Encyclopedia of

ELECTRONIC

CIRCUITS

Volume 7

Rudolf F. Graf
and
William Sheets

McGraw-Hill

New York   San Francisco   Washington, D.C.   Auckland   Bogotá
Caracas   Lisbon   London   Madrid   Mexico City   Milan
Montreal   New Delhi   San Juan   Singapore
Sydney   Tokyo   Toronto

Library of Congress Cataloging-in-Publication Data
(Revised for vol. 7)
Graf, Rudolf F.
    The encyclopedia of electronic circuits
    Authors for v. 7–   : Rudolf F. Graf & William
Sheets.
    Includes bibliographical references and indexes.
    1. Electronic circuits—Encyclopedias. I. Sheets,
William. II. Title.
TK7867G66   1985   621.3815   84-26772
ISBN 0-8306-0938-5 (v. 1)   ISBN 0-07-011077-8 (pbk. : v. 5)
ISBN 0-8306-1938-0 (pbk. : v. 1)   ISBN 0-07-011076-X (v. 5)
ISBN 0-8306-3138-0 (pbk. : v. 2)   ISBN 0-07-011275-4 (v. 6)
ISBN 0-8306-3138-0 (v. 2)   ISBN 0-07-011276-2 (pbk. : v. 6)
ISBN 0-8306-3348-0 (pbk. : v. 3)   ISBN 0-07-015115-6 (v. 7)
ISBN 0-8306-7348-2 (v. 3)   ISBN 0-07-016116-4 (pbk. : v. 7)
ISBN 0-8306-3895-4 (pbk. : v. 4)
ISBN 0-8306-3896-2 (v. 4)

McGraw-Hill

A Division of The McGraw·Hill Companies

    4 5 6 7 8 9 0   DOC/DOC   9 0 3 2 1

ISBN 0-07-015115-6 (HC)
ISBN 0-07-015116-4 (PBK)

*The sponsoring editor for this book was Scott Grillo, the editing
supervisor was Bernard Onken, and the production supervisor
was Sherri Souffrance. It was set in ITC Century Light by Michele
Pridmore and Michele Zito of McGraw-Hill's Professional Book
Group composition unit, Hightstown, N.J.*

Printed and bound by R. R. Donnelley & Sons Company.

McGraw-Hill books are available at special quantity discounts to use as
premiums and sales promotions, or for use in corporate training programs. For
more information, please write to the Director of Special Sales, McGraw-Hill,
Two Penn Plaza, New York, NY 10121-2298. Or contact your local bookstore.

Contents

Encyclopedia of

ELECTRONIC
CIRCUITS

UHI
Millennium
Institute

Please return/renew this item by the last date shown

Tillibh/ath-chlaraidh seo ron cheann-latha mu dheireadh

To Scott
With much love from Popsi

Patent notice

Preface

This latest volume of *The Encyclopedia of Electronic Circuits* contains approximately 1000 new electronic circuits that are arranged alphabetically into more than 100 basic circuit categories, ranging from "Active Antenna Circuits" to "Weather-Related Circuits." When taken together with the contents of the previously published six volumes, we provide instant access to more than 7000 circuits that are meticulously indexed and cross referenced. This represents, by far, the largest treasure trove of easy-to-find, practical electronic circuits available anywhere.

We wish to express our sincere gratitude and appreciation to the industry sources and publishers who graciously allowed us to use some of their material. Their names are shown with each entry and further details are given at the end of the book under "Sources."

Our thanks also go out to Ms. Tara Troxler, whose skill at the word processor and dedication to this project made it possible for us to deliver the manuscript to the publisher in a timely fashion.

Rudolf F. Graf and William Sheets
January 1998

Encyclopedia of

ELECTRONIC
CIRCUITS

Volume 7

1

Active Antenna Circuits

The sources of the following circuits are contained in the Sources section, which begins on page 1043. The figure number in the box of each circuit correlates to the entry in the Sources section.

Active Dipole Antenna
High-Frequency Loop Antenna
Active Antenna, 100 kHz to 30 MHz
Dc Block for Active Antenna and Downconverters
Splitter Circuit for Active Antenna to Enable Multiple Short Wave Receiver Usage

ACTIVE DIPOLE ANTENNA

R1	R2	-3 dB BW		Diff.	Max.	Max.
		Low	High	V$_G$	In	Out
(Ω)		(kHz)	(MHz)	(dB)	(mV rms)	
100	75	50	39	12.8	233	510
100	150	56	36	15.3	230	670
27	75	50	31	23.9	62	490
27	150	56	29	26.4	61	640

Table title: **BANDWIDTH, GAIN, AND SIGNAL LEVELS**

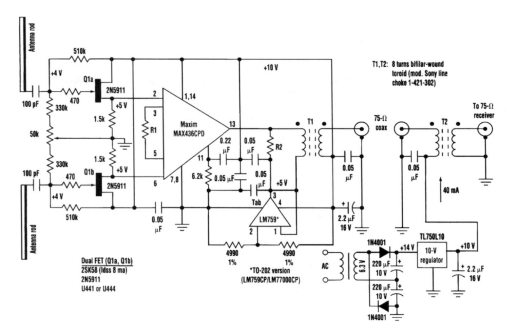

Fig. 1-1

This active antenna acts as a dipole and consists of a dual FET source follower feeding a differential amplifier. The LM759 acts as a virtual ground, splitting the 10-V supply. R1 is a gain set resistor (see table). Dc is fed in via coax cable. The frequency range is 100 kHz to 40 MHz.

HIGH-FREQUENCY LOOP ANTENNA

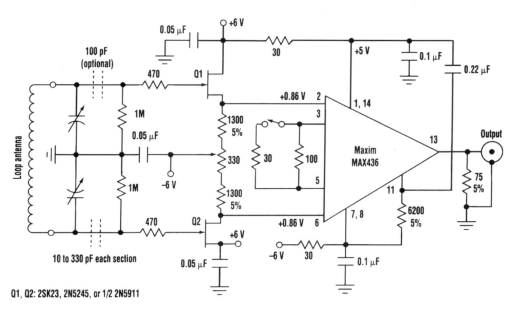

ELECTRONIC DESIGN

Fig. 1-2

Good performance in the 5- to 30-MHz range requires an amplifier with extremely high input impedance and low noise that can drive 75-Ω loads at high signal levels at frequencies over 30 MHz. Combining dual FET source followers and the new Maxim 436 wideband transconductance amplifier can produce just such an amplifier. A balanced configuration is used for the tuned loop to preserve the symmetry of the figure-eight polar antenna pattern. As a result of using FET source followers on the amplifier's front end, only the 1-MΩ gate resistors load the tuned circuit, so tuning is very sharp and resistance to off-frequency interference is very high. The FETs drive the differential inputs of the MAX436, which amplifies the balanced signal and converts it to a single-ended output. Voltage gain for this amplifier is switch selectable at either 8 dB or 20 dB into a 75-Ω load. Since the amplifier is designed to work into a 75-Ω load, the device can be connected to the receiver with a length of RG-59U cable. Maximum undistorted output is 1500 mV into 75 Ω, so overloading is unlikely even at the high-gain setting.

ACTIVE ANTENNA, 100 kHz TO 30 MHz

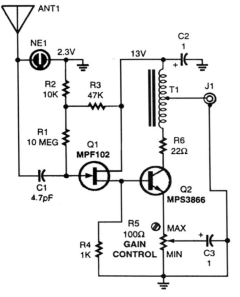

This circuit uses a FET and bipolar transistor combination to achieve high performance with a short pickup antenna (24 in. or 70 cm). R5 is a gain control and should be set for the minimum gain necessary for good results. T1 is 24 turns of #32 trifilar wire on a Ferroxcube P/N 768T188-3E2A core. J1 is used for dc feed and RF output. A dc block unit can be used at the receiver end. Dc power is 12 V at 30 mA. The pickup wire and pre-amp should be enclosed in a weatherproof assembly so that the antenna can be mounted outdoors in an electrically quiet area.

A complete kit of parts, including the PC board, is available from North Country Radio, P.O. Box 53, Wykagyl Station, New Rochelle, NY 10804-0053A.

POPULAR ELECTRONICS   *Fig. 1-3*

DC BLOCK FOR ACTIVE ANTENNA AND DOWNCONVERTERS

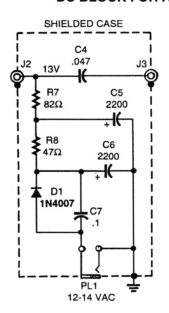

A suitable dc block and power source for active antennas and downconverters can be made from a few components as shown. This circuit is suitable for frequencies from 50 kHz up to VHF. A shielded case should be employed to prevent stray signal pickup. Dc power supplied to the active antenna or downconverter, via the coaxial signal cable, is 12 V at 30 mA.

POPULAR ELECTRONICS   *Fig. 1-4*

SPLITTER CIRCUIT FOR ACTIVE ANTENNA TO ENABLE MULTIPLE SHORTWAVE RECEIVER USAGE

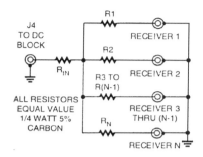

NUMBER OF RECEIVERS	50Ω CABLE	75Ω CABLE
2	16Ω	24Ω
3	24Ω	36Ω
4	30Ω	43 OR 47Ω
5	33Ω	51Ω
N	$50\left(\dfrac{N-1}{N+1}\right)\,\Omega$	$75\left(\dfrac{N-1}{N+1}\right)\,\Omega$

RESISTOR VALUES
(NEAREST 5% TOLERANCE)

Identical-resistor-value star networks can be used to connect multiple receivers to a single active antenna. Although the loss is higher with this method than with conventional ferrite transformer splitters, the frequency response goes down to dc, and the extra loss is not usually a problem at LF, MF, and HF, as active antennas have plenty of gain and the received signal levels are limited by atmospheric and ambient noise signal strengths. The splitter should be mounted in a shielded box with suitable connectors to minimize stray signal pickup.

POPULAR ELECTRONICS

Fig. 1-5

2

Alarm and Security Circuits

The sources of the following circuits are contained in the Sources section, which begins on page 1043. The figure number in the box of each circuit correlates to the entry in the Sources section.

Cut Phone Line Alarm
Freezer Sentry
Intruder Alarm
Burglar Alarm
Cut Phone Line Alarm
Digital Lock Circuit
Analog Electronic Lock
Safer Security System
Property Alarm Circuit
Tamper Alarm
Rain Alarm

CUT PHONE LINE ALARM

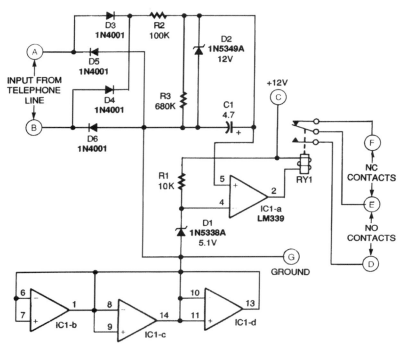

Fig. 2-1

The figure shows the schematic diagram of the phone line monitor. Note that 12-Vdc power is connected to the unit via solder pads C (positive) and G (ground) on the PC board. This power should be supplied from your home security system, since it will have battery backup, and this will be as reliable as your base security system. The input from the telephone line is fed to pads A and B. Polarity is not important to the monitor because the input is rectified by the full-wave bridge rectifier formed by diodes D3 to D6. Connecting the phone line to the phone line monitor in no way diminishes phone service, and its presence on the line will be unnoticed by you and those who call you. The rectified or polarized voltage from the bridge rectifier is fed to a long-time-constant filter formed by R2 and C1. This filter includes a zener diode (D2) that limits the voltage maximum charge on C1 to 12 V during normal operation. Resistor R3 provides a high-resistance shorting path to drain the charge from C1 when there is no input voltage. This resistor sets the time delay for activating the alarm. The trigger voltage for comparator IC1-a is generated by R1 and D1. Diode D1 is a 5.1-V zener whose output is fed to the inverting input (pin 4) of IC1-a. With the inverting input at 5.1 V, any voltage above 5.1 V on the noninverting input (pin 5) will cause the output of the comparator to go high. Since the output of IC1 is tied to the relay coil of RY1 and the other end of the relay is tied to +12 Vdc, the relay coil is not energized. When the voltage from the input filter drops below 5.1 V, the output of IC1-a goes low and energizes the relay. The contacts of RY1 (either the normally open set or the normally closed set) could also be used to trigger the home's security system.

FREEZER SENTRY

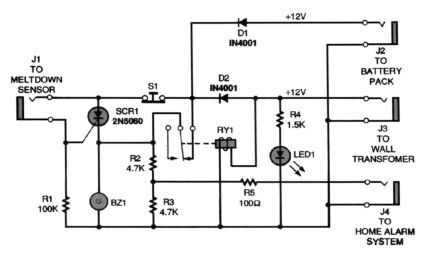

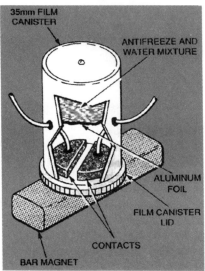

The meltdown sensor contains a liquid that melts at about 15° F. When the liquid starts to melt, it falls down to the bottom of the canister, pushing an aluminum foil disc against the electrical contacts.

FREEZER SENTRY *(Cont.)*

The heart of the circuit is SCR1, a silicon-controlled rectifier. Trigger voltage is supplied through J1 by the meltdown sensor when its contacts close. Resistor R1 is included to prevent any false triggering of SCR1 as a result of voltage fluctuations or noise from the power supply. Once SCR1 is triggered, it latches on, sounding the alarm buzzer until the reset switch (S1) is pressed. The reset switch interrupts current flow through the SCR, letting it turn off. Power for the circuit is supplied by a 12-Vdc wall-mounted transformer, which is connected to J3. As long as the wall-mounted power supply is working, relay RY1 will be activated, opening its normally closed contact. Normal power will also be indicated by LED1, with R4 limiting the current flow through the light-emitting diode. If the ac power fails, LED1 will go out and RY1 will deenergize, closing the relay's normally closed contacts. The relay contacts bypass SCR1, sounding the alarm. Backup power for that situation is supplied by a 12-V battery pack, which is connected to the circuit through J2 and D1. Diode D1 prevents the ac-derived power supply from attempting to charge the batteries, and D2 prevents the batteries from lighting the LED or energizing RY1 in the event of ac power failure. An additional feature of the circuit is that since the relay bypasses SCR1 during a loss of ac power, the reset button will not silence the alarm. Regardless of which type of failure caused the alarm to sound, R2 and R3 form a voltage divider that provides a 5-V signal to J4.

INTRUDER ALARM

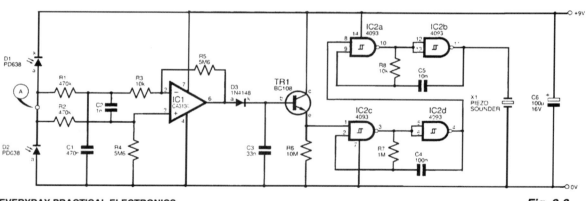

EVERYDAY PRACTICAL ELECTRONICS

Fig. 2-3

The circuit triggers a pulsed-tone alarm when anyone passes by, and several of these units can be scattered around the premises to give burglars the impression that they are being monitored! It is guaranteed to drive any burglar bananas. Two photodiodes, D1 and D2, are used as shown to ensure sensitivity over a very wide range of lightning conditions, though the circuit will be less effective in diffuse light. IC1 is connected as a differential amplifier. As the voltage at point *A* swings high, the response at pin 2 of IC1 is slowed by capacitor C1, and the voltage swings high at the output (pin 6). Capacitor C3 charges up and transistor TR1 level-shifts the signal to enable the oscillator section formed from IC2, a quad NAND Schmitt. The piezo disk X1 will pulse accordingly. C3 determines the duration of the pulsed tone and resistor R8 determines the pitch. Photodiodes D1 and D2 should be mounted a small distance apart for best effect. The circuit draws a mere 0.5 mA.

BURGLAR ALARM

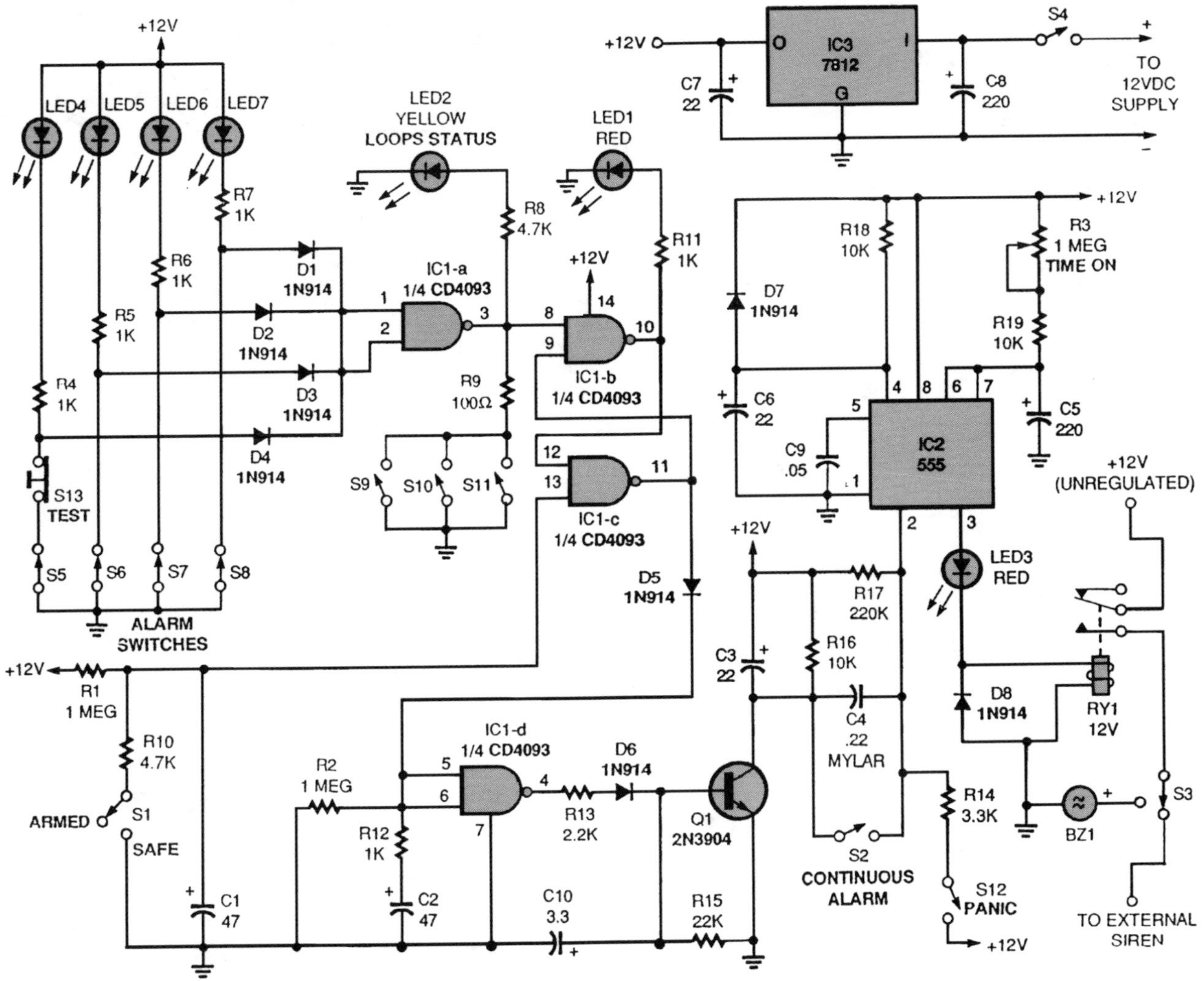

Fig. 2-4

POPULAR ELECTRONICS

10

BURGLAR ALARM *(Cont.)*

This burglar alarm uses either normally closed (S5 to S8) or normally open (S9 to S11) switches or a combination thereof. Activation sets latches IC1-b and IC1-c. S1 is used to disarm the circuit. IC1-d and related components drive switch Q1, which triggers timer IC2. IC2 generates a pulse, determined by R3 and C5, that energizes alarm relay RY1. S2 allows for a continuous alarm and S12 is a panic switch to sound the alarm immediately. Either a built-in buzzer (BZ1) or an external siren can be used as an annunciator.

CUT PHONE LINE ALARM

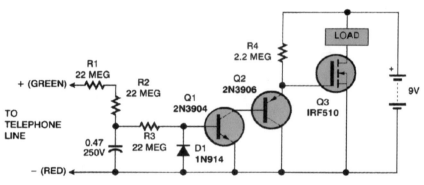

ELECTRONICS NOW

Fig. 2-5

Detecting a cut phone line can be an important function of an electronic security system. Unfortunately, detecting a cut phone line isn't easy because the voltages on a normal phone line vary so much. The voltage is typically 48 Vdc when the telephone is on the hook, 2 to 15 Vdc during a conversation, 90 Vac during ringing, and 200 Vdc when the telephone company is testing the line. Brief moments of 0 V are common; what you really want to detect is a voltage that goes to zero and stays there. A second restriction is that the cut phone line detector cannot draw any appreciable current. Its impedance has to be higher than 50 MΩ or the telephone company will think it's a leaky cable. Components R1, R2, R3, and C1 smooth out momentary variations in voltage so that the alarm doesn't trigger every time the telephone rings. If the voltage stays at zero for 30 s or more, the alarm should trigger. The load can be a piezo buzzer, an optocoupler, or a small relay.

Because of the tiny currents it must detect, this circuit should be powered by its own 9-V battery, with no direct connection to the rest of the burglar alarm. Otherwise, a slight difference in ground potentials might cause the circuit to perform somewhat erratically.

DIGITAL LOCK CIRCUIT

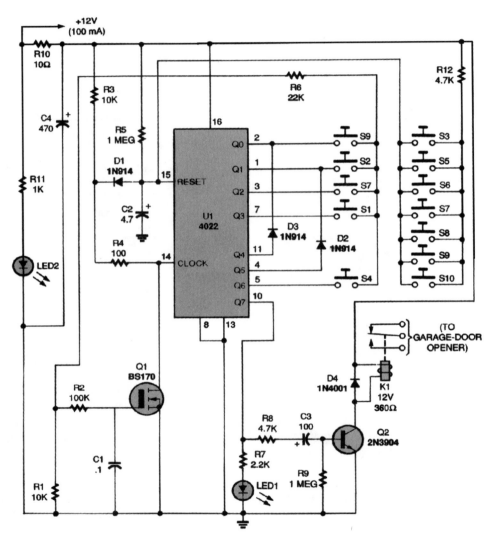

POPULAR ELECTRONICS

Fig. 2-6

This circuit depends on the entry of a correct sequence code. An incorrect number that is not part of the code causes the circuit to reset. When the correct code is entered, Q2 operates relay K1 for a short time, depending on C3.

ANALOG ELECTRONIC LOCK

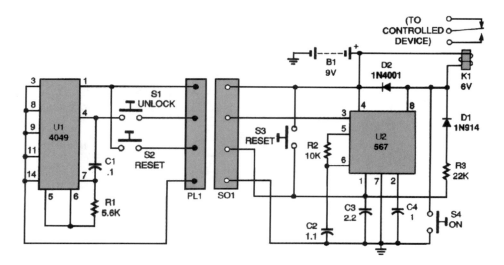

POPULAR ELECTRONICS

Fig. 2-7

The system is formed of two separate circuits—a key and a keyhole. The key engages the keyhole by means of a mating pair of connectors. The key is a tone-generator circuit consisting of a 4049 hex inverter CMOS IC (U1), switches (S1 and S2), a resistor (R1), and a capacitor (C1). The value of the tone generated by that circuit, in hertz, is determined by $1/(1.4 \times R_1 C_1)$. The keyhole is a 567 tone-decoder circuit that can be configured to detect any frequency from 0.01 Hz to 500 kHz. The frequency it detects (f_o) is set by resistor R2 and capacitor C2, according to $f_o = 1.1/(R_2 C_2)$. When the key is inserted into the lock and switch S1 is pressed, a tone is supplied to the input of the keyhole circuit. If the tone frequency is close enough to f_o the 567 IC turns on the relay (K1), which should be connected to the electronic locking device. Components R3 and D1 are used to latch the circuit, so that the output stays on even after the input tone is removed. When S2 is pressed, the system is reset. Switch S3 resets the circuit from inside. *Note:* The accuracy with which the keyhole circuit detects f_o can be set by changing the values of three components. First, R2 should be between 2000 and 20,000 Ω. Second, the value in microfarads of capacitor C4 should be n/f_o, where n is a value between 1300 and 0000 (which gives a detection accuracy of up to 14 percent of f_o). Finally, capacitor C3 should have about twice the capacitance of capacitor C4. Battery B1 supplies both circuits.

SAFER SECURITY SYSTEM

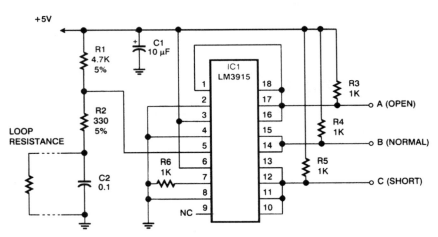

Fig. 2-8

Many security systems use a closed loop of wires and switches, arranged so that whenever a door or window is opened, the loop will be broken and the alarm will sound. An obvious problem is that someone can tamper with the system, short out the loop, and later on come back and burglarize the premises without sounding the alarm. Hiding a known resistance in the loop is a very good idea. That way, the alarm can distinguish a short circuit from a correctly functioning closed loop. The figure shows a circuit that does the job. It is a somewhat unusual application for a National Semiconductor LM3915 IC, normally used to drive LED bar-graph displays. That chip happens to contain the right combination of comparators and logic circuits to do what you need. Step 1 is to translate the loop resistance into a voltage. That is done by putting a voltage divider with resistors R1 and R2. Capacitor C2 protects the circuit against electromagnetic noise—important because burglar alarms use long wires, often running near heavy electrical equipment. Step 2 is to translate the voltage into a logic signal indicating whether it's in the correct range. Normally, the LM3915 would drive 10 LEDs, one for each of the small ranges of voltage. To obtain logic-level outputs, we have it driving 1-kΩ resistors instead of LEDs. Since we need to distinguish only three situations, we tie some of the outputs together. The LM3915 has open-collector outputs that can be paralleled in that way. Note that they use negative logic (0 V for "yes," +5 V for "no"), the opposite of ordinary logic circuits. You can use inverters, such as the 74HC04, to produce positive logic signals, if that's what you need. Finally, note that the circuit will actually work with any supply voltage from 3 to 25 V.

PROPERTY ALARM CIRCUIT

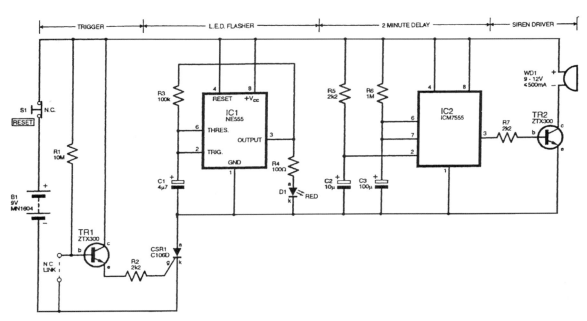

Fig. 2-9

The circuit is designed to protect by means of a wire loop or mercury switch. TR1 and associated components form a switch that is biased off by the normally closed (N.C.) link. When the link is broken, TR1 conducts and triggers the thyristor (CSR1) into conduction. Consequently, the LED flasher centered around the 555 astable of IC1 causes D1 to blink. Additionally, IC2 is a monostable timer that triggers through R5/C2 and drives an audible warning device via TR2 for a period of approximately 2 min before resetting. When the audio alarm has timed out, the LED will continue to flash, and this can be cancelled by opening S1, which resets the thyristor. The quiescent current of the circuit is very low because of the high value of resistor R1. A mercury switch can be used in place of the wire link to act as an antitamper alarm, in which case care must be taken to avoid accidental contact with the mercury bead itself because it is highly toxic.

TAMPER ALARM

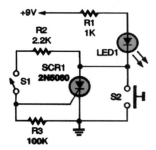

The silicon-controlled rectifier (SCR1) operates as a memory device to indicate a security breach in a room, desk drawer, safe, etc. Switch S1 can be a mechanical or magnetic switch. Position S1 in an object that you want to keep protected, making sure that the switch will close when the object is tampered with. When S1 closes, SCR1 turns on, lighting LED1. Pressing S2 resets the circuit.

POPULAR ELECTRONICS

Fig. 2-10

RAIN ALARM

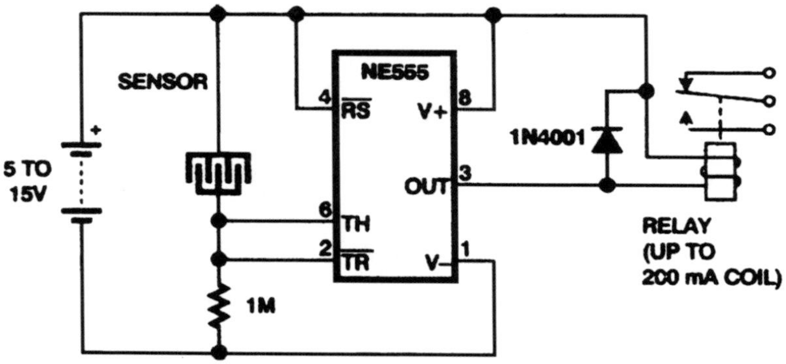

ELECTRONICS NOW

Fig. 2-11

This rain-detector circuit closes the relay when water bridges the gap between the metal electrodes.

3

Amateur Radio Circuits

The sources of the following circuits are contained in the Sources section, which begins on page 1043. The figure number in the box of each circuit correlates to the entry in the Sources section.

Automatic Voltage Controller
10-dB 50-W Attenuator
10-W CW Transceiver
2-m HT Base Station Adapter
WWV to 75-m Band Converter
Repeater Emergency Power Supply
Dummy Load and Detector
10-GHz Wavemeter Amplifier

AUTOMATIC VOLTAGE CONTROLLER

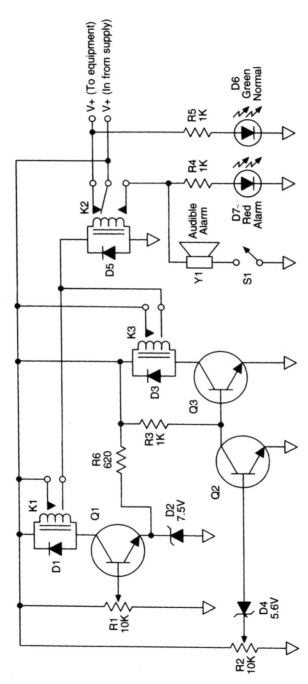

Fig. 3-1

Station dc voltage, a nominal 113.8 Vdc, is fed through the normally closed contacts of relay K2. This voltage also provides power for the protective circuit illustrated. The voltage appears across R1 and R2, 10-kΩ trimpots, the wipers of which are set to exactly midrange, measuring a nominal 16.9 V dc. As the station dc varies between 111.2 and 115 V, the voltage at the wipers will vary from 5.6 to 7.5 Vdc. Zener diode D2 controls the high-voltage limit of 115 Vdc. R5 and zener diode D4 control the low-voltage limit of 111.2 Vdc. A voltage is fed from the wiper of R1 directly to the base of Q1, relay K1 is not energized, and its normally open contacts prevent relay K2 from being energized, allowing station dc to flow through K2's normally closed contacts. Above 115 V, the voltage at the base of Q1 also rises to or above 17.5 V, energizing K1, whose normally open contacts close, energizing K2, whose normally closed contacts open, removing the voltage from the station equipment. Should the voltage fall to or below 111.2 Vdc, zener diode D4 ceases to conduct, cutting off Q2, which causes Q3 to conduct, energizing K3. K3 applies operation voltage through its contacts to the coil of relay K2, opening its normally closed contacts, removing voltage from D6, the station equipment, and applying power to LED D7 and the audible alert.

10-dB 50-W ATTENUATOR

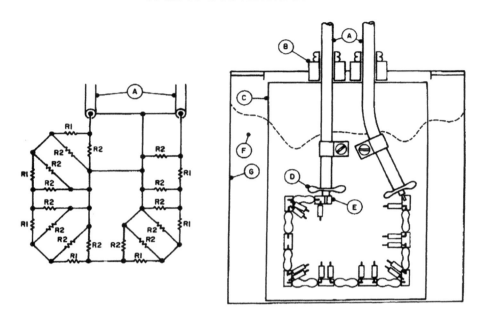

73 AMATEUR RADIO TODAY

Fig. 3-2

The 10-dB attenuator has an attenuation of 10 dB at lower frequencies at 2 m and 70 cm; circuit losses increase to 10.8 dB and 12.0 dB, including interconnecting RG-58/U cable losses of 0.4 dB and 0.6 dB, respectively. Establishing an SWR of less than 1.1:1 at 70 cm was the most crucial design consideration. Changes occur with different coolants. If the assembly is tailored for minimum SWR in air, its SWR increases to about 1.3:1 when you put it in vegetable oil. The SWR changes are caused by increased circuit capacitance due to the dielectric. This condition was improved by using household wax, paraffin, instead of oil. *Caution: Paraffin has a relatively low flash temperature; it can be used to make candles.* Next, to decrease the circuit distributed capacitance, increase the distance of the components from the circuit board by stacking two layers of PC boards at the tie-down pads. This raises the parts of connecting positions to about 0.120 in. The completed assembly has an SWR of less than 1.1:1 at both 70 cm and 2 m.

Schematic and layout of the 10-dB attenuator: R1, 10-Ω 1/2-W (7); R2, 1-kΩ 1/4 W (14). **A.** Input/output, 3 ft RG-58/U. **B.** Cable bushing, $\frac{1}{4}$ in. clearance hole. **C.** Component mounting board, 3$\frac{1}{2}$ in×2$\frac{3}{8}$ in×$\frac{1}{16}$ in PCB. **D.** Fan out end cable braid, twist into two segments, and solder to PCB with minimum lead lengths. **E.** Component tie-down pads (9), double thickness glass-epoxy PCB, $\frac{1}{4}$ in×$\frac{1}{8}$ in×$\frac{1}{8}$ in pieces cemented together and into position with clear household cement (Elmer's). **F.** Coolant, household wax (paraffin). Fill container with melted wax to $\frac{1}{4}$ in from the top. To melt wax, insert the container in hot water (about 200°F). **G.** Container, one-pint can.

10-W CW TRANSCEIVER

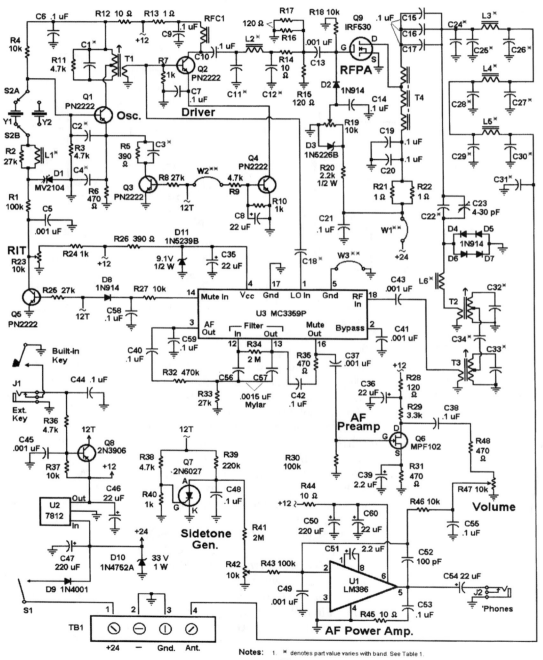

Notes: 1. ✳ denotes part value varies with band. See Table 1.

2. ✳✳ W1-3 represent bare wire jumpers needed with the accompanying pcb layout.

3. Both Q9 and U2 require a heatsink. See text and photographs for details.

4. Component ratings: Unless specified, resistors are 1/4 W, electrolytic capacitors are 35 V and ceramic disk capacitors are 100 V.

20

10-W CW TRANSCEIVER *(Cont.)*

Table 1 Band Data

Component	80-M	40-M	30-M
C1	390 pF	68 pF	Not used
C2	18 pF	5 pF	Not used
C3	680 pF	270 pF	Not used
C4	820 pF	680 pF	390 pF
C11, C12	820 pF	390 pF	270 pF
C18, C34	39 pF	18 pF	10 pF
C22	18 pF	Not used	Not used
C25, C31	390 pF	270 pF	180 pF
C26, C28	390 pF	Not used	390 pF
C27, C29	820 pF	680 pF	68 pF
C30	68 pF	Not used	Not used
C32, C33	390 pF	68 pF	Not used
L1 (FT37-61)	40T #30	23T #28	17T #28
L2 (T50-2)	22T #24	14T #24	12T #24
L3, L5 (T50-2)	22T #24	17T #24	14T #24
L4 (T50-2)	25T #24	19T #24	16T #24
L6 (FT-37-61)	30T #28	25T #28	15T #28

Capacitors are 100v ceramic disk type. For inductors, wind turns using the enamel wire gauge given on the toroid core specified.

Fig. 3-3

This transceiver can be operated on the 80-, 40-, or 30-m CW ham bands. A crystal oscillator drives a driver and RF PA delivering about 10 W output. The receiver uses an MC3359 IC and is fed LO from the crystal oscillator used for transmitting, and a varactor diode is used for incremental tuning (RIT) of the receiver. The unit also has a CW sidetone generator for ease of monitoring the transmitted CW signal. The power supply is 18 to 28 Vdc at 1 A.

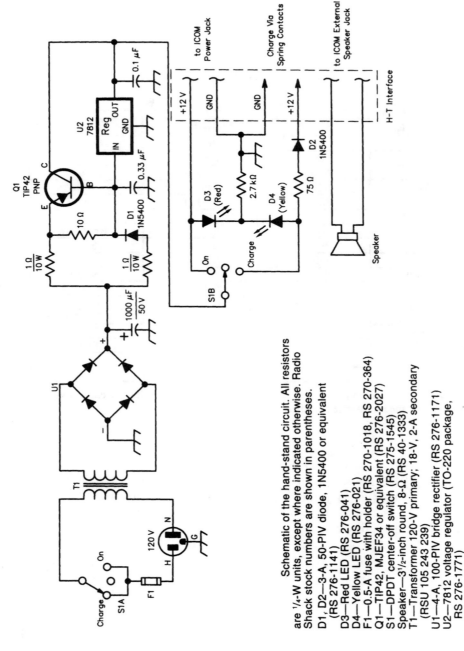

2-m HT BASE STATION ADAPTER

Schematic of the hand-stand circuit. All resistors are 1/4-W units, except where indicated otherwise. Radio Shack stock numbers are shown in parentheses.

D1, D2—3-A, 50-PIV diode, 1N5400 or equivalent (RS 276-1141)
D3—Red LED (RS 276-041)
D4—Yellow LED (RS 276-021)
F1—0.5-A fuse with holder (RS 270-1018, RS 270-364)
Q1—TIP42, MJEF34 or equivalent (RS 276-2027)
S1—DPDT center-off switch (RS 275-1545)
Speaker—3¹/₂-inch round, 8-Ω (RS 40-1333)
T1—Transformer 120-V primary; 18-V, 2-A secondary (RSU 105 243 239)
U1—4-A, 100-PIV bridge rectifier (RS 276-1171)
U2—7812 voltage regulator (TO-220 package, RS 276-1771)

Fig. 3-4

This device mates with a small handie-talkie (HT) and provides power and speaker interface as well as charging capability. It was designed for an Icom 02-AT, but can be adapted for other HTs as well.

WWV TO 75-m BAND CONVERTER

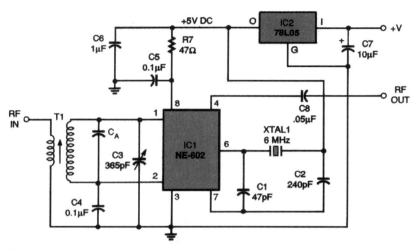

POPULAR ELECTRONICS

Fig. 3-5

This circuit for an HF-band converter will convert the 10-MHz WWV signal to a frequency in the 75-m ham band. The local oscillator section of NE-602 is available on pins 6 (base) and 7 (emitter). In this circuit, a 6.00-MHz crystal oscillator is provided by the NE-602. Capacitors C1 and C2 form the feedback network. The junction of C2 and XTAL1 can be connected to either the 15-Vdc line or ground (the former is shown here). The difference frequency between WWV's 10 MHz and the 6.00-MHz crystal frequency is 4.00 MHz, which is located at the top end of the 75-m ham band. Crystals with other frequencies will produce other output sum or difference frequencies, so tune the receiver appropriately if something other than 6.00 MHz is used.

REPEATER EMERGENCY POWER SUPPLY

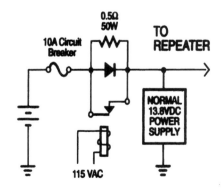

NUTS AND VOLTS MAGAZINE

Fig. 3-6

If your repeater needs emergency power, this circuit might do. Connect a 12-V storage battery, a relay, and a 0.5-Ω 50-W resistor as shown. Normally, the battery charges at a low rate (less than 1 A) through the 0.5-Ω resistor. When power fails, current can flow from the battery through the diode for a few milliseconds, until the relay drops out and closes the contacts, completely eliminating any voltage drop. If your repeater doesn't use a 13.8-V power supply, you could purchase a 2-A battery charger and connect a 0.5- to 1-Ω resistor (50 W) in series with the battery. This will charge at approximately $\frac{1}{2}$ A and maintain the battery.

DUMMY LOAD AND DETECTOR

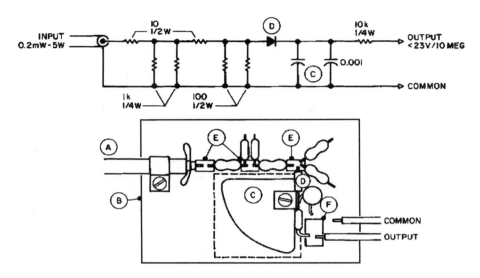

Fig. 3-7

Two 100-Ω, ½-W resistors in parallel connected across a 50-Ω cable with near-zero lead lengths will be very close to a 50-Ω termination with the SWR less than 1.1:1. However, adding the diode introduces capacitor loading that results in an SWR of 1.5:1 or more. In the circuit, the components on the 2-dB resistor pad preceding the diode/load compensate for this, as they are tailored to reduce the SWR to less than 1.1:1. Although the power capability of this assembly is 5 W, forced-air cooling is required at the higher power levels, depending on the measurement period.

Schematic and layout of the dummy-load/detector assembly: **A.** Input, 1 ft RG-58/U. Fan out braid on connecting end, twist in two segments, and solder to PC board with minimum lead lengths. **B.** Base, 3 in×2 in×$\frac{1}{16}$ in PCB. **C.** Capacitor—90° circular sector, 1 in radius, 0.21-in-thick Reynolds sheet aluminum. Surface polish with 220-grit sandpaper to remove burrs. Dielectric, 2.7-mil polyethylene (Ziploc heavy-duty freezer bag). Feed through 2-56 screw with a plastic insulator on the back side. Hole is reamed on both sides with a large drill to prevent shorting to the foil. **D.** Peak readout diode, 1N34A selected to have a reserve resistance of less than 5 MΩ (RS 276-1123). **E.** Component tie-down pads, $\frac{1}{4}$ in×$\frac{1}{8}$ in×$\frac{1}{16}$ in glass-epoxy PC board. Cemented to base with clear household cement (Elmer's). **F.** Output tie-down pad, $\frac{3}{8}$ in×$\frac{1}{4}$ in×$\frac{1}{16}$ in PC board.

10-GHz WAVEMETER AMPLIFIER

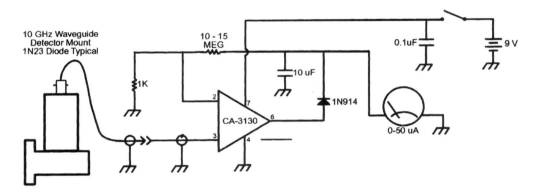

Fig. 3-8

This wavemeter amplifier is connected to the microwave waveguide detector. This circuit operates from a single 9-V transistor radio battery for simplicity.

4

Amateur Television (ATV) Circuits

The sources of the following circuits are contained in the Sources section, which begins on page 1043. The figure number in the box of each circuit correlates to the entry in the Sources section.

ATV Line Sampler
UHF ATV Downconverter for 900 and 1300 MHz
1.3-GHz ATV Transmitter

ATV LINE SAMPLER

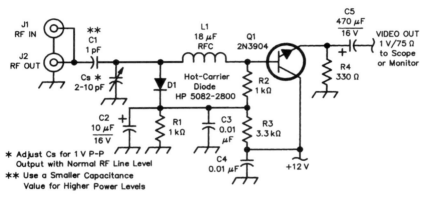

QST

Fig. 4-1

This device can be used to recover demodulated video from the output of an ATV transmitter. A small sample of the signal on the antenna feeder is tapped off and detected, and the resultant video is fed to an emitter follower. C1 is chosen for 1 V p-p under normal transmit conditions. This circuit was intended for 440-MHz AM TV signals. It will not work for FM or for 900-MHz or 1300-MHz signals. A striplike directional coupler can be used to sample the RF line without introducing an impedance bump.

UHF ATV DOWNCONVERTER FOR 900 AND 1300 MHz

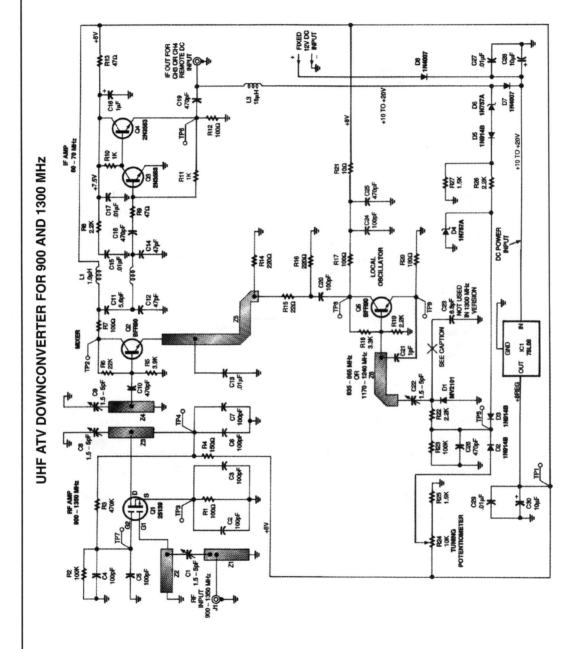

Fig. 4-2

UHF ATV DOWNCONVERTER FOR 900 AND 1300 MHz *(Cont.)*

This downconverter is tunable and can cover either 902 to 928 MHz or 1240 to 1300 MHz for receiving amateur television transmissions or other wideband signals. An NEC 25139 low-noise RF amplifier feeds a BFR90 mixer driven by a BFR tunable LO operating 61 or 67 MHz below the received frequency. A pair of 2N3563 transistors are used as a post-IF amplifier. Overall gain is 37 to 40 dB with around 1.5-dB noise figure. AGC control voltage can be used on gate 2 of the RF stage to reduce RF gain on strong signals; about -3 V is needed to reduce RF gain by >30 dB. Either on-board tuning via pot R24 or remote tuning via 10- to 20-V variable dc supply on the IF coaxial cable can be used. Note that striplines are different for 900 and 1300 MHz, and this requires separate models for each band, although the physical layout and circuit diagrams are identical.

A complete kit of parts, including PC board, is available from North Country Radio, P.O. Box 53, Wykagyl Station, New Rochelle, NY 10804-0053A.

1.3-GHz ATV TRANSMITTER

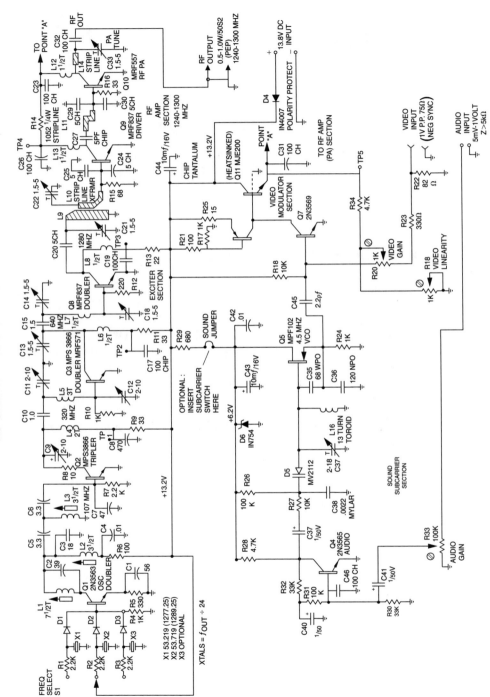

Fig. 4-3

WILLIAM SHEETS

30

1.3-GHz ATV TRANSMITTER *(Cont.)*

This TV transmitter operates from a +12-Vdc source and produces about 1 W peak power on sync tips. Output frequency is 1240 to 1300 MHz, but operation is possible to 1325 MHz (for U.K. band). Video interface is 1 V p-p/75 Ω negative sync; audio is 5 mV to 1 V @ Z55 kΩ. For PAL operation, reduce the number of turns on L16 to 11 for 5.5-MHz sound.

A kit of parts for this transmitter is available from North Country Radio, P.O. Box 53, Wykagyl Station, New Rochelle, NY 10804-0053A.

5

Amplifier Circuits—Audio

The sources of the following circuits are contained in the Sources section, which begins on page 1043. The figure number in the box of each circuit correlates to the entry in the Sources section.

22-W Amplifier for 12-V Systems
60-W Switching Amplifier
Stable LM386 Audio Amplifier Circuit
Stereo Preamp
Miniature Audio Power Amplifier
Simple Audio Output Amplifier
Simple Guitar Amplifier with Dual Inputs
2-W Stereo Amplifier
Headphone Amplifier for Guitars
High-Gain Amplifier
Mini Megaphone
Audio Power Booster
Multipurpose Mini Amplifier
Push-Pull Audio Line Driver
Electret Mike Preamp with PTT Circuit
Crystal Radio Amplifier

22-W AMPLIFIER FOR 12-V SYSTEMS

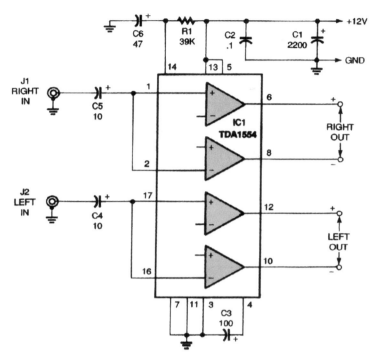

POPULAR ELECTRONICS

Fig. 5-1

Power for the circuit (+12 V) is provided by a connection to the host vehicle's battery. A connection is also provided for the vehicle/s ground. Capacitors C1 and C2 provide decoupling of any signal riding on the supply voltage, while capacitor C3, working in conjunction with IC1, provides ripple rejection. The incoming audio signal is coupled to IC1 by capacitors C4 and C5. Those 10-μF capacitors are used to avoid rolling off of the low audio frequencies. Resistor R1 and capacitor C6 feed the mute switch circuit (included in IC1), providing the delay that eliminates turn-on pop. Their *R/C* time constant is about 1.4 s. None of the component values external to IC1 are crucial, but major value substitutions should not be made. Pin 14 (the mute switch) of IC1 must have at least 8.5 V for the amplifier to be ON, or be held below 3.3 V to ensure that the chip stays in the mute condition. Current requirements at this pin are on the order of 40 μA in the ON condition, and 100 μA for standby. The R1/C6 combination used here (47μF and 39,000 Ω) provides enough delay to eliminate turn-on pop without having an excessive wait for normal operation. In addition to this slight delay, pin 14 gradually comes up above the 8.5-V threshold as C6 charges up, rather than coming on instantly, as it would if a simple switch were used. Values for C6 and R1 are not crucial, but R1 should be no larger than 100,000 Ω, and the R1/C6 time constant should be on the order of a second or two. Too short a time constant may not eliminate the turn-on pop; too long a time constant does not harm, except causing an irritating delay.

60-W SWITCHING AMPLIFIER

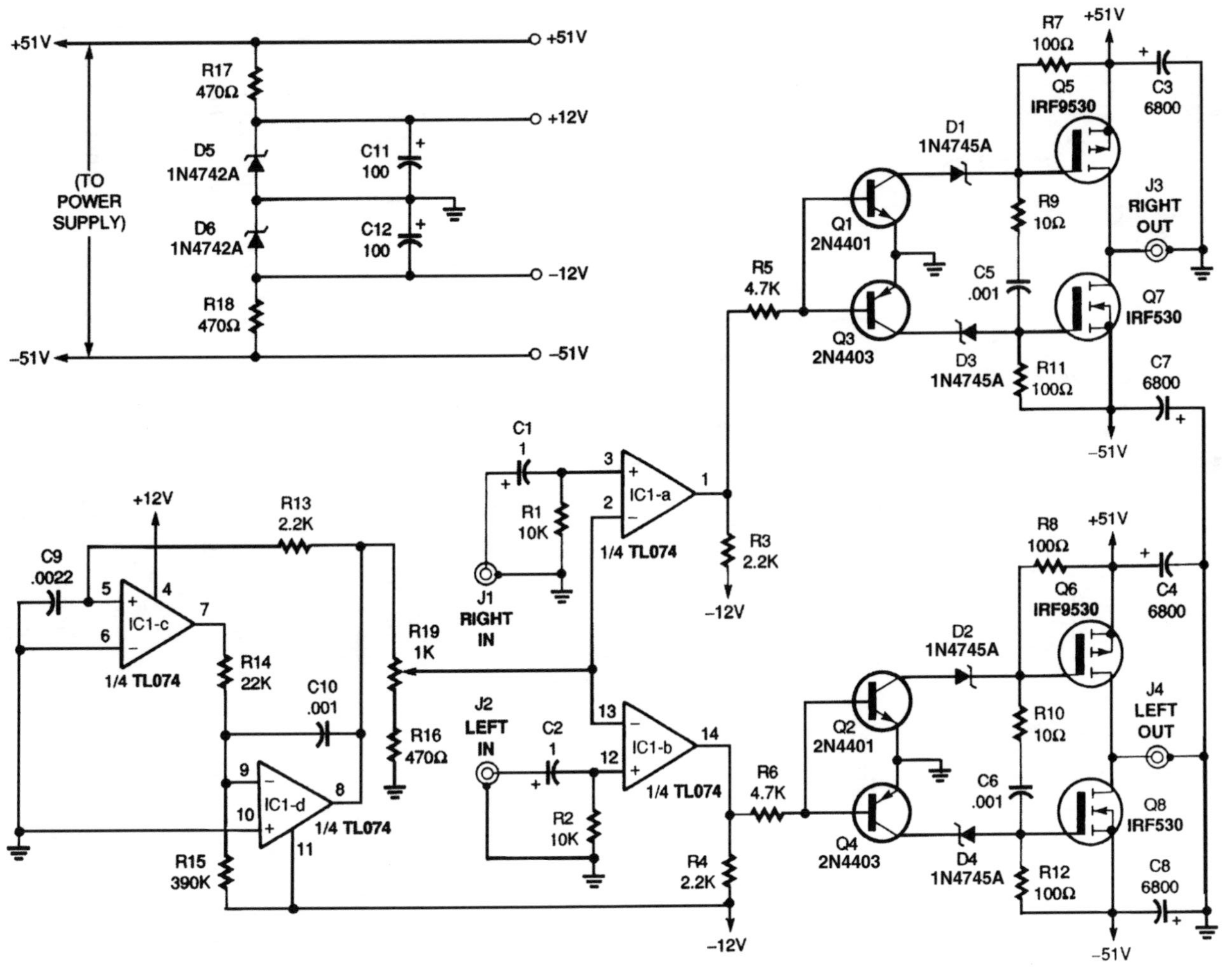

POPULAR ELECTRONICS

Fig. 5-2

60-W SWITCHING AMPLIFIER *(Cont.)*

The schematic for the switching amplifier is shown. A separate 51-Vdc source is required to power the amplifier circuit. The 51-Vdc source is fed to a pair of zener diodes, D5 and D6, and is filtered by capacitors C11 and C12 to provide a 12-Vdc source for part of the circuit. Also, part of the 51-Vdc source bypasses the zeners to power the sections of the circuit that require a high voltage. The right and left signals are input to the amplifier through jacks J1 and J2, respectively. Two sections of a TL074 op amp, IC1-c and IC1-d, generate a 4-V p-p, 50-kHz triangular reference waveform. The generated waveform is then fed to potentiometer R19. That enables the amplifier to use input signals with amplitudes ranging from 1 V p-p to 4 V p-p. The other two op-amp sections, IC1-a and IC1-b, function as comparators to produce the pulse-width-modulating output for the left and right channels of the amplifier. In the right channel of the amplifier, the output of the voltage comparator is coupled to the bipolar translating circuit through a current-limiting resistor, R5. The translating circuit has a positive and a negative "leg;" Q1, D1, and R7 make up the positive leg, and Q3, D3, and R11 make up the negative leg. Both legs are tied to ground through the emitters of Q1 and Q3, providing a reference point for the translator. The translator arrangement results in 17 V being present across Q1, Q3, and zener diodes D1 and D3. Sufficient current is then present to overcome the power MOSFET gate capacitance; that rapidly switches on and off the power MOSFET complementary push-pull output stage, composed of Q5 and Q7. Resistor R3 keeps the output swing centered at the midpoint of the supply voltage. Without R3, the square-wave output drifts down toward the negative rail. The *RC* network, composed of R9 and C5, which connects to both N- and P-channel gates, minimizes switching noise and sharpens the square-wave output. Note that both channels contain power-supply elements to split the incoming single-polarity voltage in half. Capacitors C3, C4, C7, and C8 make up a series-parallel circuit that converts the 51-Vdc supply to 25.5 Vdc. The output can feed full-range 60-W rms speakers, which demodulate the signal and produce an amplified audio output. (At peak output power, the current draw for an 8-Ω dynamic load is approximately 1.2 A at 51 Vdc.)

STABLE LM386 AUDIO AMPLIFIER CIRCUIT

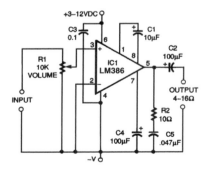

The circuit shown has components installed to improve the stability of the LM386 circuit. R2, C3, C4, and C5 are sometimes omitted, leading to instability with certain layouts. These components should be used to ensure stability. Output is up to 1 W, depending on supply voltage and load impedance.

Fig. 5-3

STEREO PREAMP

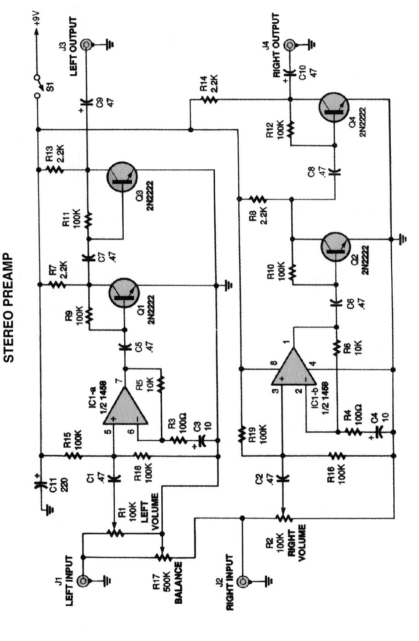

POPULAR ELECTRONICS

Fig. 5-4

The output of the audio source is fed to the left and right inputs of the circuit (J1 and J2). Potentiometers R1 and R2 control the volumes of the input signals, while potentiometer R17 is a balance control. The incoming signals are coupled through capacitors C1 and C2 to the noninverting inputs of op amps IC1-a and IC1-b. Because IC1 operates as a single-supply amplifier, its output signal fluctuates above and below half of the supply voltage. The output signals of the op amps are coupled to the bases of two 2N2222 transistors (Q1 and Q2), which further amplify the left and right signals. Then, the outputs of the transistors are coupled to the bases of two more 2N2222 transistors (Q3 and Q4), further boosting the left and right signals. The transistor pairs also act as buffers.

MINIATURE AUDIO POWER AMPLIFIER

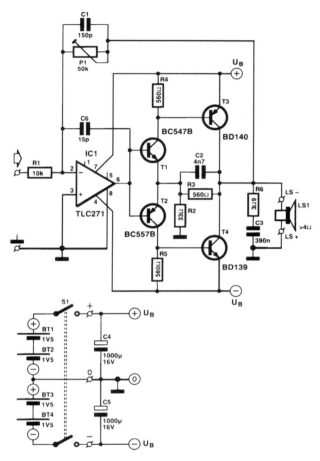

Fig. 5-5

A compact audio power amplifier with low current drain has many applications. These are the design basis for the mini amplifier. It continues working satisfactorily with a battery voltage down to 1.5 V. Its quiescent current drain is about 1 mA, and its efficiency is a worthwhile 70 percent. It provides an output power of 500 mW into 8 Ω (or 800 mW into 4 Ω), has a sensitivity of 400 mV, and its distortion is never higher than 1.2 percent. Because the output transistors have no emitter resistor, the voltage is determined solely by the knee voltage of T3 and T4. With a load of 4 to 8 Ω, these voltages are limited to 0.2 to 0.3 V so that the transistors can be driven virtually up to the supply voltage. The overall bandwidth of the amplifier is limited to not less than 21 kHz at the maximum amplification of ×5. With a 4-Ω load, the peak output current is 700 mA. A 315-mA fuse in series with the output is, therefore, a simple but effective short-circuit protection. At maximum drive with a music signal, the average current is only 50 mA. In operation the drive will never be continuously maximum, so the actual current drain will be much lower. A set of four penlight batteries should last about 200 hours.

SIMPLE AUDIO OUTPUT AMPLIFIER

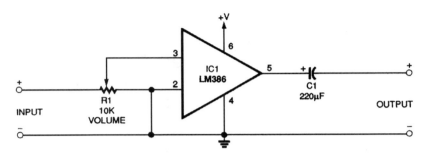

ELECTRONIC EXPERIMENTERS HANDBOOK

Fig. 5-6

This audio amplifier circuit has a gain of about 20. A supply of 3 to 12 V can be used. Output is up to about 1 W, depending on load impedance and supply voltage.

SIMPLE GUITAR AMPLIFIER WITH DUAL INPUTS

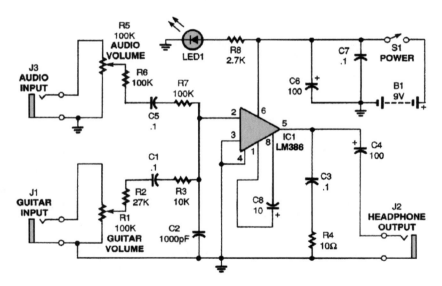

POPULAR ELECTRONICS

Fig. 5-7

This amplifier uses an LM386 IC and has two inputs, an input for guitar and a separate audio input for a second guitar, a mike, etc. Power is supplied by a 9-V battery. While headphones are indicated, the amplifier will drive a small speaker. Output is around 300 to 500 mW, depending on load and battery voltage.

2-W STEREO AMPLIFIER

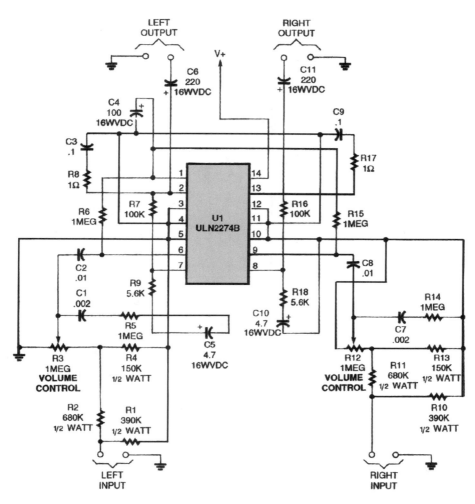

POPULAR ELECTRONICS

Fig. 5-8

The amplifier is built around a ULN2274B dual audio power amplifier and provides a maximum of 2 W of quality sound. Because the amplifier is composed of two identical subcircuits, only one subcircuit will be covered. Resistor R5 sets the tone and it can be replaced with a variable unit; lower values of that resistor produce more bass. The bias is set by R6 and C4, and R7, R9, and C5 are feedback elements. For a more realistic sound, R8 and C3 are used to roll off the high frequencies. Capacitor C6 is a dc-blocking capacitor. The circuit chip requires from 6 to 26 Vdc without distortion.

HEADPHONE AMPLIFIER FOR GUITARS

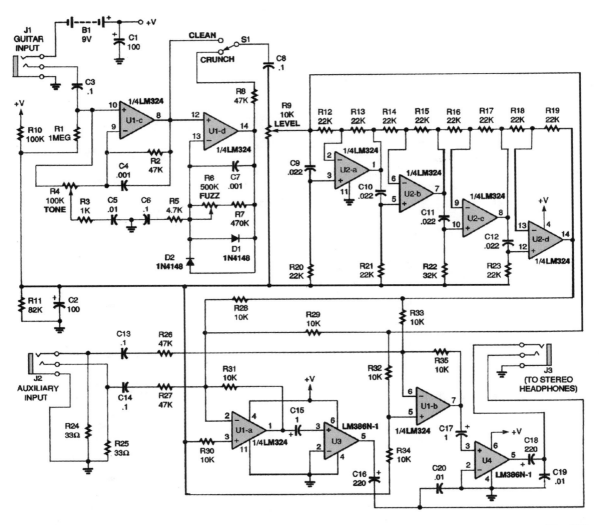

Fig. 5-9

From the hard distortion provided by U1-d and D1 and D2 to the stereo imaging accomplished by U2, this is a guitarist's dream come true. Note that there is no power switch in the circuit; J1 turns on the unit whenever an instrument cable is plugged in.

The main components in the circuit are two LM324 quad op amp ICs (U1 and U2) and two LM386 power amp ICs (U3 and U4). The inputs to U1 and U2 are biased to a little less than half the power-supply voltage by resistors R10 and R11. Capacitors C1 and C2 filter the power-supply and bias voltages. J1 turns on the amplifier when the input plug is inserted. When an audio signal from an instrument is input through J1, the signal is fed through coupling capacitor C3 to the tone-control

HEADPHONE AMPLIFIER FOR GUITARS *(Cont.)*

circuit composed of U1-c, R2, R4, and C4. Frequencies above 1 kHz are amplified or attenuated depending on the position of potentiometer R4, which is the tone control. Resistor R2 and capacitor C4 filter unwanted high frequencies. Audio level and overdrive are controlled by potentiometer R9; with that level control adjusted to full volume, the circuit's final amplifiers are overdriven to produce a soft distortion effect. To prevent any unwanted dc "swishing" noise, a coupling capacitor, C8, is used. Switch S1 toggles between the clean and distorted signals. When S1 is on the CRUNCH setting, diodes D1 and D2 and U1-d produce a distortion effect by clipping the amplified signal at 0.7 V. Frequencies below 160 Hz are attenuated by R5 and C6. The amount of gain or "fuzz" is controlled by R7 and potentiometer R6, and resistor R8 adjusts the distortion level to match the tone-control level.

HIGH-GAIN AMPLIFIER

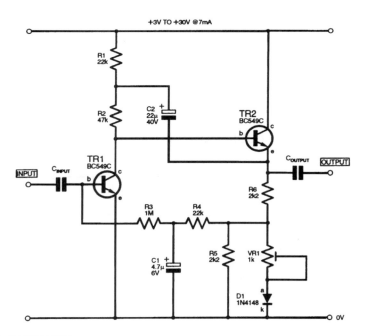

EVERYDAY PRACTICAL ELECTRONICS

Fig. 5-10

This high-gain inverting amplifier stage was designed to operate with a rail of between 3 and 30 V. It includes a bootstrap network (R1, C2), which serves to increase the gain of the stage to approximately 3000, as well as offering a low-distortion output waveform at maximum amplitudes. The input impedance is approximately 80 kV at 200 Hz, the output level is 80 percent at 20 kHz, and noise at the output is 14 mV p-p. R5 and D1 were included to proportionately lower the bias to TR1 to compensate for any increases in the rail voltage; R5 can be omitted and D1 shorted out if the rail is constant. VR1 is adjusted to give symmetrical clipping of the maximum output signal.

MINI MEGAPHONE

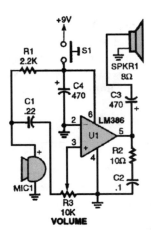

The mini megaphone is composed of an electret microphone (MIC1), an LM386 low-voltage audio-power amplifier (U1), a horn speaker (SPKR1), and a few other components.

POPULAR ELECTRONICS

Fig. 5-11

AUDIO POWER BOOSTER

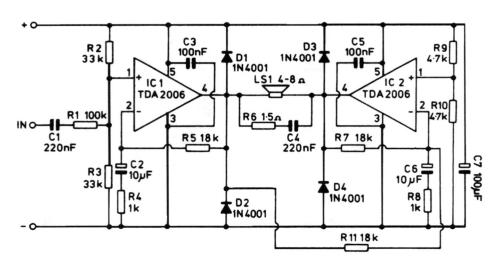

ELECTRONIC EXPERIMENTERS HANDBOOK

Fig. 5-12

The audio power booster is based on two TDA2006 audio IC power amplifiers. Power amplifier IC1 is used as what is virtually a noninverting amplifier, with the noninverting input of the device being biased to half the supply voltage by R2 and R3. R5 provides 100-percent negative feedback from the output to the inverting input of IC1 at dc so that the circuit has unity voltage gain and the output

AUDIO POWER BOOSTER *(Cont.)*

is biased to the required level of half the supply voltage. C2 and R4 remove some of the feedback at audio frequencies, and this gives a voltage gain of about 18 times at these frequencies. R1 is used at the input of the amplifier to reduce the sensitivity to a more suitable level, and C1 simply provides dc blocking at the input. IC2 is used in virtually the same configuration, but its noninverting input is not fed with an audio signal and it only receives the dc bias signal from R9 and R10. Resistor R11 couples the output signal of IC1 to the inverting input of IC2, and the value of R11 is chosen to give IC2 an effective voltage gain of unity. However, as the input signal is coupled to IC2's inverting input, there is a phase inversion through this section of the amplifier, giving the required antiphase relationship at the two outputs. Diodes D1 to D4 are protection diodes for the two ICs, while R6 plus C4 helps to prevent instability. Components C3, C5, and C7 are all supply decoupling capacitors.

MULTIPURPOSE MINI AMPLIFIER

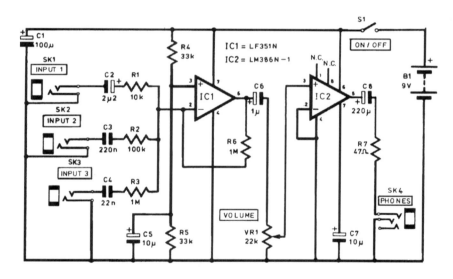

Fig. 5-13

This circuit uses two audio IC amplifiers, one for a preamp and one for a power output stage. Three audio inputs are provided—low (1 mV), medium (10 mV), and high (100 mV) input—to drive a pair of headphones or a small speaker. The LM386 is good for 0.5 W or more output, and a small speaker can be connected directly to the junction of C8 and R7. A series combination of a 22-Ω resistor and a 0.1-mF capacitor should be connected between this point and ground to suppress possible high-frequency oscillations in this case.

PUSH-PULL AUDIO LINE DRIVER

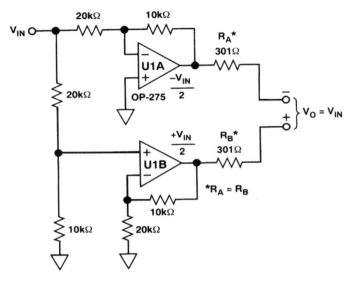

ANALOG DIALOGUE

Fig. 5-14

A line driver for 600 Ω can be configured from two IC op amps. U1A and U1B are Analog Devices OP-275. These devices feature low power consumption, low THD, and high slew rate.

ELECTRET MIKE PREAMP WITH PTT CIRCUIT

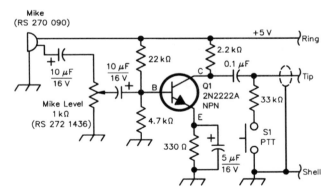

QST

Fig. 5-15

The schematic of electret mike element and amplifier circuit. All resistors are $\frac{1}{4}$-W units. The mike connector is an RS 274-284.

CRYSTAL RADIO AMPLIFIER

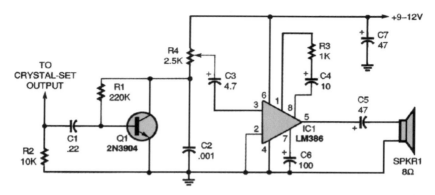

POPULAR ELECTRONICS

Fig. 5-16

If you'd like to hear your crystal set at a comfortable, room-filling volume, hook it up to this amplifier.

6

Amplifier Circuits—Miscellaneous

The sources of the following circuits are contained in the Sources section, which begins on page 1043. The figure number in the box of each circuit correlates to the entry in the Sources section.

WIDEBAND CURRENT FEEDBACK AMPLIFIER

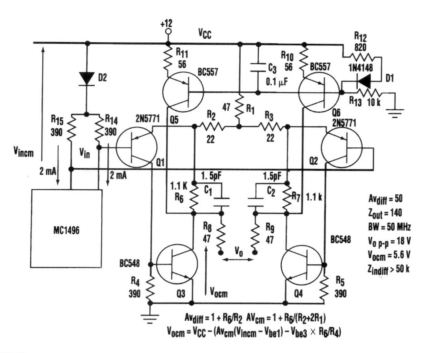

ELECTRONIC DESIGN

Fig. 6-1

When using multipliers such as the LM1496 or MC1495, low-value output resistors are necessary to obtain maximum bandwidth. This reduces the output swing available, which necessitates a differential high-gain, wideband amplifier. The amplifier has a differential gain ($A_{V_{i\,diff}}$) of 50 and a bandwidth of about 50 MHz, giving it a total gain-bandwidth product of 2.5 GHz. It also provides an output swing of 18 V p-p from a 112-V supply. Transistors Q1 and Q2 form a differential pair that drives Q3 and Q4. Feedback is provided to the emitters of Q1 and Q2 by R6, C1 and R7, C2, which bootstraps the input impedance and sets the overall gain. Q5 and Q6 provide about 15 mA of current each, providing most of the sourcing current for the load. The basic design equations are provided to modify component values to suit different applications.

Simply envision the circuit is as a differential current-feedback amplifier, with the low-impedance port being the emitters of Q1 and Q2, and the current output as the collectors of Q3 and Q4. Because the output is a bridge circuit and the maximum positive current is set by R6, R7 and Q5, Q6, the output is short-circuit-protected between resistors R8 and R9. R1 can be replaced by a current source to reduce the common-mode gain. The main criterion is to balance the currents at the emitters of Q1 and Q2 to give a common-mode output voltage of $V_{cc}/2$. Some care should be taken when driving capacitive load because the circuit can oscillate under such conditions.

47

ACCURATE LOG AMPLIFIER

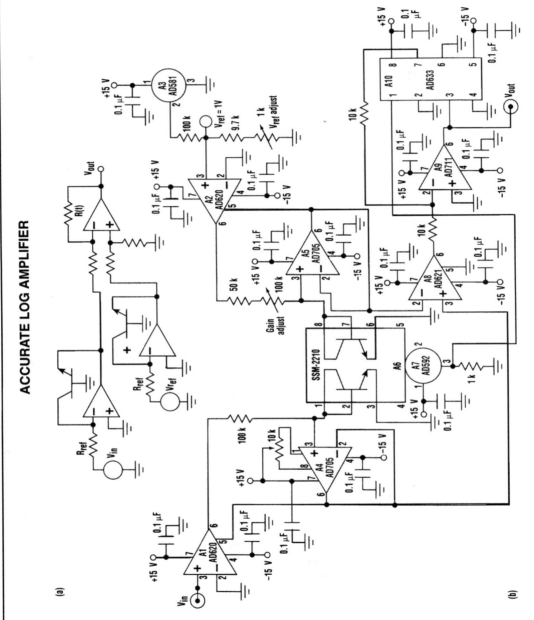

(a)

(b)

ELECTRONIC DESIGN

Fig. 6-2

ACCURATE LOG AMPLIFIER *(Cont.)*

The log circuit consists of an instrumentation amp and an op amp together with a diode-connected transistor that produces a voltage proportional to the logarithm of the current. A circuit consisting of a voltage reference, an instrumentation amp, and an op amp, together with a diode-connected transistor, acts as a reference circuit. A thermometer IC, a fixed-gain instrumentation amp, and a divider circuit provide the necessary temperature compensation and scaling for a transfer function:

$$V_{out} = 51.985 \log_{10}(V_{in}/1 \text{ V})$$

V_{ref} must be set to 1.000 V and, with $V_{in} = 5V_{ref}$, the gain adjust has to be set so that $V_o = 50$ V. Calibration at low input voltage is done by changing buffer A4's offset voltage.

DIFFERENTIAL AMPLIFIER

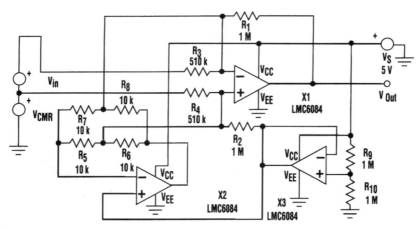

Fig. 6-3

The differential amplifier shown handles common-mode voltages up to ±24 V on a 3.3-V supply, or up to ±40 V on a 5-V supply. It is handy when interfacing analog-to-digital converters (ADCs) or data-acquisition systems (DASs) in 3.3-V or 5-V single-supply systems to inputs with a wide common-mode range. Differential amplifier X1 is the actual differential amplifier, and R1, R2, R3, and R4 are the gain-setting resistors. X2 forces the common-mode voltage at X1's inputs to zero with respect to a quiescent biasing point provided by X3. The wide availability of dual and quad low-voltage op amps permits implementations with a single IC package. Biasing at one-half supply, or at some reference well within the supply rails, is necessary because positive-going common-mode inputs require X2's output to swing negative. R5, R6, R7, and R8 should have impedance at least an order of magnitude less than the gain-setting resistors of X1. The impedance relationship of R6 and R8 to gain-setting resistors limits the use of this circuit to high-impedance differential-amplifier circuits. Therefore, it is not well suited for wide-bandwidth circuits. The high impedances also favor the use of JFET or CMOS input op amps.

WIDER-BANDWIDTH PHOTODIODE AMPLIFIER

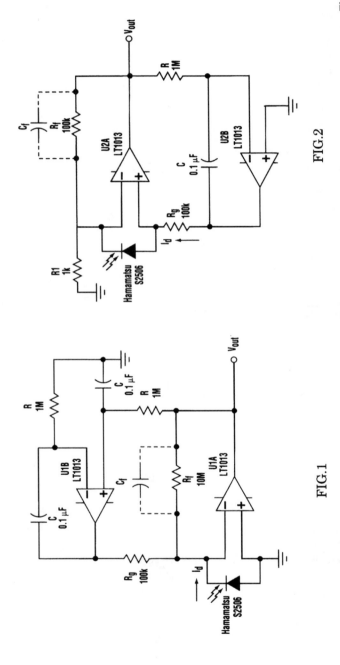

FIG.1

FIG.2

Fig. 6-4

ELECTRONIC DESIGN

Many photodiode applications require high gain and wide bandwidth. A dc-restored photodiode amplifier (Fig. 1) is useful in situations where a time-varying light signal of interest competes with unwanted ambient background light. To extend the amplifier's bandwidth, the design can be modified to provide current-to-voltage conversion with subsequent voltage gain (Fig. 2). Current from the low-capacitance photodiode flows through R_g and establishes a voltage at the noninverting terminal. This voltage then receives a gain equal to $11R_f/R1$. With the values shown in Fig. 2, the equivalent transimpedance $R_g(11R_f/R1)$ is equal to 10 Ω. The design uses an invertingintegrator to drive a restoration current through R_g. This low-frequency current cancels current from the photodiode at frequencies below the low-frequency cutoff pole. The low-frequency pole in Fig. 2 is given by $f_{-3dB}5(1/2\pi RC)(11R_f/R1)$. For the values shown, this is equal to 159 Hz. R_g helps determine the equivalent transimpedance and provides the path for the low-frequency restoration current. The measured high-frequency cutoff for the circuit in Fig. 1 is 4.9 kHz, with a total in-band noise of 370 μV rms. In Fig. 2, the high-frequency cutoff is extended to 11.7 kHz, with a total in-band noise of 589 μV rms.

COMPRESSION AMPLIFIER

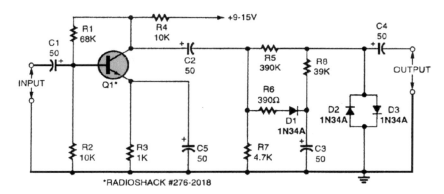

POPULAR ELECTRONICS

Fig. 6-5

Here's a compression-type amplifier that can be used to keep the volume level of an organ constant. Unlike compressors used for public-address-system applications, the organ leveler can respond to the entire range of frequencies generated by the organ without coloring the voice. It can handle large fluctuations in input signal without clipping. It also works well as a microphone leveler.

OP-AMP BIAS-PRECAUTION CIRCUIT

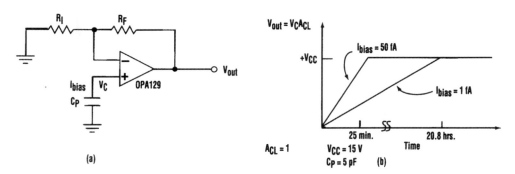

ELECTRONIC DESIGN ANALOG APPLICATIONS

Fig. 6-6

All amplifiers must have a dc path for input-bias current to return to ground (a). Without this path, an amplifier will eventually drift to the output rail (b).

OP-AMP ERROR-SOURCE CIRCUIT

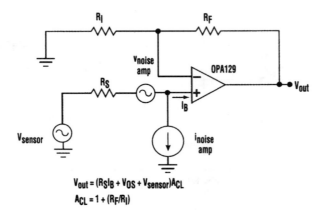

$$V_{out} = (R_S I_B + V_{OS} + V_{sensor})A_{CL}$$
$$A_{CL} = 1 + (R_F/R_I)$$

Fig. 6-7

The basic error sources for an op amp are shown referenced to the output. In high-source-impedance applications, the input bias current terms can become quite dominant.

OP-AMP ERROR-CURRENT MEASUREMENT CIRCUIT

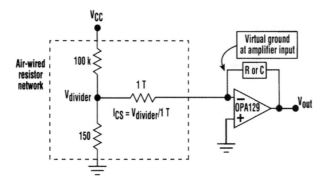

Fig. 6-8

Using an air-wired resistor-divider network, designers can make a precision femtoampere source that can inject a known error current into the amplifier to check the circuit's accuracy.

PROGRAMMABLE AMPLIFIER

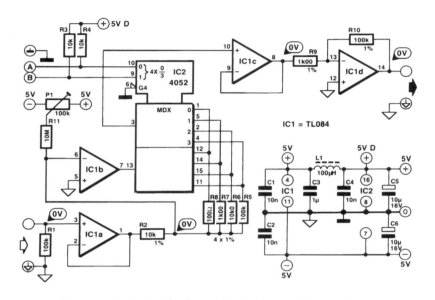

A	B	Amplification	
0	0	1	$(R_8/R_2) \cdot (R_9/R_{10})$
0	1	10	$(R_7/R_2) \cdot (R_9/R_{10})$
1	0	100	$(R_6/R_2) \cdot (R_9/R_{10})$
1	1	1000	$(R_5/R_2) \cdot (R_9/R_{10})$

ELEKTOR ELECTRONICS

Fig. 6-9

The programmable gain amplifier (PGA) is ideal for applications, such as data loggers and automatic measuring instruments. The amplification can be set anywhere between unity and ×1000. The bandwidth at any amplification is >30 kHz. The current drain does not exceed 7 mA. The input signal is buffered by IC1a and then applied to amplifier IC1b. The amplification of this stage depends on how feedback resistors R5 to R8 are switched into the circuit by IC2. How the resistors are intercoupled depends on the logic levels at inputs A and B. The interrelation among the resistors, the total amplification, and the logic levels is shown in the table. After the output signal of IC1b has passed through the multiplexer, it is applied at ×100 amplifier IC1d via buffer IC1c. The overall amplification is set with logic levels at inputs A and B as shown in the table. The amplification of the first stage has been kept low purposely to ensure that the value of the feedback resistor does not have to be high, which makes the effect of the leakage current of the multiplexer negligible.

CLAMP AMPLIFIER

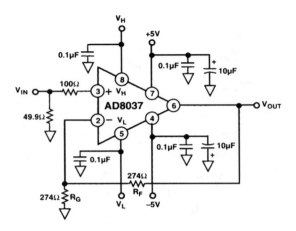

ANALOG DIALOGUE

Fig. 6-10

A clamp amplifier is a limiting or bounding circuit. For input voltages between two levels, V_H and V_L, the output is proportional to the input. For inputs greater than V_H or less than V_L (V_H or V_L × the amplifier gain, A_v), irrespective of the input. The threshold voltages, V_H and V_L, can be fixed or variable. If the amplifier can handle positive and negative input/output voltages, V_H and V_L can have any positive or negative value within the specified range, as long as $V_H > V_L$.

LM3900 AC AMPLIFIER

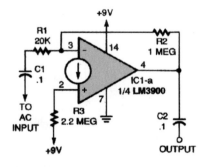

Although the LM3900 current difference amplifier (IC1) used in this ac amplifier really isn't a true op amp, it does simulate one in its performance; however, the IC requires only a one-polarity power source.

Gain is R2/R1, and R3 is 2×R2.

POPULAR ELECTRONICS

Fig. 6-11

CURRENT DIFFERENCE AMPLIFIER

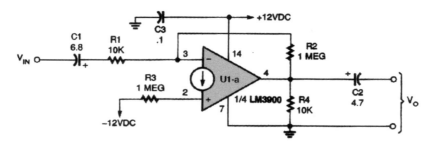

Fig. 6-12

This practical 20-dB gain ac amplifier uses the LM3900 CDA.

CURRENT DIFFERENCE AMPLIFIER USAGE CIRCUITS

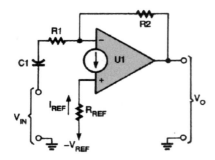

Here a current difference amplifier is used much like an op amp in an inverting follower circuit.

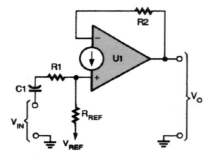

Here is a noninverting follower built around a current difference amplifier.

Fig. 6-13

POWERED SUBWOOFER

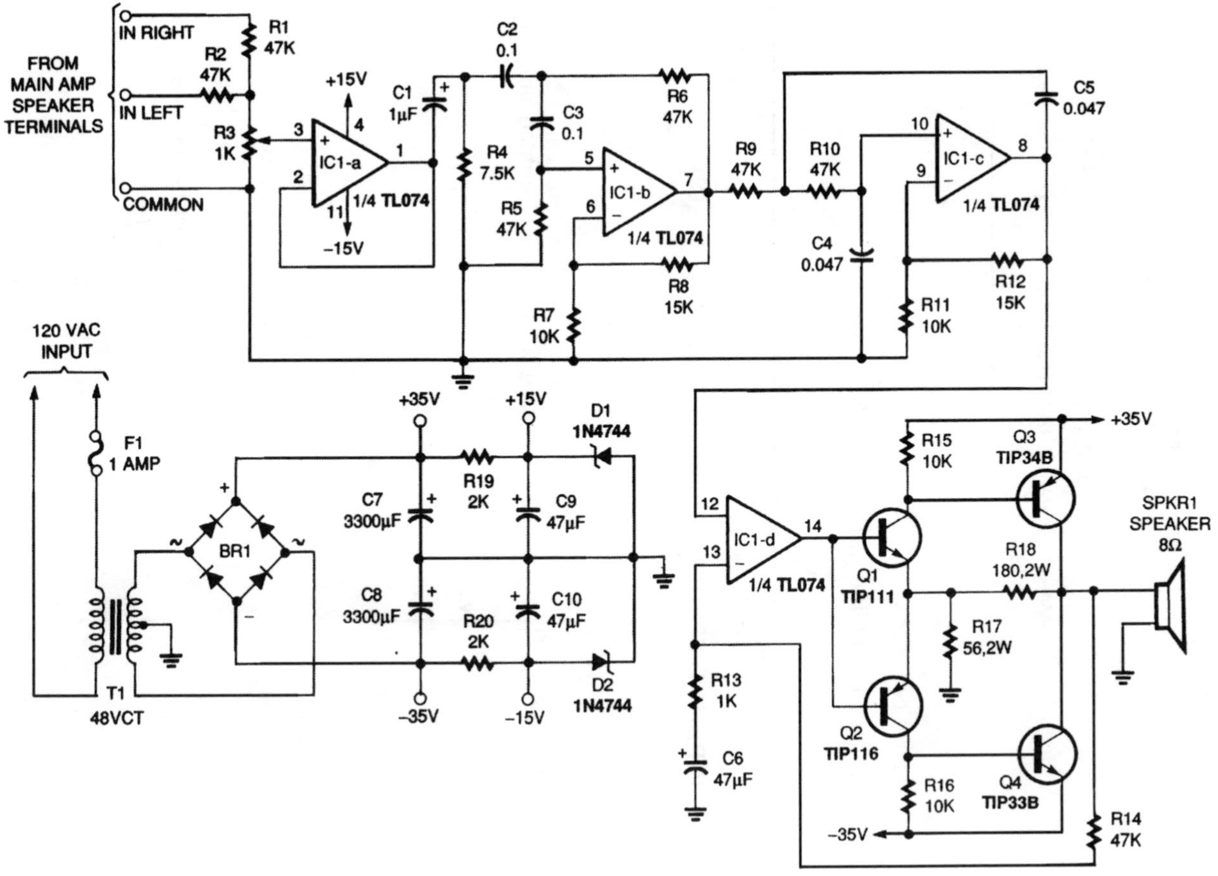

Fig. 6-14

POWERED SUBWOOFER *(Cont.)*

The power supply consists of a center-tapped 48-V transformer, a bridge rectifier, and filter capacitors. The rectified and filtered output is about ±35 V. The power supply for op amp IC1 is regulated to ±15 V by zener diodes D1 and D2 and resistors R19 and R20. The input circuit consists of a mixer and voltage divider formed by resistors R1 and R2, potentiometer R3, and unity-gain buffer IC1-a. Potentiometer R3 is provided to adjust the output of the subwoofer to the desired level. Op amp IC1-b provides a 12-dB-per-octave high-pass filter with capacitors C2 and C3 and resistors R5 and R6. The cutoff frequency for this filter is $1/2\pi RC$, or about 34 Hz with the values shown. Resistors R8 and R7 set the gain and Q of the filter. Capacitor C1 and resistor R4 form an additional 6-dB-per-octave high-pass filter at about 20 Hz. A 12-dB-per-octave low-pass filter is formed by IC1-c, C4 and C5, and R9 and R10. The values shown set the low-pass cutoff at 72 Hz. The gain and Q of this stage are set by R11 and R12. These two filters, connected back-to-back, form a bandpass filter. When operated with ±15-V supplies, the output of op amp IC1-d can swing about 10 V peak to drive transistors Q1 and Q2. Resistors R17 and R18 provide negative feedback, and set the gain of the output stage at about 3. Hence, the output can swing to about 30 V peak. As long as the transistors are the high-beta types specified, the peak power output into an 8-Ω load is (30×30/8)/2556 W rms. The overall gain of the amplifier is set by resistor R13 and feedback resistor R14.

HIGH-GAIN AMPLIFIER FOR PHOTODETECTORS

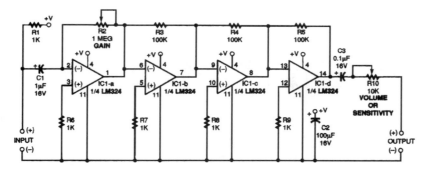

Fig. 6-15

Suitable for photodetectors, laser experiments, or general use, this four-stage amplifier uses a single LM324 quad op amp. A 9-V supply is recommended.

CAR STEREO SUBWOOFER CROSSOVER

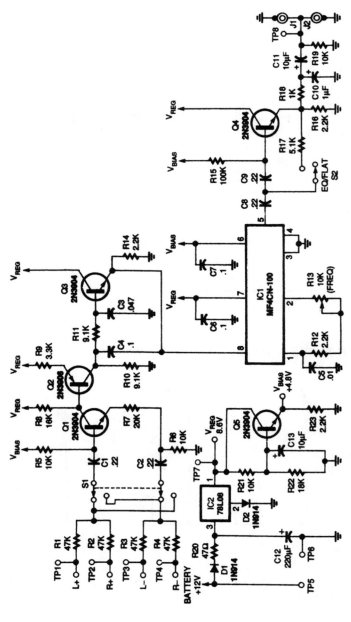

Fig. 6-16

ELECTRONICS NOW

This figure is the schematic for the subwoofer crossover. The inputs through Q1 form a differential summing amplifier, with switch S1 functioning as a polarity inverter. A 24-dB-per-octave switched capacitor filter (IC1) is the heart of the continuously variable filter. Potentiometer R13 controls the cutoff frequency of IC1 by controlling its sampling frequency. Because of the inherent sampling action of switched capacitor filters, an antialiasing filter is required at the input of IC1. Transistors Q2 and Q3, along with the surrounding components, form this second-order low-pass antialiasing filter. The subsonic filter with a boost stage follows the output of IC1 at 5. When switch S2 is closed, the boost is added. Additional subsonic filtering action is provided by C1 and C2 at the inputs of the crossover circuit. A "reconstruction filter" that eliminates sampling artifacts is formed by R18 and C10

58

CAR STEREO SUBWOOFER CROSSOVER *(Cont.)*

at the output of IC1. The power-supply circuit, based on the 78L08 voltage regulator (IC2), provides both an 8.6-V main supply and a 4.8-V bias supply. Diode D1 protects against negative voltage spikes and incorrect hookup. Diode D2 biases the 78L08 regulator reference pin at 0.6 V to provide an output of 8.6 V, rather than 8 V.

CURRENT LIMITERS

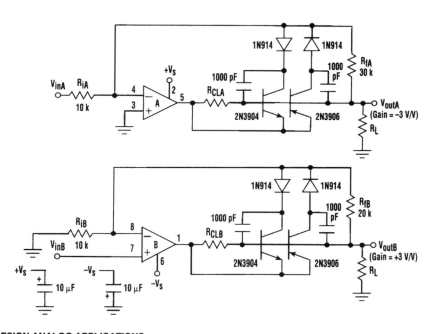

ELECTRONIC DESIGN ANALOG APPLICATIONS

Fig. 6-17

A user-controlled internal current limit on power op amps isn't always provided by manufacturers. However, the current limit can be set externally by using the technique shown, in which the output current is sensed with a single resistor. The resistor activates a complementary transistor switch that reduces inverting gain, limiting output current. The technique is demonstrated with 8-pin OPA2541 or OPA2544 dual-power op amps that are lacking an internal user-controlled current limit. The external components add an adjustable current limit to these amplifiers. The PNP transistor controls the positive current limit, and the NPN transistor controls the negative current limit. Both transistors are OFF until the voltage drop across R_{CL} reaches the current-limit set point. At the current-limit set point, current from the controlling transistor will sum with the input current (through R_i) at the op amp's inverting input summing junction. This will limit the output. The diodes in series with the collectors of both transistors prevent forward base-collector bias. High-frequency oscillation during current limit is damped out with the 1000-pF capacitors. These capacitors only nominally affect the configuration's closed-loop bandwidth because the diodes, which are normally off, isolate them.

STEREO AUDIO COMPRESSOR

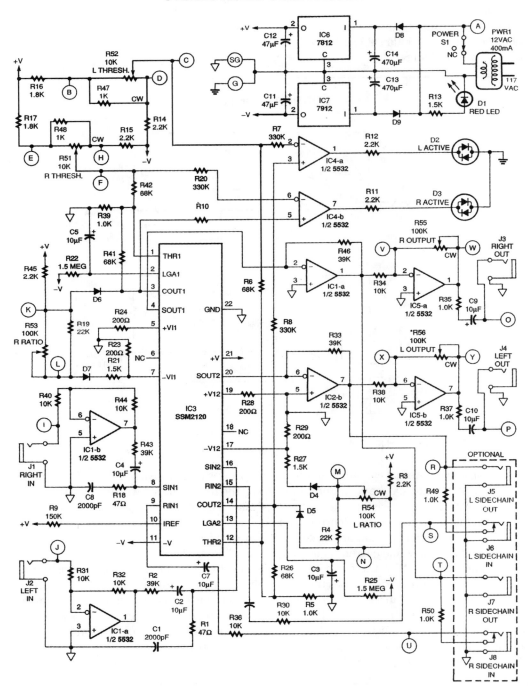

STEREO AUDIO COMPRESSOR *(Cont.)*

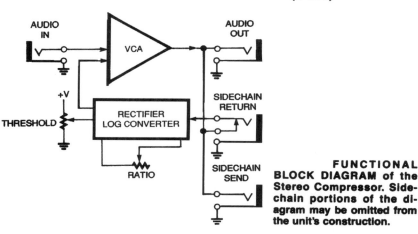

FUNCTIONAL **BLOCK DIAGRAM of the Stereo Compressor. Sidechain portions of the diagram may be omitted from the unit's construction.**

Fig. 6-18

The block diagram of the stereo compressor is shown. The input signal is fed to a voltage-controlled amplifier (VCA) that has a nominal gain of unity. Some of the output signal is fed to a precision rectifier followed by a logarithmic converter circuit. The output of this block is a dc voltage proportional to the log of the average level of the input signal. By sending some of this dc control voltage to the VCA, the gain of the VCA is automatically reduced when the input signal exceeds a user-determined threshold level. It is important to note that the signal level is determined after the VCA and not before. This allows the output level to increase and sound normal, but not increase as much as the input signal does. By varying the amount of feedback, the compression ratio is adjusted, which, in conjunction with the THRESHOLD control, determines the operating characteristics of the compressor. The optional sidechain jacks permit external processing of the audio signal or substitution of a completely different audio signal as the control signal. This add-on circuitry lets the user experiment and achieve some useful audio effects.

OP-AMP RESPONSE NULL CIRCUIT

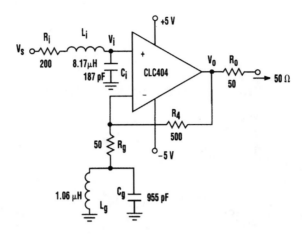

ELECTRONIC DESIGN ANALOG APPLICATIONS

Fig. 6-19

A sharp null can be achieved in a current feedback op amp's frequency response by adding a sharp cutoff filter. The response is modified by adding a resonant circuit in series with the gain-setting resistor R_g.

7

Amplifier Circuits—RF

The sources of the following circuits are contained in the Sources section, which begins on page 1043. The figure number in the box of each circuit correlates to the entry in the Sources section.

1.8- to 7.2-MHz POWER AMPLIFIER

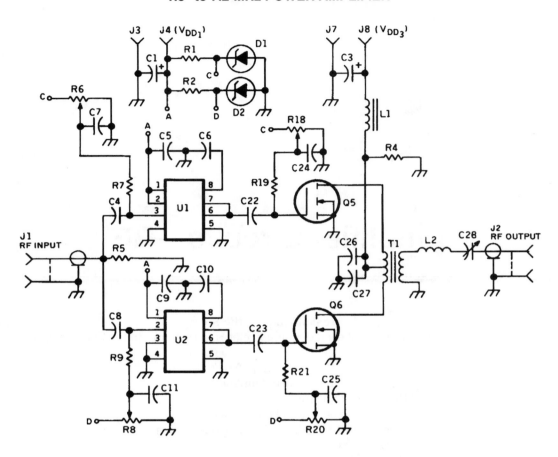

Fig. 7-1

A novel approach to a 250-W power amplifier (PA) is shown. The driver uses a pair of low-cost ICs, rather than the conventional RF transformer to provide out-of-phase driving signals for the two final MOSFETs. It also provides hard limiting (sine-wave–to–square-wave conversion) of the input signal. The Elantec EL7144C is intended for use as a gate driver. The internal Schmidt trigger allows it to serve as a hard limiter, and the presence of both inverting and noninverting inputs allows a pair to serve as a phase splitter. The RF input is ac-coupled to the noninverting input of U1 and the inverting input of U2. Adjustment of the biases via R6 and R8 allows the transition points to be selected to produce the desired duty ratio (50:50). The phase error between the two EL7144s is about 0.5 ns. (If an oscilloscope is not available, use a voltmeter with a 1-kΩ series resistor and set

Reference Designator	Part Description
C1	33-µF, 50-V electrolytic, Mallory
SKR33OM1HE11V	
C3	20-µF, 250-V electrolytic, Mallory TT25OM20A
C4-C27	0. 1-µF, 50-WV chip, ATC 200B 104NP50X
C28	
DI, D2	5.1-V, 0.25-W Zener, 1N751A
J1, J2	BNC female bulkhead
	Recommended: Amphenol 31-5538
	Used: RF Industries RFB-1116S/U
J3, J4, J7, J8	European-style binding post, Johnson 111-0104
Ll	3.5-µH, 7 tums #24 enameled wire on Ferroxcube
L2	768XT188 4C4 toroid.
Q5	*n*-channel MOSFET, APT ARF440
Q6	*n*-channel MOSFET, APT ARF441
Rl, R2	330-Ω RC07
R4	1O-kΩ RC07
R5	51-kΩ RC07
R6, R8, RI8, R20	1-kΩ trimpot, Bournes 3299X-1-102
R7, R9, R19, R21	4.7-kΩ RC07
T1	One Ceramic Magnetics 3000-4-CMD5005 block of CMD5005 ferrite
U1, U2	Schmidt trigger/limiter, Elantec EL7144C
Heatsink for DIP IC	Aavid 5802 clip-on
Heatsink for Q5 and Q6	Aavid 61475, 6.5-in wide by 4-in long
IC socket, 8-pin DIP (4)	Augat 208-AG 190C
PC board	Approximately 6.5-in wide by 8-in long
Plastic bracket for L2	Cut from plastic L
Plastic bracket for C28	Cut from plastic L
Feet (6)	Aluminum threaded spacer, 4-40 × 2 in (Keystone 2205)

the average output voltage to 6 V.) The EL7144s have high input impedances, so R5 provides a 50-V input impedance for the signal source. Input signals in the range of 10 to 100 mW are satisfactory, allowing this PA to be driven directly from a laboratory signal generator or oscillator with buffer. The best switching speed is obtained with VDD1=12 V. Output transistors are ARF440 and ARF441. See parts list.

25-W SOLID-STATE LINEAR AMPLIFIER

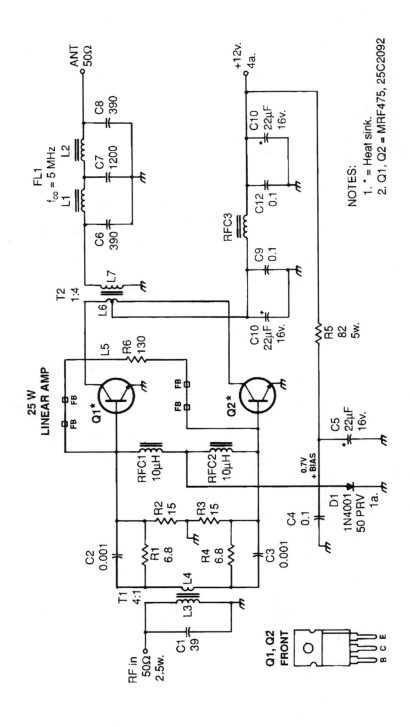

NOTES:
1. * = Heat sink.
2. Q1, Q2 = MRF475, 25C2092

CQ MAGAZINE

Fig. 7-2

This shows a schematic diagram of the 25-W linear amplifier. Decimal-value capacitors are in picofarads; others are in microfarads. Polarized capacitors are electrolytic or tantalum, 16 V or greater. Resistors other than R5 are 1/4-W carbon-composition or carbon-film units.

BIDIRECTIONAL RF AMPLIFIER

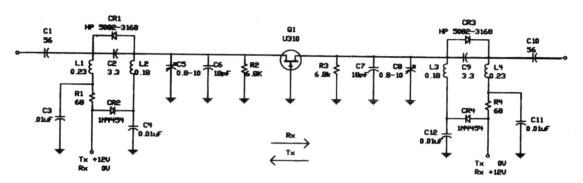

RF DESIGN

Fig. 7-3

JFETS can often be used with the source and drain interchanged. This interesting circuit makes use of this fact as a bidirectional RF amplifier at 70.0455 MHz. It is useful for transceiver applications.

GENERAL-PURPOSE WIDEBAND PREAMP

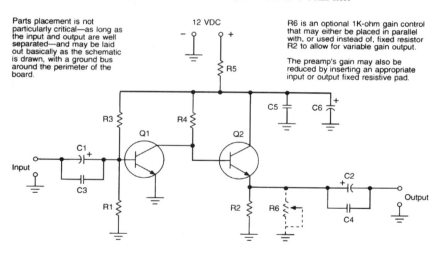

Parts placement is not particularly critical—as long as the input and output are well separated—and may be laid out basically as the schematic is drawn, with a ground bus around the perimeter of the board.

R6 is an optional 1K-ohm gain control that may either be placed in parallel with, or used instead of, fixed resistor R2 to allow for variable gain output.

The preamp's gain may also be reduced by inserting an appropriate input or output fixed resistive pad.

73 AMATEUR RADIO TODAY

Fig. 7-4

This preamp has a gain of 35 dB at 100 kHz, 30 dB at 10 MHz, and 17 dB at 100 MHz and draws 15 mA at 12 V.

LOW-NOISE 9-MHz AGC-CONTROLLED AMPLIFIER

	16 dB	13 dB
	gain	gain
Noise figure	0.6 dB	.6 dB
Input impedance	50 ohms	50 ohms
Output impedance	50 ohms	50 ohms
Third order intercept		
$V_s = +12V$ (max gain)	23 dBm	26 dBm
$V_s = +20V$ (max AGC)	28 dBm	30 dBm
Input for 1 dB compression		
$V_s = +12V$ (max gain)	0 dBm	+3 dBm
(max AGC)	7 dBm	+11 dBm
$V_s = +20V$ (max gain)	+5 dBm	+8 dBm
(max AGC)	+11 dBm	+14 dBm

Amplifier input and output impedances are 50 ohms regardless of AGC-controlled gain. Gain range 45 dB.

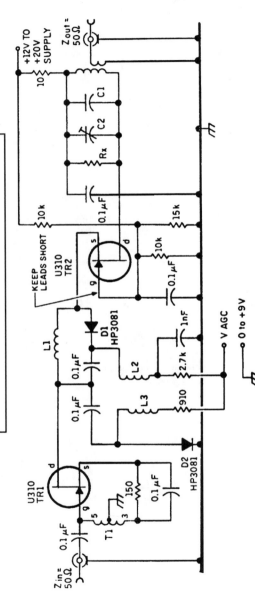

Fig. 7-5

LOW-NOISE 9-MHz AGC-CONTROLLED AMPLIFIER *(Cont.)*

The circuit diagram of the 9-MHz amplifier is given. Note that the AGC amplifier must be capable of sinking the current through D1 at 0 V (i.e., maximum gain). The warning to keep leads short in the drain circuit of the second U310 FET arises from experience in which the initial IP3 measurements proved to be poor because of this stage's oscillating around 400 MHz.

Component notes: The two U310s are Siliconix low-noise JFETs. C1 is an 82-pF ceramic, C2 is a 60-pF ceramic trimmer (Cirkit), and all other capacitors are monolithic ceramic (RS components). Resistors are 1/8 W metal film (RS components). D1 and D2 are HP3081 PIN diodes (Farnell). T1 is 1513 turns of 0.224-mm diameter Bicelflux enamel on Fairite Balun core 28-43002402 (Cirkit). T2 (primary) is 2.81 μH, 31 turns of 0.314-mm Bicelflux enamel on Micrometals toroid T37-6 (Cirkit). T2 (secondary) is (1) for 16-dB gain, 3 turns, Rx 6k2; (2) for 13-dB gain, 4 turns, Rx 3k9. Note that (1) and (2) could be relay switched for use with an SSB or CW filter (loss 10 dB or 3 dB). L1, L2, and L3 have 7 turns of 9.314-mm enamel on balun core 28-43002402 (Cirkit).

RECEIVER PREAMPLIFIER

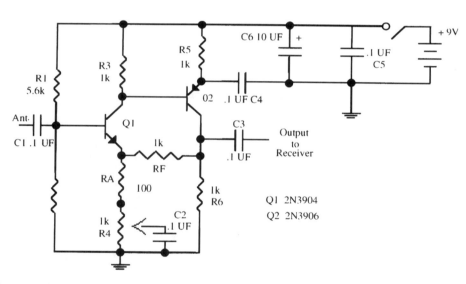

Fig. 7-6

This preamplifier lets you use a short antenna over the range from 100 kHz to well over 55 MHz, with excellent sensitivity, using a vertical 30-in piece of #12 wire or a few feet of wire laying on the floor. Transistor Q1 (NPN) is directly coupled to PNP transistor Q2. Feedback from the collector of Q2 to the emitter of Q1 is accomplished by resistors RF and RA. Because of the high open-loop gain of the amplifier, the gain of the amplifier is R_F/R_A, or 20 dB maximum. A 1-kΩ potentiometer changes the effective value of R_A, resulting in a minimum gain of near unity.

RECEIVER RF PREAMP

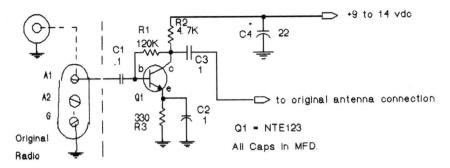

Fig. 7-7

The RF broadband preamplifier uses an NPN VHF transistor as an untuned broadband (0.5 to 30 MHz) RF amplifier. Input impedance is 50 Ω, allowing usage for all receiver inputs, and the unit has a 600-Ω output to match virtually all RF input circuits. The preamp delivers 30 dB of gain at 10 MHz, with a noise factor below 1 dB. Power for the preamp can be obtained from a variety of sources. The assembly requires from 9 to 14 Vdc and draws 8 mA of current. This makes it ideal for use in battery-operated portable equipment.

RF PREAMP SUPPLY

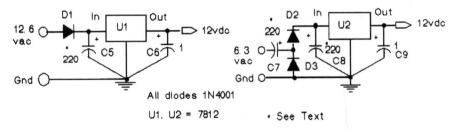

Fig. 7-8

Early tube-type receivers used filament voltages of 6.3 and 12.6 Vac, but had no low-voltage dc power supplies. The figures show how to "borrow" a little of the filament voltage, which is rectified and regulated to provide 12 Vdc for the preamp. In the 6.3-Vac version, diodes D2 and D3 act as a simple voltage doubler to step the input voltage up to approximately 15 Vdc. While the 220-μF capacitors used for the voltage doubler are sufficient for the load presented by the preamp, they will not provide the 115 V dc to the input of the regulator at more than about a 25-mA load. If other circuits will be used with these dc power sources, then the 220-μF capacitors should be increased accordingly (2200 μF will provide about 65 mA regulated output).

COMBINED MMIC AMPLIFIERS

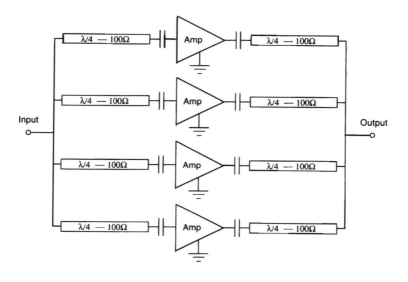

RF DESIGN

Fig. 7-9

Quadrature combining, practical in microstrip, can be used to obtain a medium power output from several MMIC amplifiers.

CASCADED MMIC AMPLIFIER CIRCUIT

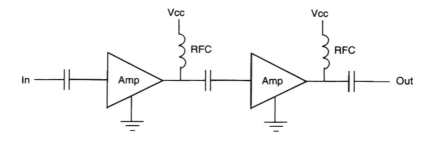

RF DESIGN

Fig. 7-10

Most MMIC amplifliers can be cascaded, with dc blocking as the only interstage connection requirement.

MMIC AMPLIFIER CIRCUIT

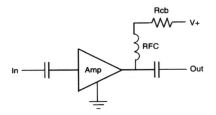

The addition of a series dropping resistor allows operation at higher supply voltages.

Fig. 7-11

SIMPLE MMIC AMPLIFIER CIRCUIT

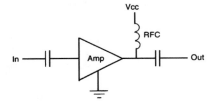

This is the simplest implementation of an MMIC amplifier.

Fig. 7-12

8

Amplifier Circuits—Vacuum Tube

The sources of the following circuits are contained in the Sources section, which begins on page 1043. The figure number in the box of each circuit correlates to the entry in the Sources section.

Single-Ended Hi-Fi Amplifier
Voltage-Controlled Amplifier Stage
Audio Amplifier

SINGLE-ENDED HI-FI AMPLIFIER

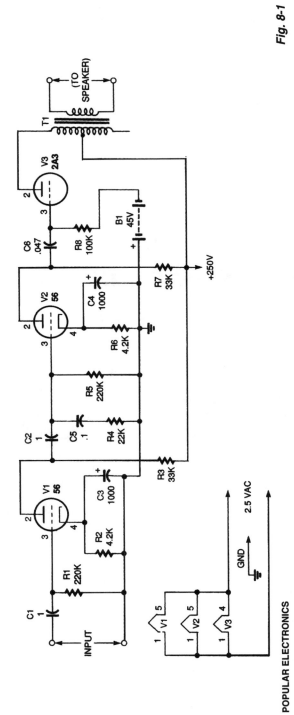

POPULAR ELECTRONICS

Fig. 8-1

The circuit as shown will handle up to about 0.5 V; above that, distortion will be present. The input stage amplifies the signal voltage about seven times, and the second stage amplifies the voltage about 10 times. Both stages use type 56 triodes (V1 and V2). The output stage uses a type 2A3 tube, V3. Two 22.5-V batteries wired in series (B1) were used to provide 45 V on the grid of V3. In this circuit (and other single-ended amplifiers), direct current flows in the primary of the output transformer (T1) to use the circuit modification shown in Fig. 2. That change keeps dc out of the primary at some sacrifice in power. With the modification, the amplifier will be flat (within 1 dB) from 20 to 20,000 Hz. (The modified circuit requires that a high-voltage transformer be used in the power-supply circuit.) Note that transformer T1 is shown connected from one end of the primary to the center tap. That was done because it is assumed that the transformer used will have a primary impedance of about 8000 Ω. The recommended load for a 2A3 tube (V3) is 2500 Ω for maximum output, but increasing the impedance lowers the distortion while only slightly lowering the power.

74

VOLTAGE-CONTROLLED AMPLIFIER STAGE

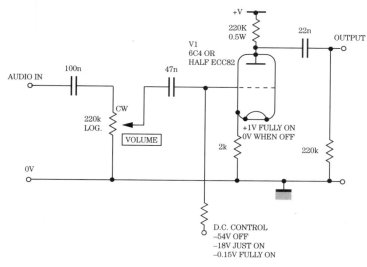

Gain is controlled by a variable negative grid bias in this amplifier stage.

Fig. 8-2

AUDIO AMPLIFIER

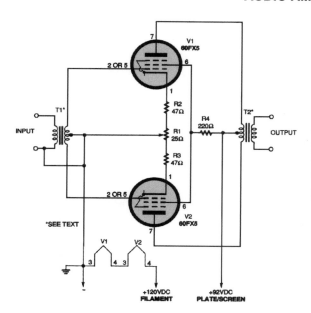

Some hobbyists still prefer to use older vacuum tube technology. A push-pull audio amplifier using a pair of 60FX5 tubes is shown in the figure. T1 is a 1:3 audio interstage, while T2 is a universal output transformer of about 5000 Ω to voice coil (4 or 8 Ω) impedance.

Fig. 8-3

9

Analog-to-Digital Converter Circuits

The sources of the following circuits are contained in the Sources section, which begins on page 1043. The figure number in the box of each circuit correlates to the entry in the Sources section.

ADC CONTROLLING DIGITAL POTENTIOMETER

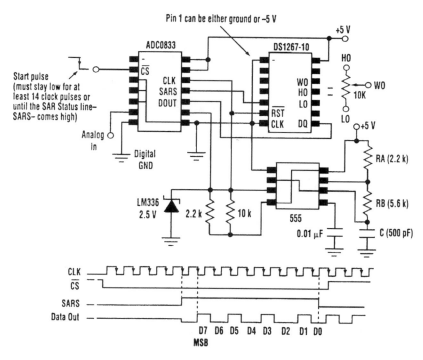

Fig. 9-1

This design was created to control a DS1267 digital potentiometer with an analog signal. The DS1267 is a dual-pot chip, but this design will enable control of only one section. Here, the analog-to-digital converter (ADC) used is an ADC0833 8-bit serial I/O converter with a four-channel multi-plexer. The analog input to channel 3 of the ADC is employed. The timing diagram illustrates the operation of the circuit. A negative start pulse on the chip select of the ADC starts the sequence (the pulse must stay low for at least 14 clock pulses or until the ADC's SAR Status line comes high). The next five clock pulses perform various housekeeping in the ADC. The Data Out line comes out of tris-tate on the negative edge of the fifth pulse, and the SAR Status line comes high to signal a conversion in progress. The first bit on the Data Out line is a leading zero for one clock period. Data is clocked into the DS1267 on the positive edge of the clock pulse. The input format for the DS1267 requires that the first bit determine the stack select (used in the DS1267 when the two pots are combined) and the following 8 bits provide data. Following transmission of these 9 bits, the SAR Status line goes high, disabling further input to the DS1267 (the ADC0833's output format continues transmission of 8 more bits of the conversion in reverse order, but the DS1267 ignores these). Input range for the ADC is 0 to 5 V. Pin 1 of the DS1267 is shown tied to ground (for pot connections referenced to ground; however, −5 V can be used to provide a range of 15 to −5 V on the pot).

ANALOG-TO-DIGITAL CONVERTER CIRCUIT

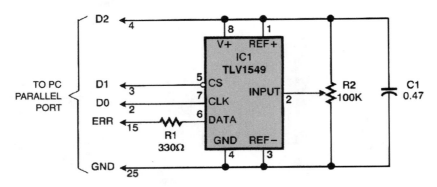

Fig. 9-2

A working circuit that uses a PC parallel port to receive data from a 10-bit analog-to-digital converter (ADC) (Texas Instruments TLV1549) is shown. A fourth wire for reset can be powered from a logic-level signal. In fact, if your printer port produces 5 V (not all do), you can use the more-common TLC1549. Resistor R1 reduces transmission errors by isolating the ADC from the cable capacitance; it might be unnecessary if the cable is short.

10

Antenna Circuits

The sources of the following circuits are contained in the Sources section, which begins on page 1043. The figure number in the box of each circuit correlates to the entry in the Sources section.

Antenna Tuning Indicator
Low-Power Antenna Tuner and SWR Meter
Op-Amp Antenna Amplifier
10-GHz Waveguide Test Antenna Setup
Balun Box
Microwave Reflection Antenna
Antenna Noise Bridge Detector
Inverted Vee Antenna
Swept Oscillator for Ham-Band Antenna Tuning
40-m Loop Antenna
Undercover Scanner Antenna
800-MHz Antenna
10-W Dummy Load
Coaxial Line Balun
Toroidal Transformer
Simple Antenna Tuner

ANTENNA TUNING INDICATOR

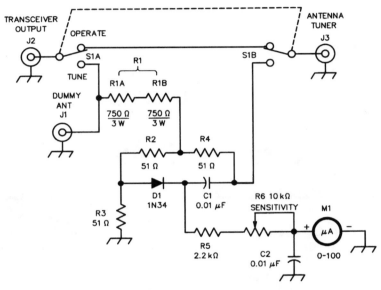

C1, C2—0.01-μF general-purpose ceramic
  capacitor
D1—Germanium diode (1N34, 1N60,
  1N270 or equivalent)
J1-J3—Coaxial jacks
M1—0-50 or 0-100 μA dc meter
R1A, R1B—750-Ω, 3-W, 5%-tolerance
  metal-oxide film resistor

R2-R4—51-Ω, ½-W, 5%-tolerance carbon-
  film resistor
R5—2.2-kΩ, ½-W, 5%-tolerance carbon-
  film resistor
R6—10-kΩ, linear taper potentiometer
S1A—Double-throw, double-pole (DPDT)
  toggle switch (I used one rated for
  120-V service)

QST MAGAZINE

Fig. 10-1

The stealth antenna tuning indicator consists of a sensitive reflected-power bridge and indicator (R2 to R6 and associated components) and a switch (S1) that lets you route your transceiver's
power into a dummy antenna (TUNE) or your antenna system (OPERATE). One more component,
R1 (a resistance consisting of two separate 3-W resistors, R1A and R1B), leaks just enough of your
transmitter's power to the bridge and your antenna system to allow low-interference tune-up when
S1 is set to TUNE.

LOW-POWER ANTENNA TUNER AND SWR METER

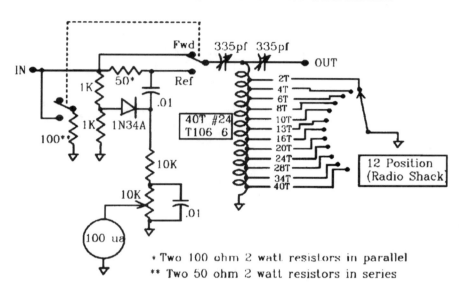

Fig. 10-2

The SWR meter is a variation of a resistive bridge. This circuit has the advantage of providing high sensitivity at low power inputs. Its disadvantage is that the maximum power that can be applied to the circuit is limited by the power dissipation of the 50-Ω bridge resistors because 75 percent of the transmit power is absorbed by these resistors under perfectly matched conditions. In this circuit, a DPDT switch adds in a 100-Ω resistor only for SWR readings, allowing the circuit to provide a reasonable SWR for your transmitter even during severe tune-up conditions. Additionally, you have minimum insertion loss in the FORWARD position because the 1-kΩ bridge resistors permit you to always indicate forward relative power and calibrate the full-scale reading for SWR measurements. The antenna tuner itself is a standard T-type tuner utilizing a pair of 335-pF miniature transistor radio variable capacitors and a toroidal inductor, tapped as shown in the figure. To actually make the taps, simply wind the toroid with 40 turns of #24 wire, and then scrape the outer surface of each appropriate turn with a hobby knife to clean off the enamel insulation. Then tin each wire turn, and tack-solder tap wires to them.

OP-AMP ANTENNA AMPLIFIER

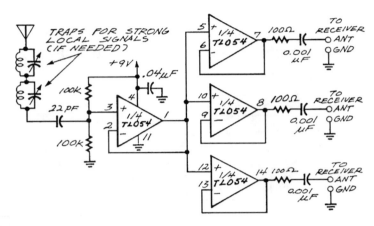

ELECTRONICS NOW

Fig. 10-3

The impedance of a random-wire antenna at broadcast-band frequencies will be several kilohms, a poor match for the 50-Ω input of a receiver. Also, the inputs of multiple receivers can't be connected directly to a single antenna, or they'll detune each other. The circuit solves both problems. The first op amp overcomes the impedance mismatch, strengthening the signal greatly even though it has no voltage gain. The rest of the op amps feed the signals to the separate receivers. No low-pass filter is needed because the gain of these op amps drops off sharply above 2 or 3 MHz.

10-GHz WAVEGUIDE TEST ANTENNA SETUP

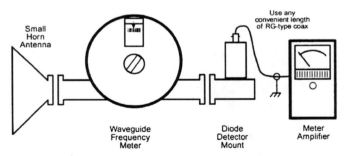

73 AMATEUR RADIO TODAY

Fig. 10-4

This waveguide test antenna, wavemeter, and meter amplifier setup uses a small horn antenna and a variable commercial waveguide frequency meter to determine RF frequency. RF is sampled with a detector and displayed on an amplified meter for sensitive meter indications.

BALUN BOX

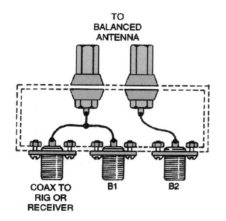

TO
BALANCED
ANTENNA

COAX TO
RIG OR
RECEIVER

B1

B2

Here's the connection box for making a 4:1 coaxial balun. It is intended for mounting on the antenna center isolator.

POPULAR ELECTRONICS

Fig. 10-5

MICROWAVE REFLECTION ANTENNA

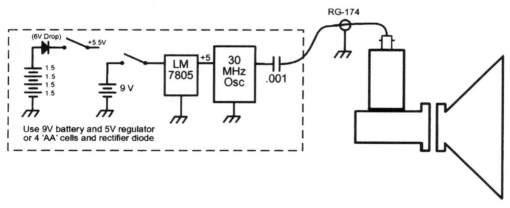

RG-174

(6V Drop)   +5.5V

1.5
1.5
1.5
1.5

9 V

LM
7805   +5

30
MHz
Osc

.001

Use 9V battery and 5V regulator
or 4 'AA' cells and rectifier diode

73 AMATEUR RADIO TODAY

Fig. 10-6

Originally used for 10-GHz microwave transmitter testing, this antenna modulates a received signal with 30 MHz and reflects it, producing a signal component 30 MHz offset from the incident signal for reception and dish-aiming tests.

ANTENNA NOISE BRIDGE DETECTOR

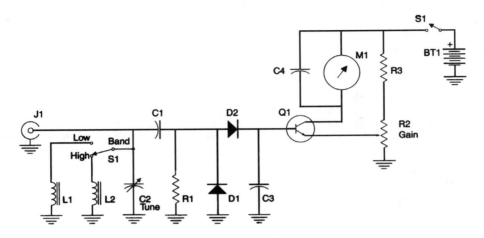

Parts List

BT1 9-Volt alkaline battery
C1   100 pF mica or poly capacitor
C2   140-150 pF variable capacitor

C3   0.1 µF disc capacitor
C4   0.01 µF disc capacitor
D1,D2 Germanium diode: 1N34,
        1N60, 1N90, 1N270, etc.
J1   SO-239 UHF female connector
        (or builder's choice)
L1   33 turns #26 enam. wire on
        T50-2 toroid (40-30 meters)
L2   11 turns #26 enam. wire on
        T50-2 toroid (20 through 10
        meters)
M1   100 or 200 µA meter (see
        text)
Q1   NPN small signal transistor:
        2N2222, 2N3904, 2N4124,
        etc.
R1,R3 10k 5% 1/4 Watt resistor
R2   1000 Ohm linear taper poten
        tiometer, with switch S2
S1   SPDT toggle or slide switch
        (Bandswitch)

ANTENNA NOISE BRIDGE DETECTOR *(Cont.)*

The circuit is illustrated in the figure. The noise output from your antenna noise bridge is applied though a coax jumper cable to J1, an SO-239. This noise, which usually will peak slightly below 1.0 V, is broadband white noise and is fed though C1, a 100-pF capacitor, to a pair of small-signal diodes connected as a rectifier/voltage doubler. The rectified dc voltage, filtered by C3, an 0.1-μF disk capacitor, is then applied to the base of a small-signal NPN transistor (Q1), which serves as a meter amplifier. Meter M1 is a 200-μA meter. It monitors collector current through Q1. On/off switch S2 is mounted on the GAIN control, which, in series with R3, forms a voltage divider across battery BT1, a 9-V battery, which powers this instrument. The GAIN control is wired so that the wiper travels from the end of R3 to ground as the knob is rotated clockwise. This sets the emitter bias and the point at which Q1 will go into conduction as rectified noise voltage is applied to the base of Q1. The current drain from the battery is approximately 8 μA with no input, increasing to slightly over 200 μA with the meter at full scale. With such low current drain, an alkaline battery should last for years, even if you forget to turn the instrument off! This instrument covers the range from below 40 m to above 10 m in two bands: 40 and 30 m, and 20–17–15–12–10 m. Bandswitch S1, an SPDT toggle or slide switch, selects the frequency range. The tuning capacitor C2 is a small 150-pF air variable.

INVERTED VEE ANTENNA

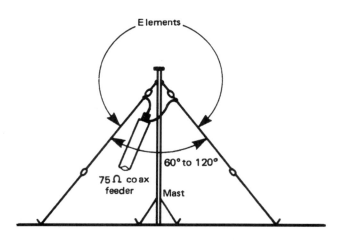

ELECTRONIC EXPERIMENTERS HANDBOOK

Fig. 10-8

The inverted vee antenna is widely used by SWLs and radio amateurs. It has the advantage of needing only one support and can be fit into somewhat smaller spaces in some instances. A balun having a 1:1 impedance designed for 50 Ω can be installed at the feed point for improved performance, if desired.

SWEPT OSCILLATOR FOR HAM-BAND ANTENNA TUNING

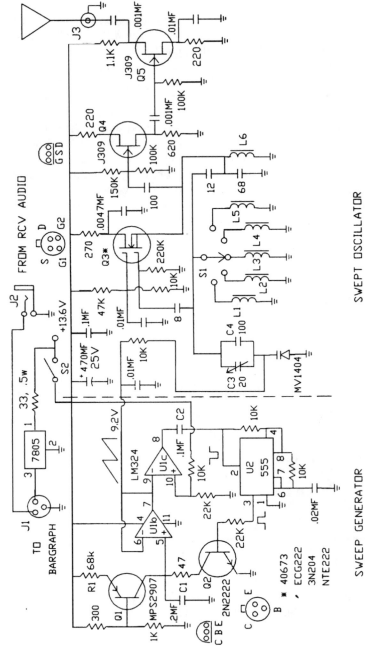

Fig. 10-9

Transistor Q1 acts as a constant-current generator, charging capacitor C1 at a rate determined by the value of R1. The high input impedance of U1b ensures that the constant-current generator is not excessively loaded and forwards the voltage level of C1 to U1c. U1c is configured as a comparator, and output pin 8 goes low when pin 9 rises above 9 V. This switching action applies a trigger pulse via capacitor C2 to activate the monostable multivibrator, U2. The resulting high pulse from pin 3 of U2 provides a discharge path for capacitor C1 through turned-on transistor Q2. A continuous sweep ramp with reset is produced at pin 7 of U1b at a rate determined by R1C1 (approximately 75 Hz). The swept oscillator section consists of a dual-gate MOSFET transistor in a Colpitts configuration, followed by a buffer amplifier that provides the RF from the selected frequency range to an output amplifier and short antenna. As the output sawtooth waveform from U2, pin 7, is applied to the varactor, the capacitance of the tank circuit decreases from approximately 62 pF at the low end of the ramp to a very low value at the maximum sweep level of about 9.2 V. The on-off switching action of the sweep waveform modulates the generated RF to produce an obnoxious buzz, which is easy to differentiate from other low-level signals at the selected receiver frequency when adjusting the tuner or antenna.

40-m LOOP ANTENNA

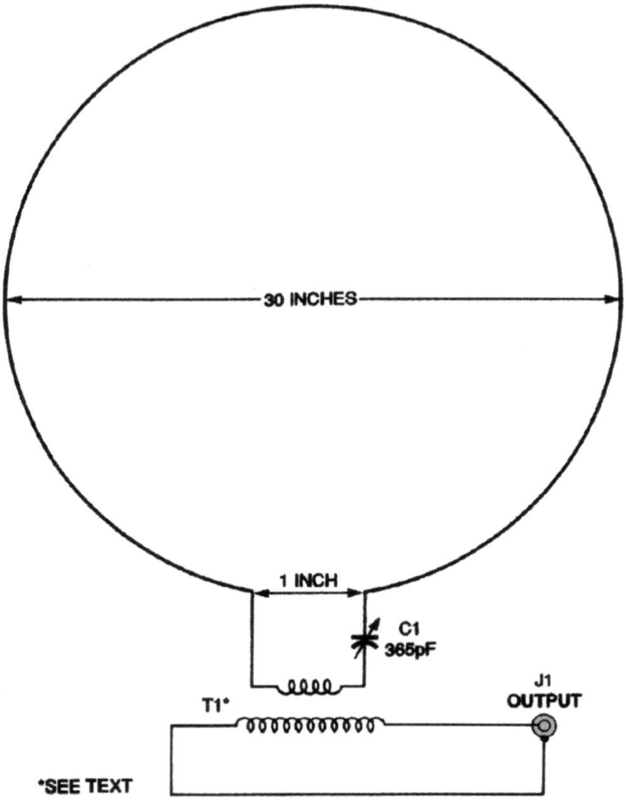
30 INCHES
1 INCH
C1
365pF
T1*
J1
OUTPUT
*SEE TEXT

POPULAR ELECTRONICS

Fig. 10-10

A cliff-dweller's dream, this 40-m loop can get apartment-based hams on the air. It also makes a great receiving antenna. The matching transformer, T1, is wound on T50-2 toroid core. The primary consists of 4 turns of 18-gauge enameled copper wire, and the secondary consists of 12 turns of 20-gauge enameled copper wire. Close-wind the secondary and the primary, and make sure that the two windings do not overlap.

UNDERCOVER SCANNER ANTENNA

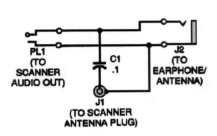

PL1
(TO
SCANNER
AUDIO OUT)

C1
.1

J2
(TO
EARPHONE/
ANTENNA)

J1
(TO SCANNER
ANTENNA PLUG)

The figure shows the schematic for the antenna circuit. Because the circuit is passive, no power supply is needed. Plug PL1 connects to the audio output or earphone jack of a scanner. The audio signal is then fed to J2, the earphone/antenna jack of the circuit. A 0.1-μF capacitor, C1, connects the center ("hot") terminal of J2 to a BNC jack, J1. When an earphone is plugged into J2, C1 feeds RF from the earphone wire to the front end of the scanner, through J1, without directly connecting the audio circuit to the receiver. An earphone connected to J2 will therefore both provide audio and act as an antenna.

POPULAR ELECTRONICS   *Fig. 10-11*

800-MHz ANTENNA

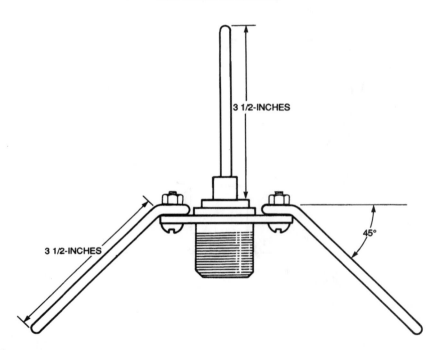

3 1/2-INCHES

3 1/2-INCHES

45°

POPULAR ELECTRONICS   *Fig. 10-12*

This simple quarter-wave antenna can improve performance on the 800-MHz band and can be built in just a couple of hours.

10-W DUMMY LOAD

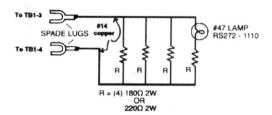

SPADE LUGS

To TB1-3

To TB1-4

#14 copper

#47 LAMP
RS272 - 1110

R = (4) 180Ω 2W
OR
220Ω 2W

This dummy load was intended for tune-up of a 10-W CW ham rig. It should be useful for general-purpose applications in the HF range. The spade lugs shown can be replaced with a UHF connector for more versatility. The lamp acts as a relative power indicator.

73 AMATEUR RADIO TODAY   *Fig. 10-13*

COAXIAL LINE BALUN

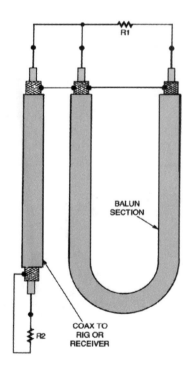

R1

BALUN
SECTION

COAX TO
RIG OR
RECEIVER

R2

A coaxial cable 4:1 balun transformer can be made from 75-Ω coax cable.

Remember that the electrical length of the balun section is a half wavelength at the operating frequency. This is about equal to the physical length times the velocity factor of the cable, plus any effect the connections might have on these lengths.

POPULAR ELECTRONICS   *Fig. 10-14*

TOROIDAL TRANSFORMER

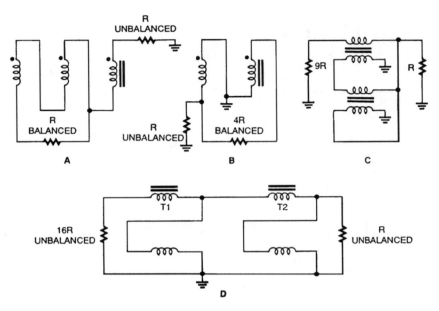

Fig. 10-15

The several types of impedance-matching transformers are: 1:1 balun (A), 4:1 balun (B), 9:1 un-un (C), and 16:1 un-un (D).

SIMPLE ANTENNA TUNER

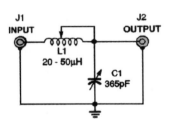

This simple tuner can properly match anything from a bedspring to a long-wire antenna. Be sure to use a voltage-transmitting-type unit for C1 if more than 5 W of output power is to be used.

Fig. 10-16

11

Automotive Circuits

The sources of the following circuits are contained in the Sources section, which begins on page 1043. The figure number in the box of each circuit correlates to the entry in the Sources section.

Air Conditioner Monitor
Oil-Pressure-Actuated Running Lights
Electrical System Analyzer
Battery Cranking Tester
Car Presence Detector
Automatic Headlight Dimmer
Car Battery Voltmeter with Bar Graph Display
Auto Stethoscope
Daytime Running Light Circuit
Radio Automatic Level Control
Automotive Neon Driver
Intermittent Windshield Circuit
Capacitor-Discharge Ignition Circuit
Visual and Audible Headlight Monitor
Lights-On Indicator
Windshield Wiper Circuit
Headlight Monitor
SW Converter
Automatic Vehicle Door Unlock Circuit

AIR CONDITIONER MONITOR

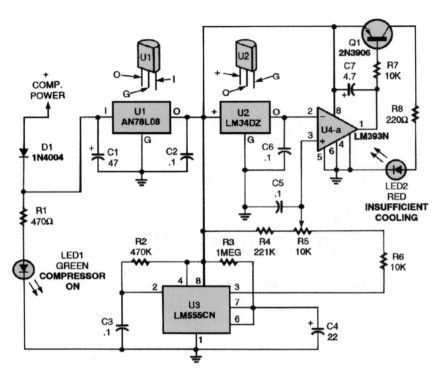

Fig. 11-1

Power for the air conditioner monitor is derived from the +12-V power lead that feeds the magnetic clutch of the air conditioning compressor. Thus, the circuit is active only when cooling is demanded of the A/C system. The heart of the circuit is U2, a factory-calibrated three-terminal temperature-sensing IC that is designed to deliver an output voltage of 10 mV per degree Fahrenheit. The sensor, which is driven by the regulated 8-V supply, is physically attached to the A/C evaporator-coil return pipe so that it accurately senses the operating temperature. Its temperature-dependent output voltage, which is proportional to the sensed temperature, is fed to negative input terminal U4-a (half of an LM393N voltage comparator) at pin 2. The output of U4-a can thus be used to determine the operating temperature of the evaporator return pipe. The positive input of the comparator allows the trigger voltage level of U4-a to be set slightly higher than the normal operating temperature of the A/C evaporator return pipe, or about 0.4 to 0.5 V, representing 40 to 50°F.

In order to avoid a false alarm caused by the warm evaporator return pipe each time the cycling A/C compressor starts, a time delay is provided by U3, a 555 oscillator/timer connected as a one-shot or monostable multivibrator that has an output pulse width of about 25 s as determined by R3 and C4. When compressor power is applied to the monitor, U3 is automatically triggered. Once triggered, U3 generates a positive output voltage at pin 3. That causes the open-collector output tran-

AIR CONDITIONER MONITOR *(Cont.)*

sistor within U4-a to be cut off. No base current flows to Q1, and LED 2 will not light. At the end of the timed cycle of U3, U4-a is ready to monitor the evaporator temperature.

Should the temperature be above the limit set by R5, the output of U4-a is pulled low, illuminating the warning LED (LED2). As mentioned earlier, a warm evaporator return pipe is a symptom of loss of refrigerant or other problems with the air-conditioning system. Another light-emitting diode, LED1, has been included in the circuit as a visual indication as to the operation of the air conditioner compressor. Should the refrigerant charge be extremely low, the compressor will cycle rapidly, alerting the driver to an almost total loss of refrigerant.

OIL-PRESSURE-ACTUATED RUNNING LIGHTS

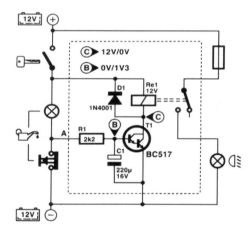

ELEKTOR ELECTRONICS

Fig. 11-2

Motorcycles and cars that are not equipped with (automatic) day running lights can be so equipped with the present circuit. The circuit is connected to the oil-pressure indicator. When the engine is not running, the contacts of the oil-pressure sensor in the engine block are closed. When the ignition is then switched on, the oil-pressure light comes on. The potential at A is then low, and nothing happens. When the engine is running, oil-pressure builds up, whereupon the contacts open and the indicator light goes out. The potential at A is then high so that T1 comes on and the relay becomes energized. The relay contact in series with the headlights closes so that the headlight is switched on. When the engine is switched off, the relay is deenergized and the headlights go out.

ELECTRICAL SYSTEM ANALYZER

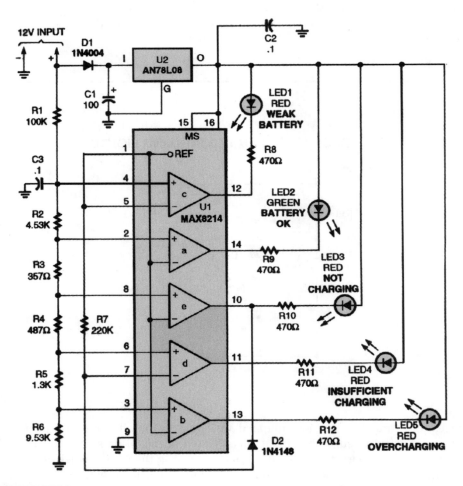

Fig. 11-3

ELECTRICAL SYSTEM ANALYZER *(Cont.)*

AUTOMOTIVE ELECTRICAL FAULTS

Condition	Normal Voltage	Possible Fault
Vehicle at rest	12.6 volts	<12.4 volts: bad cell or severly undercharged battery
Cranking	>9 volts	<9 volts: weak battery
Idling	>12.8 volts	<12.8 volts: not charging; bad alternator or wiring
Running minimum load	>13.4 volts	<13.4 volts: defective alternator or voltage regulator
Running minimum load	<15.2 volts	>15.2 volts: overcharging; defective regulator
Running maximum load	>13.4 volts	<13.4 volts: alternator defective or belt slipping

The automotive electrical diagnostic system is built around a Maxim MAX8214ACPE five-stage voltage comparator, which contains a built-in 1.25-V precision reference and on-board logic that allows the outputs of two of the comparators to be inverted.

BATTERY CRANKING TESTER

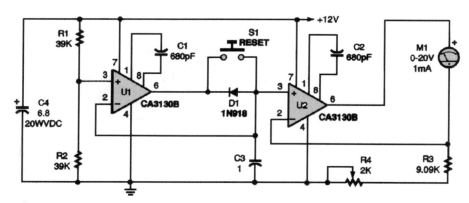

POPULAR ELECTRONICS

Fig. 11-4

Need to discover if your battery is weak or if your starter's windings are shorted? This meter will display how low your battery voltage drops during a start. The circuit is a negative-peak reading voltmeter.

CAR PRESENCE DETECTOR

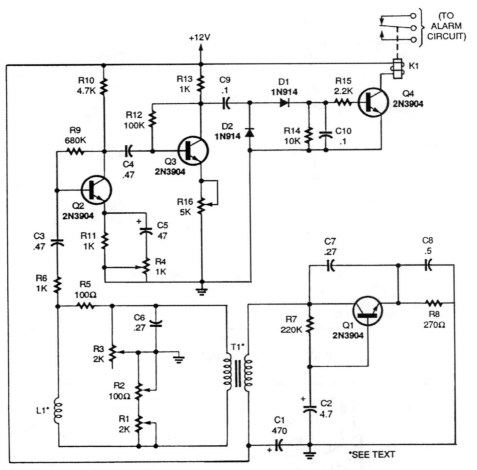

POPULAR ELECTRONICS

Fig. 11-5

Here's a circuit that will, if calibrated carefully, detect the presence of a large metal object. Coil L1 is made by winding 50 turns of 26-gauge wire on a 50-inch form. If L1 is buried in the ground, the circuit can be used to detect any car or truck that is driven over the coil. Transistor Q1 and its surrounding components make up a Colpitts oscillator, operating at 10 kHz. The output is coupled to a Maxwell bridge through the secondary of T1, which is a 600-Ω-to–600-Ω audio coupling transformer. With R1, R2, and R3 set properly, the bridge is balanced when no metal is close to L1, producing a near-zero ac signal at the bridge's output (the R5/L1 junction and ground). Should a large enough ferromagnetic object pass over L1, the bridge becomes unbalanced, producing a signal at the junction of R5 and L1. That signal is fed to the base of Q2, part of a common-emitter amplifier stage.

CAR PRESENCE DETECTOR *(Cont.)*

Resistor R4 adjusts to the gain of that stage. The signal from the amplifier is fed to the base of Q3, which drives a voltage doubler that supplies current to the relay driver, Q4. The relay, K1, is a 1000-Ω, sensitive, 12-V unit that can drive a normally open or closed alarm circuit to indicate the presence of a vehicle. Use shielded cable to attach L1 to the circuit, and do not bury L1 deeper than 8 inches for best results. Remember to weatherproof all components that will reside outside.

AUTOMATIC HEADLIGHT DIMMER

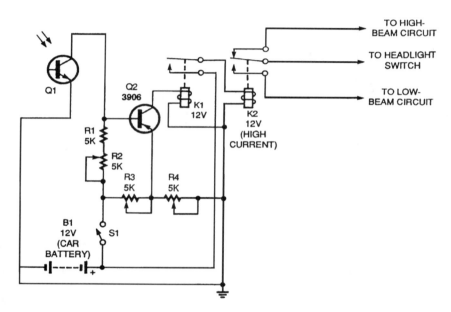

POPULAR ELECTRONICS

Fig. 11-6

The circuit is designed to switch your car's lights from its high beams to its low beams when traffic approaches. With Q1 in moderate darkness, the variable resistors should be adjusted to produce no base current through Q2. When there are no approaching headlights to trigger the system, K1 is not energized, and its contacts remain closed. Relay K2 is energized, and the high-beam lights are on. When light from an approaching car shines on the phototransistor, the base current and collector current of the PNP transistor increase substantially, and K1 is energized. That automatically deenergizes K2 (a power-type relay fed directly by the car battery), shutting off the high-beam headlights and turning on the low-beam lamps. Switch S1 should be conveniently located on the dash of the vehicle. Keep the switch turned off during the day so that the circuit will not leave your lights on in sunlight.

CAR BATTERY VOLTMETER WITH BAR GRAPH DISPLAY

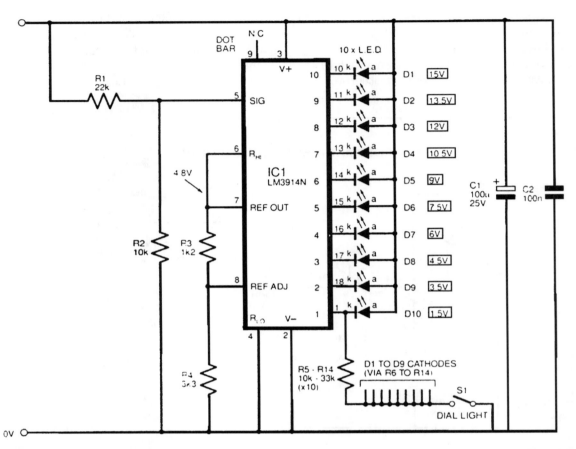

EVERYDAY PRACTICAL ELECTRONICS

Fig. 11-7

This uses an LM3914 bar-graph display, IC1, adapted to measure its own supply, so you simply wire it directly across the 12-V supply voltage (right way round—watch pins 2 and 3 carefully!). Ten LEDs, D1 to D10, will indicate the applied battery voltage, ranging from 1.5 to 15 V. The IC contains 10 internal comparators connected totem-pole-like, each sinking current through an LED. They compare an internal reference voltage against the chip's input voltage, at pin 5. Set the IC reference voltage with resistors R3 and R4 to just under 5 V. This means that one LED will light for every 500 mV (5 V reference/10 stages) increase in the signal. To enable this to be used to read a 12-V battery (the chip's own supply rail, in this case), resistors R1 and R2 are included as a divider: An input of roughly +15 V will cause D1 to light. When the voltage gradually falls, LEDs D2 to D10 will progressively illuminate. The circuit is set as a "moving dot" display. Connect pin 9 to the positive rail for a bar-graph display (not recommended because of the current consumption). Because the 10

CAR BATTERY VOLTMETER WITH BAR GRAPH DISPLAY *(Cont.)*

outputs are effectively constant-current sinks, the LEDs will glow at a level independent of the changes in the supply rail; they won't dim when the rail drops. However, the first two or three display LEDs (D8 to D10) are superfluous in this application, because the LM3914 won't operate correctly below a rail of about +5 V.

AUTO STETHOSCOPE

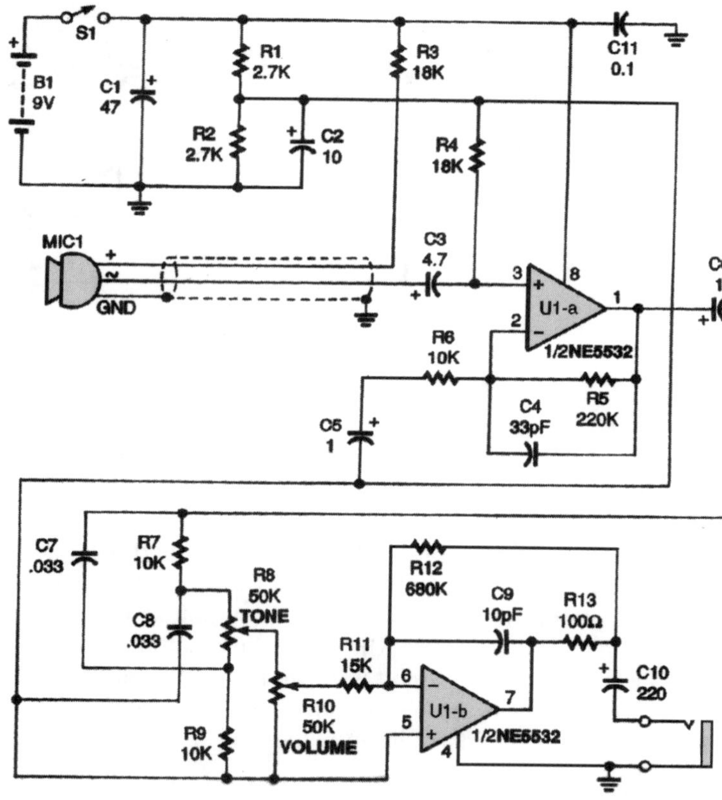

Fig. 11-8

The heart of the stethoscope is the NE5532 audio op amp, U1. That component directly drives low impedances and allows the use of headphones without adding another amplifier.

This circuit uses an electret microphone and an audio amplifier using a NE5532 audio op amp as an automotive diagnostic tool. The mike is mounted in a probe or piece of tubing and placed near or on various parts of the engine or other components as an aid in diagnosis; the sound generated by the suspected part is used to determine possible problems.

DAYTIME RUNNING LIGHT CIRCUIT

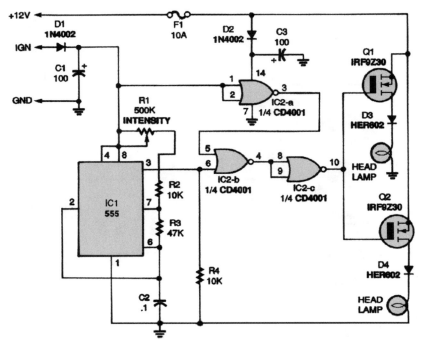

POPULAR ELECTRONICS

Fig. 11-9

The DRL circuit is a variable-duty-cycle power oscillator that allows pulses of current to be applied to the headlights whenever the ignition of the vehicle is turned on. The 12-Vdc power that drives the circuit is taken from two sources in the vehicle. Power for the headlights is provided by the vehicular battery and generator system, and is input through fuse F1 to the circuit. The power source for the digital or logic part of the circuit is the +12-V line that feeds the ignition system of the vehicle. When the ignition is turned on, the circuit is on, and when the ignition is turned off, so is the circuit. When the ignition is on, power is applied through D1 to operate IC1, a 555 timer that is connected as a free-running multivibrator. Pin 3 of IC1 provides the output of the oscillator; the duty cycle is determined by the following ratio:

$$(R_1 + R_2) / R_3$$

Potentiometer R1 allows the duty cycle of the negative part of the waveform to be adjusted from about 10 percent to almost 50 percent. Adjustment capability is provided so that the intensity of illumination can be set as desired. A set of logic gates (IC2), powered by the battery and generator, is used to control the operation of the circuit. When the ignition is off, the logic output of pin 3 of IC2-a is high. Under that high-logic condition, the transistors are not forward-biased and the headlights remain off. When the ignition is on, the oscillator is powered through D1. The pulse train output of IC1 appears simultaneously at the gates of Q1 and Q2. Those two transistors then feed +12-V pulses

DAYTIME RUNNING LIGHT CIRCUIT *(Cont.)*

through their respective diodes, D3 and D4, to the low-beam headlights of the vehicle. The diodes provide isolation between the DRL circuit and the electrical system of the vehicle. The design of the circuit makes it possible to add second pairs of hex-FET transistors and corresponding diodes to the circuit and have them driven by IC2-c. Thus, if desired, the taillights of the vehicle may also be used for DRL operation.

RADIO AUTOMATIC LEVEL CONTROL

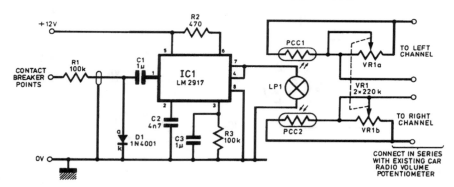

EVERYDAY PRACTICAL ELECTRONICS

Fig. 11-10

This experimental circuit automatically increases the volume of a car radio when the vehicle's speed (and consequently road and wind noise) rises, and returns the volume to a pleasant level at lower speeds. This relieves the driver of the distracting chore of continually fiddling with the radio's volume control. In the figure, a special tachometer IC type LM2917 takes its input from the ignition system's contact breaker points and converts this into a voltage that is directly proportional to engine speed. This voltage will always be in the range of 0 to 5 V and is used to illuminate a miniature filament lamp LP1. On either side of this lamp is mounted a pair of light-dependent resistors (LDRs) that are connected in series with the wipers of the radio's volume control potentiometers. A ganged potentiometer across each of the LDRs sets the minimum volume level, the radio's own volume being used for the high level. The value of capacitor C2 determines the range over which the unit will function satisfactorily. A suggested value of 4n7 would be a suitable starting point for further experimentation. The 100-kΩ input resistor (R1) should be placed as near as possible to the ignition coil to prevent RFI (radio frequency interference) and to ensure that the contact breakers are not shorted out should the wire short to earth (chassis) for any reason.

AUTOMOTIVE NEON DRIVER

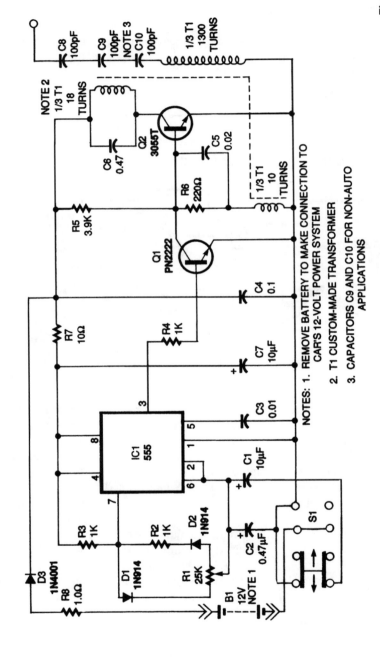

NOTES: 1. REMOVE BATTERY TO MAKE CONNECTION TO CAR'S 12-VOLT POWER SYSTEM

2. T1 CUSTOM-MADE TRANSFORMER

3. CAPACITORS C9 AND C10 FOR NON-AUTO APPLICATIONS

Fig. 11-11

ELECTRONICS NOW

A 555 timer drives driver transistor Q1 and switching transistor Q2 at 25 to 30 kHz. T1 is wound on a ferrite core (try using a core from an old B/W TV flyback) and has the windings arranged as shown in the schematic. About 2500 Vac p-p is produced, so be careful to avoid any possibility of contact with this voltage, as it can cause a bad shock or RF burn. The neon tube is a custom-size item that can be made up at any neon sign shop or possibly purchased commercially. It is not legal to illuminate certain parts of your vehicle in some parts of the United States and in some nations.

INTERMITTENT WINDSHIELD CIRCUIT

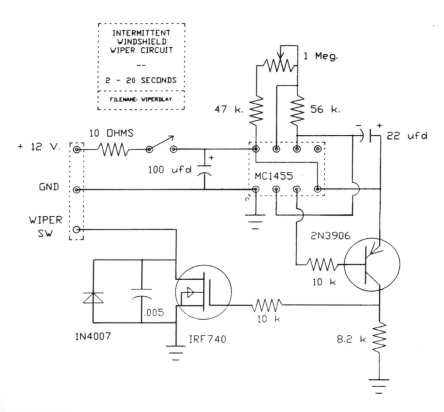

Fig. 11-12

Basic timing is accomplished by the 555 IC. The resistor and capacitor network on pins 6, 7, and 8 sets the minimum delay at 2 s and the maximum at 20 s. The 555 output on pin 3 drives a 2N3906 PNP transistor to provide the proper input to the IRF740 power MOSFET. The MOSFET does the work and switches the wiper motor on at the end of the delay time. The wiper circuit uses a cam and switch arrangement on the drive assembly to cause the wiper blades to "park" in the proper place when the wipers are turned off. This cam switch is open when the blades are parked; the dash-mounted wiper switch is in parallel with the cam switch. The intermittent circuit is wired to the cam switch as shown. When activated, and with the dash switch in the OFF position, the power MOSFET puts a ground on one side of the wiper motor, and the motor starts running. Shortly afterwards, the cam switch closes and forces the motor to continue running (regardless of the MOSFET's on/off status) until the PARK position is reached and the cam switch opens. The motor then stops, and one cycle is complete.

CAPACITOR-DISCHARGE IGNITION CIRCUIT

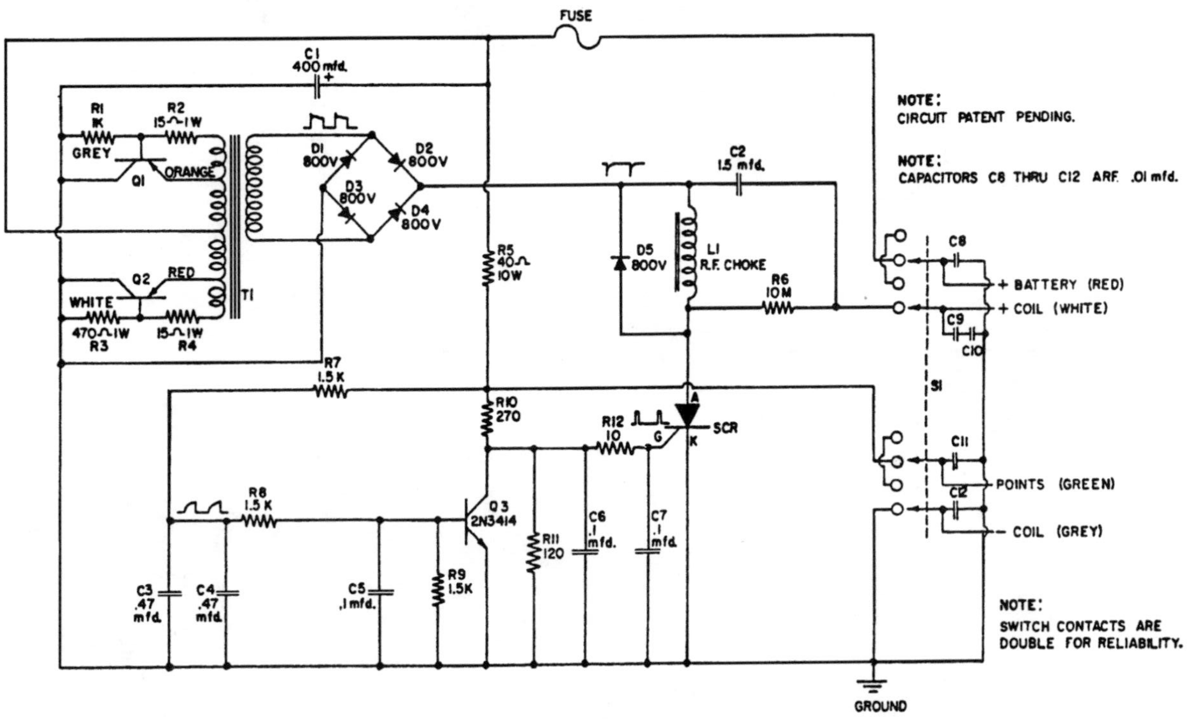

NOTE:
CIRCUIT PATENT PENDING.

NOTE:
CAPACITORS C8 THRU C12 ARE .01 mfd.

+ BATTERY (RED)
+ COIL (WHITE)

POINTS (GREEN)
− COIL (GREY)

NOTE:
SWITCH CONTACTS ARE
DOUBLE FOR RELIABILITY.

GROUND

Fig. 11-13

NUTS AND VOLTS MAGAZINE

This ignition circuit consists of a dc-to-dc converter delivering 250 to 300 V to a 1.5-mF capacitor, and discharging it through the ignition coil primary upon the opening of the distributor points. SCR is triggered by the opening of the distributor points via inverter and pulse shaper Q3 and associated components. This circuit is typical of CD ignition systems. T1 is a special transformer for this circuit, but any suitable dc-to-dc converter transformer and circuit that delivers 250 Vdc (or more) can be used.

VISUAL AND AUDIBLE HEADLIGHT MONITOR

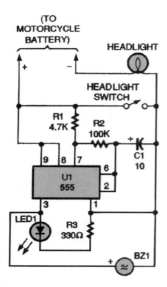

If the LED in this circuit is flashing and the piezo sounder is buzzing, then your headlights are not on.

POPULAR ELECTRONICS

Fig. 11-14

LIGHTS-ON INDICATOR

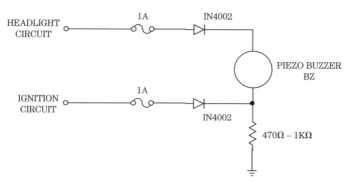

WILLIAM SHEETS

Fig. 11-15

The buzzer (BZ) will sound if the ignition is off and the headlights are on.

WINDSHIELD WIPER CIRCUIT

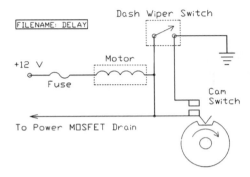

NUTS AND VOLTS MAGAZINE

Fig. 11-16

When the wipers are turned on, the switch in parallel with the cam-operated points keeps the motor running. When the switch is turned off, the motor keeps running until the cam opens the points, stopping the motor when the wipers are in the PARK position.

HEADLIGHT MONITOR

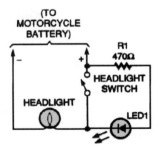

With this simple circuit, you'll never again forget to turn your headlights on.

POPULAR ELECTRONICS

Fig. 11-17

SW CONVERTER

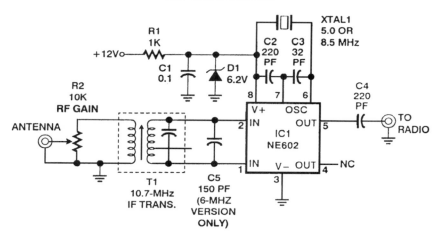

ELECTRONICS NOW

Fig. 11-18

This simple circuit, built around a NE602, allows you to receive either the 6- or 9.5-MHz short-wave band on your car radio.

AUTOMATIC VEHICLE DOOR UNLOCK CIRCUIT

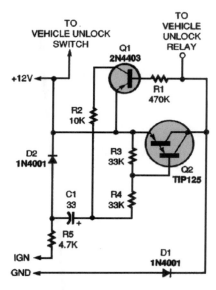

POPULAR ELECTRONICS

Fig. 11-19

When the ignition switch is on, the +12V and IGN lines connected to the circuit are within half a volt or so of each other. Resistor R3 reverse-biases Darlington Q2 and keeps it cut off. Transistor Q1 is in parallel with Q2's emitter and collector; Q1 is ON as a result of the large voltage drop across Q2's emitter and collector. When the ignition switch is turned off, voltage is removed from the IGN line, effectively placing R5 at ground level. Capacitor C1 provides a path for current to flow through R3 to R5 and the vehicle components to ground. That flow will forward-bias Q2, driving it into saturation and lowering its emitter/collector voltage to less than 0.5 V. As a result, almost full battery voltage is applied to the vehicle unlock relay; transistor Q1 is cut off at the same time. The voltage divider of R3 and R4 prevents false triggering during voltage sags on the IGN line.

12

Automotive Security Circuits

The sources of the following circuits are contained in the Sources section, which begins on page 1043. The figure number in the box of each circuit correlates to the entry in the Sources section.

Auto Signal Minder
Auto Guardian Transmitter
Starter Cutoff
Auto Guardian Receiver
Auto Security Device

AUTO SIGNAL MINDER

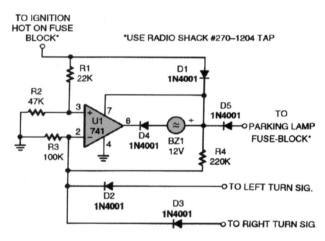

POPULAR ELECTRONICS

Fig. 12-1

If the ignition is on and the parking lights are on or off, the noninverting input of U1 is at 8 V and the inverting input is either 0 or 4 V. That causes pin 6 to go high, keeping the buzzer off. If the ignition is off and the parking lights are on, the noninverting input is at 0 V and the inverting input is at 4 V. That brings pin 6 low, activating the buzzer. With the ignition and turn signal on and the parking lights on or off, the noninverting input is at 8 V and the inverting input pulses at 12 V, causing the buzzer to sound in step with the blinker.

AUTO GUARDIAN TRANSMITTER

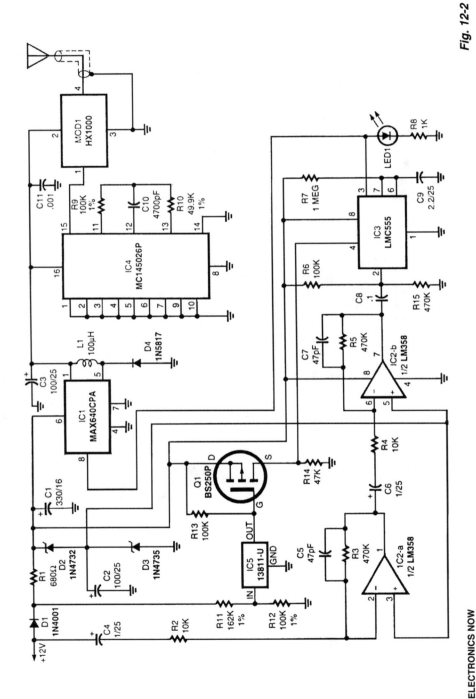

Fig. 12-2

The circuit is powered by the vehicle's 12-V battery. A pair of zener diodes, D2 and D3, create two regulated voltage sources of 10.9 and 6.2 V. Integrated circuit IC2 contains a pair of identical op amps, which are cascaded. The negative input of IC2-a is ac-coupled to the 12-V bus so that it can detect a sudden sag in voltage caused by the current draw of the vehicle's dome lamps

110

AUTO GUARDIAN TRANSMITTER *(Cont.)*

when a door is opened. A change in battery voltage of 3 mV will cause a −6-V swing at the output of IC2-b, which is great enough to trigger IC3, a CMOS 555 timer chip. That IC is wired to operate as a one-shot pulse generator. An output pulse about 2[fr1/2] s long appears on pin 3 of IC3 when triggered by IC2-b. The output of IC3 is connected to the enable input of IC1, a switching-type voltage regulator. When IC1 is enabled, it outputs a regulated 3.3-Vdc power source for the transmitter portion of the circuit, encoder IC4 and transmitter module MOD1. When power is applied to the encoder, a series of pulse trains is generated that contain the address of the encoder. Those pulse trains are applied to pin 1 of the hybrid module to produce an RF signal with an on-off pulse modulation, sometimes called *amplitude-shift keying.* When the 2½-s pulse time of IC3 is completed, the transmitter shuts down, returning to its dormant state.

In order to prevent RF transmission while the vehicle is running, voltage detector IC5 uses R11 and R12 as a voltage divider to sense when the supply voltage rises above 12 V as a result of an alternator charging when the engine is running. Under that condition, the output of IC5 is open, turning off Q1. That holds the reset input of IC3 low, preventing the timer chip from responding to any trigger pulses from the amplifier. As a result, no transmission takes place. The HX1000 hybrid-transmitter module, MOD1, is a four-terminal device that contains a surface-acoustic-wave-stabilized UHF oscillator. The oscillation frequency, 433.92 MHz, is stabilized and controlled by the resonant frequency of the internal surface-acoustic-wave filters, which also filter undesirable harmonics. The hybrid module is capable of delivering about 1 mW of power (0 dBm) into a 50-Ω load. It is connected to the transmitting antenna through a 50-Ω coaxial cable.

STARTER CUTOFF

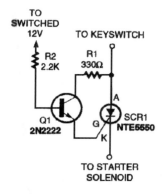

POPULAR ELECTRONICS · *Fig. 12-3*

In this auto-starter cutoff circuit, two conditions are required to start the car: The key switch must be turned to the starting position, and Q1 must be turned on. When Q1 is switched to the ON state, current flows through the gate, allowing current to pass through the SCR, provided that the key switch is also turned to the starting position. Transistor Q1 can be turned on by connecting R2 through a separate switch to a 12-V power source, or by using an existing switched source. Examples of existing sources might be a brake light, turn signal, parking light, or anything that most people would not normally activate while starting a vehicle. With regard to using a turn-signal indicator, it doesn't matter that the power applied to Q1 is not constant. Only a pulse is required to latch the SCR on, provided that the key switch is in the starting position. The SCR is a 25-A, 50-V unit.

AUTO GUARDIAN RECEIVER

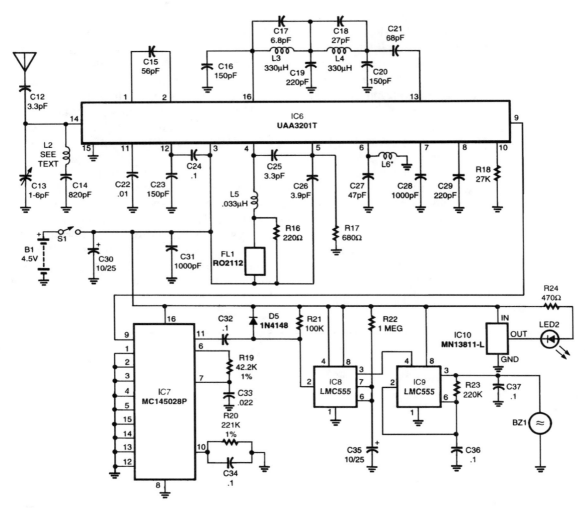

Fig. 12-4

The decoder IC7 checks the incoming pulse train for the correct address. If that address is present, the valid output terminal, pin 11, goes high for the 2½-s duration of the transmitted pulse trains. When the valid output signal returns to a low state, timer IC8 is triggered into operation. The output terminal of IC8, pin 3, goes high for a time period of about 11 s, as determined by R22 and C35. That output, in turn, enables IC9, a timer configured as a 50-percent duty-cycle oscillator with a period of about ¼ s. The output of IC9 drives a piezo buzzer, which produces an attention-getting audio signal that alerts the user that a door of the vehicle has been opened. Power to

AUTO GUARDIAN RECEIVER *(Cont.)*

operate the receiver is obtained from a set of three AAA alkaline cells connected in series to produce 4.5 V. Current draw is only 3 mA during standby, so battery life will be in the hundreds of hours. IC10 is a voltage-detector IC that allows current to flow through LED2 when the battery voltage falls below 3 V. That alerts the user that battery replacement is necessary.

AUTO SECURITY DEVICE

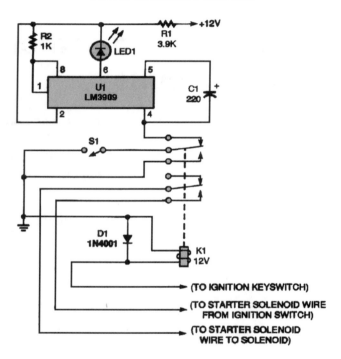

POPULAR ELECTRONICS

Fig. 12-5

Combining two functions in one, this circuit is both a passive cutoff device and a fake car alarm. Depressing S1 with the key in the RUN position disables the flasher and allows the car to start.

13

Battery Backup and Switchover Circuits

The sources of the following circuits are contained in the Sources section, which begins on page 1043. The figure number in the box of each circuit correlates to the entry in the Sources section.

Single-IC Battery-Backup Manager
Low-Voltage Battery Switchover System

SINGLE-IC BATTERY-BACKUP MANAGER

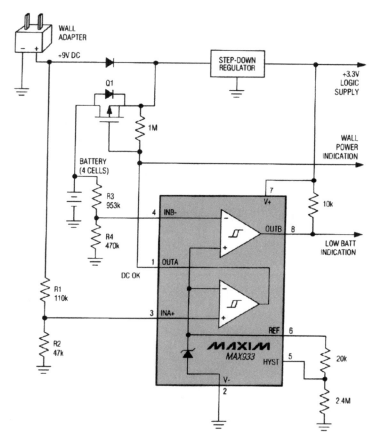

Fig. 13-1

Instruments powered by a "wall adapter" with battery backup typically diode-OR the battery and wall-adapter connections. That arrangement carries a penalty, however—the diode in series with the battery limits the minimum voltage at which the battery can supply power. One alternative is a dual-comparator/reference IC, which monitors the battery and wall-adapter voltages with respect to its internal reference voltage. The open-drain output of comparator B (with pull-up to 3.3 V) provides a low-battery warning in the form of a low-to-high transition when battery voltage drops to 3.6 V. The open-drain output of comparator A (with pull-up to 9 V) flags low wall-cube voltage in the same way, with a warning threshold of 3.9 V. Comparator A also controls the PMOS switch Q1, which replaces the OR-connection diode in a conventional circuit. When wall power is removed, Q1 turns on and provides a low-resistance path from battery to regulator.

LOW-VOLTAGE BATTERY SWITCHOVER SYSTEM

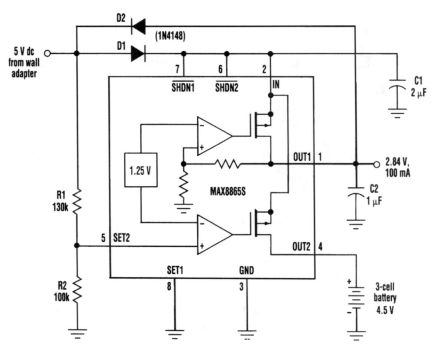

ELECTRONIC DESIGN

Fig. 13-2

Portable systems often include the flexibility to operate either from an internal battery or from an ac-dc wall adapter. Many such systems include circuitry that switches automatically between an internal battery and an external source as the user connects and disconnects the wall adapter. The circuit shown implements this idea with a dual linear regulator, one side of which is preset for a regulated output of 2.84 V (other versions of the IC offer 2.80-V and 3.15-V outputs). The other side of the regulator is configured to allow user-adjustable outputs, but, in this case, it monitors the wall-adapter voltage. When that voltage is removed by unplugging the adapter, the regulator's pass transistor routes battery current into the IC for support of the 2.84-V output (current flow in this transistor is counter to that of most applications). The input bypass capacitor (C1) provides enough holdup time for seamless transitions between the battery and adapter voltages. Resistors R1 and R2 sense the wall-adapter voltage and determine the switchover threshold, V_{sw}:

$$V_{sw} = V_{set}[(R_1 + R_2)/R_2] = 1.25 \text{ V}[(130 \text{ k}\Omega + 100 \text{ k}\Omega)/100 \text{ k}\Omega] = 2.875 \text{ V}$$

Diode D1 isolates the wall-adapter voltage so that the battery can't cause limit cycling by retriggering the switchover. D2 holds the IC's dual-mode input in the external feedback mode by maintaining a minimum voltage at the SET2 input. The wall-adapter voltage should be equal to or greater than the maximum battery voltage.

14

Battery Charger Circuits

The sources of the following circuits are contained in the Sources section, which begins on page 1043. The figure number in the box of each circuit correlates to the entry in the Sources section.

Precision Li-Ion Battery Charger
300-mA Step-Down Battery Four-Cell NiCd Charger
90-Percent-Efficient Four-Cell NiCd Charger
Battery Charger
More-Efficient Battery Charger
Battery Trickle Charger
Step-Up Supply Battery Charger
3-A Battery Charger for Li-Ion or NiCd
Timed NiCd Charger
2.5-A Battery Charger
Constant-Current NiCd Charger
Float-Charging Circuit
Simple Charger
Lead-Acid Battery Temperature-Compensated Charger

PRECISION Li-ION BATTERY CHARGER

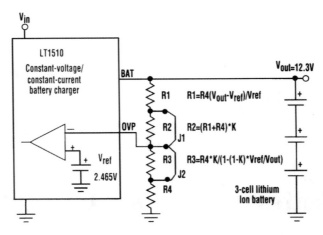

Fig. 14-1

An alternative to using expensive 0.25-percent precision resistors is presented in this battery-charger design, which adds two 1-percent resistors and two jumpers to the charger.

In constant-voltage-mode charging, a lithium-ion cell requires 4.1 V ± 50 mV. The 1.2-percent requirement represents a tight tolerance. In a regulation loop where a voltage divider is compared against a reference, the accuracy is typically achieved by selecting a 0.7-percent reference and a voltage divider with 0.25-percent tolerance resistors. Unfortunately, 0.25-percent precision resistors cost three times as much as 1-percent resistors and have very long lead times. One solution for moderate-volume production involves adding two 1-percent resistors and two jumpers to the charger circuit (see the figure). The jumpers are removed as necessary to bring the constant voltage to the required accuracy of 1.2 percent. The charger selected for this example is the LT1510. There are three lithium-ion cells in the battery. After a value is selected for R_4, the values for R_1, R_2, and R_3 can be calculated using the equations given. K is the relative change required for a circuit with all of its tolerances in one direction. For example, in the case of a 0.5-percent reference and two 1-percent resistors, the total tolerance is 2.5 percent. To bring it back to 1.2 percent, the percentage change required is 2.5 percent − 1.2 percent = 1.3 percent, and $K = 0.013$. The jumpers (J1 and J2) must be opened based on the following:

If V_{out} is $K/2$ below nominal, remove J1.
If V_{out} is $K/2$ above nominal, remove J2.

The following values were calculated: $R_1 = 4.99$ kΩ, $R_2 = 324$ Ω, $R_3 = 80.6$ Ω, and $R_4 = 20$ kΩ. The voltage below which jumper J1 should be opened is 12.34 V − 1.3 percent / 2 = 12.22 V. The voltage above which jumper J2 should be opened is 12.34 V + 1.3 percent / 2 = 12.42 V.

300-mA STEP-DOWN BATTERY FOUR-CELL NiCd CHARGER

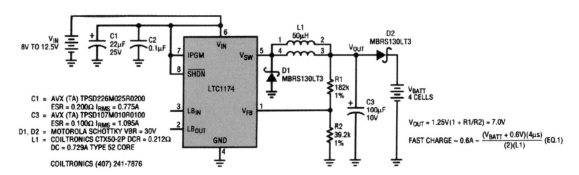

C1 = AVX (TA) TPSD226M025R0200
ESR = 0.200Ω I_{RMS} = 0.775A
C3 = AVX (TA) TPSD107M010R0100
ESR = 0.100Ω I_{RMS} = 1.095A
D1, D2 = MOTOROLA SCHOTTKY VBR = 30V
L1 = COILTRONICS CTX50-2P DCR = 0.212Ω
DC = 0.729A TYPE 52 CORE

COILTRONICS (407) 241-7876

V_{OUT} = 1.25V(1 + R1/R2) = 7.0V

$$\text{FAST CHARGE} \approx 0.6A - \frac{(V_{BATT} + 0.6V)(4\mu s)}{(2)(L1)} \quad (EQ.1)$$

LINEAR TECHNOLOGY POWER SOLUTIONS

Fig. 14-2

Low-current battery-charger circuits are required in hand-held products such as palmtop, pen-based, and fingertip computers. The charging circuitry for these applications must use surface-mount components and consume minimal board space. The LTC1174 circuit shown provides both of these features. The LTC1174 is a current-mode, step-down switching regulator with a low-loss internal P-channel MOSFET power switch. In this circuit, the peak switch current is programmed to 600 mA. The average charging current is determined by the choice of inductor value and the actual battery voltage, according to the equation. The voltage feedback resistor network is set for an output voltage of 7 V. NiCd cells have a nominal voltage of 1.2 V, a discharged voltage of 0.9 V, and a fully charged voltage of 1.5 V. A four-cell NiCd pack's voltage will range from 3.6 to 6 V, depending upon its state of charge. When it is attached to the charger, it will pull the output voltage below 7 V and place the LTC1174 into current-limited operation at about 300 mA with the 50-µH inductor shown. Diode D2 prevents the batteries from discharging through the divider network when the charger is shut down by bringing the SHUTDOWN pin low. Less than 10 µA of current is drawn from the supply when the charger is shut down.

90-PERCENT-EFFICIENT FOUR-CELL NiCd CHARGER

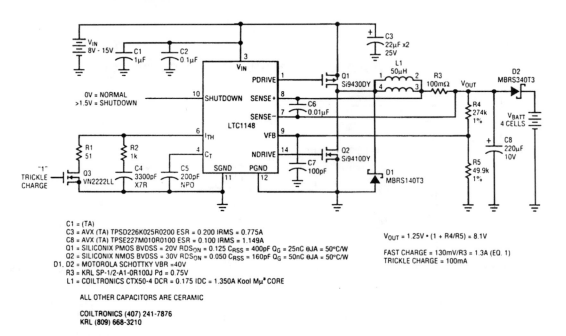

C1 = (TA)
C3 = AVX (TA) TPSD226K025R0200 ESR = 0.200 IRMS = 0.775A
C8 = AVX (TA) TPSE227M010R0100 ESR = 0.100 IRMS = 1.149A
Q1 = SILICONIX PMOS BVDSS = 20V RDS_{ON} = 0.125 C_{RSS} = 400pF Q_G = 25nC θJA = 50°C/W
Q2 = SILICONIX NMOS BVDSS = 30V RDS_{ON} = 0.050 C_{RSS} = 160pF Q_G = 50nC θJA = 50°C/W
D1, D2 = MOTOROLA SCHOTTKY VBR =40V
R3 = KRL SP-1/2-A1-0R100J Pd = 0.75V
L1 = COILTRONICS CTX50-4 DCR = 0.175 IDC = 1.350A Kool Mμ² CORE

ALL OTHER CAPACITORS ARE CERAMIC

COILTRONICS (407) 241-7876
KRL (809) 668-3210

V_{OUT} = 1.25V • (1 + R4/R5) = 8.1V

FAST CHARGE = 130mV/R3 = 1.3A (EQ. 1)
TRICKLE CHARGE = 100mA

LINEAR TECHNOLOGY POWER SOLUTIONS

Fig. 14-3

An embedded battery charger requires extremely low power dissipation in order to minimize heat buildup in compact portable systems. This schematic shows a charger that can charge four NiCd cells and is selectable for either a 1.3-A fast charge or a 100-mA trickle charge with up to 90 percent efficiency. The LTC1148 is a step-down synchronous switching regulator; it controls the charge rate, monitoring the output current via external current-sense resistor R3. Fast charge current is determined by the value of R3, according to the fast charge equation. In this case, it is set to 1.3 A. The resistor divider network R4 and R5 sets the output voltage to a nominal 8.1 V under no-load conditions, such as when the battery is removed. A four-cell NiCd pack's voltage will range from 3.6 to 6 V, depending upon its state of charge. When installed, the battery will pull the output below 8.1 V and place the LTC1148 into current-limited operation at 1.3 A. This constant current will be delivered until trickle charge is enabled by an external charge termination circuit or the battery is removed. Q3 enables trickle charge operation with charge current set by choice of R1. Diode D2 prevents the battery from being drained by the feedback resistor network when the LTC1148 is shut down.

BATTERY CHARGER

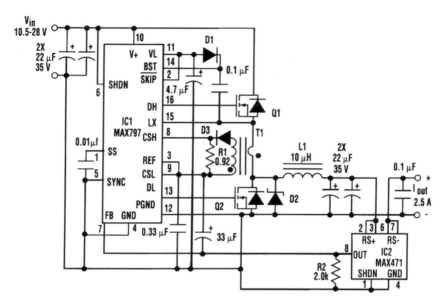

ELECTRONIC DESIGN

Fig. 14-4

This current-source battery charger can source 2.5 A with up to 96 percent efficiency. It can operate from an ac adapter or directly from a car battery. By sensing current on the high side of the battery being charged, it preserves the low impedance of the automobile's ground-return system. The charger handles battery stacks of 5 to 15 cells, and its input voltage can range from 28 V down to a level 1.5 V above the terminal voltage of a fully charged battery. Charging current is generated by the current-mode, buck-regulator controller IC1, operating with an external power switch (Q1) and a synchronous rectifier (Q2). Both are N-channel MOSFETs, whose low on-resistance (vs. P-channel types) contributes to high efficiency in the circuit. The IC includes a charge pump for generating the required positive gate voltage for Q1. It also senses the Q1 current (via R1) and shuts down if this current becomes excessive.

The current-sense transformer (T1) saves power by diverting a fraction of Q1's current through R1. Operating on the output's high side is the current-sense amplifier IC2. Its OUT current (1/2000 of the current flowing internally from RS+ to RS−) flows through R2 to produce a feedback signal for IC1. Digital control of the charging current can be introduced by switching among different-switching FETs like the 2N7002; its 7.5-Ω on-resistance is not a problem because its drain current—no greater than 1.25 mA—introduces an error no greater than 0.5 percent. Circuit efficiency ranges up to 96 percent. Efficiency and power output both increase with output voltage, because the circuit's power consumption (associated mainly with IC1 and the switching MOSFETs) is almost constant. This buck regulator's output cannot rise about V_{in}, in most cases, overvoltage protection is not required. V_{out}, which supplies power to IC2, must not go below 4 V.

MORE-EFFICIENT BATTERY CHARGER

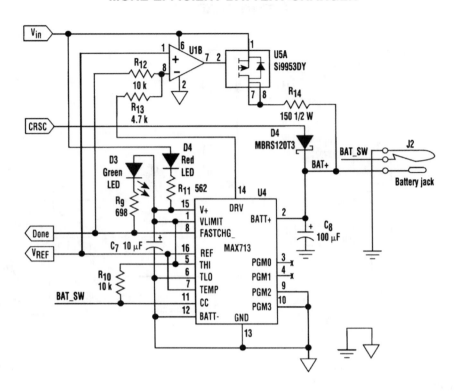

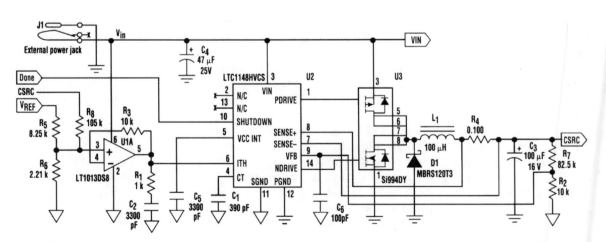

Fig. 14-5

MORE-EFFICIENT BATTERY CHARGER (Cont.)

The project at hand was to build a small, efficient, inexpensive, full-function battery charger that could charge 2 to 10 NiCd or NiMH cells. A switching regulator that could be set up as a constant-current source was needed. Connecting the negative end of the battery directly to ground provides more voltage and reduces IR losses. Linear Technology's LTC1148HV synchronous step-down switching regulator fills this role because it is more than 90 percent efficient, and it features two current sense inputs (Sense+ and Sense−) and a current control pin (I_{th}) that has a dc input linearly related to the maximum coil current (Fig. 1). For example, with a low, commonly available 0.1-Ω sense resistor and I_{th} connected to the 2-JV reference output of the MAX713, the peak coil current is set to 1.55 A. The average current will still vary with output voltage, but this can be compensated for by feeding back some of the output voltage to I_{th}. The constant-voltage regulation loop of the LTC1148 is disabled once the voltage divider (R2 and R7) for V_{FB} is set above the highest voltage that the battery is going to reach during charge. With the battery above the constant voltage regulation point, the switching regulator will supply no current. If a trickle charge current is desired, a switch (U5A) and a resistor (R14) can be added that supply the desired current directly from the primary dc source (V_{in})—a simple wall cube—when the MAX713 controller terminates fast charge or during battery undervoltage condition at startup.

BATTERY TRICKLE CHARGER

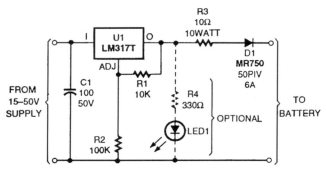

Fig. 14-6

The circuit is built around an LM317T adjustable voltage regulator. Its output voltage (V_{OUT}) is determined by the formula $V_{OUT}=1.25 \ (1+R_2/R_1)$. I chose a 10-k$\Omega$ resistor for R1 and a 100-kΩ resistor for R2 to allow for a little wasted current flow (around 125 μA). This will develop 13.75 V between the output terminal and ground. R3 and a diode prevent burnout if the terminals are shorted or reversed, and prevent battery discharge in the event that the regulator is disconnected from the main power supply. The capacitor is optional but will make life easier on the chip under an extreme charging cycle. With proper heat sinking, the IC is capable of carrying 1.5 A to the battery. The circuit can be left hooked to the battery all of the time and will not overcharge it.

STEP-UP SUPPLY BATTERY CHARGER

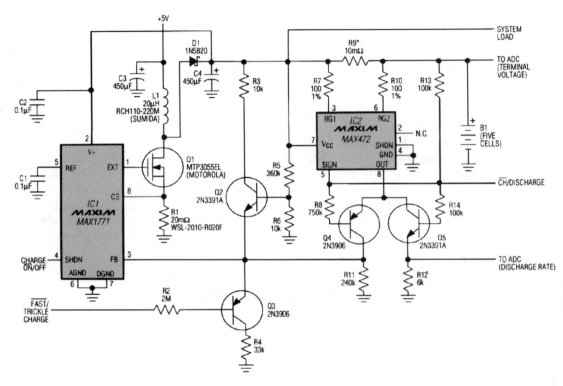

Fig. 14-7

IC1 is a step-up switching regulator that boosts V_{in} (nominally 5 V) as necessary to supply the combination of charging current and load current. The 5-V source must include short-circuit protection. IC2 is a high-side current-sense amplifier that monitors the charging current. Commands from the processor include CHARGE ON/OFF and FAST/TRICKLE CHARGE. IC2 produces an output current (OUT) equal to 10^{-4} of the current through sense resistor R9. Q3 and Q4 are on during a fast-charge operation, so this output current flows through the parallel combination of R11 and (approximately) R4. The resulting feedback to IC1 (pin 3) maintains the fast-charge current through R9 at 500 mA. This feedback also enables the regulator to supply as much as 500 mA of load current in addition to the fixed 500 mA of charging current. Q2 limits the battery voltage to 10 V (2 V/cell). During the fast charge, an external processor and multichannel A/D converter (ADC) must monitor the battery's terminal voltage. When the ADC senses a change of slope in this voltage, the processor terminates the fast charge by asserting a high on FAST/TRICKLE CHARGE. Q3 turns off, causing a rise in the feedback (FB) that lowers the charging current to the trickle charge rate (approximately 60 mA). If IC1 shuts down or if the load current plus charging current exceeds the capability of IC1, the R9 current reverses as current flows out of the battery. IC2 indicates the reversal by allowing R13

STEP-UP SUPPLY BATTERY CHARGER *(Cont.)*

to pull its open-collector SIGN output high, turning off Q4 and turning on Q5. Current through R12 then produces a voltage proportional to the battery's discharge current (5 A through R9 produces a full-scale response of 3 V across R12). By integrating this voltage over time (sampling at fixed intervals and multiplying by the time interval), the A/D-processor system can monitor energy removed from the battery. Based on this measurement and the terminal-voltage measurement, the processor can then reinitiate a fast charge (by asserting FAST/TRICKLE CHARGE low) before the battery reaches its end of life.

3-A BATTERY CHARGER FOR Li-ION OR NiCd

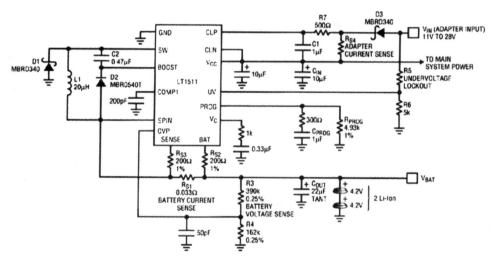

Fig. 14-8

The LT1511 is a high-efficiency switched-mode current source designed for battery charging in portable applications. The LT1511 implements the constant-voltage/constant-current profile required for lithium-ion (Li-ion) batteries. It can also charge nickel-cadmium (NiCd) and nickel-metal-hydride (NiMH) batteries by using an external charge-termination method. Full charging current can be programmed by resistors or a DAC. An input current regulating loop on the LT1511 allows simultaneous equipment operation and battery charging without overloading the wall adapter. The charging current is automatically reduced to keep the wall adapter current within specified levels.

TIMED NiCd CHARGER

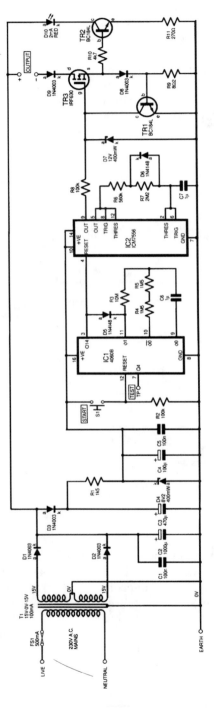

Fig. 14-9

EVERYDAY PRACTICAL ELECTRONICS

The primary supply is provided by 230 V through transformer T1, diodes D1 and D2, and reservoir capacitor C2. A supply for the timing circuit is provided through diode D3, with resistor R1 and zener diode D4 to keep it to a safe voltage for the CMOS devices used. These CMOS devices, IC1 and IC2, use so little current that the timer will keep going for up to 30 s if the power fails, preventing the charge period from being restarted by every minor glitch in the supply. IC1 is a 4060B 14-stage divider with built-in oscillator. Pressing switch S1 sets all its outputs low, so diode D5 is reverse-biased and the oscillator operates. With the component values shown, it runs at about 0.17 Hz, so the last output of the divider, pin 3, goes high after about 14 h. This applies forward bias to D5, which stops the oscillator. The first timer of the 7556 dual CMOS timer IC2 is connected to operate as an oscillator with a frequency of about 0.5 Hz with a duty cycle of about one to five. Pin 4 is an "active low" reset for this timer, so although the input to this from IC1 is low, its output is also low. The second timer is used as an inverter to convert this to a high output, which, via resistor R8, activates the output constant-current generator. When IC1 times out, the oscillator in IC2 starts running, and the current generator is then pulsed for about 400 ms every 2 s. NiCds have quite a high self-discharge rate: about 10 percent of capacity per week. If left on this charger, they will be kept fully topped up ready for use without overcharging, and by flashing in time with the current pulses, LED D10 will let the user know that the main charge period is complete.

2.5-A BATTERY CHARGER

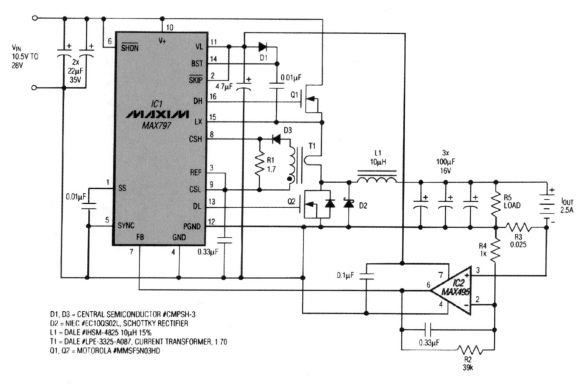

D1, D3 = CENTRAL SEMICONDUCTOR #CMPSH-3
D2 = NIEC #EC10QS02L, SCHOTTKY RECTIFIER
L1 = DALE #IHSM-4825 10μH 15%
T1 = DALE #LPE-3325-A087, CURRENT TRANSFORMER, 1 70
Q1, Q2 = MOTOROLA #MMSF5N03HD

MAXIM ENGINEERING JOURNAL

Fig. 14-10

Battery chargers are usually designed without regard for efficiency, but the heat generated by low-efficiency chargers can present a problem. For those applications, the charger shown in the figure delivers 2.5 A with efficiency as high as 96 percent. It can charge a battery of one to six cells while operating from a car battery. IC1 is a buck-mode switching regulator that controls the external power switch, Q1, and the synchronous rectifier. IC1 includes a charge pump for generating the positive gate-drive voltage required by Q1. The battery-charging current develops a voltage across the 25-MΩ resistor (R3) that is amplified by the op amp and presented as positive-voltage feedback to IC1. This feedback enables the chip to maintain the charging current at 2.5 A. While charging, the circuit can also supply current to a separate load, up to a limit set by current-sense transformer T1 and sense resistor R1. T1 improves efficiency by lowering power dissipation in R1. The transformer turns ratio (1:70) routes only 1/70 of the total battery-plus-load current through R1, creating a feedback voltage that enables IC1 to limit the overall current to a level compatible with the external components.

CONSTANT-CURRENT NiCd CHARGER

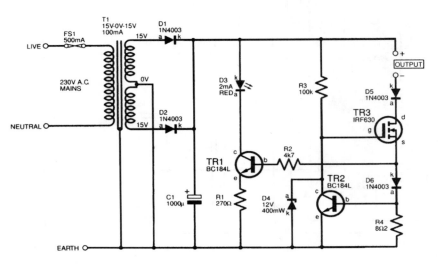

EVERYDAY PRACTICAL ELECTRONICS

Fig. 14-11

A tapped transformer (T1) with two rectifier diodes (D1 and D2) and capacitor C1 provides the primary dc supply from the 230 Vac. The power MOSFET device TR3 is initially biased into conduction by a positive gate voltage from resistor R3. Current flowing through the load flows through current-sensing resistor R4, developing a voltage. When this reaches about 0.5 V, transistor TR2 begins to conduct, reducing the gate (g) voltage of TR3. The circuit stabilizes at the load current that gives 0.5 V across R4. Zener diode D4 prevents excessive gate voltage from being applied to power MOSFET TR3, which might happen if the circuit was powered without the load connected. Diode D6 compensates for the base-emitter voltage drop of transistor TR1 so that the voltage across R4 is duplicated across resistor R1, controlling the current flow through LED D3. The apparent brightness of LED D3 therefore depends on the charging current, so any problems with connection to the cells on charge will be immediately apparent to the user. Diode D5 prevents any "back feeding" of current from the batteries power supply fails while the batteries are still connected.

FLOAT-CHARGING CIRCUIT

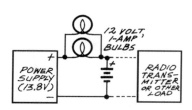

To float-charge a 12-V lead-acid battery, all you need is an accurately regulated 13.8-Vdc power supply. You could connect the power supply directly to the battery, but it is safer to make the connection through a couple of 12-V, 1-A automotive light bulbs. The light bulbs serve as current limiters; they guarantee that no matter what happens, the power supply will never have to deliver more than 2 A. Resistors can also serve as current limiters, but light bulbs are better because their resistance varies with current in a useful way. It's nearly zero at low current, but as the current through each bulb approaches 1 A, the bulb lights and the resistance increases. Thus, the bulb limits the current, but doesn't waste energy when the current is low. This setup is a good way to power a heavy load that operates intermittently, such as a ham radio transmitter. The battery delivers heavy current when necessary and the power supply keeps the battery charged.

ELECTRONICS NOW   *Fig. 14-12*

SIMPLE CHARGER

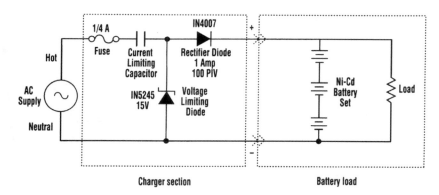

Charger section   Battery load

ELECTRONIC DESIGN   *Fig. 14-13*

Built with just four components, this battery charger for small NiCd batteries is useful as a compact travel charger or in "floating" simple battery-backed projects. It is small enough to fit in a 35-mm film canister.

Warning: This circuit is not isolated from the ac line and presents a shock hazard if there is any contact between it and a person or another device.

LEAD-ACID BATTERY TEMPERATURE-COMPENSATED CHARGER

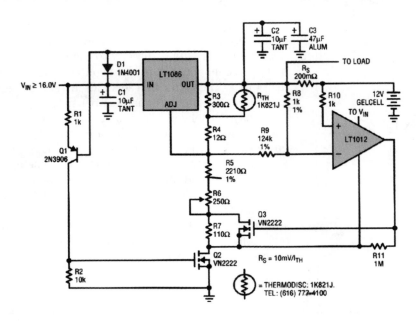

Table 1. The Regulator Should Be Chosen to Provide at Least C/4 Charging Current

Battery Capacity	Device	Maximum Charging Current	Float Current Threshold	Sense Resistor (Shunt)
≤3Ah	LT1117	0.8A	20mA	500mΩ
3–6Ah	LT1086	1.5A	50mA	200mΩ
6–12Ah	LT1085	3.2A	100mA	100mΩ
12–24Ah	LT1084	5.5A	200mA	50mΩ
24–48Ah	LT1083	8.0A	400mA	25mΩ

LINEAR TECHNOLOGY POWER SOLUTIONS

Fig. 14-14

The charging characteristics of lead-acid cells are strongly linked to the ambient temperature. To prevent over- or undercharging of the battery during periods of extended low or high temperatures, temperature compensation is desirable. This circuit incorporates a low-dropout linear regulator, temperature compensation, dual-rate charging, and true negative ground, and consumes zero standby current. The LT1086 linear regulator is used to control the charging voltage and limit maximum charging current. If the input supply is removed, Q1 and Q2 turn off and all charger current paths from the battery to ground are interrupted, resulting in zero shutdown current draw. Diode D1

LEAD-ACID BATTERY TEMPERATURE-COMPENSATED CHARGER *(Cont.)*

provides reverse current flow protection for the regulator should the input fall below the battery voltage or be shorted. The temperature compensation employed in this circuit follows the true curvature of a lead-acid cell. Temperature compensation is provided by R_{TH}, which is a Tempsistor, in parallel with R3. Changes in temperature alter the resistor divider ratio of the regulator. The match is within 100 mV for a 12-V battery over a range of -10 to $+60°C$. The best location for the Tempsistor is directly under the battery, with the battery resting on a pad of Styrofoam. Dual-rate charging is implemented by comparator LT1012, which senses the charging current through current-sense resistor R_S. When the current is greater than 10 mV/R_S, the high-rate charging voltage is 14.4 V at 25°C; when the current is less than this threshold, the float charging voltage is 13.8 V.

15

Battery Test and Monitor Circuits

The sources of the following circuits are contained in the Sources section, which begins on page 1043. The figure number in the box of each circuit correlates to the entry in the Sources section.

TRICOLOR LED NiCd CHECKER

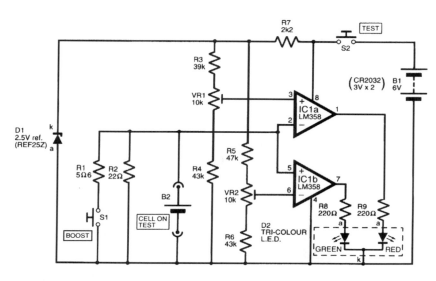

EVERYDAY PRACTICAL ELECTRONICS

Fig. 15-1

This circuit uses a window comparator and a bicolor LED. VR1 and VR2 set the high and low trip points of the comparators, respectively. The LED shows green for highest voltage (good), yellow for intermediate, and red for low voltage (bad). The circuit can be calibrated for other voltages and is generally useful as a voltage indicator.

LED BATTERY MONITOR FOR 12-V SYSTEMS

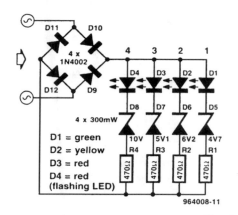

ELEKTOR ELECTRONICS

Fig. 15-2

This monitor is handy in cars and is particularly suitable for radio amateurs who power their equipment from a car battery. Bridge rectifier D9–D12 ensures that the battery cannot be connected with incorrect polarity. The zener voltage of diodes D5–D8 increases in standard steps, so the battery voltage needs to be higher to cause successive LEDs to light. In other words, the higher the battery voltage, the more LEDs will light. The component values are chosen so that at a battery voltage of 9 V—battery poor—only D1 lights; when it is about 11 V—battery doubtful—D1 and D2 light; when it is 13 V—battery fine—D1, D2, and D3 light. Diode D4 is a flashing LED. The value of zener diode D8 is such that the LED begins to flash when the battery voltage approaches 15 V—that is, an overvoltage situation.

BATTERY SIMULATOR CIRCUIT

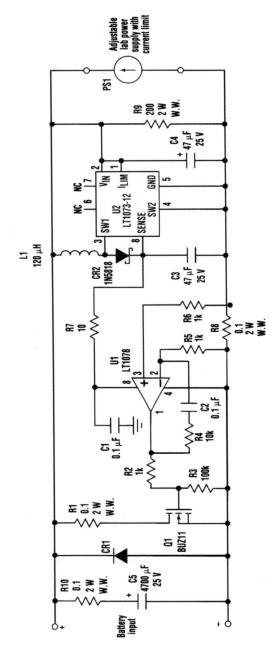

ELECTRONIC DESIGN

Fig. 15-3

When developing a battery charger, using a real battery might be inconvenient. The battery simulator circuit described here is an alternative. The battery input positive and negative terminals should be connected in place of the battery in the charger circuit. Also, a current-limited lab-type power supply must be connected to the simulator circuit (as shown). In Discharge mode, the battery simulator uses the current-limited lab power supply PS1 as a source, and the simulator is inactive. In Charge mode, charge current is forced through the battery input terminals. Low voltage that develops across R8 is amplified by U1 and causes Q1 to shunt the charge current while maintaining the power-supply (PS1) voltage. U2, L1, CR2, C3, and C4 produce an internal 12-V power supply that is required to operate U1 and drive Q1. The PS1 voltage range is 1.5 to 15 V. Diode CR1 protects the circuit from reverse polarity and is needed if the maximum charge current is too high for the body diode of Q1 (Q1 must be heat sunk). R1 is a sense resistor for measuring the charge current. R10 and C5 simulate the ac characteristics of the battery.

DUAL-MODE BATTERY-LIFE EXTENDER

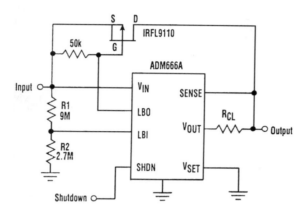

ELECTRONIC DESIGN

Fig. 15-4

Most voltage regulators stop functioning when the differential voltage between the input and output is reduced below a certain level. At this level, the regulator's output voltage begins to drop at an accelerated rate, resulting in a much lower voltage's being available to the circuit. To prevent this from happening, the battery voltage is monitored and a low-battery warning signal is issued, indicating the approaching end of battery life. In the circuit presented here, the voltage regulator is bypassed when a low battery level is detected—the battery is connected directly to the circuit. At high battery voltage levels, the output voltage is regulated by the linear regulator, and the low-battery indicator (LBI) comparator output is off. Also, the LBO pin is pulled to V_{batt} (input), keeping the power MOSFET off. As the input voltage drops and approaches the regulator's dropout voltage, the base current to the regulator's pass transistor reaches its maximum, turning it on very hard. Lowering the input voltage any further will result in an increase in the pass transistor's saturation voltage and a reduction in the regulator's output voltage. The input level detector's crossover voltage should be set to $V_{out} + V_{sat}$ of the pass transistor. When the input voltage falls below the crossover point, the comparator output is turned on. This pulls the gate voltage to ground, which turns on the power MOSFET. From this point, the voltage available to the circuit is the battery voltage minus the voltage across the power FET, which is insignificant.

FLASH BATTERY TESTER

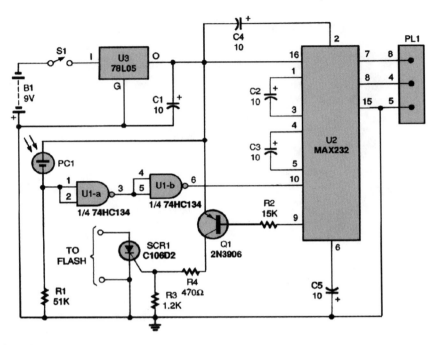

Fig. 15-5

Power from a 9-V battery (B1) is fed to a 78L05 regulator (U3). The IC then produces a regulated 5 V to power the rest of the circuit. A cadmium sulfide photocell, PC1, conducts when a connected flash's ready light comes on and illuminates it. When current flows through PC1, U1-a, and U1-b, two sections of a 74HC134 quad Schmitt trigger NAND gate produce a sharp-edged TTL high. A MAX232, U2, converts the high to an RS233 low that is compatible with the PC's serial port. The MAX232 also takes an incoming serial-port pulse and converts it to TTL levels to fire the flash via SCR1, a C106D2 silicon-controlled rectifier. The circuit works with a BASIC program. That program fires the flash by sending a brief pulse through plug PL1 to the serial port's DTR pin. The program then starts timing how long it takes for the flash to recycle. In other words, the program "looks" at the CTS pin over and over until it sees that it went low. That happens when the photocell is lit by the flash's ready light. The program records the elapsed time and continues counting until 1 minute elapses. At that time, the DTR pin is cycled again to fire the flash, and the process is repeated. The test concludes when the recycle time exceeds 45 seconds.

BATTERY CAPACITY INDICATOR

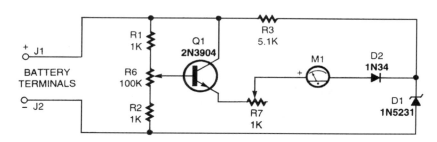

Fig. 15-6

Zener diode D1 creates a reference voltage to which the battery voltage is compared. The diode specified has a breakdown voltage of 5.1 V. That rating will work fine with most six- or seven-cell NiCd battery packs, as well as with 12-V lead-acid batteries. The circuit can be customized for a particular battery by selecting a unit for D1 that has a voltage rating about 1 V below the completely discharged voltage of the battery pack that you wish to measure. Transistor Q1, an emitter-follower amplifier, greatly increases the sensitivity of the circuit over what it would be if R7 were connected directly to the wiper of R6. A further advantage to that arrangement is that it reduces the current drain that flows through R1, R2, and R6. By amplifying the current flowing through the resistors, the resistance value can be increased to a very high value, lowering the total current draw of the circuit. Resistor R6 adjusts the meter to read 0 mA when the battery is completely discharged and R7 adjusts the meter to read 1 mA when the battery is fully charged. If the circuit were accidentally connected backward to the battery, current would flow through D1 and M1. The transistor would become reverse-biased, allowing a complete path back to the battery. That situation would allow excessive current to flow through D1, M1, and Q1, destroying them in the process. D2, R1, and R2 are included to prevent any current flow in case the battery is reversed.

BATTERY IMPEDANCE MEASURER

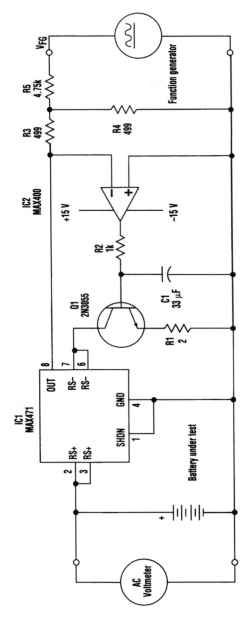

ELECTRONIC DESIGN ANALOG APPLICATIONS

By applying an ac voltage superimposed on a ten times larger negative dc voltage (at V_{FG}), the function generator determines the battery current drawn by Q1 (see the figure). The generator voltage causes the op-amp (IC2) output to go high and turn Q1 on, which allows battery current to flow through the high-side current-sensing amplifier (IC1). The output current of IC1, on pin 8, is equal to 1/2000 of this battery current. As a result, IC1, IC2, and Q1 form a loop in which the op amp forces a virtual ground at the IC1/IC2 end of R3. The op amp's extremely low-voltage offset (10 µV maximum) ensures accuracy. This virtual-ground condition enables the voltage divider (R5, and R3 in parallel with R4) and the function generator to determine the voltage across R3. The resulting current in R3 is $i_{R3} = R_P \times V_{FG}/[(R_P + R_5)R_3]$, where R_P is the parallel combination of R3 and R4. Substituting resistor values and noting that battery current (i_B) is 2000 times i_{R3}, $i_B = -V_{FG}/5$. To operate the circuit, set the function generator's ac voltage to approximately 10 percent of its dc component. The equation then gives the resulting ac current in the battery (i_B). Using an ac voltmeter, you can measure the ac voltage across the battery (v_B) and calculate the average cell impedance as v_B/Ni_B, where N is the number of cells. The circuit easily accommodates battery voltages of 3 V or more.

Fig. 15-7

NiCd BATTERY CYCLER

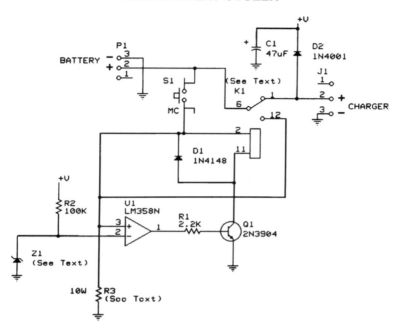

Fig. 15-8

The battery cycler works with NiCd configurations of between three and nine cells. Cell combinations are accommodated by selecting Z1, K1, and R3 based upon the number of cells in the battery pack. This circuit is basically composed of a voltage comparator (U1), a voltage reference (Z1), and a switch (Q1 and K1). The op amp (U1) compares the reference voltage generated by the zener diode (Z1) and the battery. When the battery voltage matches (or is less than) the reference, the switch is turned off and the battery is allowed to charge. While the battery voltage is above the reference voltage, it is discharged through resistor R3. The resistor has been selected to discharge the battery at a rate between 200 and 300 mA. This rate keeps the power dissipation in R3 below 3 W and reduces the risk of opening any fuses in the battery circuitry located in the portable equipment. The push-button switch (S1) initiates the discharge process. If the battery voltage is below the reference when S1 is pressed, the unit will not discharge the battery because the lower operating limit of the battery has already been exceeded. If your batteries are in this condition, charge them for 18 hours and measure the voltage. If the batteries are still below the cutoff voltage, the problem is with one or more cells in the battery pack.

LOW-BATTERY MONITOR

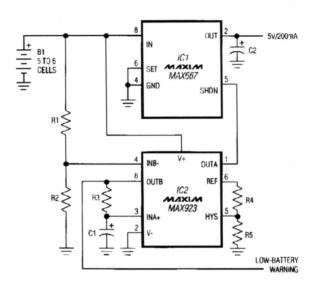

MAXIM ENGINEERING JOURNAL

Fig. 15-9

This circuit gives an early warning of declining battery voltage. Then, to allow a controlling processor time for emergency housekeeping chores, such as the storage of register data, the circuit delays system shutdown by a specified time interval (rather than waiting for battery voltage to decline further, to a specified lower level). Circuit components are chosen for low quiescent current, which protects discharged cells by minimizing the battery drain during shutdown: IC1 draws 1 μA, IC2 draws 3 μA, and R1 and R2 draw 3 μA, for a total shutdown current of about 7 μA. R1 and R3=1 MΩ, R2=280 kΩ, R4=49.9 kΩ, R5=2.4 MΩ, and C1=3.9 μF.

LOW BATTERY DETECTOR

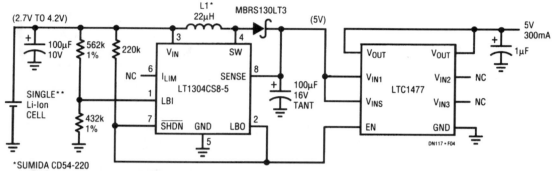

*SUMIDA CD54-220
**PRIMARY Li-Ion BATTERY PROTECTION MUST BE PROVIDED BY AN INDEPENDENT CIRCUIT

LINEAR TECHNOLOGY

Fig. 15-10

Low-voltage cutoff is desirable in battery-operated systems to prevent deep discharge damage to the battery. The LT1304 micropower boost regulator contains a low-battery detector which is active even when the regulator is shut down. The output of the detector controls both the LTC1477 and the LT1304 5-V boost regulator. In this application, the LTC1477 serves to protect against short circuits (850-mA limit selected) and completely disconnects the load under a low-battery condition. In shutdown, the circuit draws less than 25 μA from the battery.

DUAL BATTERY CAPACITY INDICATOR

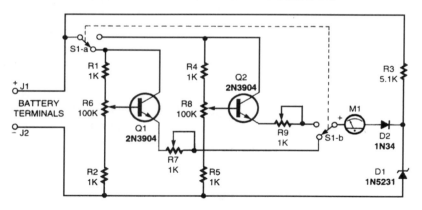

ELECTRONICS NOW

Fig. 15-11

This indicator circuit uses two separate expanded meter scale circuits and can be calibrated for use with two different types of batteries.

BATTERY TESTER

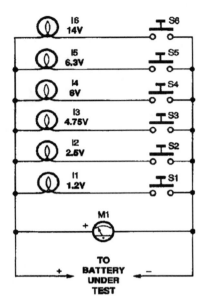

POPULAR ELECTRONICS

Fig. 15-12

Although deceptively simple, this battery tester can help you weed out marginal cells that might otherwise test good. Six incandescent lamps, chosen for their voltage and current ratings, are selected as a load for the battery under test. The meter (M1) and the lamp's brightness will give a good indication of a battery's output capacity. In addition, because the inrush current that occurs when the lamp is first connected across the battery is much higher than the rated operation value, the circuit makes it easy to spot marginal cells. To use the circuit, connect a battery and observe the measured voltage on M1. Then close the switch that corresponds to the meter's reading and/or the battery's rated voltage and observe the brightness of the appropriate lamp. Be careful to select the right lamp for testing. Never test using a lamp whose voltage rating is more than 30 percent lower than the battery's rating. Otherwise, damage to the lamp could result if the battery under test is good.

CAMCORDER BATTERY PROTECTOR

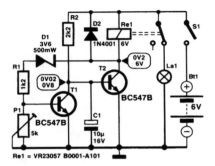

ELEKTOR ELECTRONICS

Fig. 15-13

Many camcorder enthusiasts use their spare battery for powering video lights. Many such lights give no indication when the battery has gone flat. This can be avoided by the use of this protection circuit. When it gets switched on, the potential across C1 is zero, so that T2 is cut off, the relay is inactive, and the indicator lamp lights. As long as the battery voltage remains above a certain level, T1 is on and holds the base of T2 at earth potential. In this state, only a small current is drawn. When the battery voltage is no longer higher than the sum of the zener voltage, the potential set by divider R2-P1, and the drop across the base-emitter junction of T1, this transistor is cut off, whereupon C1 is charged via R2. When the potential across C1 has risen to a value high enough for T2 to be switched on, the relay is energized, and its contact disconnects the lamp from the battery. Because the current drain ($\leq$70 mA) is then determined almost entirely by the relay, it is essential to remove, or disconnect, the battery from the light unit. The switch-off voltage level, set with P1, should be about 1 V per battery cell.

16

Bugging Circuits

The sources of the following circuits are contained in the Sources section, which begins on page 1043. The figure number in the box of each circuit correlates to the entry in the Sources section.

> Low-Power FM Telephone Bug
> FM Telephone Bug
> Telephone Bug

LOW-POWER FM TELEPHONE BUG

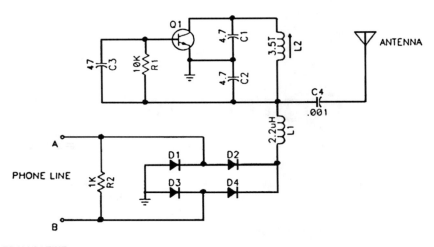

Fig. 16-1

Q1 (2N3904) is an oscillator tuned to a quiet spot in the FM broadcast band. D1 through D4 (1N914) ensure proper polarity. Dc from the line powers the bug. Audio on the line causes incidental FM, which can be heard on an FM receiver tuned to the frequency of the oscillator.

Warning: Use of this device for certain purposes could violate federal and/or state laws and subject the violator to prosecution.

FM TELEPHONE BUG

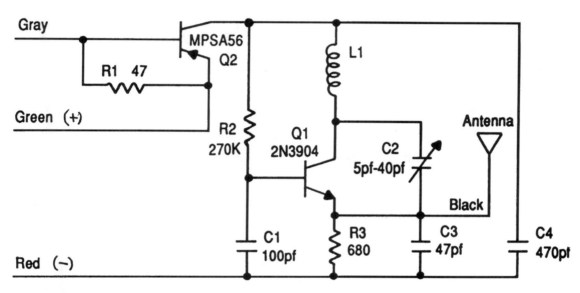

Fig. 16-2

Q1 is an oscillator tuned to a quiet spot in the FM broadcast band. Dc from the line powers the bug. Connect the gray wire to the phone in place of the green wire, and the green wire on the bug to the green line wire. The red wire goes to the red phone line wire. Audio on the line causes incidental FM, which can be heard on an FM receiver tuned to the frequency of the oscillator. Because this device "sees" the full line voltage, Q2 is a high-voltage PNP MPSA56 transistor.

Warning: Use of this device for certain purposes could violate federal and/or state laws and subject the violator to prosecution.

TELEPHONE BUG

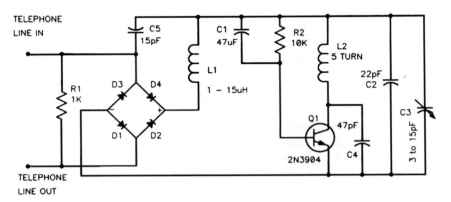

Fig. 16-3

Q1 is an oscillator tuned to a quiet spot in the FM broadcast band. Dc from the line powers the bug. Diodes D1 through D4 ensure proper polarity. R1 maintains a suitable voltage drop for the bug. Audio on the line causes incidental FM, which can be heard on an FM receiver tuned to the frequency of the oscillator.

Warning: Use of this device for certain purposes could violate federal and/or state laws and subject the violator to prosecution.

17

Clock Circuits

The sources of the following circuits are contained in the Sources section, which begins on page 1043. The figure number in the box of each circuit correlates to the entry in the Sources section.

Low-Power Wide-Supply-Range Clock
Simple 5.2-kHz Clock

LOW-POWER WIDE-SUPPLY-RANGE CLOCK

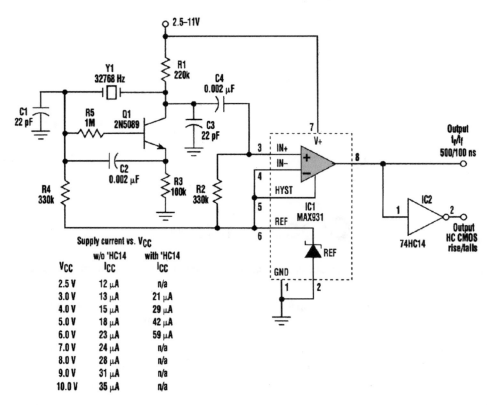

Supply current vs. V_{CC}		
	w/o 'HC14	with 'HC14
V_{CC}	I_{CC}	I_{CC}
2.5 V	12 μA	n/a
3.0 V	13 μA	21 μA
4.0 V	15 μA	29 μA
5.0 V	18 μA	42 μA
6.0 V	23 μA	59 μA
7.0 V	24 μA	n/a
8.0 V	28 μA	n/a
9.0 V	31 μA	n/a
10.0 V	35 μA	n/a

ELECTRONIC DESIGN

Fig. 17-1

This 32-kHz, low-power clock oscillator offers numerous advantages over conventional oscillator circuits based on a CMOS inverter.

Many times, 32-kHz oscillators are used to generate a system clock or an auxiliary sleep clock in low-power instruments and microcontrollers. The typical implementation uses a CMOS inverter (74HC04 or CD4049UB type). Inverter circuits present problems, though. Supply currents fluctuate widely over a 3- to 6-V supply range, and currents below 250 μA are difficult to attain. Operation can be unreliable for wide variations in supply voltage. A very low power crystal oscillator solves these problems. Drawing only 13 μA from a 3-V supply, it consists of a single-transistor amplifier/oscillator (Q1) and a low-power comparator/reference device (IC1). Q1's base is biased at 1.25 V via R5, R4, and the reference in IC1. V_{BE} is about 0.7 V, placing the emitter at approximately 0.5 V. This constant voltage across R3 sets the transistor's quiescent current at 5 μA, which fixes the collector voltage at about 1 V below V_{CC}. The amplifier's nominal gain (R_1/R_2) is approximately 2 V/V. The crystal and load capacitors (C1 and C3) form a feedback path around Q1, whose 180° of phase shift causes the oscillation. C4 cou-

LOW-POWER WIDE-SUPPLY-RANGE CLOCK *(Cont.)*

ples this signal to the comparator input; the input's quiescent voltage (1.25 V) is set by the reference via R2. The comparator's input swing is thus centered around the reference voltage. Operating at 3 V and 32 kHz, IC1 draws about 7 μA.

SIMPLE 5.2-kHz CLOCK

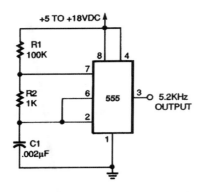

This circuit will produce a clock signal of 5.2 kHz using a NE555 timer.

ELECTRONICS NOW *Fig. 17-2*

18

Code-Practice Circuits

The sources of the following circuits are contained in the Sources section, which begins on page 1043. The figure number in the box of each circuit correlates to the entry in the Sources section.

Morse Code Oscillator
Crystal-Controlled Code-Practice Transmitter
PLL Code-Practice Oscillator
Touch-Operated Code-Practice Oscillator
Wireless FM Code-Practice Transmitter
Infrared Code-Practice Transmitter
Infrared Code-Practice Receiver
Code-Practice Oscillator

MORSE CODE OSCILLATOR

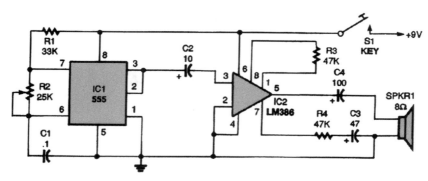

POPULAR ELECTRONICS

Fig. 18-1

The circuit is built around a 555 oscillator (IC1) and an LM386 audio amplifier (IC2). The 555 circuit is an astable oscillator, with the chip output retriggering the circuit. When the key (S1) is pressed, it activates the circuit. The 555 oscillates at a frequency determined by R1, R2, and C1. Potentiometer R2 is used to adjust the tone frequency of the oscillator. Some of the output current of IC1 is coupled to IC2 via a 10-°F capacitor, C2, so that it will be sufficient to drive a loudspeaker. Because the circuit has no gain control, the volume depends on the size and wattage of the speaker.

CRYSTAL-CONTROLLED CODE-PRACTICE TRANSMITTER

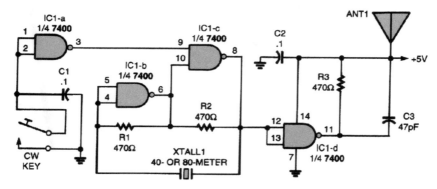

POPULAR ELECTRONICS

Fig. 18-2

When the CW key is closed, IC1-a's output goes high, allowing IC1-b and IC1-c to oscillate. The crystal supplies the feedback path setting the oscillator's operating frequency. The circuit will operate on the 40- and 80-m bands. Section IC1-d isolates the oscillator from the short antenna, ANT1. A clip lead should do here to get the signal out and about for operation. Tune your ham-band receiver to the crystal's frequency and key down. If the receiver doesn't have a CW mode, turn on the BFO and tune for the desired CW tone.

PLL CODE-PRACTICE OSCILLATOR

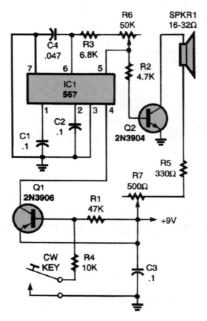

Fig. 18-3

This CPO uses a 567 phase-locked loop, IC1, as the variable tone generator. The oscillator's frequency is set by R6, and the frequency range can be changed by selecting a different-value capacitor for C5. To lower the oscillator's frequency range, make the value of C_5 larger, and to increase the frequency range, reduce C_5. A general-purpose 2N3906 PNP transistor, Q1, supplies power to the 567 through pin 4 each time that the CW key is closed. Meanwhile, Q2, a general-purpose 2N3904 NPN transistor, buffers the oscillator's output and drives the speaker. Potentiometer R7 sets the output volume.

TOUCH-OPERATED CODE-PRACTICE OSCILLATOR

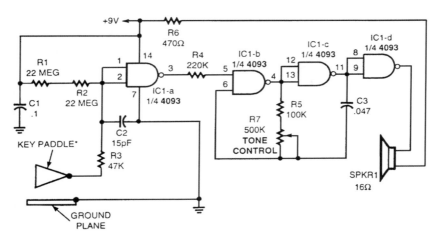

POPULAR ELECTRONICS

Fig. 18-4

A touch-operated CPO is shown in the figure. The gates of IC1-a are biased high through the two 22-MΩ resistors (R1 and R2), keeping its output low. Gates IC1-b and IC1-c are connected in an audio oscillator circuit that can operate only when pin 5 of IC1-b is high. The last gate of the 4093, IC1-d, adds isolation to the oscillator's circuit and drives the speaker (SPKR1). Touching the key paddle and ground plane lowers IC1-a's input gate voltage to near zero, allowing the output at pin 3 to go high. The tone generator then turns on and sends out an audio note. The touch key paddle and ground plane can be made from a circuit board or any other conductible material. Note that the ground plane should lie flat for a hand rest and the key paddle should be positioned for ease of touch.

WIRELESS FM CODE-PRACTICE TRANSMITTER

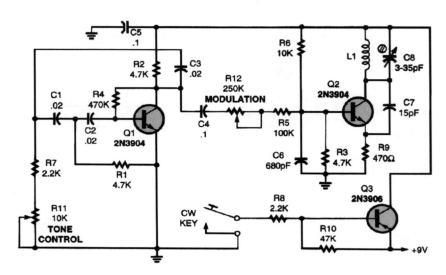

POPULAR ELECTRONICS

Fig. 18-5

The circuit shown is a low-power, tone-modulated FM transmitter that can be used with any FM broadcast receiver for code practice. Transistor Q1 and its associated components make up a phase-shift audio-frequency generator circuit. Potentiometer R11 sets the tone frequency. Transistor Q2 is connected in a high-frequency RF oscillator circuit that operates in the FM broadcast band. Potentiometer R12 sets the modulation level. Transistor Q3 operates as a switch, turning on the FM transmitter each time the CW key is closed. Coil L1 is a homemade air-wound coil. Take a 6½-in length of 20-gauge enamel-covered wire and close-wind it around a ¼-in-diameter form; leave about ¼ in free at each end. Remove the insulation from the ends and slide the coil off the form. The overall length of the finished coil should be about ¼ in. Set R11 and R12 to midposition and close the CW key. Then set your FM receiver to a clear spot on the low end of the dial and slowly adjust C8. Once the tone is heard, R11 can be set for the desired tone frequency and the tone level set by R12. If your oscillator won't tune to the top end of the band, carefully stretch the windings of L1 and retune. The circuit's operating range can be increased by adding a very short antenna to the emitter of Q2.

INFRARED CODE-PRACTICE TRANSMITTER

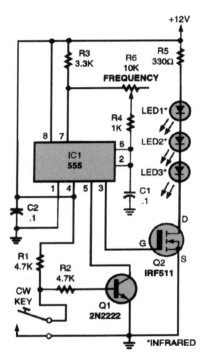

POPULAR ELECTRONICS

Fig. 18-6

A 555 timer, IC1, is connected in an audio-oscillator circuit with its frequency set by potentiometer R6. Transistor Q1, with the CW key up, is biased on, thereby holding pin 5 of IC1 low and keeping it turned off. The 555 timer's output, at pin 3, ties to the gate of a power hexFET, Q2, which drives the three IR emitters, LED1 to LED3. Placing the CW key in the down position turns the 555 oscillator on. That sends out the audio tone signal via IR.

INFRARED CODE-PRACTICE RECEIVER

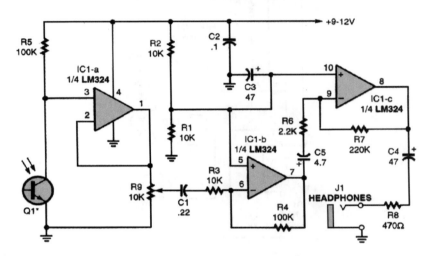

Fig. 18-7

An IR phototransistor, Q1, is directly-coupled to the input of op amp IC1-a. The output of IC1-a is fed through the gain-control potentiometer R9 to the input of op amp IC1-b, which has a voltage gain of 10. Section IC1-b's output drives IC1-c, which has a voltage gain of 100. The output of IC1-c supplies audio to the headphones via J1. The IR phototransistor can be mounted in reflectors to increase the CPO's operating range.

CODE-PRACTICE OSCILLATOR

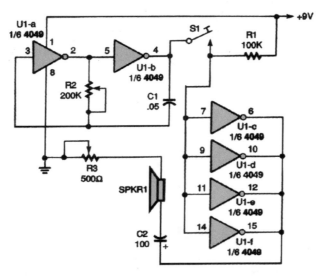

POPULAR ELECTRONICS

Fig. 18-8

This versatile code-practice oscillator has a variable frequency and volume control. The unit is especially suitable for use by small groups that are interested in learning and practicing code. A single 4049 CMOS hex inverting buffer is the heart of the oscillator, with inverters U1-a and U1-b making up the variable audio-oscillator circuit. The oscillator's output is coupled to the speaker-driver circuit through the CW (Morse code) key S1. The audio frequency of the oscillator (hence, its tone) is varied by the potentiometer. R3 is used to vary the speaker's volume.

19

Comparator Circuits

The sources of the following circuits are contained in the Sources section, which begins on page 1043. The figure number in the box of each circuit correlates to the entry in the Sources section.

DUAL-VOLTAGE COMPARATOR

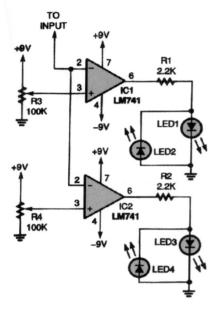

Use this dual comparator to monitor a battery while it's charging; the circuit will let you check for under- and overvoltage conditions.

R3 and R4 set the trip levels at which the LEDs are activated. Almost any standard LEDs can be used. LED current is about 3 to 4 mA, as determined by the op-amp capability and resistors R1 and R2.

POPULAR ELECTRONICS

Fig. 19-1

VOLTAGE WINDOW COMPARATOR

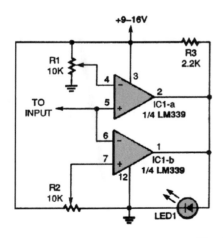

This window comparator determines if a voltage is between two limits, upper and lower.

Using a digital voltmeter, set the reference voltages in this voltage window to the same values. Then vary one to set the width of the window; when the input voltage is within the window area, LED1 will light.

POPULAR ELECTRONICS

Fig. 19-2

FOUR-LEVEL VOLTAGE COMPARATOR

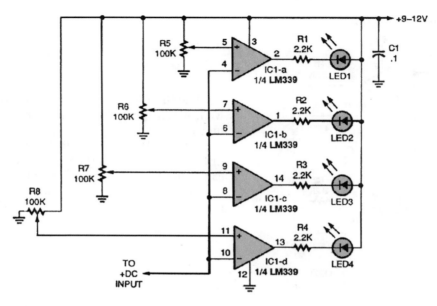

Fig. 19-3

This four-level voltage detector can be used as a bar-graph voltmeter. Simply set each potentiometer (R5 to R8) for a specific voltage.

ADJUSTABLE COMPARATOR

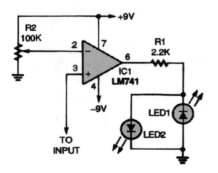

The setting of potentiometer R2 determines the level at which this comparator circuit will switch. The output can be used to drive any device needing a comparator signal, within the drive capabilities of the particular op amp used (741 is shown).

Fig. 19-4

FAST TTL-COMPATIBLE COMPARATOR

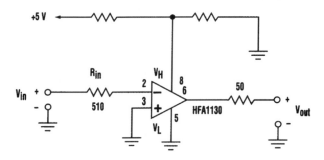

ELECTRONIC DESIGN ANALOG APPLICATIONS

Fig. 19-5

The HFA1130 by Harris Semiconductor is useful as a comparator. Depicted is an inverting 2-ns comparator with TTL-compatible output levels that are realized by using the HFA1130 output-limiting, current-feedback amplifier.

VOLTAGE COMPARATOR

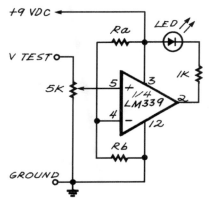

The LED turns on when the input voltage at pin 5 of the LM339 falls below the reference voltage at pin 4.

ELECTRONICS NOW

Fig. 19-6

20

Computer-Related Circuits

The sources of the following circuits are contained in the Sources section, which begins on page 1043. The figure number in the box of each circuit correlates to the entry in the Sources section.

Printer-Port D/A Converter
Isolated RS232 Interface
Computer Serial Port Relay Controller
Ultra-Simple RS232 Tester
RS232-to-Parallel Data Converter
Three-Wire RS232-to-RS485 Converter
RS232 Test Circuit
PC Power Pincher
Computer Voice
Joystick Changeover
Serial Transmitter Circuit
Flash EEPROM Communicator
PC Watchdog
SCSI Switch
Baud-Rate Generator
PC IR Card Reader
5-V Supply from Three-Wire RS232 Port
Computer-Controlled A/D Converter
PC Signal Generator

PRINTER-PORT D/A CONVERTER

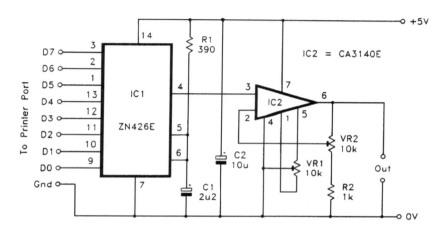

EVERYDAY PRACTICAL ELECTRONICS

Fig. 20-1

Where eight latching outputs are available, the Ferranti ZN426E probably represents the simplest means of providing an analog output. The figure shows the circuit diagram for a PC analog output port based on this chip. The full-scale output voltage is equal to the reference voltage fed to pin 5. This terminal can be fed from an external reference voltage of up to 3 V, but, in most cases, the built-in 2.55-V reference is perfectly adequate. The output voltage from IC1 (in volts) is equal to the value written to the printer port multiplied by 0.001. In most practical applications, this output-voltage range will have to be modified using an amplifier or an attenuator. In virtually all cases it will be a small amount of amplification that is required. This is the purpose of IC2, which also provides output buffering. The noninverting-mode amplifier IC2 can have its closed-loop voltage gain varied from unity to about 11 times by means of preset VR2.

The maximum output voltage of IC2 is about 2 V less than its supply potential (about 3 V) if it is powered from a 5-V supply. Therefore, maximum output voltages of more than 3 V require IC2 to be powered from a higher supply potential of up to about 30 V. This means using a separate supply for IC2 because the converter circuit must be powered from a 5-V supply. If preset VR1 is included, the best way to find the correct setting is to first write a low value to the port and adjust VR1 for the correct output voltage. Then write a high value to the port and adjust VR2 for the appropriate output voltage. Repeat this process a few times until no further adjustment is needed. If VR1 is omitted, write a value of 255 to the port and then adjust VR2 for the required maximum output voltage. Reasonable accuracy should then be obtained over the full range of output voltages. Using GW BASIC, it is just a matter of writing the values to the appropriate address using the OUT command. For example, OUT &H378,123 would write a value of 123 to a digital-to-analog converter connected to printer port LPT1.

ISOLATED RS232 INTERFACE

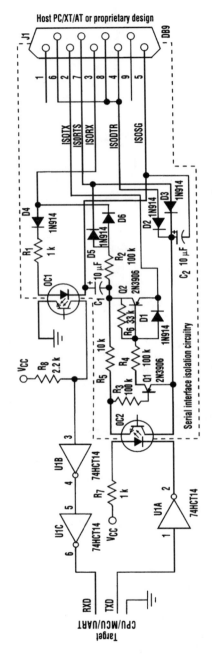

Fig. 20-2

ELECTRONIC DESIGN

Electrical isolation of 1500 V or higher is possible with this cost-effective RS232 interface. It is ideal for various data-communication designs, including medical instrumentation and devices within harsh electrical environments. The interface uses the host's TXD (transmit data), RXD (receive data), DTR (data terminal ready), and RTS (request to send) lines to provide power. For fully duplexed operation, the host's communications port should be set up so that its RTS is off (at negative voltage). Half-duplex operation can use the host's TXD negative voltage, available through D6, to run the circuitry. Sending a low bit (space state) to the host is just the opposite. This time, the required positive voltage is supplied from the host via diodes D2 and D3, and is routed through Q1 and D1 to DB9-2.

COMPUTER SERIAL PORT RELAY CONTROLLER

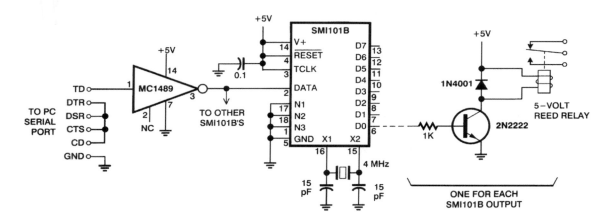

```
10 OPEN "COM1:9600,N,8,1" FOR OUTPUT AS #1   ' open serial port
20 PRINT #1,"ØN1" + CHR$(13)                 ' turn on relay 1 of processor Ø
30 FOR I=1 TO 10000: NEXT I                  ' pause a moment
40 PRINT #1,"ØF1" + CHR$(13)                 ' turn it off again
```

ELECTRONICS NOW

Fig. 20-3

Decoding RS-232 signals is a job for a microcontroller (single-chip computer). Fortunately, you don't have to program the microcontroller yourself; you can get a PIC 16C54A microcontroller already programmed for exactly this job from Stone Mountain Instruments. Shown is the circuit and the BASIC program to demonstrate how it works. Each SMI101B has eight logic-level outputs. Further, you can connect up to seven SMI101Bs to a single serial port. The three N pins give each SMI101B a distinctive identifying number, from 0 to 6. If all three are grounded, the identifier is 0; if N1 is connected to +5 V, the identifier is 1, etc. At power-up, all the data outputs are off (logic 0). To turn an output on, send a command of the form "xNy, where x is the identifier of the SMI101B and y indicates which data output you want to switch. To turn the outputs back off, use an F in place of an N (e.g., 0F3). All communication is done with 8 data bits and no parity bits. The baud rate is 9600 baud with a 4-MHz crystal, or 1200 baud with a 500-kHz ceramic resonator. As shown in the diagram, each relay requires a transistor to drive it, along with a resistor and a protective diode. To cut down the total number of components, you can use a relay driver chip, such as the Allegro UDN2987, which contains everything necessary to drive eight small relays from logic-level signals.

ULTRA-SIMPLE RS232 TESTER

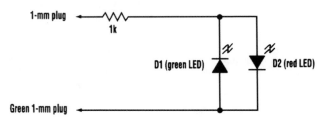

Fig. 20-4

The tester consists of no more than a two-color LED (or a red and a green LED connected in anti-parallel) in series with a 1-kΩ resistor. The free ends of the resistor and the LEDs should preferably be terminated in 1-mm plugs. One of the free ends should be covered in green sleeving, or use a green 1-mm plug. This is touched on pin 7 (signal ground) of the connector under test. The other end is touched on each pin to be tested, in turn. The LED (or LED pair) is connected so that a positive voltage emits a red glow and a negative voltage causes a green glow. Sometimes, an RS232 input will be found that has an internally connected pull-up to drive a particular default RS232 level when unconnected. Although this will cause the tester to glow as if the pin were an output, it will do so with markedly less brightness. Bearing this in mind, the tester can be used to diagnose most RS232 problems at the electrical level.

RS232-TO-PARALLEL DATA CONVERTER

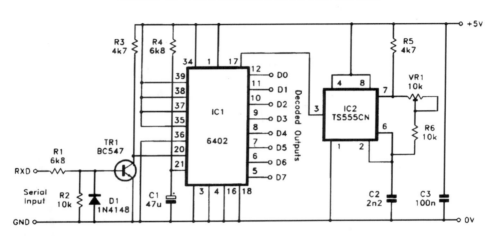

EVERYDAY PRACTICAL ELECTRONICS

Fig. 20-5

A simple serial decoder circuit based on a 6402 UART (IC1) is shown in the figure. The RS232C input signal is at signal levels of about ±12 V; these must be converted to standard 5-V logic levels before being applied to the serial input of IC1. A simple common-emitter switching stage based on transistor TR1 is used to provide the conversion to normal logic levels, and it also produces the necessary inversion of the input signal. Resistor R4 and capacitor C1 provide IC1 with a positive reset pulse at switch-on. The inputs at pins 35 to 39 program the word format; the method of connection shown in the figure provides a format on 1 start bit, 8 data bits, 1 stop bit, and no parity. Pin 34 is connected to the +5-V rail so that the binary pattern on pins 35 to 39 is loaded into IC1's control register. The decoded bytes of parallel data are available at pins 5 to 12, and pin 4 is connected to the 0-V rail so that these outputs are permanently enabled. In a stand-alone application, the tristate capability of these outputs is not of great value, but, if necessary, this facility can be utilized by applying a control signal to pin 4 of IC1.

THREE-WIRE RS232-TO-RS485 CONVERTER

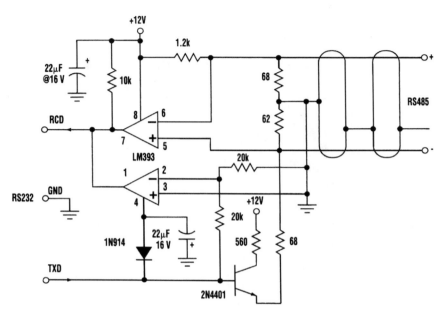

Fig. 20-6

This circuit needs only a minimal three-wire RS232 implementation plus one +10- to +15-V supply voltage to provide a transparent link capable of sending data at transmission rates up to tens of kbaud. Circuit operation is as follows: When both the 232 and the 485 are idle (232 port in MARK state and no 485 device active), the 485 link is held in the 1 state by the 1200-Ω pullup. This causes the top comparator to hold the 232 RCD line negative and therefore in MARK state. When a character is transmitted by the 232 port, it begins with a positive-going (SPACE state) START bit on the TXD line. In response, the 2N4401 pulls the "−" 485 conductor more positive than the "+" wire, thus transmitting the START down the 485 cable. Meanwhile, the bottom comparator holds the "wire-or" (LM393s have open-collector outputs) 232 RCD low, blocking the 232 port from "hearing" its own transmission. The receive side of the 232 port remains idle. The rest of the bits of the character follow along in the same fashion. When a character originates somewhere along the 485 bus, it begins with a 485 transceiver going active and driving the "−" line above the "+." This causes the upper half of the LM393 comparator to release RCD, and this time the bottom comparator doesn't prevent it from being pulled high. The data bits are thus allowed to arrive at the RS232 port, where they appear at standard RS232 bipolar voltage levels. The common-mode voltage range and noise-rejection capabilities of this circuit are compatible with standard 485 specifications. The converter's speed is mainly limited by the loading of the comparator outputs because of cable capacitance.

RS232 TEST CIRCUIT

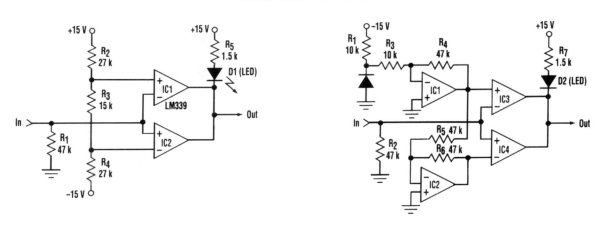

ELECTRONIC DESIGN

Fig. 20-7

The tester is basically a window comparator, in which the low and high levels are set at +3.0 V and −3.0 V, respectively, by resistors R2, R3, and R4. Resistor R1, when not driven by an RS232 output, will have a low voltage across it (approximately 0 V), and the LED D1 at the output of the comparators is turned off. If the unknown wire of the cable that is tested is an RS232 output, then it will drive the In point to a voltage either between +3 and +12 V or between −3 and −12 V. In both cases, one of the two comparator outputs will be driven low. This turns the LED on, indicating the presence of a wire connected to an RS232 output. The comparator should be an LM339 type or equivalent (with an open-collector output). The disadvantage of this scheme is that the thresholds are very sensitive to the supply variations. To eliminate this problem, the thresholds at the inputs of the comparators can be created using the normal forward drop on a simple diode and then be brought to the necessary levels by IC1 (+3 V at its output) and IC2 (a simple inverter).

PC POWER PINCHER

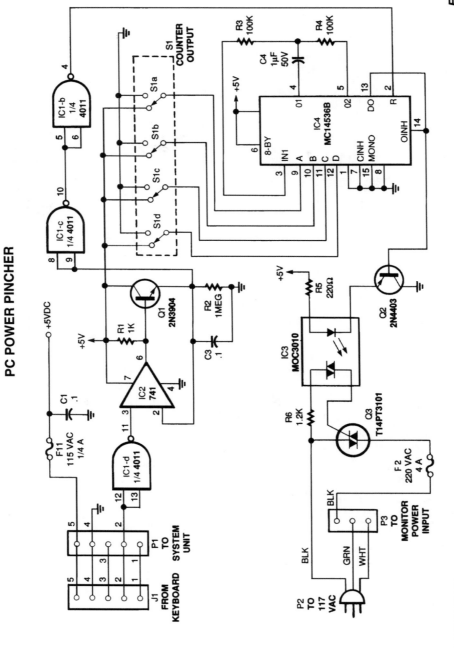

PC POWER PINCHER *(Cont.)*

The figure is a diagram of the circuit, in which a low-frequency oscillator continually drives the input of a multistage binary counter. Whenever the count reaches the setting selected by DIP switch S1, the circuit turns triac Q3 off, thereby interrupting the flow of 120-Vac to the monitor. A keyboard-monitoring circuit keeps the video monitor powered up during active use by resetting the counter every time a key is pressed. As long as a key-press occurs before the time delay expires, the counter keeps resetting. Hence, it never times out, and the monitor continues to receive power. When the computer turns on, a routing in its basic input/output system (BIOS) polls the keyboard. The keyboard, in turn, sends a series of data pulses back to the microprocessor to indicate its status. The data line is normally high (+5 V), and the pulses are low-going transitions. The first stage of the power pincher inverts the sense of the logic to normally low with high-going transitions.

COMPUTER VOICE

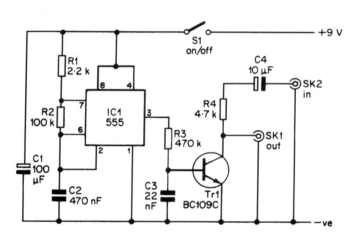

ELECTRONIC EXPERIMENTERS HANDBOOK

Fig. 20-9

This circuit will enable one to simulate the "computer voice" effect commonly heard in films, ads, and TV programs. It consists of two sections: an oscillator to provide the modulation signals, and the modulator itself. The oscillator uses a 555 timer chip in the astable multivibrator mode, and the frequency of operation has been set at about 10 Hz by the values given to R1, R2, and C1. A very simple modulator is used, but this is quite acceptable. Distortion produces new frequencies that help to change the voice signal and make it sound less like the original. A large amount of distortion is obviously not desirable because it would severely impair the intelligibility of the output signal. Transistor TR1 is used as a sort of voltage-controlled resistor, and, in conjunction with R4, it forms a voltage-controlled attenuator.

JOYSTICK CHANGEOVER

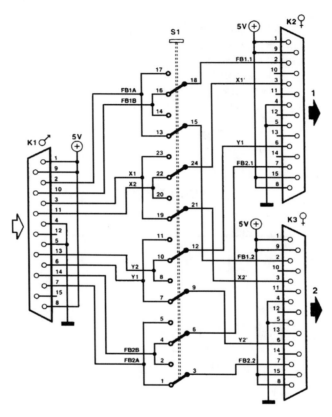

ELEKTOR ELECTRONICS

Fig. 20-10

Many I/O cards and sound cards have a standard provision of a 15-way connection for two joysticks. Unfortunately, many programs use the connections for only one joystick. Because often several kinds of joystick are used (in particular, modern flight simulators have provision for very advanced, specialized joysticks), it is frequently necessary to change over connectors. As the joystick connectors are invariably found at the back of the computer, this can be a tedious operation. Moreover, in the long term, it does not do the connectors any good. The present circuit replaces this changing over of connectors by a simple push on a button. In this way, two joysticks can be connected to the computer in a simple and user-friendly way. An eight-pole switch arranges the interconnection of controls X and Y and fire button 1 and fire button 2.

SERIAL TRANSMITTER CIRCUIT

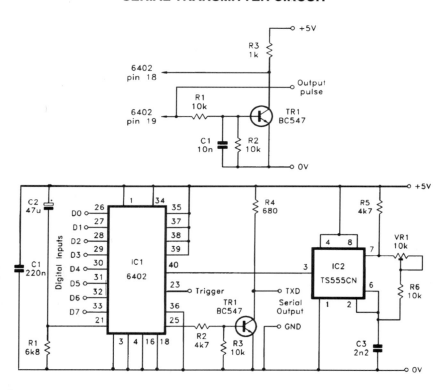

Fig. 20-11

This circuit provides basic parallel-to-serial conversion, with the control inputs connected to provide a word format of 1 start bit, 8 data bits, 1 stop bit, and no parity. VR1 should provide operation at 1200 baud. Adjusting VR1 is just a matter of using trial and error to locate a setting that enables the computer to reliably decode the serial data. If a frequency meter is available, adjust VR1 for an output frequency of 19.2 kHz at pin 3 of IC2.

FLASH EEPROM COMMUNICATOR

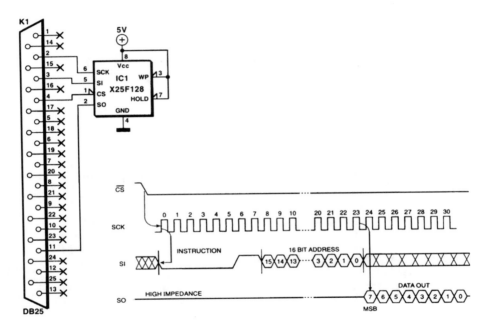

ELEKTOR ELECTRONICS

Fig. 20-12

Modern equipment frequently uses serial EEPROMS to store data that must not be lost. With the circuit shown, it is possible to program or read such an IC via the Centronics port of a computer. Because the serial interface of a flash EEPROM is identical to that of an EEPROM, the two devices can communicate with each other. This means that the present circuit can be used for either device. The IC is enabled via CS (Chip Select). The command "read" followed by the address that needs to be read is set on to the high level is a little low, some 4.7 kΩ pull-up resistors can be added between the Centronics outputs and the positive supply line (not the SO line).

PC WATCHDOG

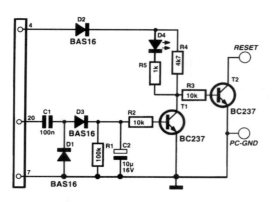

BAS16

The watchdog is intended to monitor a microprocessor and determine whether it functions correctly or not. It and its software are suitable for use in the background with any PC that runs under DOS or Windows. The hardware is linked to the serial interface, which is then controlled by the software. After the computer has been switched on, the data connections at the serial interface are low. The watchdog software is started at the same time as the selected application program. It provides a permanent rectangular signal at pin 20, which results in C2 being charged and T2 conducting. Pin 4 is made high so that the computer cannot receive a RESET signal. This condition is stable as long as the program runs. If the computer fails (crashes), the rectangular signal falls out and C2 is discharged via R1. This causes T1 to be switched off, whereupon the base of T2 goes high. This transistor is then on and pulls the RESET line of the computer to ground. The computer then restarts. Note that the circuit works only if a computer RESET also causes the serial interface to be reset because that is essential for the high level at pin 4 to be removed. It is only when this pin is at ground level that the RESET pulse is terminated and the computer can reboot.

Fig. 20-13

SCSI SWITCH

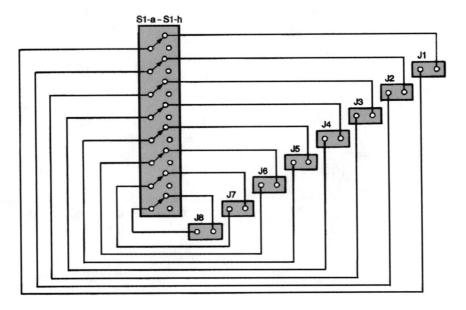

POPULAR ELECTRONICS

Fig. 20-14

The SCSI switch consists of an eight-position DIP switch, about 24 inches of ribbon cable, and eight female header blocks. A schematic of the switch is shown. The header blocks plug onto the ID pins of the SCSI hard disks. Through the DIP switch, those IDs can be reassigned at will. In any given configuration, whichever physical drive has the lowest SCSI ID becomes the boot drive. The others are allocated in order by SCSI ID.

BAUD-RATE GENERATOR

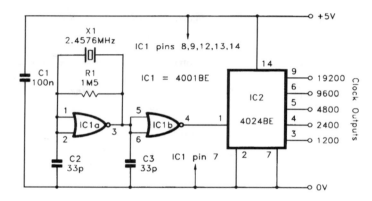

EVERYDAY PRACTICAL ELECTRONICS

Fig. 20-15

Gate IC1a is used in a conventional crystal oscillator circuit and IC1b acts as a buffer stage. IC2 is a CMOS 4024BE seven-stage binary counter and its clock input is fed with the 2.4576-MHz signal from IC1b. The first two stages of IC2 are not used, but the other five outputs provide baud rates from 1200 to 19,200 baud. Of course, the clock frequency is 16 times the baud rate and output frequencies from IC2 range from 19.2 to 307.2 kHz.

PC IR CARD READER

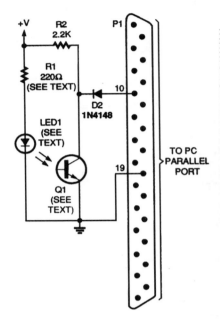

LISTING 1—MAIN PROGRAM

```
REM*******************************************************
REM** SWIPE.BAS V950121 (c) 1995, JJ Barbarello *****
REM** NOTICE: This is a non-compilable version *****
REM*******************************************************
CLEAR : CLS : DEFINT A-X: DEFSTR Y-Z: DIM x(16)
DEF SEG = 64: ON ERROR GOTO errortrap
OPEN "R", 1, "BITPORT.DAT": FIELD 1, 4 AS a$
IF LOF(1) = 0 THEN
a1 = PEEK(8) + 256 * PEEK(9) + 1
ELSE
GET 1, 1: a1 = VAL(a$) + 1
END IF
CLOSE 1
REM************* MAIN PROGRAM LOOP
start1:
GOSUB screenlayout
WHILE (INP(a1) AND 64) = 0
a$ = INKEY$: IF a$ <> "" THEN GOTO readytoend
WEND
x = 0: j = 0: start! = TIMER
readholes:
WHILE (INP(a1) AND 64) = 64: WEND
x = 0: WHILE (INP(a1) AND 64) = 0: x = x + 1: WEND
j = j + 1: x(j) = x
IF x = 0 OR (TIMER - start!) > 2 THEN ERROR 6
IF j < 16 THEN GOTO readholes
done1:
VIEW PRINT 3 TO 24: CLS : VIEW PRINT: BEEP
stat = 0: ttl = 0
FOR i = 2 TO 16
SELECT CASE stat
CASE IS = 0
IF x(i) > 1.5 * x(i - 1) THEN
ttl = ttl + 2 ^ (i - 2): stat = 1
ELSE
stat = 0
END IF
CASE IS = 1
IF x(i) < .667 * x(i - 1) THEN
stat = 0
ELSE
ttl = ttl + 2 ^ (i - 2): stat = 1
END IF
CASE ELSE
ERROR 6
END SELECT
NEXT
LOCATE 14, 3
LOCATE 10, 35: PRINT "ID SENSED:"; ttl
GOSUB screenlayout
GOTO start1
readytoend:
IF a$ = CHR$(27) THEN CLS : LOCATE 18, 1, 1: END
BEEP: GOTO readholes
REM**
REM** SCREEN LAYOUT
REM**
screenlayout:
LOCATE 1, 34, 0: PRINT "Pc SWIPE CARD";
LOCATE 2, 1: PRINT STRING$(79, 220)
LOCATE 18, 35: COLOR 23, 0: PRINT "Waiting........"; :
COLOR 7, 0
LOCATE 21, 33: PRINT "(Press ESC to end)"
RETURN
REM**
REM** ERROR TRAP
REM**
errortrap:
IF ERR = 6 THEN
```

LISTING 1—MAIN PROGRAM

```
SOUND 500, 1
CLS : LOCATE 1, 34: PRINT "Pc SWIPE CARD";
LOCATE 2, 1: PRINT STRING$(79, 220): COLOR 0, 7
LOCATE 9, 25: PRINT SPACE$(34)
LOCATE 10, 25: PRINT " Error In Reading Swipe Card. "
LOCATE 11, 25: PRINT " Wait For The Beep and Try Again.
"
LOCATE 12, 25: PRINT SPACE$(34): COLOR 7, 0
start! = TIMER
WHILE (TIMER - start!) < 1: WEND: CLS
END IF
BEEP
RESUME start1
```

LISTING 2—TEST PROGRAM

```
REM************************
REM** SWIPETST.BAS 1/20/95 *
REM************************
CLEAR : CLS: DEFINT A-X: DEF SEG = 64
a1 = PEEK(8) + 256 * PEEK(9) + 1
LOCATE 1, 34, 0: PRINT "PcSWIPE TEST"
LOCATE 2, 1: PRINT STRING$(79, 220)
LOCATE 4, 31: PRINT "(Press ESC To End)"
previous = (INP(a1) AND 64) / 64
LOCATE 10, 39
IF previous = 1 THEN PRINT "HI" ELSE PRINT "LO"
loop01:
a = (INP(a1) AND 64) / 64
a$ = INKEY$: IF a$ <> "" THEN GOTO endit
LOCATE 10, 39
IF a = 1 AND previous = 0 THEN
SOUND 600, 1
PRINT "HI"
previous = 1
ELSEIF a = 0 AND previous = 1 THEN
SOUND 100, 1
PRINT "LO"
previous = 0
END IF
GOTO loop01
endit:
END
```

PC IR CARD READER *(Cont.)*

LISTING 3—DECIMAL TO BINARY CONVERSION

```
REM*****************************
REM** SWIPENOS.BAS 1/20/95 **
REM*****************************
CLEAR : CLS : DIM n$(14)
LOCATE 1, 23: PRINT "PC SWIPE DECIMAL TO BINARY
CONVERSION"
LOCATE 2, 1: PRINT STRING$(79, 220)
loop1:
LOCATE 6, 23: INPUT "Enter Decimal Number (0 to
32767).."; n
IF n < 0 OR n > 32767 THEN
BEEP
LOCATE 6, 20: PRINT SPACE$(50)
GOTO loop1
END IF
number = n
FOR i = 14 TO 0 STEP -1
bin = 2 ^ i
IF bin <= n THEN n = n - bin: n$(i) = CHR$(79) ELSE
n$(i) = CHR$(248)
NEXT
LOCATE 10, 1
LOCATE 10, 23: PRINT CHR$(218); STRING$(33, 196);
CHR$(191)
FOR i = 11 TO 15
LOCATE i, 23
PRINT CHR$(179); SPACE$(33); CHR$(179)
NEXT i
LOCATE 16, 23: PRINT CHR$(192); STRING$(33, 196);
CHR$(217)
LOCATE 13, 25: PRINT "Ref"; : LOCATE 14, 25: PRINT
CHR$(179);
LOCATE 15, 25: PRINT CHR$(248); " ";
FOR i = 0 TO 14
PRINT n$(i); " ";
NEXT i
LOCATE 12, 35: PRINT USING "ID: #####"; number
LOCATE 20, 23: PRINT "Press a key to try again, ESC to
end...";
LOCATE 6, 23: PRINT SPACE$(50)
a$ = INPUT$(1)
IF ASC(a$) = 27 THEN END
LOCATE 20, 23: PRINT SPACE$(50)
GOTO loop1
```

Fig. 20-16

The LED is a high-output infrared emitter. It receives its power through current-limiting resistor R1. With a 9-V power supply and a value of 220 Ω for R1, the diode will receive about 25 mA of current. With a 5-V supply, R_1 should be 150 Ω to keep diode current in the 25-mA range. The LED energizes NPN phototransistor Q1, which is configured as a simple inverting amplifier. As more light shines on Q1, the output voltage at its collector decreases. With a value of 2.2 kΩ for R2, the circuit provides TTL-compatible logic levels. The output of Q1 feeds one bit of a PC's parallel port. Diode D2 allows the use of power sources greater than 5 V, thereby maintaining TTL level compatibility—even with high supply voltages. If the voltage at the collector of Q1 ever exceeds 5 V, D2 will block the voltage, thereby protecting the port. On the other hand, when Q1 goes logic low, D2 becomes forward-biased, so the low level can be sensed by the port.

5-V SUPPLY FROM THREE-WIRE RS232 PORT

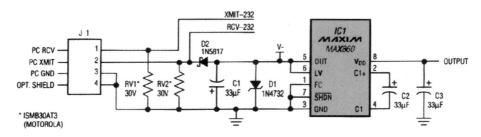

Fig. 20-17

This circuit produces a semiregulated 5-V output from an RS232 port. The output current—about 8 mA—is sufficient for CMOS microcontrollers and other low-power circuits. IC1 is a switched-capacitor, charge-pump voltage converter that can either invert an input voltage or double it. The connections shown provide a doubler configuration in which the normal input voltage is reversed: A positive input voltage normally connects between GND and OUT, but this circuit connects a negative input between OUT and GND. The IC then doubles the negative V_{in} in the positive direction, producing a positive output (at V_DD) equal to $[V_{in}]$. The zener diode D1 acts as a shunt regulator that "semiregulates" V_{in} to -5 V (actually to -4.7 V). The 33-μF capacitor values shown are larger than normal to support the output voltage during worst-case (all-zero) patterns of transmission. At 9600 baud, for example, an all-zeros character causes an output droop of about 0.2 V. For lower baud rates, substitute a proportionally higher value for C_1.

COMPUTER-CONTROLLED A/D CONVERTER

TABLE 2: ADC COMMANDS

Parallel port data	ADC operation
0 0 0 0 0	All off
0 0 1 0 0	Select ADC
0 0 1 0 1	Clock low
0 0 1 1 1	Clock high
0 1 1 0 1	Deselect ADC

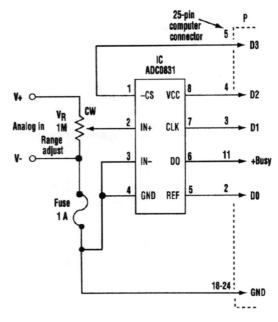

```
'********************************************************************
'program EDCCADC written in BASIC  Barry Voss
'for use with the computer-controlled ADC pc board
'********************************************************************
'define variables
    DIM addout AS INTEGER
    DIM addin AS INTEGER
'graph
    DIM graph(3000)
'variables used
    'bitin input bit for building a digital word
    'word is the binary built from bitin
    'lastvoltage is the previous voltage value. used for determining
        'the number of characters in the previous display
    'voltage is the calculated voltage to be displayed
'********************************************************************
'SETUP                                                              *
'********************************************************************
'setup display
    SCREEN 2: WIDTH 80: CLS
'header
    LOCATE 1, 10: PRINT "Computer Controlled Analog to Digital Converter"
    LOCATE 2, 58: PRINT "voltage ="
'footer
    LOCATE 24, 1: PRINT "<ESC> to quit";
'draw the graph box
    LINE (1, 59)-(601, 161),,B
'********************************************************************
'address of parallel port                                          *
'the next line should read:                                        *
'addout = 888 if you have a PC clone                               *
'addout = 956 if you have an IBM (type computer)                   *
'addout = 632 if you are using LPT2                                *
'********************************************************************
    addout = 888
    addin = addout + 1
```

```
'***************************************************************
'main program                                                 *
'***************************************************************
DO UNTIL INKEY$ = CHR$(27): 'do until escape key pressed
'get data from ADC
   'clear the old data
       REDIM bitin(8) AS INTEGER
       word = 0
   'select the ADC
       OUT addout, 5
   'clock the chip this starts the conversion
       OUT addout, 7
       OUT addout, 5
   'get the 8 bit word
       FOR a = 7 TO 0 STEP -1
   'clock the chip high
       OUT addout, 7
   'the bit is present after the negative clock edge
       OUT addout, 5
   'get the status word
       bitin(a) = INP(addin)
       NEXT a
   'de-select the ADC
       OUT addout, 13
   'reconstruct the word MSB is first
       FOR r = 7 TO 0 STEP -1
       IF bitin(r) < 128 THEN word = word + (2 ^ r)
       NEXT r
   'calculate the voltage
       voltage = 50 * word/255
   'printing the voltage with formatting
       IF LEN(lastvoltage) > LEN(STR$(INT(voltage * 100)/100) THEN LOCATE 2, 69: PRINT " "
       voltage = INT(voltage * 100)/100
       LOCATE 2, 69
       PRINT voltage
       lastvoltage = voltage
   'update the graph
       GET (3, 60)-(600, 168), graph
       PUT (2, 60), graph, PSET
       PSET (599, (160 - (100 * (voltage/50)))
LOOP
'escape pressed end the program
   'clear the parallel port
       OUT addout, 0
END
```

ELECTRONIC DESIGN   *Fig. 20-18*

This simple, inexpensive computer-controlled A/D converter (ADC) plugs into a PC parallel port. The 8-bit peripheral device requires only seven components to implement and is completely controlled by a short BASIC program.

PC SIGNAL GENERATOR

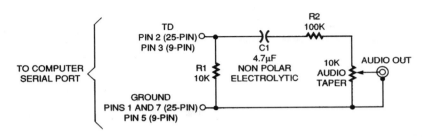

ELECTRONICS NOW

Fig. 20-19

This circuit will produce audio from your PC. The trick is U—the ASCII character "U," that is. The hexadecimal value of U is 55, which in binary is 01010101 (with 8 data bits and no parity, or 7 data bits and even parity). The RS232 protocol specifies that the bits of an ASCII character are transmitted from least to most significant, preceded by a start bit (always 0) and followed by a stop bit (always 1). So, after adding the requisite start and stop bits, the result is 1010101010. Now, suppose a string of Us is generated at the serial port at some steady rate. The result is a continuous series of alternating 1s and 0s—a square wave. The frequency of the signal will be half the baud rate, which by definition is the number of transitions per second. Each cycle of a square wave comprises two transitions, so, for example, a 9600-bps baud rate produces a 4800-Hz square wave. In practical terms, just about any computer should be able to deliver frequencies of 55, 150, 300, 600, 1200, 2400, and 4800 Hz, corresponding to the standard baud rates from 110 to 9600. The output of a serial port is nominally 24 V p-p, which is much too high a voltage to feed to the input of an audio amplifier. The circuit attenuates the signal to a more useful level, a variable 2-V p-p. The circuit also protects the computer from static electricity and voltage surges. Capacitor C1, a nonpolar unit, blocks dc because the serial port, when idling, outputs approximately −12 V.

21

Controller Circuits

The sources of the following circuits are contained in the Sources section, which begins on page 1043. The figure number in the box of each circuit correlates to the entry in the Sources section.

LOW-VOLTAGE POWER CONTROLLER

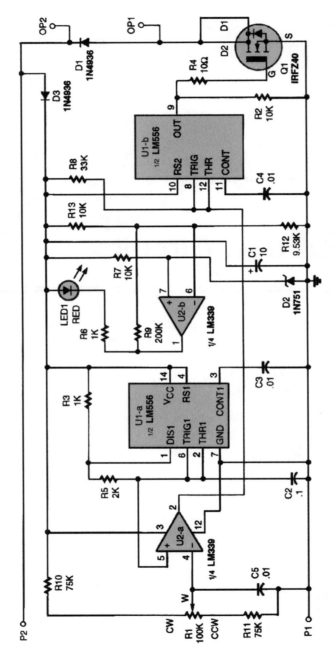

Fig. 21-1

POPULAR ELECTRONICS

The circuit has a duty-cycle generator that produces an output varying from fully off to fully on and pulses of any duty cycle in between the two extremes. The circuit can be fed from any dc supply source of between 10 and 15 V. U1-a and U2-a combine to form a voltage-to-pulse-width converter. The first half of the dual oscillator/timer is configured as an astable oscillator, generating a continuously oscillating ramp voltage. Op amp U2-a compares the voltage at its noninverting input to the voltage at its inverting input. The op amp will give a low output if R1's wiper voltage is higher than the instantaneous voltage present at pins 2 and 6 of U1-a. The output of U2-a at pin 2 will have an on/off ratio that is proportional to the voltage at R1's wiper. Because the output of U2-a does not have enough power-handling capacity to drive the MOSFET, its output is fed to U1-b, which is used to buffer the signal. The low-impedance, pulsed

LOW-VOLTAGE POWER CONTROLLER *(Cont.)*

output of U1-b at pin 9 is fed to the gate of MOSFET Q1, driving it hard on or off. Diode D1 is used to suppress the reverse-voltage spikes generated by inductive loads during turn-off. If the circuit will not be used to drive inductive loads (motors), D1 can be eliminated. *Note:* This controller can be used only with incandescent lamps and permanent-magnet dc motors.

TEN-STEP COUNTER FOR CONTROLLERS

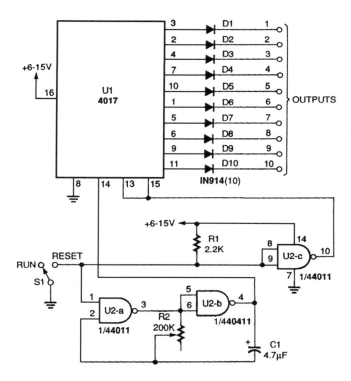

POPULAR ELECTRONICS

Fig. 21-2

A 4017 divide-by-10 counter IC is the heart of this simple 10-output controller circuit. Two gates of a 4011 quad two-input NAND gate IC are connected in an astable oscillator circuit to clock the divide-by-10 counter U1. The step time is set by R5. With the RUN/RESET switch in the RUN position, U1 takes 10 equal steps and then stops. Momentarily switching S1 to RESET starts the cycle over.

FOUR-OUTPUT CONTROLLER

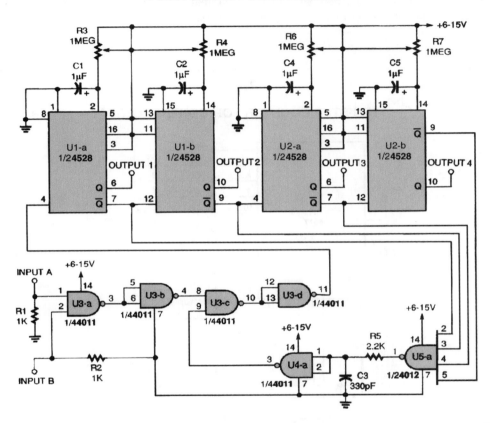

POPULAR ELECTRONICS

Fig. 21-3

This controller offers up to four timed outputs that can be used to operate motors, air valves, solenoids, relays, etc. The timer sections are cascaded so that as one timer times out, it triggers the next timer, and so on, until the last timer times out. The intended application for the controller required a two-sensor input that would start only when both inputs occur simultaneously. Note, however, that the start-signal logic could be modified to accommodate a combination of any number of input sensors, or even a single switch closure. The controller also includes an inhibit circuit that keeps the sequence from restarting before a cycle is completed.

TWO-FUNCTION CONTROLLER

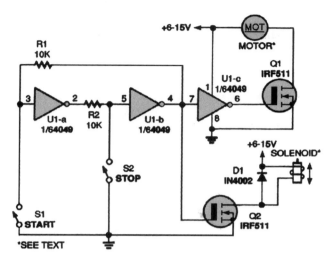

POPULAR ELECTRONICS

Fig. 21-4

This two-function controller operates a motor when started, then energizes a solenoid, relay, etc., when stopped.

22

Converter Circuits

The sources of the following circuits are contained in the Sources section, which begins on page 1043. The figure number in the box of each circuit correlates to the entry in the Sources section.

Linear DAC with Nonlinear Output
8-Bit Binary-to-Decimal Converter
Audio-to-Dc Converter Circuit
Ratiometric 20-kHz V-F Converter
Multiswitch Charge Pump Boost Converter
Isolated 3-V–to–5-V DAC
Voltage Converter

LINEAR DAC WITH NONLINEAR OUTPUT

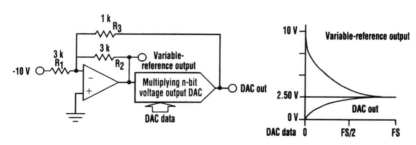

ELECTRONIC DESIGN

Fig. 22-1

When controlling a nonlinear device, such as an incandescent lamp, it is desirable to have fine resolution at the high end, where a small change in current can cause a large change in brightness. At the low end, coarser resolution is quite adequate. Using the circuit shown, any desired compression can be produced using just about any multiplying DAC. A negative 10-V reference is fed through R1 to inverting amplifier A1, which has an initial gain set to unity by R2. A1's output supplies a positive variable reference to the DAC. The DAC output provides additional feedback through R3, reducing the amplifier's gain as the DAC data increase. The variable reference is gradually reduced so that each step is progressively smaller than the one before. With the values shown, as the DAC data approaches full-scale, the reference approaches $\frac{1}{4}$ of its original value. This produces output with four times as much resolution at the high end as at the low end. By decreasing the value of R_3, greater compression and higher resolution can be achieved. The variable-reference output also can be useful in some applications.

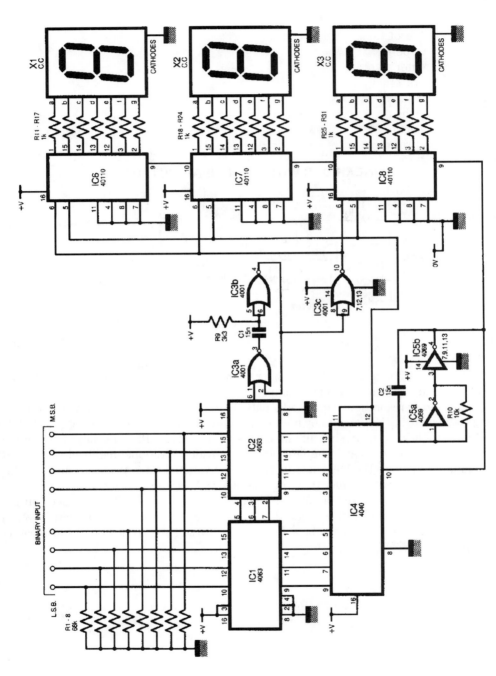

8-BIT BINARY-TO-DECIMAL CONVERTER

Fig. 22-2

EVERYDAY PRACTICAL ELECTRONICS

8-BIT BINARY-TO-DECIMAL CONVERTER *(Cont.)*

The decimal value of the eight input lines is displayed on three seven-segment displays. Although chips exist to perform this function for single displays, none include the possibility of displaying numbers greater than 9. No EPROM is necessary in this design. It operates by using two synchronized counters, one that generates an 8-bit binary output, and a second that drives the display. If the display is updated only when the binary counter is at the same value as the input, then the display will show the decimal value of the input. IC4 is the binary counter, and IC6 to IC8 form the display/counter section. IC5a and IC5b form the astable, which clocks both counters at 5 kHz. The minimum display refresh rate is, therefore, 20 times per second. IC1 and IC2 are two 4-bit comparators, ganged to form an 8-bit comparator. This compares the output of the binary counter (IC4) with the binary input taken from the circuit under test. Resistors R1 to R8 pull down the input lines to prevent them from floating when no input is connected. When the comparator inputs are equal, pin 6 of IC2 goes high, triggering the monostable (IC3ab), which outputs a brief pulse. This latches the value of the display counters (IC6 to IC8) to the display. When IC4 reaches a value of 256, the link between pins 11 and 12 resets the chip, and this resets the display counter also.

AUDIO-TO-DC CONVERTER CIRCUIT

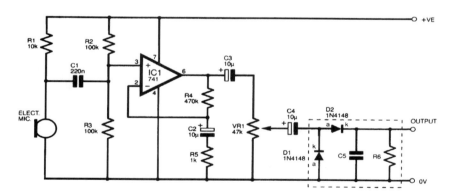

Fig. 22-3

The circuit diagram for the audio input module, which produces a dc output voltage relative to the input signal amplitude.

RATIOMETRIC 20-kHz V-F CONVERTER

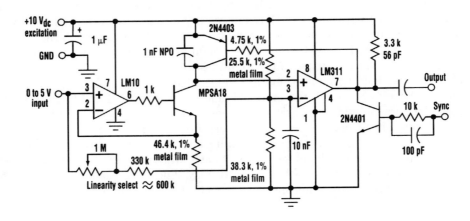

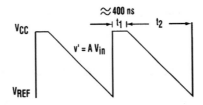

Fig. 22-4

This ratiometric 20-kHz voltage-to-frequency converter (VFC) provides superior performance with strain gauges and other ratio-responding transducers—even with noisy, unregulated excitation voltages. Feedback isn't used to achieve the excellent 4-Hz linearity, so there is very-low-frequency jitter—period measurements can be used to get several digits of resolution—even when operating at a fraction of full scale. An optional synchronizing transistor starts the VFC with zero charge at the beginning of each count cycle, eliminating the characteristic digit-jumping often encountered with VFC designs. Good linearity is attained by making the comparator's reference voltage vary with the input voltage, which precisely compensates for the finite capacitor reset time:

$$Period = t_1 + t_2$$
$$= t_1 + (V_{CC} - \Omega_{pe\phi})/A\Omega_{\iota\nu}$$
$$= (t_1 A V_{in} + V_{CC} - \Omega_{pe\phi})/A\Omega_{\iota\nu}$$

where $AV_{in} = \Delta V/\Delta t$. If V_{ref} is made to include the amount $t_1 A V_{in}$, then the effect of t_1 is eliminated:

RATIOMETRIC 20-kHz V-F CONVERTER *(Cont.)*

$$Period = [t_1 AV_{in} + V_{CC} - (\tau + A\Omega_{LV} + V_{ref})]/AV_{in} = (V_{CC} - \Omega_{pe\phi})/A\Omega_{LV}$$

The MPSA18 is a remarkably high-gain transistor, even at low currents, producing good current-source linearity down to 0 Hz. In addition, bipolar transistors work well with the low collector voltages encountered in this single-supply 10-V design. Moreover, most single-supply op amps will work in place of the LM10. But the LM10 also has a reference amplifier that could be used to construct a 10-V excitation regulator. The LM311 propagation delay gives a reset pulse width near 400 ns, which gives the transistor time to discharge the capacitor. Also, the 311's bias current produces a small negative offset that ensures a 0-Hz output for 0 V_{in}.

MULTISWITCH CHARGE PUMP BOOST CONVERTER

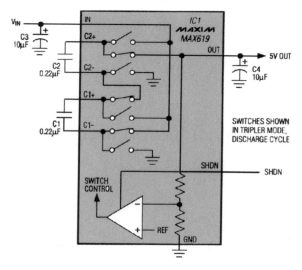

Fig. 22-5

This device is useful both in main supplies and in backup supplies. It generates a regulated 5-V output for load currents to 20 mA and inputs ranging from 1.8 to 3.6 V. For input voltages no lower than 3 V, the output current can reach 50 mA. The circuit accomplishes regulation without a linear pass element, but its losses are the same as those of an unregulated doubler or tripler feeding into a linear regulator.

ISOLATED 3-V-TO-5-V DAC

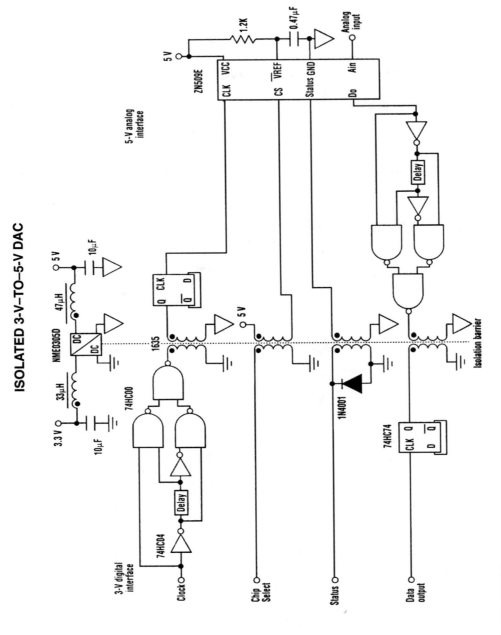

3-V digital interface

5-V analog interface

Isolation barrier

Fig. 22-6

ELECTRONIC DESIGN

ISOLATED 3-V–TO–5-V DAC *(Cont.)*

The circuit uses a low-power (1 W) dc-to-dc converter for the power isolation and a transformer isolator/translator for the data interface. The transformer isolator not only provides the galvanic isolation required, but also converts the data signals between 3-V and 5-V levels. As a result, no additional level conversion is required. The ADC can run at 1-MHz clock rates. To allow the 50 percent duty cycles (clock and data signals) to pass through a transformer isolator requires an edge-detection and conversion technique. The edge detector is built from simple logic gates (two inverters and three NAND gates) and a short delay (either a delay line or a passive RC circuit can be used). The signal is rebuilt with a D-type flip-flop. Using a four-channel isolator allows full control over the ADC to be exercised. Conversion is requested by pulling the Chip Select pin low, and a Status high for one clock cycle is reported back to acknowledge conversion start. The first data bit (MSB) then is presented onto the data line, and all eight bits are transferred with a further Status signal at the end of conversion. Conversion can be requested asynchronously with the system clock, if necessary, and the Status flag can be used to poll the controlling logic circuitry. Filtering is placed on either side of the dc-to-dc converter to reduce power-supply ripple and prevent noise on the logic power supply from affecting the analog system. Although not shown, all ICs have 0.22-µF decoupling capacitors.

VOLTAGE CONVERTER

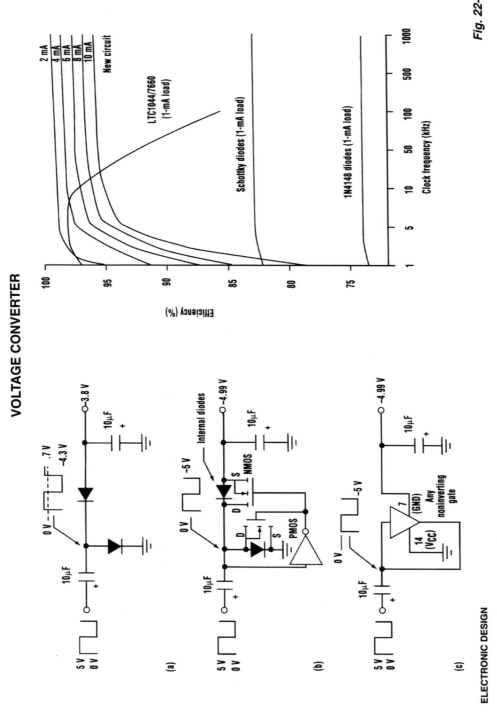

Fig. 22-7

ELECTRONIC DESIGN

This bootstrap voltage converter begins with the basic diode inverter circuit (a). Placing a MOSFET across each diode will improve efficiency (b). The final step involves using a single noninverting CMOS digital gate to replace the diodes (c).

The performance curves depict the bootstrap voltage converter's efficiency, which is 99 percent from 5 to 500 kHz using 10-µF capacitors and a 1-mA load. A typical commercial unit operates at only 94 percent efficiency.

23

Counter Circuits

The sources of the following circuits are contained in the Sources section, which begins on page 1043. The figure number in the box of each circuit correlates to the entry in the Sources section.

Simple 25-MHz Counter
Digital Counter Circuit
Four-Mode Frequency Counter
Up/Down Counter with XOR Gates
Four-Digit Counter

SIMPLE 25-MHz COUNTER

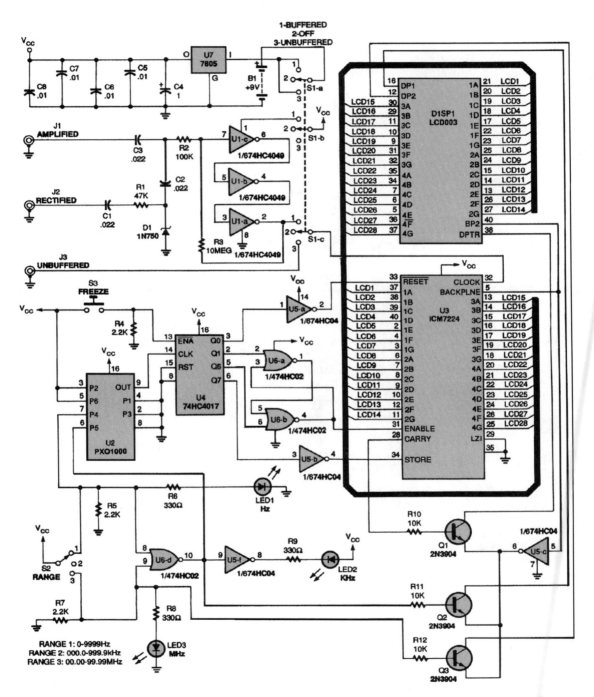

Fig. 23-1

SIMPLE 25-MHz COUNTER *(Cont.)*

An Intersil ICM7224 is used to drive an LCD003 four-digit LCD display unit. Three inputs are provided, which add signal conditioning, either gain, rectification, or unbuffered input. A Statek PX0-1000 1-MHz clock is used as a frequency reference. S2 selects gate time and hence range, with LED1, LED2, and LED3, indicating Hz, kHz, or MHz, respectively.

DIGITAL COUNTER CIRCUIT

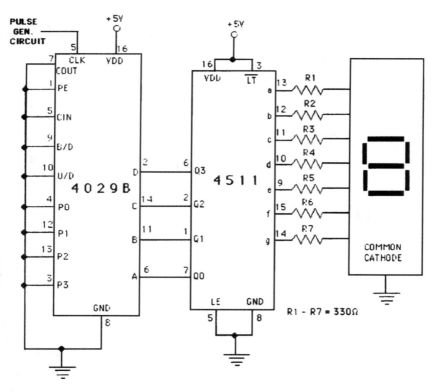

NUTS AND VOLTS

Fig. 23-2

This circuit shows how a simple digital counter can be implemented. A CD4029B drives a DC4511 decoder and LED driver. A common-cathode LED display is used.

FOUR-MODE FREQUENCY COUNTER

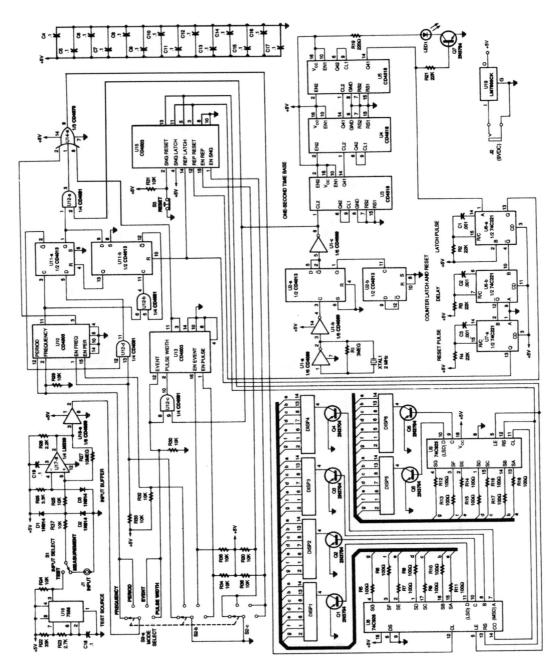

Fig. 23-3

FOUR-MODE FREQUENCY COUNTER *(Cont.)*

This counter can measure frequencies from 2 Hz to 1 MHz, time interval, and period, and count random events. A 74C926 and 74C925 are used as the counter, and these will drive a multiplexed LED display. A 2-MHz time base is used, and a divider chain is used to derive a 1-s gate. An LM339 op amp serves as the input buffer, and CD4000 series logic is used for gating and switching functions.

UP/DOWN COUNTER WITH XOR GATES

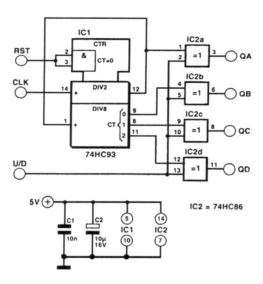

ELEKTOR ELECTRONICS

Fig. 23-4

This circuit shows how a regular 4-bit binary counter can be extended with an up/down function just by adding four XOR (exclusive-OR) gates. The principle is simple: The level at the common inputs of the XOR gates determines whether the gates invert the counter's Q_A to Q_D output levels or not. In this way, the outputs of the XOR gates can be made to cycle from 1111 down to 0000 instead of from 0000 to 1111. The disadvantage of this circuit over a real up/down counter is the jump, which occurs when the level on the U/D control input is changed. The sum of the "old" state and the "new" state is always 15. For example, if the counter is at state 3 in count-up mode, the state becomes 12 when the U/D line is made logic high to initiate down counting.

FOUR-DIGIT COUNTER

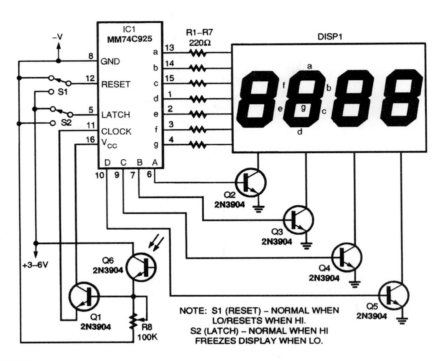

Fig. 23-5

A general-purpose four-digit counter can be made from a 74C925 IC and a few external components. That IC contains the counter, multiplexer, and seven-segment digit drivers, all in one convenient package. A couple of important notes about the circuit: The seven 220-Ω resistors (R1 to R7) are used as dropping resistors between the driver outputs and the segments of the display to protect the LEDs and keep the 74C925 from overheating. The four 2N3904 transistors (Q2 to Q5) act as amplifiers for the digital drivers to keep the display at full illumination. Switches S1 and S2, for reset and latch, respectively, are not crucial, but are highly recommended. The latch will freeze the count if needed. If the switches are not used, pin 12 needs to go to ground, and pin 5 to $+V$ (the circuit's voltage supply). If more than four digits are needed, use the MM74C926 IC. The functions are the same, but it has a "carry out" to cascade more than one chip and four-digit display.

24

Crystal Oscillator Circuits

The sources of the following circuits are contained in the Sources section, which begins on page 1043. The figure number in the box of each circuit correlates to the entry in the Sources section.

Versatile Wideband Crystal Oscillator
CMOS Crystal Oscillator
Pierce Crystal Oscillator
NE602 Overtone Crystal Oscillator
NE602 Third-Overtone Crystal Oscillator
NE602 Adjustable Crystal Oscillator
Basic NE602 Colpitts Crystal Oscillator
32.768-kHz Micropower Oscillator
TTL Crystal Oscillator
Colpitts Crystal Oscillator

VERSATILE WIDEBAND CRYSTAL OSCILLATOR

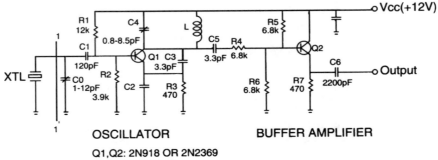

OSCILLATOR   BUFFER AMPLIFIER

Q1,Q2: 2N918 OR 2N2369
C2,L: SEE TABLE

freq. MHz	L µH	C2 pF
6.4	5.6	4700
12.0	3.4	4700
20.0	1.98	470
34.29	0.78	470
45.454	0.78	47
73.0	0.39	47
104.0	0.16	47
120.0	0.1	47

RF DESIGN

Fig. 24-1

The crystal oscillator operates from 6 to 120 MHz by changing only C2 and L. The table lists component vaues for crystal oscillator at different frequencies.

CMOS CRYSTAL OSCILLATOR

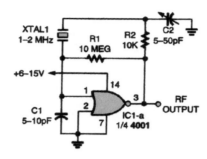

POPULAR ELECTRONICS

Fig. 24-2

This single CMOS two-input NOR-gate crystal oscillator circuit has one major limitation: It lacks high-frequency performance. Otherwise, it is a solid performer.

PIERCE CRYSTAL OSCILLATOR

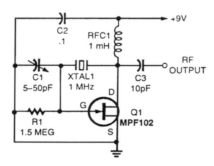

POPULAR ELECTRONICS

Fig. 24-3

NE602 OVERTONE CRYSTAL OSCILLATOR

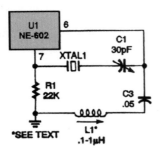

For higher frequencies, use an overtone crystal oscillator like the one shown here. The circuit is a Butler oscillator. The overtone crystal is connected between the oscillator emitter of the NE602 (pin 7) and a capacitive voltage divider that is connected between the oscillator base (pin 6) and ground. An inductor is also in the circuit (L1), and it must resonate with C1 to the overtone frequency of crystal XTAL1. The circuit can use either third- or fifth-overtone crystals up to about 80 MHz.

POPULAR ELECTRONICS   *Fig. 24-4*

NE602 THIRD-OVERTONE CRYSTAL OSCILLATOR

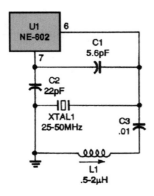

This overtone crystal oscillator uses third-overtone crystals and will work from 25 to 50 MHz.

POPULAR ELECTRONICS   *Fig. 24-5*

NE602 ADJUSTABLE CRYSTAL OSCILLATOR

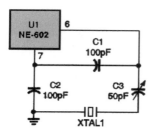

Here, a variable capacitor is added to the circuit to make it easier to obtain the desired frequency.

POPULAR ELECTRONICS   *Fig. 24-6*

BASIC NE602 COLPITTS CRYSTAL OSCILLATOR

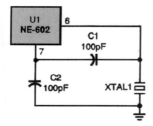

This basic Colpitts crystal oscillator will work with fundamental-mode crystals up to 20 MHz.

POPULAR ELECTRONICS   *Fig. 24-7*

32.768-kHz MICROPOWER OSCILLATOR

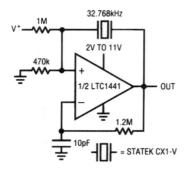

Using an LTC1441, this oscillator pulls 9 μA at a supply voltage of 2 V. The circuit has no spurious modes.

LINEAR TECHNOLOGY   *Fig. 24-8*

TTL CRYSTAL OSCILLATOR

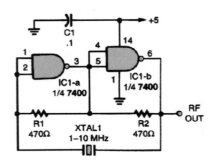

POPULAR ELECTRONICS

Fig. 24-9

Here is an oscillator circuit that uses no L/C components. It uses two sections of a 7400 TTL IC, two resistors, and a crystal to make up a simple and stable oscillator circuit.

COLPITTS CRYSTAL OSCILLATOR

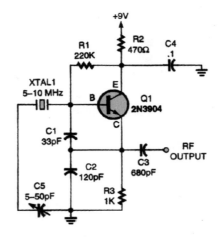

POPULAR ELECTRONICS

Fig. 24-10

This circuit is commonly called a *Colpitts oscillator*. C_1 and C_2 determine the feedback ratio that maintains oscillation. To obtain maximum frequency stability and output level, C_1 and C_2 should be selected for a given frequency.

25

Crystal Radio Circuits

The sources of the following circuits are contained in the Sources section, which begins on page 1043. The figure number in the box of each circuit correlates to the entry in the Sources section.

Full-Wave Detector Crystal Set
Tunable Dual-Coil Crystal Radio
Three-Coil Crystal Receiver
Antenna-Matched Crystal Radio
Crystal Radio Coils
Single-Coil Crystal Radio

FULL-WAVE DETECTOR CRYSTAL SET

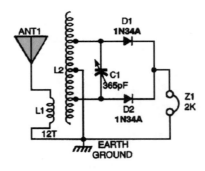

POPULAR ELECTRONICS

Fig. 25-1

The 12-turn coil of L1 couples the RF signal to the large coil, L2. Connect the center tap of L2 to ground, the fourth tap up from the center to diode D1, and the fourth tap down from the center to diode D2. The combined audio output drives the headphones (Z1). If you change tap positions, keep the same number of turns on each side of center. That will balance the RF that feeds each detector diode. The circuit's sensitivity and audio output can be increased by placing L1 inside of L2 (the forms specified for the coils should make that possible). For maximum selectivity, L1 should be loose-coupled to L2.

TUNABLE DUAL-COIL CRYSTAL RADIO

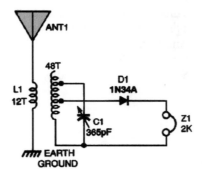

POPULAR ELECTRONICS

Fig. 25-2

The 12-turn primary winding couples the RF signal from the antenna/ground system to the 48-turn secondary winding. Here C1, a 365-pF variable capacitor, tunes the L/C circuit to the desired radio-frequency signal. A 1N34A germanium diode, D1, detects the audio and feeds it to the headphones (Z1). The various taps on L1's secondary allow impedance matching of the antenna/ground system and the detector diode, as well as the inductance value needed to tune to the desired RF signal.

THREE-COIL CRYSTAL RECEIVER

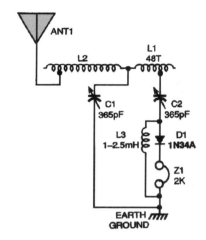

POPULAR ELECTRONICS

Fig. 25-3

This circuit uses three inductors to increase the receiver's selectivity and sensitivity. Components L2 and C1 are used in an antenna impedance-matching circuit, while L1 and C2 operate in a series-tuned low-output impedance circuit that matches the impedance of the diode detector. A 1- to 2-mH inductor (L3), as in the previous circuit, offers dc continuity to the detector circuit.

ANTENNA-MATCHED CRYSTAL RADIO

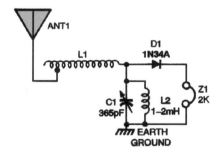

POPULAR ELECTRONICS

Fig. 25-4

This receiver uses a tuning circuit that is, in some ways, similar to an antenna-matching device used by amateur-radio operators to impedance-match their receiver/transmitter input/output circuitry to the impedance of the antenna for maximum signal transfer. Inductor L2 provides a dc-signal return path for D1's output. The inductance of L2 is too large to affect the circuit's tuning function.

CRYSTAL RADIO COILS

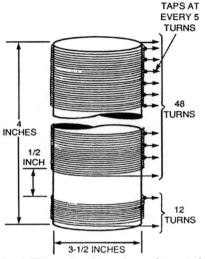

TAPS AT EVERY 5 TURNS

48 TURNS

4 INCHES

1/2 INCH

12 TURNS

3-1/2 INCHES

Fig. 1. This small coil is made up of two windings. Note that the 48-turn winding has taps at every five turns. The 12-turn coil will be used for RF coupling.

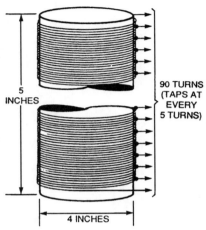

5 INCHES

90 TURNS (TAPS AT EVERY 5 TURNS)

4 INCHES

Fig. 2. The large coil has one 90-turn winding with taps at every five turns.

POPULAR ELECTRONICS

Fig. 25-5

These coils are suitable for crystal sets and broadcast receiver experiments. They are wound with 19- or 20-gauge wire and should be wound on a material having low loss at AM broadcast frequencies. PVC or polystyrene tubing would be suitable, but cardboard or fiberglass will do as well.

SINGLE-COIL CRYSTAL RADIO

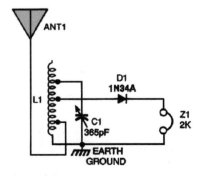

ANT1

L1

D1
1N34A

C1
365pF

Z1
2K

EARTH GROUND

POPULAR ELECTRONICS

Fig. 25-6

A starting setup for this circuit is as follows: Connect the antenna to the second tap up from the bottom of the coil (that's the end of the coil that's connected to ground). The diode should connect to about the fourth tap up from the bottom, and C1 should be attached to the seventh tap or so up from the bottom. Those tap positions might not be the best starting point for your antenna/ground arrangement. That doesn't matter, however, because to obtain the best results with the receiver at your location, you should experiment with all variables anyway.

26

Crystal Test Circuits

The sources of the following circuits are contained in the Sources section, which begins on page 1043. The figure number in the box of each circuit correlates to the entry in the Sources section.

Crystal Activity Tester
Quartz Crystal Specifications
Meter Indicator for Crystal Activity Tester

216

CRYSTAL ACTIVITY TESTER

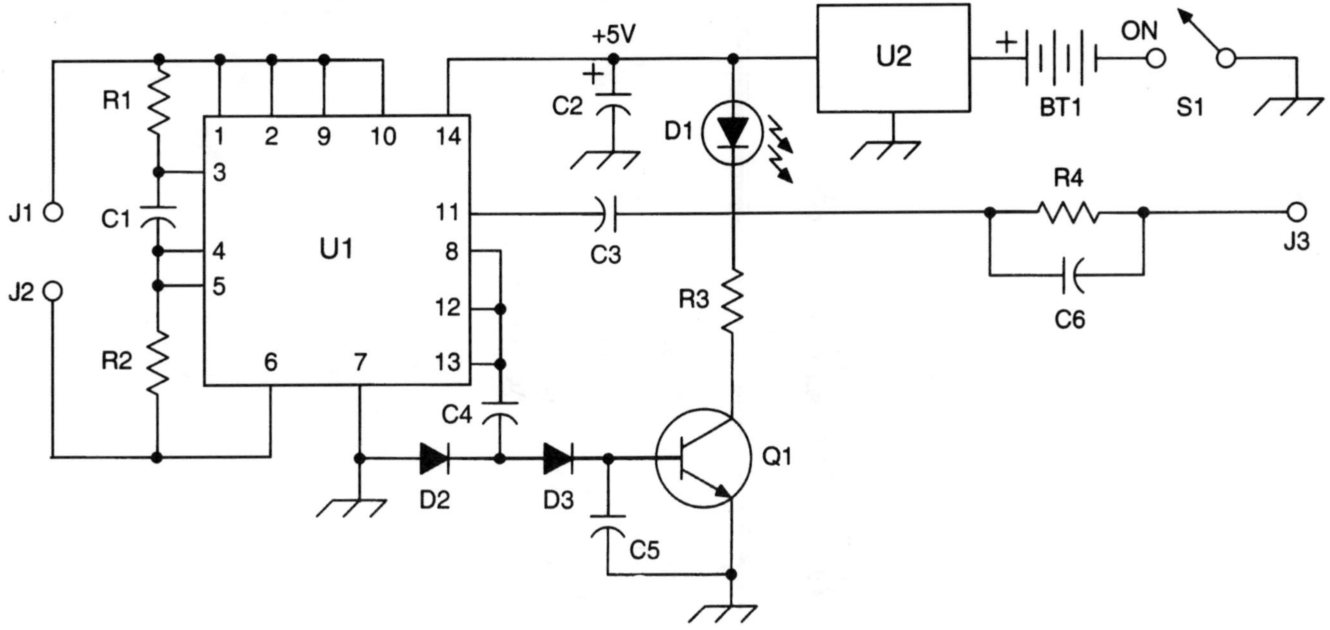

Fig. 26-1

U1 is a 74LS00 quad two-input NAND-gate logic chip. The LS version was selected because of its low current requirement. Two of its internal gates are connected as inverters and, in conjunction with R1, R2, and C1, form a crystal oscillator when a crystal is connected between J1 and J2. The third gate, also connected as an inverter, is used as a buffer to provide RF at the crystal frequency from pin 11, through isolation capacitor C3 and high-pass filter R4 and C6, to J3 so that the crystal frequency can be monitored by an external-frequency counter. The fourth gate, also connected as an inverter-buffer, provides RF at the crystal frequency from pins 8, 12, and 13 connected in parallel. The RF is fed through isolation capacitor C4 to voltage-doubling rectifiers D2 and D3, and the resulting dc voltage is filtered by C5 and provides a positive bias to the base of Q1, an NPN transistor, which is normally cut off. Q1 goes into conduction, and its collector current flows through LED D1 and R3 in series, both forming the

CRYSTAL ACTIVITY TESTER *(Cont.)*

collector load circuit. The greater the activity of the crystal, the higher the positive bias on the base of Q1 will be, and the higher its collector current will be. This current illuminates D1, and its relative brightness is indicative of the level of crystal activity.

QUARTZ CRYSTAL SPECIFICATIONS

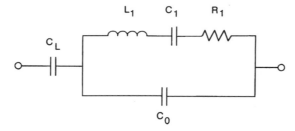

RF DESIGN   *Fig. 26-2*

Five main parameters control the characteristics of a crystal, as noted in the equivalent circuit for the figure. These parameters are:

- C_1, the motional capacitance
- L_1, the motional inductance
- R_1, the equivalent series resistance (ESR)
- C_0, the parallel capacitance resulting from the electrodes and crystal packaging
- C_L, external load capacitance of the circuit

C_1 and L_1 are interdependent because they determine the resonant frequency of the crystal. If we know one of the parameters, we can readily compute the other if we know the series resonant frequency. R_1 is the resistance determined by the motional (piezoelectric) behavior of the crystal. If it is too high, the crystal might not start oscillation. C_0 is a physical capacitor, created by the electrodes plated onto the crystal surface, along with some additional capacitance from the package. Generally, larger C_0 contributes to better pullability. C_L is the load capacitance of the user's circuit. The crystal must operate at the right frequency in the intended circuit, so this value needs to be included in the crystal purchase specification.

METER INDICATOR FOR CRYSTAL ACTIVITY TESTER

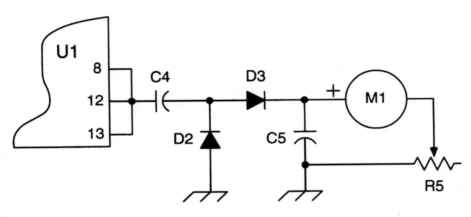

Fig. 26-3

A meter can be added as shown, replacing the LED in the original circuit. The meter movement should be a 0- to 1-mA type.

27

Current Source Circuits

The sources of the following circuits are contained in the Sources section, which begins on page 1043. The figure number in the box of each circuit correlates to the entry in the Sources section.

Constant-Current Source Converter
Programmable Current Source

CONSTANT-CURRENT SOURCE CONVERTER

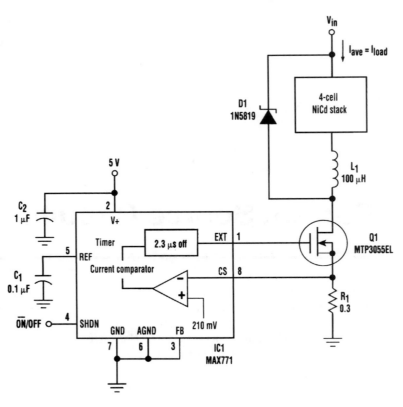

ELECTRONIC DESIGN

Fig. 27-1

To maintain regulation, the switching voltage regulator shown includes independent loops of current and voltage feedback. If the voltage loop is disabled, the current loop can be used to implement a general-purpose current source. The first step in obtaining a current source is to apply 5 V to V^+. Because the chip expects 12 V of feedback at that terminal, it assumes a loss of regulation and shifts control to the current loop. This mode of operation allows an increasing ramp of current through Q1, causing the voltage at pin 8 to increase until it reaches the internal comparator threshold (210 mV). Timing circuitry then turns off Q1 for a fixed 2.3 µs, and the cycle repeats. The result is a relatively constant inductor current, which also happens to be the load current. With a proper choice of component values, the circuit generates constant current over a wide range of input voltages. The circuit (with component values shown) is a fast charger for NiCd batteries that provides 60-mA charging currents.

PROGRAMMABLE CURRENT SOURCE

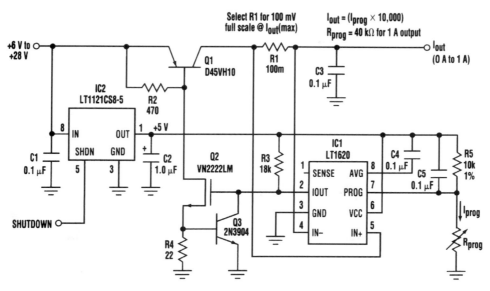

ELECTRONIC DESIGN

Fig. 27-2

Constant-current sources are required in many applications, particularly when it comes to battery charging. In such applications, it is desirable that the output current be accurate, temperature-stable, and adjustable. The controller also can be successfully employed as the control element in a low-cost linear current source. Output current is sensed by resistor R1, with the value selected so that 100 mV is full-scale. The voltage across R1 is amplified by a factor of 10 and averaged across capacitor C4. An internal transconductance error amplifier compares the voltage on pin 8 against the programming voltage at pin 7. The error-amplifier output is present on pin 2 (I_{out}), and is level-shifted by Q2 to control the PNP pass transistor. Output current is programmed by adjusting the voltage across R5 (1 V full-scale). The LT1121 LDO regulator provides a 5-V, ±1.5 percent reference voltage so that current can be accurately programmed by simply connecting different values for R_{prog}. The input voltage can range from +6 to +28 V, with output current changing less than 0.3 percent. Proper heat sinking must be provided for Q1, especially when operating with large input-to-output voltage differentials. Transistor Q3 and R4 limit the magnitude of Q1's base drive during dropout, preventing excessive dissipation in driver transistor Q2. Using voltage regulator IC2, the constant-current source operates directly from the unregulated input voltage. Pulling IC2's shutdown pin low turns off V_{CC} to the entire circuit, and limits the reverse current drawn from the output to less than 25 μA.

28

Dc-to-Dc Converter Circuits

The sources of the following circuits are contained in the Sources section, which begins on page 1043. The figure number in the box of each circuit correlates to the entry in the Sources section.

QUICK-SWITCHING PNP REGULATOR

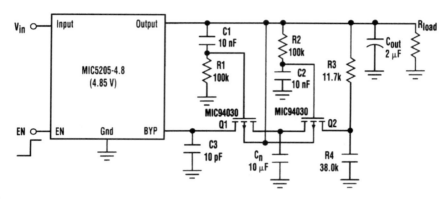

ELECTRONIC DESIGN

Fig. 28-1

The MIC5205 is a low-dropout PNP regulator that incorporates a noise bypass pin for additional noise reduction. A single 10-nF capacitor, connected from the bypass pin to ground, reduces output noise by $V_{out}/1.24$ V (12 dB for the 5-V part) and creates a noise pole below 100 Hz. Switch-on time is increased from 80 µs to 15 ms. With the addition of a few components, the following circuit preserves a low switch-on time. A low-to-high signal on EN switches the output on quickly. This allows R1 and C1 to hold Q1 off while R2 and C2 hold Q2 on. C_n is also quickly charged to the bypass pin voltage through Q2 by the voltage divider R3, R4. C1 and C2 then charge to the output voltage, turning Q1 on and Q2 off. C_n is now switched from the voltage divider to the bypass pin. The regulator is now in the low-noise configuration, which takes about 100 µs. When EN goes low, C1 and C2 discharge through R1, R2, R3, R4, and R_{load}. This resets the circuit for the next turn-on cycle. C3 helps prevent overshoot on the output. Ratio R_3/R_4 can be found empirically: First, set the ratio close to the ratio 1.24 V/(V_{out} −1.24 V), then adjust the value so the output turns on quickly without overshoot. The final tolerance needs to be 1 percent. Switch-off time is determined only by output capacitor size and the load.

±20-V CHARGE PUMP BOOST CONVERTER

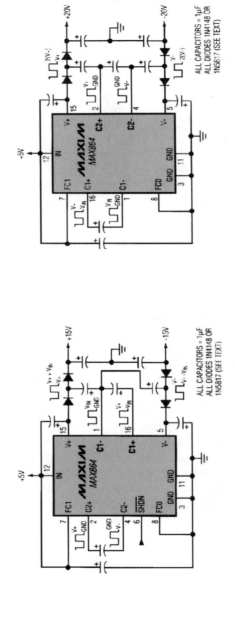

Fig. 28-2

MAXIM ENGINEERING JOURNAL

A low-power converter of 5 to ±20 V can be made surprisingly small by enhancing a dual-output charge-pump IC with an extra boost stage composed of discrete diodes. Such supplies are useful for CCD power supplies, LCD bias, and varactor tuners. The MAX864 on its own can generate ±10 V (minus load-proportional losses) from a 5-V input, or ±6.6 V from a 3.3-V input. Using additional diode-capacitor stages, these outputs can be doubled again to approximately ±4V_{in}, or multiplied by 1.5 to approximately ±3V_{in}. Note that the external diode/capacitor network connects to C1 for ±15-V outputs or to C2 for ±20-V outputs.

500-kHz 3.3-V–TO–12-V DC CONVERTER

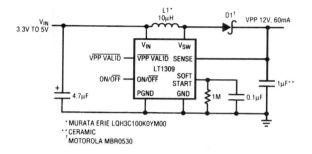
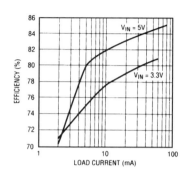

LINEAR TECHNOLOGY POWER SOLUTIONS

Fig. 28-3

The LT1309 500-kHz micropower dc-to-dc converter circuit shown provides a compact 12-V supply. High-frequency operation permits the use of low-inductance and low-capacitance values for surface-mount parts. The LT1309 provides 60 mA of output current at 12 V, required for 8-bit-wide flash memory chips. In addition, when the flash memory card is removed, the LT1309 can be shut down by the system, reducing current draw to 6 μA. A soft-start feature allows the output voltage to ramp up to 12 V over a period of time, minimizing inrush current needed from the 3.3-V supply to charge the PCMCIA input capacitance. An active-low VPP VALID output signals the system that the 12-V supply is within regulation after being switched on.

BASIC BOOST CONVERTER CIRCUIT

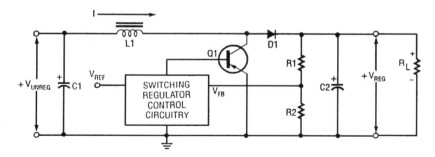

ELECTRONICS NOW

Fig. 28-4

This boost converter with a single switching transistor depends on the transformer for energy storage.

MICROPOWER STEP-UP CONVERTER

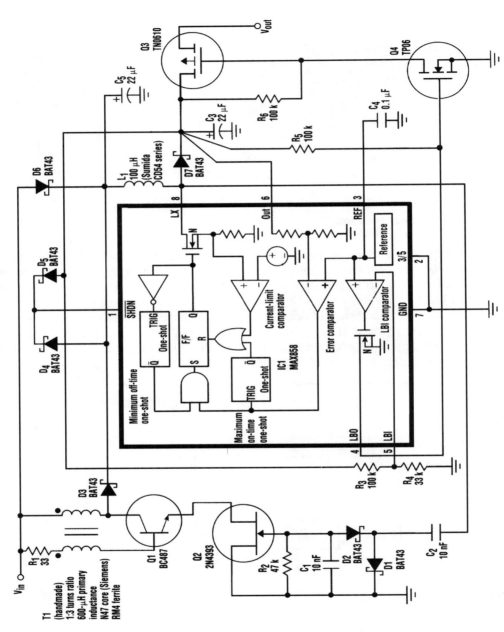

Fig. 28-5

ELECTRONIC DESIGN ANALOG APPLICATIONS

This circuit can start with inputs as low as a diode drop, i.e., 0.6 V. It provides a 5-V output with 15-mA current capability, but for this application, the load current must be limited. The circuit powers a meter from the drop across a single diode in a 4- to 20-mA current loop, and provides 77 percent efficiency at a current of 0.5 mA (with diode D6 shorted). Load current available to the meter (I_{load}) is limited by the 4-mA minimum loop current. Transistor Q1 and transformer T1 form a resonant tank circuit that self-oscillates at a frequency of approximately $1/L_1R_1$. R1 is a current-limiting resistor. At 50 kHz, with an input of 0.6 V, the value shown for resistor R1 (33 Ω) limits the maximum current into the base of transistor Q1 to 20 mA. The primary inductance of transformer T1, although not a crucial parameter, should be 660 μH for this value resistor of R1.

MICROPOWER POSITIVE-TO-NEGATIVE CONVERTER

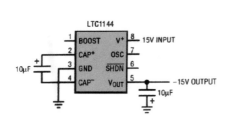

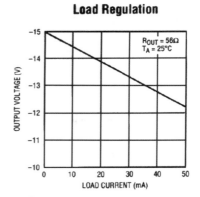

Load Regulation

LINEAR TECHNOLOGY POWER SOLUTIONS

Fig. 28-6

Switched capacitor voltage converters are a convenient way to generate a local negative supply for biasing special circuitry, but have been limited by CMOS processes to 10 V of supply or less. The LTC1144 voltage converter overcomes this limitation, extending the maximum input voltage to 20 V. Still, the part retains the low power of CMOS operation. The LTC1144 circuit shown here generates a negative supply voltage of -13.8 V typ. (-12.6 V min.) from a 15-V input at a maximum load current of 20 mA. Higher load currents are possible at slightly lower output voltages. The low-cost circuit includes two surface-mount capacitors, minimizing board space. A supply current of 1.2 mA (max.) results in high conversion efficiency, while just 8 μA of supply current is consumed in shutdown, making the LTC1144 excellent for use in battery-powered systems.

INDUCTORLESS −5-V CONVERTER

Voltage Loss

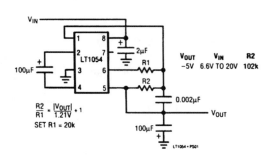

V_{OUT}	V_{IN}	R2
−5V	6.6V TO 20V	102k

$$\frac{R2}{R1} = \frac{|V_{OUT}|}{1.21V} + 1$$

SET R1 = 20k

Fig. 28-7

Switched capacitor voltage converters are great for supplying an unregulated negative voltage from a positive supply. To provide a regulated negative voltage, a linear regulator is normally required. The LT1054 eliminates the extra voltage regulator by adding a pair of feedback resistors, as shown here. An internal feedback circuit allows full regulation of the output voltage with changes in input voltage and load current. With a minimum input of 6.5 V, the LT1054 can produce a regulated −5-V output at loads of up to 100 mA max. External components required include four capacitors and two resistors.

TWO-CELL 5-V 500-mA CONVERTER

2-Cell to 5V Converter Efficiency

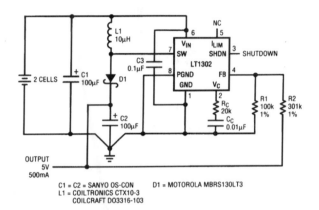

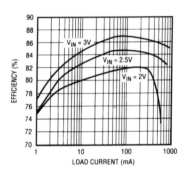

C1 = C2 = SANYO OS-CON   D1 = MOTOROLA MBRS130LT3
L1 = COILTRONICS CTX10-3
    COILCRAFT DO3316-103

LINEAR TECHNOLOGY POWER SOLUTIONS

Fig. 28-8

Hand-held instruments, PCMCIA cards, and portable communications gear often require high current for their operation, although only for short periods at a time. Current of 500 mA at 5 V can be obtained from a two-cell battery using a switching regulator controller IC, an external switch transistor, and some discrete components, but the solution is inefficient, space-intensive, and cumbersome. The LT1302 was designed to provide higher output currents by using an integrated 2-A low-loss switch. An output current of 500 mA at 5 V with 85 percent efficiency is possible from two AA cells. Component size is minimized by using a fixed-frequency 220-kHz PWM architecture. The circuit maintains high efficiencies at low load currents by automatically switching to Burst mode operation at lower switch currents.

9-V DC-TO-DC CONVERTER

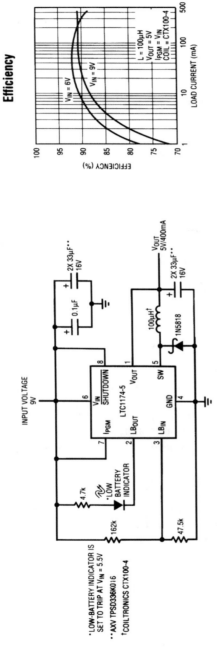

Efficiency

Fig. 28-9

LINEAR TECHNOLOGY POWER SOLUTIONS

In portable applications, 9-V batteries are frequently used. For low-power dc-to-dc step-down conversion, the LTC1174 is a simple integrated solution. With its internal MOSFET power switch, the LTC1174-5 requires only four external parts to achieve 90-percent efficiency. All surface-mount assembly is standard, with high-frequency operation at up to a 200 kHz switching frequency. Extremely low dropout operation is possible because the LTC1174 operates at 100-percent duty cycle with low input voltages. A single logic-level input selects an output voltage of 3.3 V or 5 V.

3- TO 7-V DC-TO-DC CONVERTER

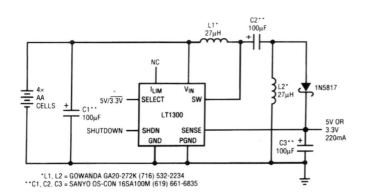

*L1, L2 = GOWANDA GA20-272K (716) 532-2234
**C1, C2, C3 = SANYO OS-CON 16SA100M (619) 661-6835

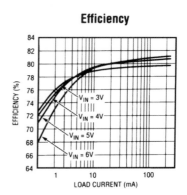

Efficiency

LINEAR TECHNOLOGY POWER SOLUTIONS

Fig. 28-10

    Converting the voltage from a four-cell battery pack to 5 V while using the full capacity of the batteries requires two modes of operation: step-down from an input voltage of 6 V and step-up from an input of 4 V (or less). A flyback topology can accomplish this, but uses a costly custom transformer. The LT1300 circuit shown utilizes the simple SEPIC topology, and is capable of 220 mA of output current at 5 V from a minimum 3-V input. The two inductors specified are available off the shelf. The circuit uses a boost section and a buck, or step-down, section, with the two inductors (L1 and L2) and two capacitors (C2 and C3) all acting as energy-storage elements. Efficiency is slightly less than that of a direct step up (see graph), but is better than that of an equivalent flyback configuration. Other features of the circuit include shutdown (10 μA max. supply current) with full input-to-output isolation, which allows the output to go to zero volts, yet present no load to the batteries. Also, either a 3.3-V or 5-V output can be selected by using the logic-select pin of the LT1300.

DUAL-OUTPUT 500-kHz ±15-V DC-TO-DC CONVERTER

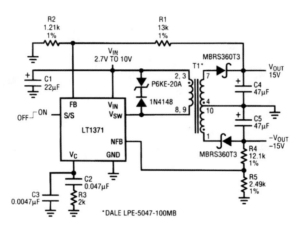

Fig. 28-11

The LT1371 dual-output flyback converter circuit shown generates ±15 V at 200 mA each from a 5-V supply. The LT1371 is a 500-kHz high-efficiency switching regulator with 3-A power switch, yet it uses only 4 mA typical quiescent current. The high operating frequency allows a small surface-mount, dual-output winding flyback transformer to be used. Each output is monitored by separate positive and negative feedback inputs on the LT1371 to be sure that neither output rises above its set point. This removes a common problem in dual-output designs using a single controller: the tendency of the voltage of the least loaded output to fly high. Output voltage regulation is best when both outputs are evenly loaded.

500-kHz 5-V–TO–12-V 400-mA DC-TO-DC CONVERTER

Efficiency

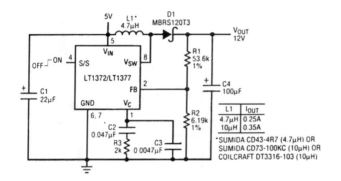

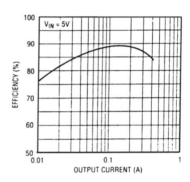

Fig. 28-12

In some 5-V-only systems, it is necessary to generate a local 12-V supply to power op amps, data-acquisition circuitry, or devices, such as small motors. The LT1372 is a 500-kHz high-efficiency switching power regulator that provides this capability, as shown in this circuit. The 500-kHz switching frequency reduces the size of the magnetics significantly over that in lower-frequency designs; they consume just 0.5 in^2 of total board space. The internal switches are current-limited to 1.5 A. In addition, the quiescent current of LT1372 is just 4 mA, which, along with very low switch losses, provides up to 89 percent efficiency in this application. Other features of the LT1372 include synchronization of the switching action to a system clock source and shutdown capability, reducing supply currents to just 12 μA. Also, the LT1372 has two feedback inputs that allow regulation of either positive or negative output voltages.

HIGH-CURRENT 5-V–TO–12-V 2.5-A DC-TO-DC CONVERTER

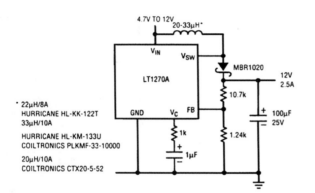

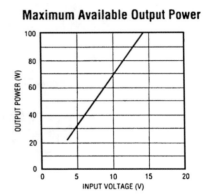

Maximum Available Output Power

LINEAR TECHNOLOGY POWER SOLUTIONS

Fig. 28-13

Many applications require a 12-V supply for control, data storage, interface, or driver functions. These designs often require high peak currents, and a 12-V supply might not be available with high enough current. The LT1270A circuit shown here will provide a minimum of 2.5 A at 12 V from a 5-V ±5 percent supply. The LT1270A has a 10-A high-efficiency switch and a low 10-mA (maximum) supply current, which provides excellent efficiency in high-current-output dc-to-dc conversion circuits. The 60-kHz switching frequency has been optimized for best efficiency.

5-V–TO–12-V 1-A DC-TO-DC CONVERTER

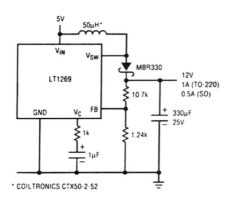

* COILTRONICS CTX50-2-52

**Maximum Available Output Power
(TO-220 Package)**

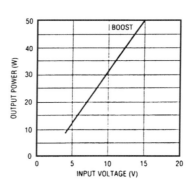

Fig. 28-14

Local on-board conversion from 5 V to 12 V for use in amplifier, signal conditioning, or bus driver circuitry normally entails the use of a module or complex dc-to-dc converter design. The LT1269 100-kHz PWM switching regulator can provide a minimum of 1 A at a regulated 12 V in a surface-mountable DD package (500 mA in 20-lead, small-outline SMT). Included on the chip is a low-ON-resistance switch (0.33 Ω) with a 4-A current limit for high-efficiency conversion in the boost converter shown here. The high switching frequency permits the use of small inductors and capacitors in this converter. The device can be placed in a micropower shutdown mode (100 μA typical supply current) by activating a clamp on the V_C pin.

70-V-INPUT 5-V 700-mA TELECOM CONVERTER

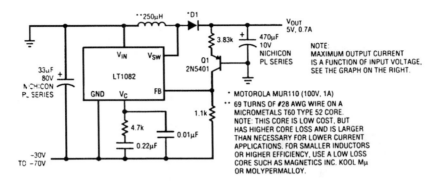

Maximum Available Output Current

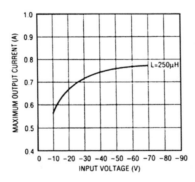

Fig. 28-15

Telecom dc-to-dc conversion applications are usually complex because of the wide input voltage range of −30 to −70 V. Either big, expensive converter modules or space- and component-intensive discrete solutions are normally required to handle these higher voltages. The LT1082 contains a 1-A switch that can handle 100 V, enabling a −48-V–to–5-V converter to be designed with minimal size and cost. Features include foldback of the 60-kHz switching frequency under short-circuit conditions, protecting the LT1082 and power components from excessive power dissipation. The LT1082 dc-to-dc converter circuit provides up to 750 mA of output current, and the solution costs less than a modular supply of similar capabilities.

40-V-INPUT 5-V 10-A DC-TO-DC CONVERTER

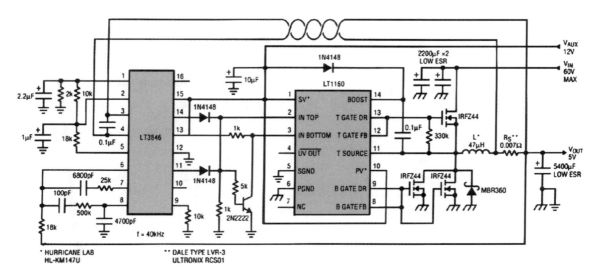

Fig. 28-16

Generating 5 V at 10 A from a 40-V supply with up to 90 percent efficiency is possible using a synchronous switching regulator. The synchronous switching regulator circuit shown achieves this high efficiency by using external MOSFETs with low R_{DS}(ON) and by shunting the Schottky diode after the topside MOSFET has completely turned off to minimize the Schottky diode's conduction losses. The LT1160 half-bridge driver alternately drives the high side and low side MOSFETs ON and OFF during switching. The loop is controlled by the LT3846 current-mode PWM controller and operates at 40 kHz. A key feature of this circuit is the ability to drive the external high-current MOSFETs safely. Internal logic in the LT1160 prevents both MOSFETs from being on at the same time, and its unique adaptive protection against shoot-through currents eliminates all matching requirements for the external MOSFETs. These protection features, in combination with the LT1160's ability to drive up to 10,000 pF of gate capacitance, make paralleling power MOSFETs for high-current applications an easy task. The high-side gate voltage is provided by a floating supply, which is boosted above the HV rail by bootstrap capacitance C_{boost}. An undervoltage detector in the LT1160 can sense an undervoltage condition at either the input supply or the floating supply and turn off both MOSFETs to prevent excessive power dissipation.

9-V-INPUT ±5-V OUTPUT DC-TO-DC CONVERTER

Negative Supply Load Regulation

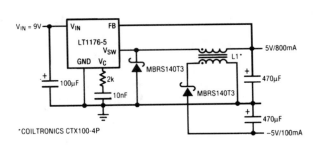

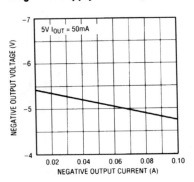

Fig. 28-17

Many dc-to-dc converter applications require regulated complementary output voltages from a single loosely regulated source. A common use for this is supplying 5 V and −5 V to a video op amp for amplification or cable driving. Accomplishing this task in a space-effective manner with a minimum of components is a challenge for a designer. The LT1176-5 circuit shown here uses a single integrated switching regulator and an off-the-shelf inductor with an extra winding to generate ±5 V from an 8- to 12-V input. The circuit is designed to supply 5 V as the main output with up to 800 mA of load current, and −5 V as a secondary output with up to 100 mA load current. Regulation is adequate for most op-amp circuits: 5-V regulation will be ±3 percent and −5-V regulation is about ±10 percent for loads between −10 and −100 mA. The LT1176-5 provides a complete 100-kHz switching regulator with a 1.2-A on-chip switch in a thin 20-lead SO package. The enhanced thermal characteristics of the fused-lead SO package allow higher power outputs than were previously possible with SOs.

6- TO 25-V-INPUT 5-V 1.25-A DC-TO-DC CONVERTER

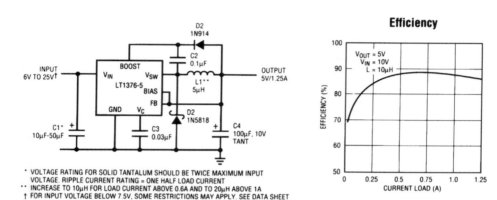

Efficiency

* VOLTAGE RATING FOR SOLID TANTALUM SHOULD BE TWICE MAXIMUM INPUT
  VOLTAGE. RIPPLE CURRENT RATING = ONE HALF LOAD CURRENT
** INCREASE TO 10µH FOR LOAD CURRENT ABOVE 0.6A AND TO 20µH ABOVE 1A
† FOR INPUT VOLTAGE BELOW 7.5V, SOME RESTRICTIONS MAY APPLY. SEE DATA SHEET

LINEAR TECHNOLOGY POWER SOLUTIONS

Fig. 28-18

One of the keys to success for many portable devices is small size. In many cases, the requirement for a wide battery supply voltage range and high-current, regulated 5-V output seems incompatible with the size requirement. The 500-kHz LT1376, shown in this circuit, provides a powerful, compact power supply. Operating at such a high frequency permits the use of a very small 5-µH surface-mount inductor and a surface-mount output capacitor. In addition, the internal switch has just 0.4 Ω of ON resistance, which reduces power loss and boosts efficiency to 88 percent. A special boost pin and circuitry reduces the minimum operating supply voltage in step-down applications. The maximum current rating of the switch is 1.5 A. The input voltage range extends from 6 to 25 V and is well matched for many battery-pack assemblies. The typical supply current is 4 mA, whereas the shutdown current is just 20 µA.

EFFICIENT 5-V–TO–3.3-V CONVERTER

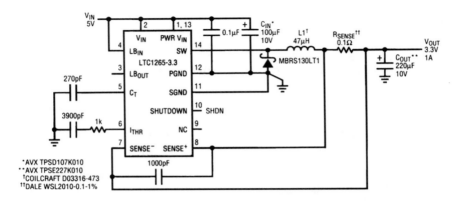

Efficiency

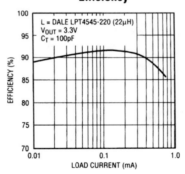

Fig. 28-19

An increasing number of portable devices need 5-V–to–3.3-V converters. High efficiency at up to 1 A is needed to power 3.3-V ICs, such as power-hungry high-speed microprocessors. The LTC1265-3.3 high-efficiency switching regulator circuit shown generates up to 1 A at 3.3 V. The LTC1265-3.3 utilizes a constant OFF-time current-mode architecture for excellent line and load regulation and contains an internal P-channel power MOSFET with 0.3 Ω ON resistance, as well as a low-battery detector. The output current is user-programmable by selection of the current sense resistor R_{sense} according to the formula $I_{out} = 100 \text{ mV} / R_{sense}$. Short-circuit protection is inherent in the current-mode architecture and limits the maximum current. The LTC1265 draws only 160 μA quiescent current under no load and just 5 μA when placed in shutdown.

240

95-PERCENT-EFFICIENT 5-V-INPUT 3.3-V-OUTPUT DC-TO-DC CONVERTER

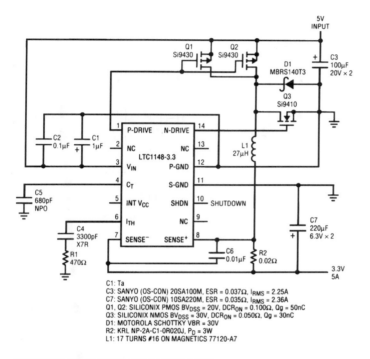

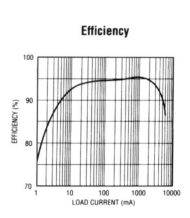

Efficiency

C1: Ta
C3: SANYO (OS-CON) 20SA100M, ESR = 0.037Ω, I_{RMS} = 2.25A
C7: SANYO (OS-CON) 10SA220M, ESR = 0.035Ω, I_{RMS} = 2.36A
Q1, Q2: SILICONIX PMOS BV_{DSS} = 20V, DCR_{ON} = 0.100Ω, Qg = 50nC
Q3: SILICONIX NMOS BV_{DSS} = 30V, DCR_{ON} = 0.050Ω, Qg = 30nC
D1: MOTOROLA SCHOTTKY VBR = 30V
R2: KRL NP-2A-C1-0R020J, P_D = 3W
L1: 17 TURNS #16 ON MAGNETICS 77120-A7

LINEAR TECHNOLOGY POWER SOLUTIONS

Fig. 28-20

Computing equipment increasingly requires both 3.3-V and 5-V logic supplies, with only the 5-V being readily available. As 3.3-V current demands increase to include microprocessors, coprocessors, DSP processors, and memory, the total power demands on the 3.3-V supply rule out the relatively low efficiency and high dissipation of a linear regulator. The circuit shown here supplies 5 A at 3.3 V with over 94-percent efficiency by using the LTC1148-3.3 synchronous switching regulator. This surface-mount solution (L1 is through-hole) requires no heatsinks and very little board space. High efficiency is also maintained at low output currents by implementing Burst mode operation, making the LTC1148-3.3 solution ideal for supplies using normal/sleep modes, such as those in "Green PCs." The LTC1148-3.3 can be placed in a shutdown mode, reducing supply current to a mere 22 μA.

TWO-CELL–TO–5-V BOOST CONVERTER

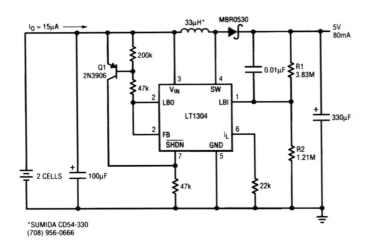
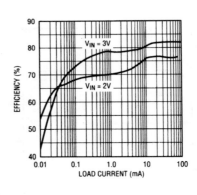

LINEAR TECHNOLOGY POWER SOLUTIONS

Fig. 28-21

Extending the battery life of a portable device that spends most of its life time in standby mode is critical in two-cell applications. The standard LT1304 boost circuit requires two capacitors, one diode, and one inductor and provides 5 V at 200 mA from a two-cell battery with 80 percent typical efficiency. For improved efficiency at very light loads, the LT1304 switching regulator circuit shown achieves an efficiency of 50 percent at just 10 µA of load current. As indicated in the graph, high efficiency over a very wide range of load current is obtained by using the extra circuitry to control Burst Mode operation. The LT1304 is a micropower step-up dc-to-dc converter with an internal comparator that is operational in shutdown. The peak switch current limit can be set up to 1 A by the resistor at the I_{LIM} input. In this circuit, it is set to 500 mA. The input voltage range extends down to 1.65 V, ensuring operation—even as the two-cell battery voltage drops during discharge. The on-board comparator shuts down the LT1304 micropower regulator when the output voltage is higher than the target 5-V output. In shutdown, the LT1304 consumes 10 µA of current, which is less than one-tenth of its active quiescent current of 120 µA. When the output voltage begins to droop below the target 5-V output, the comparator switches the LT1304 on again to recharge the output capacitor.

SIMPLE 5-V–TO–3.3-V CONVERTER

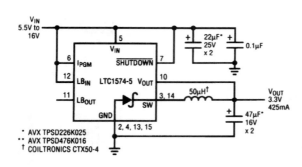

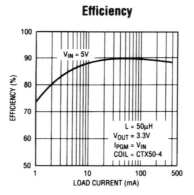

Fig. 28-22

In portable logic systems requiring mixed 5-V and 3.3-V supplies, space and efficiency are paramount. Enter the LTC1574-3.3, a small and simple solution that provides over 90 percent efficiency. The LTC1574-3.3 features an internal Schottky diode, reducing the external component count to just three parts. This high-efficiency circuit uses all surface-mount components, and employs Burst Mode operation to extend high efficiency to low current levels (see graph). A low-loss internal power MOSFET ($R_{DS} = 1.2\ \Omega$ is typical for this circuit) switch and constant off-time architecture are key in achieving this high efficiency. The LTC1574 can be shut down, limiting the supply current to 25 μA (max).

FOUR-CELL–TO–5-VDC CONVERTER

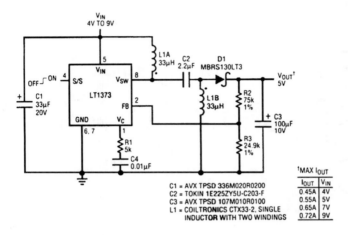

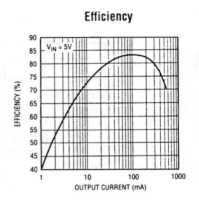

Efficiency

C1 = AVX TPSD 336M020R0200
C2 = TOKIN 1E225ZY5U-C203-F
C3 = AVX TPSD 107M010R0100
L1 = COILTRONICS CTX33-2, SINGLE
    INDUCTOR WITH TWO WINDINGS

†MAX IOUT

IOUT	VIN
0.45A	4V
0.55A	5V
0.65A	7V
0.72A	9V

LINEAR TECHNOLOGY POWER SOLUTIONS

Fig. 28-23

Generating a regulated 5-V output from a four-cell NiCd battery pack requires step-down operation when the battery voltage is above 5 V (fully charged pack at 6 V) and step-up operation when the battery voltage drops from 5 V to 3.6 V during discharge. The LT1373 converter circuit shown achieves up to 83 percent efficiency at high current (100- to 200-mA range), better than a flyback approach. The 250-kHz switching frequency minimizes inductor values, and both 33-μH inductors are wound on the same core, requiring less board area. With a quiescent current of 1 mA typically, the LT1373 offers high efficiency at high frequency, extending the battery lifetime in applications.

5-V–TO–4-V CONVERTER

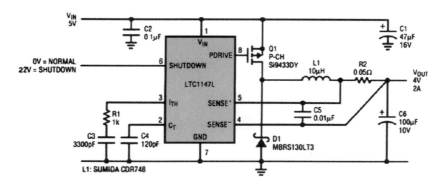

LINEAR TECHNOLOGY POWER SOLUTIONS

Fig. 28-24

Generating a 4-V supply voltage for some of the new microprocessors from a 5-V main requires a low-voltage-loss capability in an efficient switching power supply. The LTC1147L circuit shown provides this capability in only 0.6 in^2 of board space. The LTC1147L is a high-efficiency step-down switching regulator controller that provides gate drive control for an external P-channel MOSFET switch. Up to 100 percent duty cycle is possible with the LTC1147L, allowing the output voltage to be close to the input voltage. This device uses a constant off-time architecture and can operate at switching frequencies exceeding 400 kHz. The LTC1147L is an adjustable device, with the output voltage set by an external resistor divider network. Maximum load current is set by the value of the current sense resistor R2.

LOW-NOISE DC-TO-DC CONVERTER

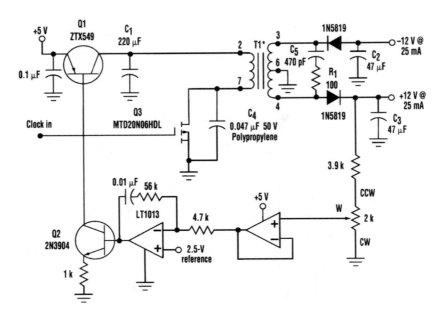

*Notes on T1:
Core: Philips 813E187-3C85
Bobbin: Philips E187PBB1-8 (Core and bobbin samples available from Elna Ferrite Labs, (800) 553-2870).
Winding: 7 turns of 20 AWG. Begin pin 2, wind clockwise, end pin 7. One layer Mylar tape. Then
12 turns, 24 AWG. Begin pin 3, wind clockwise 6 turns, pause at pin 6, continue clockwise 6 turns, end pin 4.
Gap outer legs of core at 0.002 in. Primary inductance should be about 5μH

ELECTRONIC DESIGN

Fig. 28-25

Low noise is achieved by this inexpensive and versatile 5-V–to–±12-V dc-to-dc converter. Wideband output noise appears to be well under 500 μV p-p. The converter accepts external clocks from 80 to 120 kHz. The converter operates much like a TV horizontal deflection circuit. Q3 is a logic-level power MOSFET driven by an external clock. When Q3 is switched on, current ramps up through T1's primary. When Q3 is switched off, Q3's drain flies back to 25 V as C4 resonates with T1's primary, transferring energy to the secondary. As the flyback voltage falls and attempts to go below ground, the intrinsic diode of the power MOSFET clamps it. In addition, the excess resonant energyflows backward through T1's primary, recharging C1. When this point is reached, Q3 is again switched on and the cycle repeats. Q3's gate (trace 2) is switched on and off when Q3's drain is near ground. Q3's gate driver power requirements are modest enough to be handled by any 74HC logic gate. Regulation is achieved by the error amplifier (LT1013). Despite the fact that Q1 acts as a linear pass transistor, converter efficiency can exceed 75 percent if the voltage drop across Q1 is minimized. This can be accomplished by either adjusting the clock frequency or tuning C4 to adjust the flyback voltage.

1.2-V REGULATOR FOR GTL TERMINATION

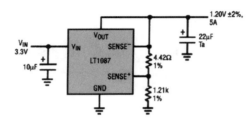

LINEAR TECHNOLOGY POWER SOLUTIONS

Fig. 28-26

A recent development in high-speed digital design has resulted in a new family of logic chips called *gunning transition logic (GTL)*. These chips use high-speed logic and require active termination for best interconnection performance. The termination voltage required is 1.2 V, and the input voltage can be as low as 3.3 V from a logic supply. The LT1087 5-A low-dropout regulator shown here addresses these requirements by providing a regulated 1.2-V ±2 percent output voltage from a minimum 2.7-V input. Note that the LT1087 has a 1.25-V reference, but provides Kelvin sensing inputs for the feedback amplifier. The 4.42-Ω resistor is inserted as a simple way to adjust the internal reference downward without sacrificing regulation. This GTL termination circuit supplies 5 A maximum load current and can handle 3.3-, 5-V, or higher supplies, although 3.3 V is recommended for minimum device dissipation.

CURRENT-LIMITING A-SERIES REGULATOR

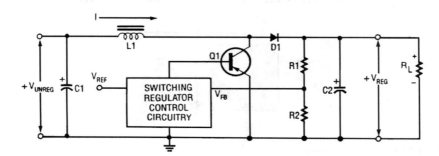

ELECTRONICS NOW

Fig. 28-27

If the current limitation of the series-pass transistor is exceeded, transistor Q1 could be damaged or destroyed. This can be prevented with the addition of a current-limiting transistor, as shown in the figure. When the current through Q1 becomes high enough, the voltage drop across R2 becomes high enough to forward-bias transistor Q2. When Q2 starts to conduct, its internal resistance decreases. When this occurs, the forward bias of Q1 is fixed, and its output is a constant current. The current-limiting transistor and resistor in the figure protect the pass transistor and rectifier diodes if the load terminals are accidentally short-circuited. However, the addition of transistor Q2 increases the already high power dissipation in pass transistor Q1 when the load demand is high.

TRANSFORMERLESS DC-TO-DC CONVERTER

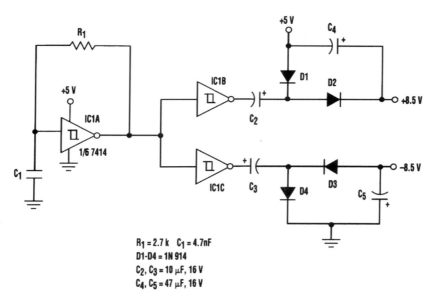

R₁ = 2.7 k   C₁ = 4.7nF
D1-D4 = 1N 914
C₂, C₃ = 10 μF, 16 V
C₄, C₅ = 47 μF, 16 V

ELECTRONIC DESIGN

Fig. 28-28

This configuration should prove handy in situations in which dual-polarity supplies are needed for a few devices on a board that has only one +5-V supply. The circuit doesn't need any dc-to-dc converter ICs, nor does it require any transformers or inductors. Three Schmitt-trigger inverters, such as the 7414, form the heart of the circuit (see the figure). One inverter is configured as a high-frequency astable multivibrator employing a single resistor and a capacitor. For the RC values shown, the frequency of the astable output is around 100 kHz. The frequency of the oscillation is given by $f = 1/T$, where $T = R_1 C_1 \ln [(1 - V_{CC}/V_{LT})/(1 - V_{CC}/V_{UT})]$ and R_1 and C_1 are the timing components of the astable multivibrator, V_{CC} is the supply voltage, and V_{LT} and V_{UT} are the lower trip point and upper trip point of the Schmitt trigger. The astable multivibrator's output drives a pair of inverters that, in turn, drive a pair of diode-capacitor voltage-doubler circuits. The outputs of the diode-capacitor circuits are around 8.5 V with the polarities shown. Diodes D1 to D4 should be fast-switching types, like the 1N914 or 1N4148. As a result, the circuit can generate ±8.5 V from a single +5-V supply making it useful in many applications.

29

Decoder Circuits

The sources of the following circuits are contained in the Sources section, which begins on page 1043. The figure number in the box of each circuit correlates to the entry in the Sources section.

Packet Radio Tuning Indicator
Alphanumeric Pager Decoder
TV Line Decoder I
TV Line Decoder II
DTMF Decoder I
DTMF Decoder II
DTMF Receiver Decoder
RTTY Tone Decoder
BCD Decoder-Driver Circuit
One-IC Tone Decoder

PACKET RADIO TUNING INDICATOR

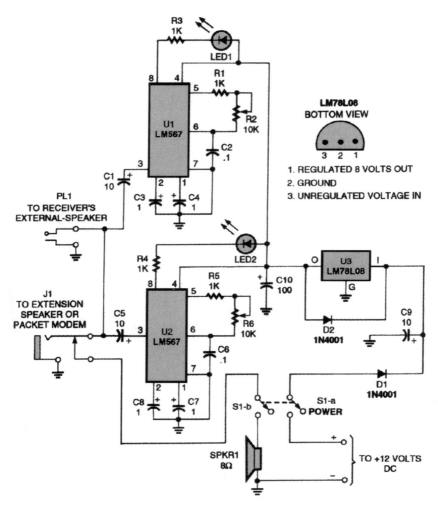

Fig. 29-1

The tuning indicator is simply two identical tone decoders adjusted to different frequencies that share a power supply. When a decoder receives a signal of the right frequency, it lights its LED. Simply tune the circuit so that both LEDs illuminate.

ALPHANUMERIC PAGER DECODER

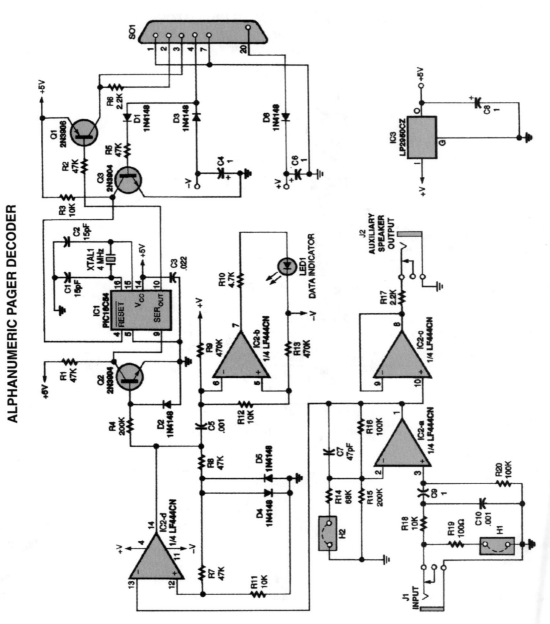

Fig. 29-2

POPULAR ELECTRONICS

252

ALPHANUMERIC PAGER DECODER *(Cont.)*

This pager decoder interfaces a scanner plugged into J1 with a personal computer's serial port via the DB25 connector. Software is necessary and can be obtained from the Internet at http://cylex-inc.com/download.htm.

TV LINE DECODER I

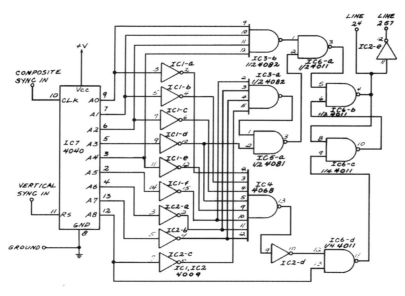

Fig. 29-3

This circuit will produce outputs on TV lines 24 and 257. It was used for a decoder circuit. It uses a CMOS counter and gate logic. Only one pin is used for the output line indicator.

TV LINE DECODER II

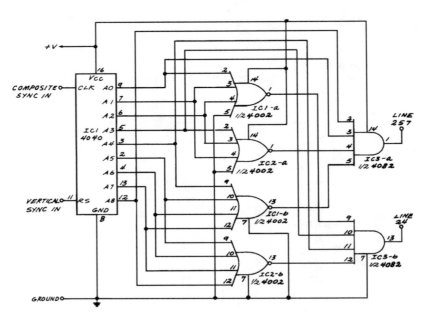

Fig. 29-4

This circuit will produce outputs on TV channels 24 and 25. It was used in a decoder circuit. It uses a CMOS counter and gate logic.

DTMF DECODER I

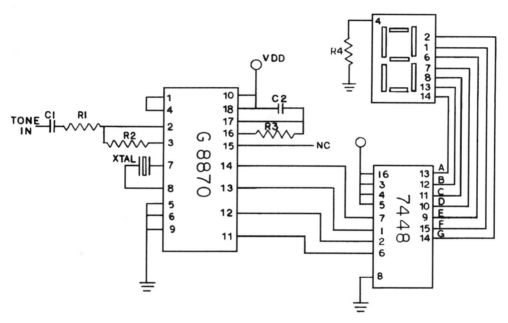

NUTS AND VOLTS

Fig. 29-5

This decoder uses a G8870 DTMF receiver decoder chip to decode DTMF signals and display them via an LED display driven by a 7448 decoder driver chip. Xtal is a 3.579-MHz TV burst crystal. $C_1=C_2=0.1\ \mu F$, $R_1=R_2=100\ k\Omega$, and $R_3=300\ k\Omega$. D1 through D4 are small red LEDs.

DTMF DECODER II

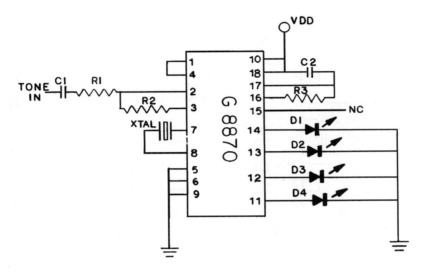

Fig. 29-6

This decoder uses a G8870 DTMF receiver decoder chip to decode DTMF signals and display them in binary via LEDs. Xtal is a 3.579-MHz TV burst crystal. $C_1=C_2=0.1$ μF, $R_1=R_2=100$ kΩ, and $R_3=300$ kΩ. D1 through D4 are small red LEDs.

DTMF RECEIVER DECODER

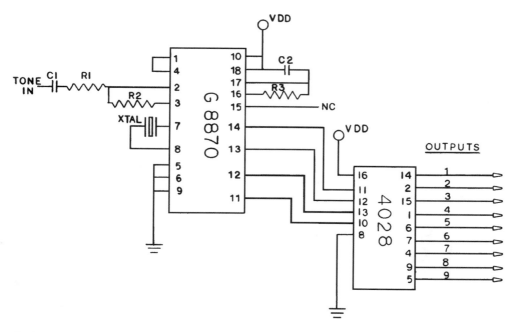

Fig. 29-7

This decoder uses a G8870 DTMF receiver decoder chip to decode DTMF signals and drive individual outputs via a 4028 binary-decimal decoder. Xtal is a 3.579-MHz TV burst crystal. $C_1=C_2=0.1$ μF, $R_1=R_2=100$ kΩ and $R_3=300$ kΩ. D1 through D4 are small red LEDs.

RTTY TONE DECODER

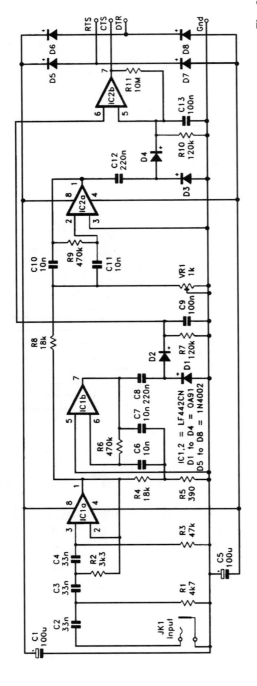

EVERYDAY PRACTICAL ELECTRONICS

Fig. 29-8

The full circuit diagram for the RTTY tone decoder is shown. The input filter is based on IC1a, and it is a third-order (18 dB per octave) high-pass type with a cutoff frequency at approximately 750 Hz. IC1b is used in the higher-frequency bandpass filter. Capacitor C8 couples the output of this filter to a conventional half-wave rectifier and smoothing network based on diodes D1 and D2. The time constant of R7 and C9 is long enough to give a well-smoothed output signal, but short enough to permit the unit to respond rapidly as the input signal alternates between one tone and the other. The second bandpass filter is based on IC2a and is essentially the same as the first, but it has a preset resistor (VR1) as one section of the input attenuator. This enables the center frequency of the filter to be adjusted, but (in operation) it is set 170 Hz lower than the center frequency of the other filter. If preferred, preset potentiometer VR1 can be set to give a center frequency 170 Hz higher than the other filter. It can also be set for shifts of other than 170 Hz. The output of IC2a (pin 1) feeds into a rectifier and smoothing circuit that is identical to the one used at the output of the other filter. The outputs of both the smoothing circuits drive the inputs of IC2b, which acts as the voltage comparator. A small amount of dc positive feedback is provided by resistor R11, and this helps to avoid problems with jitter at the output when only background noise is present at the input. Power is obtained from two of the otherwise unused handshake outputs of the PC's serial port. Diodes D5 to D8 form a bridge rectifier that ensures that the circuit is always provided with a supply of the correct polarity. Only a milliampere or two can be drawn from the handshake outputs. Accordingly, IC1 and IC2 must be low-supply-current operational amplifiers.

BCD DECODER-DRIVER CIRCUIT

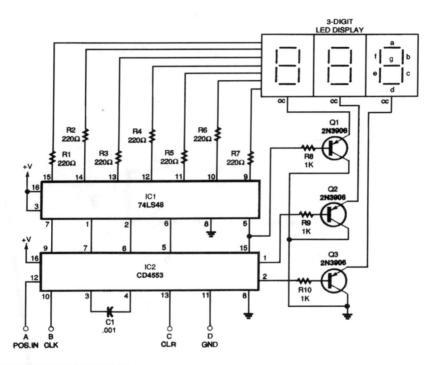

Fig. 29-9

The BCD decoder-driver circuit will interface with any standard BCD output to produce a digital display.

ONE-IC TONE DECODER

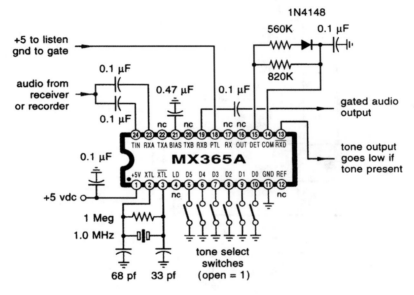

Fig. 29-10

This circuit can be used in a receiver or a repeater to require that a tone be present on a received signal so as to unsquelch the receiver.

30

Detector Circuits

The sources of the following circuits are contained in the Sources section, which begins on page 1043. The figure number in the box of each circuit correlates to the entry in the Sources section.

NE602 Product-Detector Circuit
Missing-Pulse Detector
HV Static Detector
PLL Lock Detector
Quadrature Detector Design
Basic Vacuum Tube Regenerative Detector
Vibration Detector
Peak Detector
Null Detector
Draught Detector
Carbon Monoxide Detector

NE602 PRODUCT-DETECTOR CIRCUIT

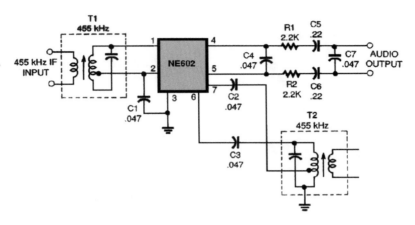

Fig. 30-1

This circuit uses an NE602 as a product detector. The component values depend on operating frequency and are typical for 455-kHz operation. Note that a passive RC filter is shown in the audio output circuit. T2 acts as an oscillator coil for the 455-kHz local oscillator (or, in this case, beat frequency oscillator).

MISSING-PULSE DETECTOR

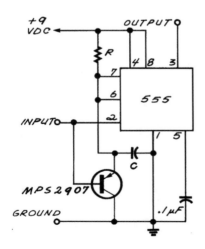

Fig. 30-2

A missing-pulse detector is a circuit whose output is triggered only when there is a loss of signal at its input. In the circuit shown, the 555 timer IC is configured as a one-shot multivibrator that produces a pulse whose width is determined by the value of resistance R and capacitance C, shown in the schematic. The values for those parts can be calculated by using the formula $T=1.1RC$. Select values that make the output pulses about twice as wide as the input trigger pulses from the control line. As long as there is a steady stream of input trigger pulses, the 555 will output a logic high. If there is only one input pulse period, the 555 will time out and its output will go logic low until the next trigger pulse arrives. The logic low at the 555 output can trigger an alarm, a time recorder, a counter, or any other device.

HV STATIC DETECTOR

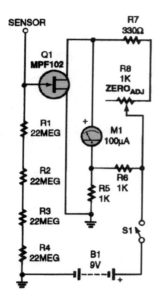

One leg of a Wheatstone bridge is replaced by an MPF102 N-channel JFET. The JFET's high gate input impedance turns the simple bridge circuit into a very sensitive high-voltage static detector. The JFET's gate is tied to ground through four 22-$M\Omega$ resistors. A 1- to 2-in piece of solid-copper wire is connected directly to the JFET's gate and serves as the input sensor. With the sensor clear, adjust R8 for a zero meter reading. Run a comb through your hair and move it toward the sensor. The meter should go off the scale. Place the static-sensor next to your computer or any static sensitive equipment and take special notice when the meter moves.

POPULAR ELECTRONICS

Fig. 30-3

PLL LOCK DETECTOR

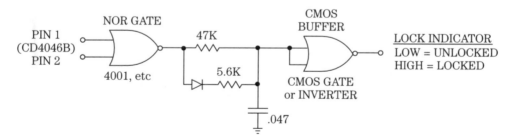

WILLIAM SHEETS

Fig. 30-4

When the loop is locked, the waveforms at pins 1 and 2 of CD4046B are almost identical, differing only in polarity. This holds the output of the NOR gate low. When the loop is not locked, the output of the NOR gate is a series of pulses. This charges the 0.047-μF capacitor, and causes a low output from the CMOS buffer.

QUADRATURE DETECTOR DESIGN

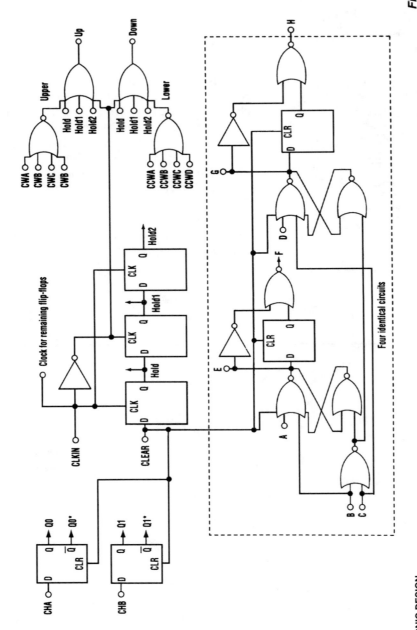

Fig. 30-5

ELECTRONIC DESIGN

One common electrical requirement when dealing with motion control is to monitor the angular position of a rotating object. Quite often, an optical encoder is attached to a rotating shaft, and the encoder's quadrature outputs provide angular displacement information. The key is to turn those quadrature signals into clock pulses, and then into a displacement count that's useful for your application. This circuit requires a reset as well as a clock for sampling the input data. The clock frequency needs to be adjusted to guarantee at least one rising clock edge for each input state change. The circuit takes a set of quadrature signals (CHA, CHB), samples them with CLKIN, and generates output clock pulses (UP or DN) for each input state change. The up/down out-

QUADRATURE DETECTOR DESIGN *(Cont.)*

put clock pulses then can be fed directly into 74XX193 counters as the up and down clocks. If CLKIN is synchronized to external timing, the 193 ripple counter's output can be latched out of phase with the updates to ensure stable counter values. The timing diagram shows the internal functionality of the design. Intervals 1 to 5 illustrate the reset function, which blanks startup instability conditions. Intervals 5 through 12 show timing for a jitter pulse that's been sampled. Interval 13 depicts a jitter pulse that missed being sampled and, therefore, caused no change in the output. The remaining intervals illustrate complete cycles for each direction of motion.

BASIC VACUUM TUBE REGENERATIVE DETECTOR

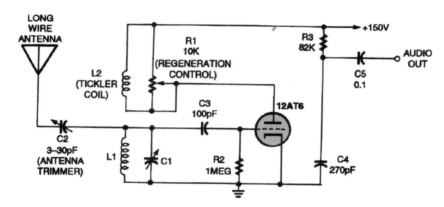

POPULAR ELECTRONICS

Fig. 30-6

The traditional tube-based regenerative receiver is essentially an Armstrong RF oscillator with variable feedback that is connected to an antenna and a resonant circuit. C1 is usually a 140-pF variable capacitor, and L1 and L2 are wound on a plug-in coil form, with L2 typically having 10 to 20 percent of the turns on L1 and wound next to the cold side of L1. The inductance depends on the desired frequency range, but 150 kHz to about 18 MHz is typical for this circuit.

VIBRATION DETECTOR

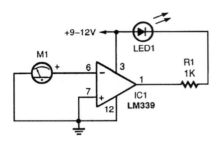

A D'Arsonval meter movement is used as a vibration detector in this novel application. The meter acts as a generator, producing a voltage when the needle moves. An inexpensive panel meter can be used, but a high-sensitivity movement (50 μA) works best.

POPULAR ELECTRONICS   *Fig. 30-7*

PEAK DETECTOR

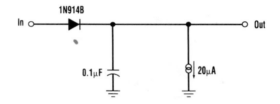

(a)

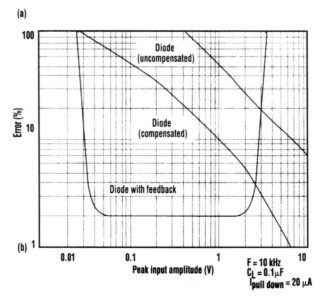

(b)

FIG 1

PEAK DETECTOR (Cont.)

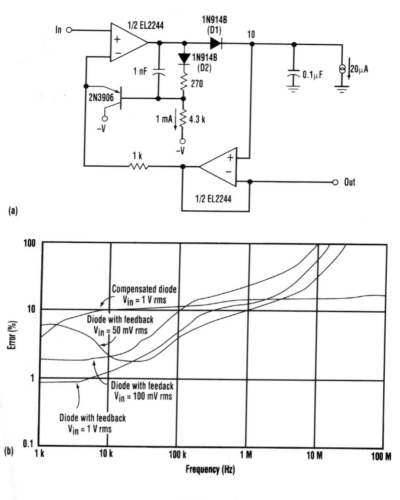

(a)

(b)

FIG 2

ELECTRONIC DESIGN

Fig. 30-8

A simple peak detector, in its most common configuration, provides a mediocre 10 percent error for very large input voltages (a). If the diode is linearized, the necessary input voltage is only reduced to 1 V peak for the same 10 percent error (b).

Feedback circuitry can provide better accuracy and more sensitivity at the input. Feedback makes possible peak detection at input voltages as small as 50 mV rms (a). If 5 percent errors can be tolerated, the circuit has a bandwidth of 100 kHz (b).

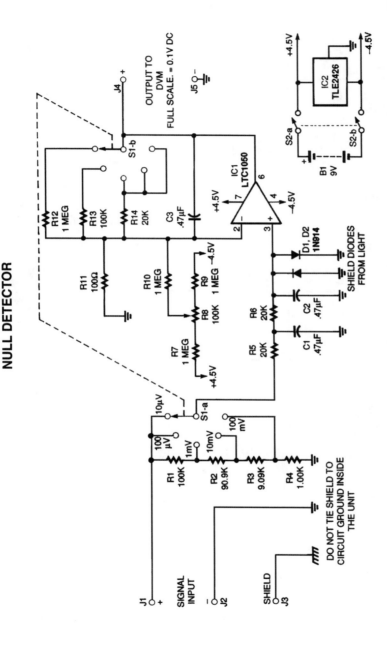

NULL DETECTOR

OUTPUT TO
DVM
FULL SCALE = 0.1V DC

Fig. 30-9

This null detector is based on the Linear Technology LTC1050 amplifier. Chopper stabilized, it provides performance that is almost drift-free. The low-frequency gain is guaranteed to be over 130 dB, so only one amplifier stage is needed. Input bias and offset currents are in the picoampere region, and, finally, the sample-and-hold capacitors have been incorporated within the amplifier, further simplifying the null detector's design. The circuit must be battery-powered, but no cells give reasonable positive and negative supply values. The answer lies in the TLE2426 "rail splitter" ground IC. Using this special divider/buffer, a standard 9-V battery will provide 4.5-V supplies—perfect for the CMOS chopper amplifier. The input of this detector is limited to a maximum of 100 mV. The most sensitive range is 10 µV, full scale. This allows the detector to resolve 1-μV differences with no difficulty. The input impedance will be 200 kΩ for normal "in-range" signals, falling to a minimm of 40 kΩ under overload conditions. The limiting factor is the voltage rating of the first input-filter capacitor, which runs at about one half of the input voltage under overload conditions.

DRAUGHT DETECTOR

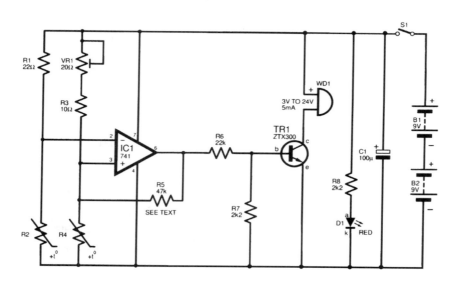

Fig. 30-10

The complete circuit diagram for the draught detector is shown. The key components are R2 and R4; these form a pair of identical positive-temperature-coefficient (p.t.c.) thermistors. It is essential to use this particular type of thermistor. With switch S1 on, current flows from the nominal 18-V supply to the rest of the circuit. Thermistors R2 and R4 are included in a pair of potential dividers. The first comprises fixed resistor R1 in the upper arm and thermistor R2 in the lower one. The second consists of potentiometer VR1, connected in series with fixed resistor R3, in the upper arm and thermistor R4 in the lower one. With room temperature below 60°C, the current in the thermistors will be 150 mA approximately, sufficient for them to self-heat. They will then approach the threshold temperature within a few seconds. The thermistors will stabilize when the temperature reaches about 75°C, corresponding to a resistance of some 500 Ω. In the absence of a draught, each thermistor will be surrounded by a blanket of warm air. However, thermistor R4 is arranged so that it does not receive any draught impinging on detector R2. When a draught is detected, the warm air around R2 is disturbed, and this thermistor is cooled slightly. This results in a lower resistance and hence a falling voltage being developed across it. This sounds the alarm. R5 supplies a little hysteresis to the alarm circuit.

CARBON MONOXIDE DETECTOR

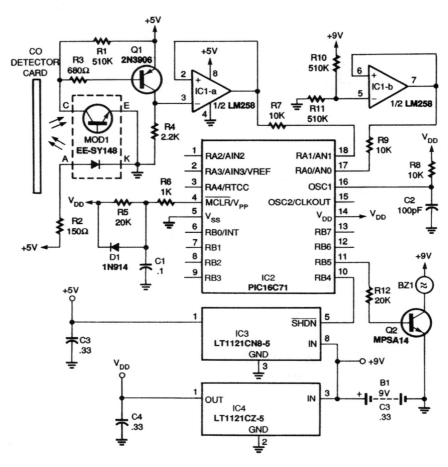

ELECTRONICS NOW

Fig. 30-11

This circuit has a focused optical sensor to transmit light to the CO reagent, and then sense the amount of light reflected. The reagent for CO detection darkens from a light yellow color when exposed to CO. According to the card's manufacturer, a concentration as low as 100 parts per million (ppm) will darken the detector after 15 to 45 min. A concentration of 600 ppm will darken it in 1 to 2 min. The reagent will return to its original color when the air freshens, usually after about 10 min. The time it takes for the reagent to return to its original color depends on the concentration of CO to which it was exposed. The "brain" of the CO detector circuit is a PIC16C71 microcontroller (IC2) that contains a built-in four-channel analog-to-digital converter. Other than the microcontroller, the main component of this circuit is MOD1, an Omron EE-SY148 optical reflector module. The module directs infrared light to the CO reagent and then receives the light reflected back to it. When CO is present, the reagent will darken, thus reducing the reflectivity. The reduction is sensed by the microcontroller, which then turns on buzzer BZ1.

31

Display Circuits

The sources of the following circuits are contained in the Sources section, which begins on page 1043. The figure number in the box of each circuit correlates to the entry in the Sources section.

SIMPLE FOUR-DIGIT DVM BASIC CIRCUIT

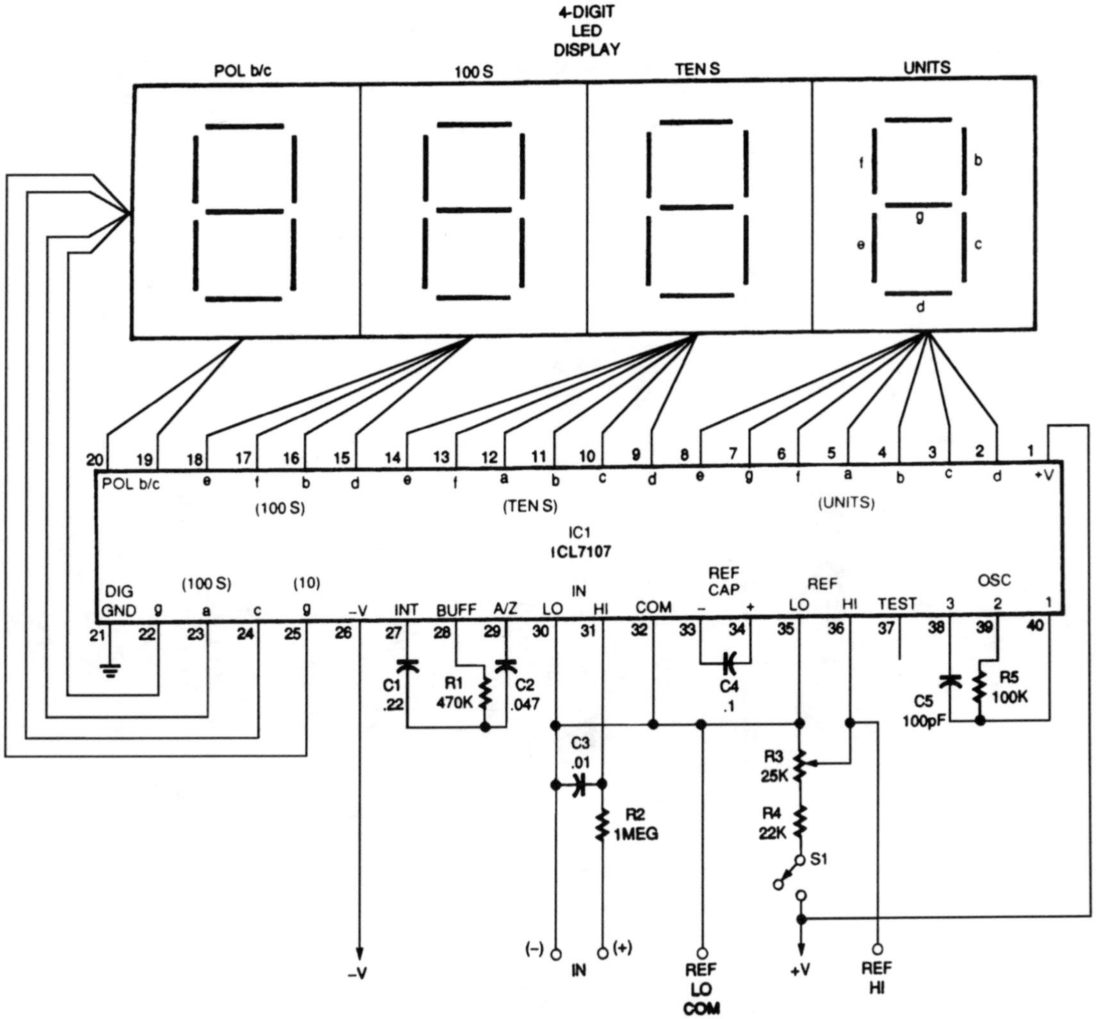

Fig. 31-1

SIMPLE FOUR-DIGIT DVM BASIC CIRCUIT *(Cont.)*

This circuit is designed around a Harris Semiconductor AD converter/LED display driver IC. The power supply is ±5 V.

HIGH-INTENSITY LED CIRCUIT

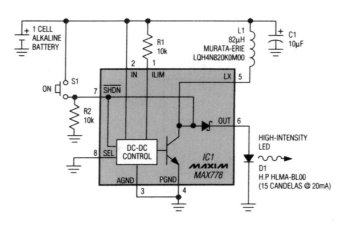

Fig. 31-2

The forward voltage for high-intensity LEDs (1.5 to 2.5 V) is too large for operation with one-cell batteries. The circuit shown overcomes this limitation with a boost-regulator technique—it drives controlled current pulses through the LED, regardless of the LED's forward voltage, and operates on input voltages from 6.2 V to below 1 V. The circuit is useful for bicycle lights, beacons, alarms, flashlights, and low-power indicators. IC1 is normally part of a regulated boost converter, but, in this case, it simply transfers energy without regulating the output. Omission of the usual rectifier and output filter capacitor makes the circuit compact, as does the high switching frequency (about 175 kHz). Programming resistor R1 sets the LED intensity by setting a peak current for the inductor and LED. A 10-kΩ value for R1 sets the approximate peak at 75 mA, and the average LED current at about 26 mA. A shutdown command turns off the OUT terminal completely, even if cell voltage exceeds the LED's forward voltage, by turning off the diode internal to IC1. (During shutdown, most step-up converters exhibit a troublesome dc path from the battery through the coil and diode to the load.) This circuit draws about 8 µA during shutdown and about 60 mA during normal operation. It operates for 35 hours continuously on one AA (or R4 size) alkaline cell.

VOLUME UNIT DISPLAY AND ALARM

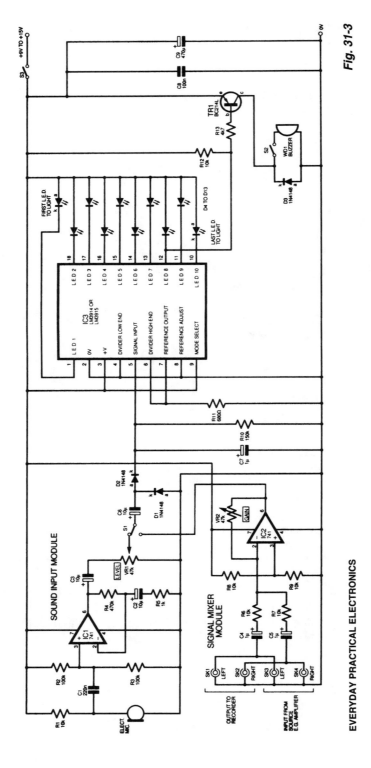

Fig. 31-3

EVERYDAY PRACTICAL ELECTRONICS

This circuit uses two audio amplifiers, one for a microphone and one for a stereo input, to drive an LED bar-graph display, rather than a meter. When the input exceeds a certain level, an LED (#8 in this circuit) will light and also cause an alarm buzzer or some other signal to actuate. A switch is provided for source selection.

274

AC-DC SUPPLY LED CIRCUIT

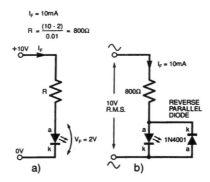

$$I_F = 10mA$$

$$R = \frac{(10 - 2)}{0.01} = 800\Omega$$

EVERYDAY PRACTICAL ELECTRONICS

Fig. 31-4

(*a*) Series resistor calculation; (*b*) using an LED on an ac supply.

EEPROM DISPLAY DRIVER

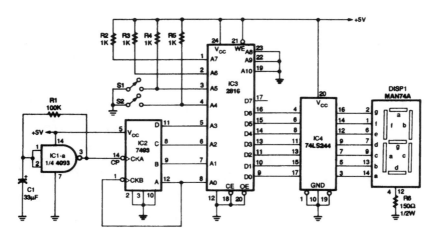

ELECTRONICS NOW

Fig. 31-5

This circuit illustrates use of an EEPROM as a decoder and display driver. IC1 and IC2 generate a binary sequence of addresses in the EEPROM, and the data out of the EEPROM drive (IC4) and the display.

MULTICOLOR LED DRIVER

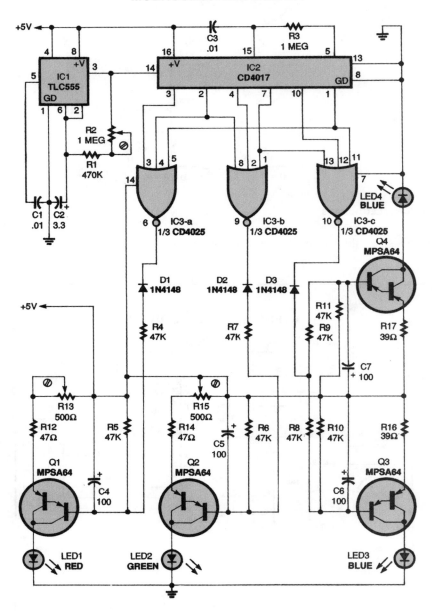

MULTICOLOR LED DRIVER *(Cont.)*

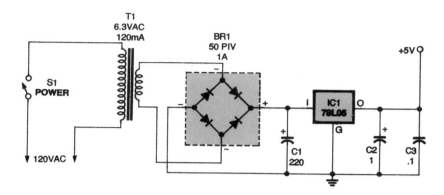

POPULAR ELECTRONICS

Fig. 31-6

A TLC555 CMOS timer, IC1, is configured as a square-wave generator that has an adjustable time period of about 2 to 7 s. That output drives IC2, a CD4017 CMOS Johnson counter, which provides the six decoded output steps necessary to trigger the three additive and three subtractive primary color combinations. A CD4025 CMOS triple three-input NOR gate, IC3, gates the six outputs to the proper LED current sources, thereby maintaining the correct color-mixing sequence. Four MPSA64 PNP Darlington transistors (Q1 to Q4) are configured as gated current sources, with capacitors C4 to C7 providing long time constants that allow LED1 to LED4 to ramp up and down in intensity. This allows the color changes to be continuous, rather than in six abrupt steps. Power for the circuit is provided by a 5-V supply. Transformer T1 steps down the voltage from a wall outlet to 6.3 Vac, rectified by BR1 and regulated by IC1, a 78L05.

FLASHING LED HV SUPPLY CIRCUIT

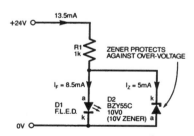

EVERYDAY PRACTICAL ELECTRONICS

Fig. 31-7

When using FLEDs on a supply voltage higher than the rating of the FLED, precautions should be taken to ensure that the device never "sees" a voltage above its maximum rating.

ECONOMICAL LED BAR DISPLAY

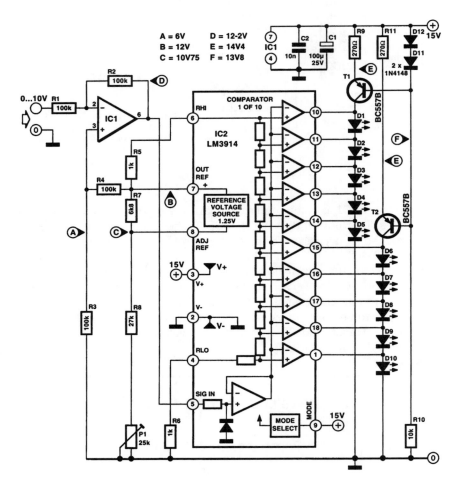

Fig. 31-8

A particular difficulty with LED bar displays is the high current drain. In the diagram, LED driver LM3914 controls a chain of LEDs instead of, as is usual, a number of parallel-connected LEDs. This means that in principle only a single diode current is needed to make several LEDs light. The cost of this, of course, is a higher supply voltage because account must be taken of a number of diode forward voltages. The economizing effect is strengthened if high-efficiency LEDs are used. In the diagram, the current drain of about 15 mA is less than half that of the standard application (32 mA).

SURROUND SOUND INDICATOR

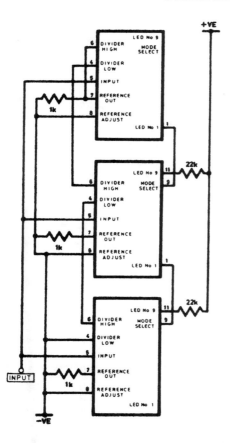

The circuit indicates, with the aid of two LEDs, whether or not the input signal contains surround data. The criterion for this is the phase difference between the two channels: If this is zero, there is no surround data. In the circuit diagram, if there is a phase difference between the two channels, the output levels of comparators IC1b and IC1c will differ. These outputs are constantly compared by XOR gate IC2c; if there is a difference, the output of the gate will go high. Depending on the output state, the red or green half of D1 will be actuated via gate IC2d, which is here connected as an inverter. If there is a pure surround signal, the red half will light brightly; if there is a mono signal, the green half will light. If the input is a standard stereo signal, the rapid changes in the output of IC2c will cause the diode to appear yellow-orange.

ELEKTOR ELECTRONICS   *Fig. 31-9*

LED AC PILOT LAMP CIRCUIT

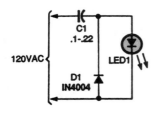

This circuit places an LED in an ac circuit, where it operates as a pilot lamp. The capacitance value sets the LED's current. For 120-Vac operation, a 0.1-μF, 400-V capacitor will result in a 5- to 10-mA current, and a 0.22-μF unit will produce approximately 20 mA of current.

POPULAR ELECTRONICS   *Fig. 31-10*

THREE CASCADED BAR-GRAPH DISPLAY DRIVERS

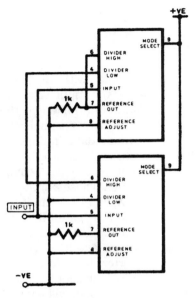

This shows how to cascade three bar-graph display drivers. The basic scheme can be extended to four or more drivers, if desired.

EVERYDAY PRACTICAL ELECTRONICS   *Fig. 31-11*

BRIDGE RECTIFIER LED LIGHT STRING

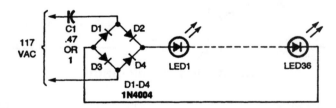

POPULAR ELECTRONICS   *Fig. 31-12*

This circuit has a full-wave bridge rectifier feeding the LED string. With a 0.47-μF capacitor, as shown, the string's current was about 12 mA. If you substitute a 1-μF unit, the current will go up to about 20 mA, allowing you to add additional LEDs without suffering too severe a loss in brightness.

TWO CASCADED BAR-GRAPH DISPLAY DRIVERS

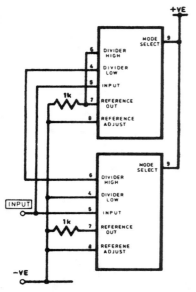

This shows how to cascade two bar-graph display drivers.

Fig. 31-13

EXTERNAL VOLTAGE REFERENCE WITH BAR-GRAPH DISPLAY DRIVER

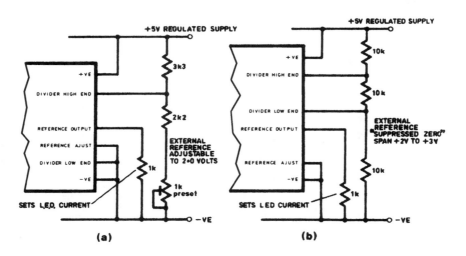

(a)　　　　　　　　(b)

　　　　　　　　Fig. 31-14

Using external reference sources to produce (*a*) an adjustable 2 V and (*b*) up to +3 V starting from +2 V.

BAR-GRAPH DISPLAY DRIVER SCALE SETTING

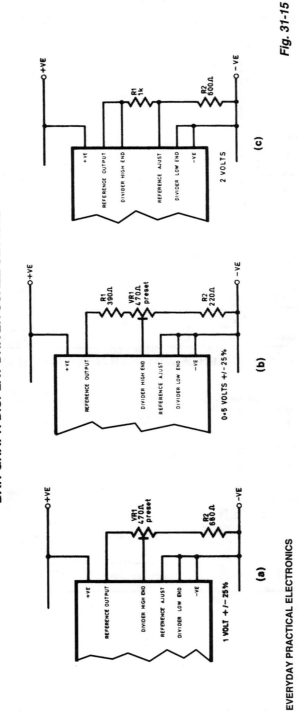

(a)

(b)

(c)

Fig. 31-15

EVERYDAY PRACTICAL ELECTRONICS

The figure shows ways of varying the reference with the internal generator of the bar-graph driver chip. (*a*) and (*b*) show two ways of using the "internal reference" to obtain full scales of 1 V and 0.5 V, respectively. In (*c*), the "reference adjust" and internal reference are used to produce 2 V.

BAR-GRAPH DISPLAY CIRCUIT

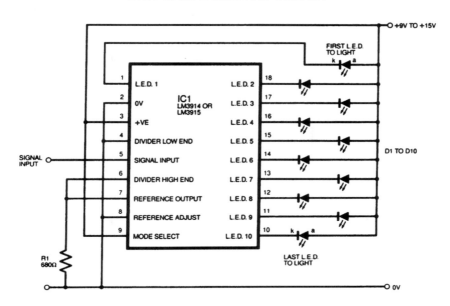

EVERYDAY PRACTICAL ELECTRONICS

Fig. 31-16

This shows how an LM3914 or LM3915 is connected for 9- to 15-V operation. D1 to D10 can be separate LEDs or part of an integral display module.

CONSTANT-BRIGHTNESS LED CIRCUIT

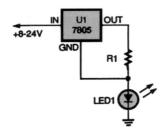

This setup ensures that the LED will glow with equal brightness as long as the input voltage remains between 8 and 24 V. $R=5/I$, where I=desired LED current.

POPULAR ELECTRONICS

Fig. 31-17

VOLUME UNIT DISPLAY AND ALARM

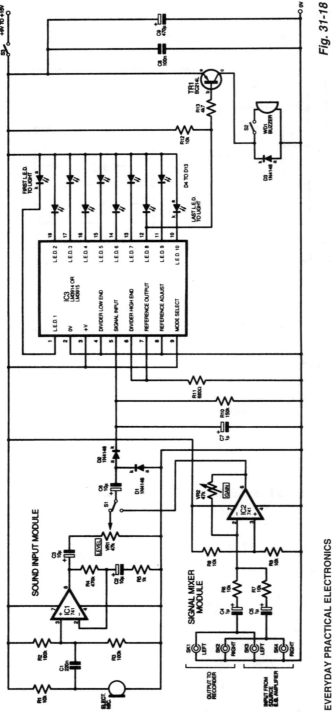

EVERYDAY PRACTICAL ELECTRONICS

Fig. 31-18

The concept is achieved by using two input modules, namely the sound input circuit and the signal mixer circuit, and selecting between them by means of a switch. The signal is then processed by an integrated circuit (IC3) and associated components, which convert a rising and falling voltage into a rising and falling LED display. The buzzer output module is connected so that when the appropriate LED lights, the buzzer sounds, indicating the excess level warning.

10-LED BAR-GRAPH INDICATOR CIRCUIT

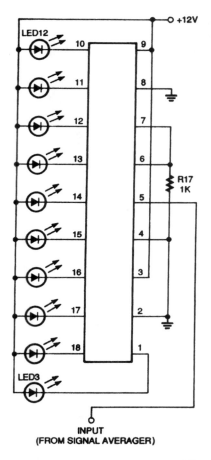

The solid-state level indicator consists of a string of 10 LEDs in one package. With a low-level signal input, all LEDs are off. As the signal increases, more LEDs light up, until finally all 10 are turned on to indicate the maximum level. The heart of the circuit shown in this figure is the National Semiconductor LM3915. The LM3915 contains a precision voltage reference, resistor divider chain, and 10 comparators to drive the LEDs. This IC also contains current-limiting circuitry that limits the brightness of the LEDs without the need for separate resistors, and logic to select either a bar graph or a moving-dot display.

ELECTRONICS NOW   *Fig. 31-19*

BICOLOR-LED SPST SWITCH

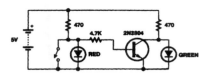

A transistor and a few resistors control two LEDs where the cathodes have a common lead.

ELECTRONICS NOW   *Fig. 31-20*

TRICOLOR LED PEAK INDICATOR CIRCUIT

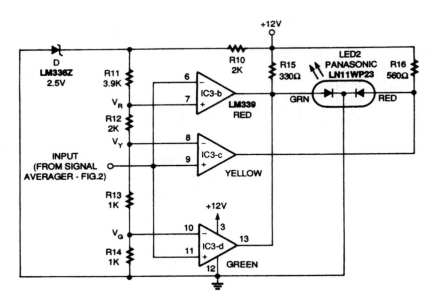

Fig. 31-21

This is a tricolor LED circuit. With a low input level, LED2 is off. As the input voltage increases, LED2 turns on and glows first green, then yellow, and finally red. Resistor R10 and 2.5-V precision voltage reference IC4 provide a precision voltage reference that is further subdivided by resistors R11, R12, R13, and R14. This chain of resistors creates three different reference voltages that set the voltage thresholds for the three LEDs, which are fed to the three comparators along with the input signal. The output of these comparators is then connected to LED2. If the output of all of the comparators is floating, the red and green dies in the tricolor LED are biased ON by R15 and R16, causing LED2 to glow yellow. However, if the output of a comparator is internally grounded, the connected color element will be pulled below 2 V, and the element will turn off. To set the gain level of the amplifier, apply a maximum-level signal to the input of the averager and reduce the level to allow for headroom.

DUAL LED LIGHT STRING

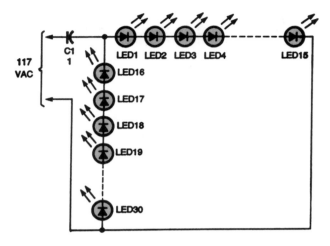

POPULAR ELECTRONICS

Fig. 31-22

This circuit lets you light up two separate 15-light strings from a single power source. On the positive half cycle of the 60-Hz ac waveform, LEDs 1 through 15 light, and on the negative half cycle, LEDs 16 through 30 light. Because the eye's response is much slower than 60 Hz, both strings appear to be on at the same time.

DUAL-COLOR LED LIGHT STRING

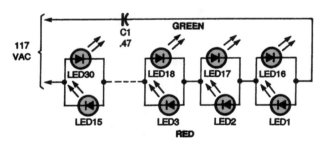

POPULAR ELECTRONICS

Fig. 31-23

The red LEDs are lit during the positive transition of the ac waveform, and the green ones are lit during the negative transition.

ELECTROLUMINESCENT PANEL

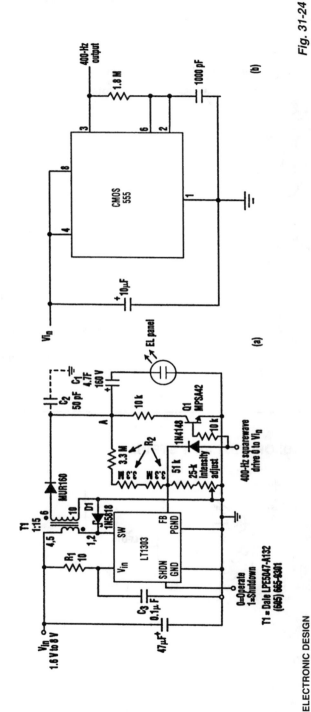

(a)

(b)

Fig. 31-24

ELECTRONIC DESIGN

Electroluminescent (EL) panels offer a viable alternative to LED, incandescent, or CCFL backlighting systems in many portable devices. EL panels are thin, rugged, lightweight, and low power; require no diffuser; and emit aesthetically pleasing blue-green light. Capacitive in nature, they typically exhibit about 3000 pF/in^2 of panel area and require a low-frequency (50 Hz to 1 kHz) 120-V rms ac drive. These problems can be solved by using a setup that includes an LT1303 micropower switching-regulator IC along with a small surface-mount transformer in a flyback topology. The 400-Hz drive signal can be supplied externally or derived from a simple CMOS 555 circuit. When the drive signal is low, flyback transformer T1 charges the panel until the voltage at point A reaches 240 Vdc. C1 removes the dc component from the panel drive, resulting in +120 Vdc at the panel. When the input drive signal goes high, the LT1303's feedback (FB) pin is pulled high as well, idling the IC and turning on Q1. Q1's collector pulls point A to ground and the panel to −120 Vdc. C2 can be added to limit voltage if the panel is disconnected or open. R1 provides intensity control by varying the output voltage. Intensity also can be modulated by varying the drive-signal frequency.

288

LED LIGHT STRING

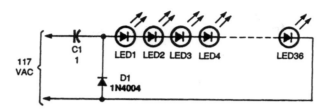

Fig. 31-25

Capacitor C1, a 1-μF, 400-V Mylar or similar unit, operates in the circuit as a no-loss, ac-limiting component connected in series with the LED string. The capacitor's reactance acts like a transistor to the ac without the losses found with a standard power resistor. The 1N4004 silicon diode protects the LEDs from reverse-voltage damage. The circuit was tested with 36 red and green LEDs with an operating current of about 15 mA. Additional LEDs might be added in series if you wish, but that will reduce the brilliance of the LED string somewhat.

LED DC-LEVEL INDICATOR

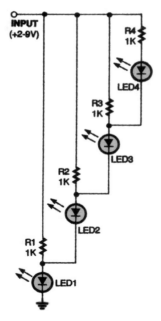

Here's a simple dc-level indicator.

POPULAR ELECTRONICS   *Fig. 31-26*

LED BAR-GRAPH CIRCUIT

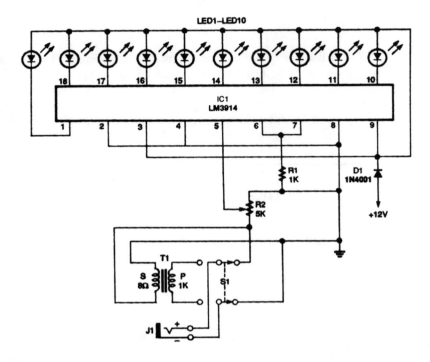

Fig. 31-27

This circuit uses an LM3914 with either dc or audio input. R2 controls the input level. The display can be either ten discrete LEDs or a bar-graph LED assembly.

ALTERNATING-COLOR LED LIGHT STRING

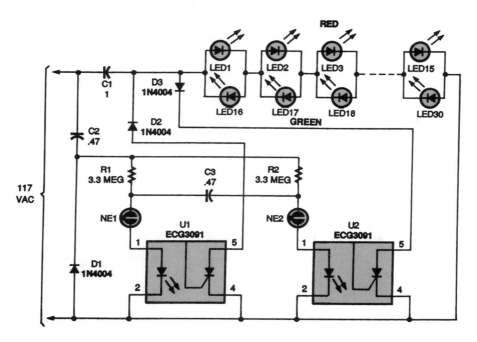

POPULAR ELECTRONICS

Fig. 31-28

This circuit produces a string that alternates between red and green at a steady 1-s rate, making it an attractive addition to any holiday display. Two neon bulbs, NE1 and NE2, are connected in a relaxation-oscillator circuit that switches back and forth at about a 1-Hz rate. The values of $R_1, R_2,$ and C_3 set the switching rate. Dc power for the neon oscillator circuit is supplied via C2 and D1. No dc filter capacitor is used or needed. When NE1 turns on, current flows through the LED in U1, an optocoupler, turning on the SCR and grounding the anode of D2. Only the positive half cycle of the ac waveform passes to the LEDs, which lights only the red ones (LED1 to LED15). After about 1 s, NE1 turns off and NE2 turns on, causing U2 to ground the cathode of D3. Now only the negative half cycle of the ac waveform passes to the LEDs, which lights only the green ones. If you'd like a change rate of less than 1 s, lower the values of R_1 and R_2; to lengthen the rate, increase the value of capacitor C_3.

TRICOLOR LED SIGN DISPLAY

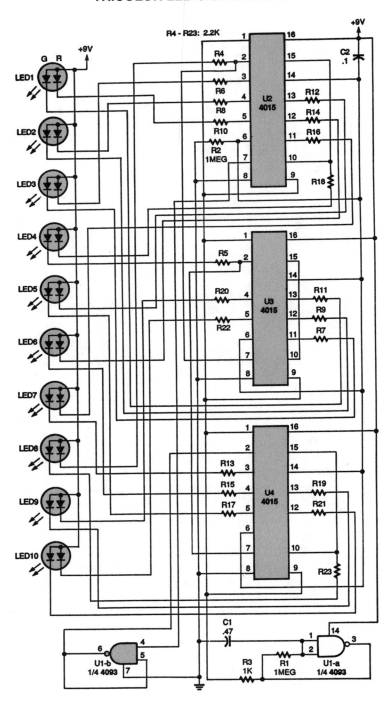

Fig. 31-29

TRICOLOR LED SIGN DISPLAY *(Cont.)*

The purpose of this circuit is to light three-terminal LEDs in sequence on a small sign for a model railroad. Capacitor C2 is connected to the resets of the three 4015 shift registers, U2 to U4. When the circuit is first turned on, C2 automatically sets all the outputs to 0, thus inputting a 1 through U1-b, a section of a 4093 NAND Schmitt trigger, to the A input (pin 7) of U2. The circuit is clocked by U1-a, which is connected as an oscillator, and U1-a inverts the 0s to 1s until the first 1 is recirculated to the output of the last shift register, U4. Then the 1s become 0s and they recirculate. That all repeats as U1-a clocks the shift registers. When the first 1 is received at the first output of U2, the LED segment connected to it lights. As U1-a clocks the shift register, the LEDs connected to the other outputs sequentially light up red. After ten clock cycles, the clock skips two, and then continues lighting up the green segments of the LEDs, giving a yellowish-orange tint to them. After all the LEDs are lit, the shift register starts turning off the red LED segments. That leaves the green segments on—the third color from the LEDs. The oscillator continues to clock the green LEDs off, leaving the sign dark. After that cycle is completed, the operation repeats, giving the effect of a sign that goes from red to yellow to green to dark.

BAR-GRAPH DISPLAY

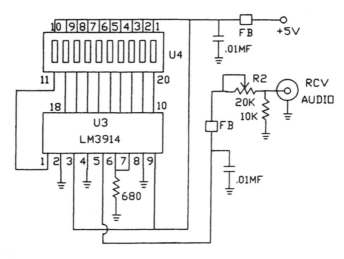

Fig. 31-30

This bar-graph display monitors received audio level and can be useful for tuning, peaking, etc.

ELECTROLUMINESCENT PANEL DRIVER

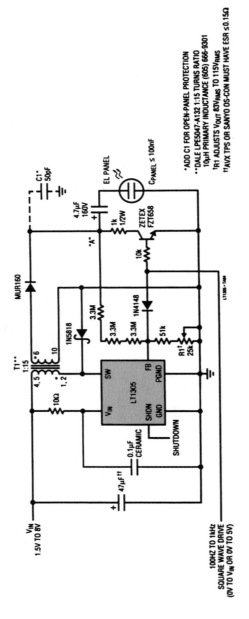

* ADD C1 FOR OPEN-PANEL PROTECTION
** DALE LPE5047-A132 1:15 TURNS RATIO
   10μH PRIMARY INDUCTANCE (605) 666-9301
† R1 ADJUSTS Vout 83Vrms TO 115Vrms
†† AVX TPS OR SANYO OS-CON MUST HAVE ESR ≤0.15Ω

Fig. 31-31

LINEAR TECHNOLOGY POWER SOLUTIONS

Driving electroluminescent (EL) panels, such as those used for backlighting LCD screens in portable devices, is a challenge because they are capacitive loads and require high-voltage ac drive signals. Typically, EL panels exhibit about 3 nF/in² capacitance. The LT1305 circuit shown can drive an EL panel with up to 100 nF total capacitance. The LT1305 micropower dc-to-dc converter contains an internal power switch capable of up to 2 A switch current. The high-power switch drives the primary of flyback transformer T1. A minimum input voltage of 1.8 V ensures operation with two-cell supplies. The input voltage is boosted to a voltage that ranges from 166 to 248 Vdc, as set by R1. The ac drive waveform is created by an external square-wave control signal turning on and off the NPN transistor, bringing point A to ground and inverting the voltage across the panel and then returning to a high positive voltage. By adjusting R1, the panel ac drive voltage can be set from 83 to 115 V_{rms} to alter the display brightness. The color or hue of the display can also be slightly varied by changing the frequency of the square-wave drive signal. This control signal can be from a system microprocessor output pin.

VARYING LCD BIAS CIRCUIT

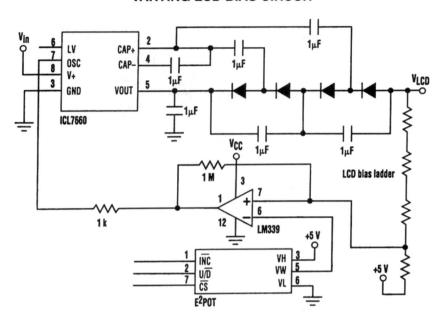

ELECTRONIC DESIGN

Fig. 31-32

Large dot-matrix LCD drivers require a negative bias voltage of up to 24 V, depending on the LCD multiplex ratio. One alternative is to combine the 7660 with a diode/capacitor voltage multiplier, using the 7660's internal square-wave generator as the signal source. This circuit will deliver an output of $3V_{in} - 4V_p$, thus generating approximately 24 V from a 9-V input. Additional stages could be added to the multiplier to allow 24 V to be generated from a 5-V regulated supply. In this circuit, the output voltage is set by the input voltage, which might be inconvenient. If a variable voltage is required for contrast adjustment, some form of microprocessor control over the output voltage will be needed. One particular configuration used a Xicor E^2POT to control the output by using a feedback circuit that pulls the oscillator pin of the 7660 low when the bias voltage exceeds the desired levels. This prevents further pumping of charge until the output returns to the required voltage. The output ripple is set by the hysteresis resistor in the comparator circuit.

BASIC BAR-GRAPH DISPLAY CIRCUIT

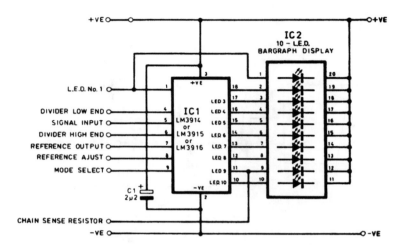

Fig. 31-33

An LM3914 is shown connected to an LED bar-graph display.

32

Doubler Circuits

The sources of the following circuits are contained in the Sources section, which begins on page 1043. The figure number in the box of each circuit correlates to the entry in the Sources section.

Simple Frequency Doubler
Digital Logic Frequency Doubler
Digital Frequency-Doubler Circuit

SIMPLE FREQUENCY DOUBLER

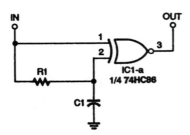

This circuit uses only a single exclusive-OR gate and a couple of passive components. The width of the output pulses is determined by the time constant of the RC network, and the maximum input frequency cannot exceed $\frac{1}{2}RC$.

ELECTRONICS NOW   *Fig. 32-1*

DIGITAL LOGIC FREQUENCY DOUBLER

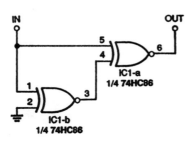

This circuit can be used if very high frequency operation is needed, or if very narrow output pulses can be tolerated. The output pulse width is determined by the propagation time delay of the exclusive-OR gate, and the input frequency cannot exceed $\frac{1}{2}$ (delay).

ELECTRONICS NOW   *Fig. 32-2*

DIGITAL FREQUENCY-DOUBLER CIRCUIT

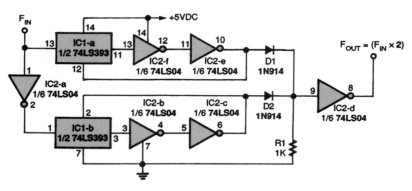

Fig. 32-3

The 74LS393 is designed to advance one count on the positive-to-negative transition of the pulse at the clock input (pins 1 and 13). An inverter in front of the B section of the counter causes the B section to advance one count on the negative-to-positive transition of the input pulse. Each section of the counter is cleared with a positive-going pulse on the master reset input (pins 2 and 12). Positive-going output pulses from pins 6 and 10 of IC2 reset the respective counters. The duration of the pulses depends upon the propagation delay of the inverters. With the 74LS04 hex inverter, delay will probably be in the vicinity of 20 to 25 ns. The output pulses are also connected to the remaining inverter gate through switching diodes and a pull-down resistor, which configures the remaining inverter as a NOR gate. The output at pin 8 of IC2 represents the input frequency multiplied by 2.

33

Driver Circuits

The sources of the following circuits are contained in the Sources section, which begins on page 1043. The figure number in the box of each circuit correlates to the entry in the Sources section.

LOW-DISTORTION LINE DRIVER

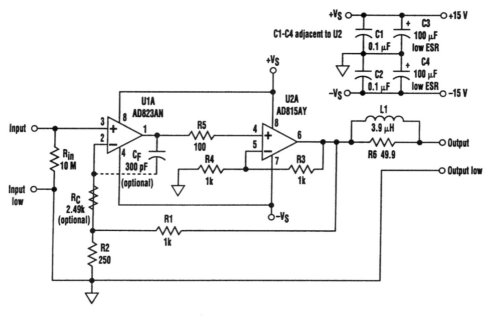

ELECTRONIC DESIGN ANALOG APPLICATIONS

Fig. 33-1

This low-distortion driver circuit delivers up to ±0.5 A and is suitable for loads of 10 Ω and up. Using a low-offset, low-bias-current input stage, the driver can be entirely direct-coupled. Gain is equal to $1+R_2/R_1$.

TWO-TERMINAL PIEZO DEVICE DRIVER

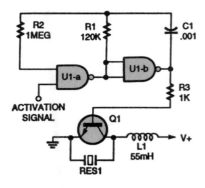

Two-terminal devices can be driven by two NAND gates. A booster coil is used to compensate for the sound-pressure attenuation caused by the case.

POPULAR ELECTRONICS

Fig. 33-2

SIMPLE LAMP DRIVER

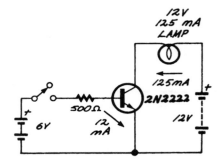

ELECTRONICS NOW

Fig. 33-3

Small current through the base controls large current through the collector. Close the switch, and the bulb lights.

R/C SERVO SWEEP DRIVER CIRCUIT

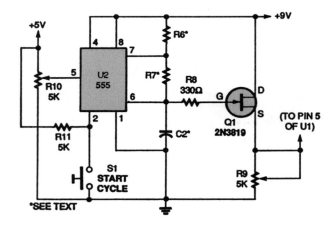

POPULAR ELECTRONICS

Fig. 33-4

The circuit shown produces a ramp that can be used to modulate the threshold of another 555 timer, which is configured to generate the pulse signal to drive the R/C servo. In this way, the servo can be slowly moved through a desired angle. Note that for long time constants, R6 and R7 will be large. Therefore, C2 should be a low-leakage capacitor, or a CMOS timer (7555) can be used to permit larger resistances and smaller capacitors to be used.

R/C SERVO DRIVER CIRCUIT

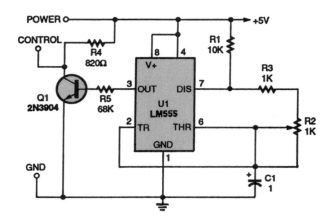

POPULAR ELECTRONICS

Fig. 33-5

The circuit shown produces the necessary pulse signal to drive R/C servos through a 90° range by adjustment of the pot R2.

12-LED SEQUENTIAL DRIVER

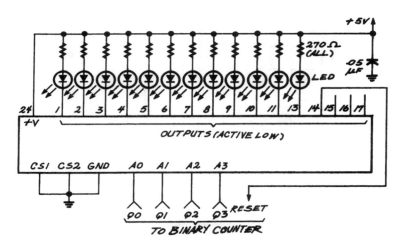

ELECTRONICS NOW

Fig. 33-6

A demultiplexer selects a different LED for each binary number at its input. The counter resets after a count of 12.

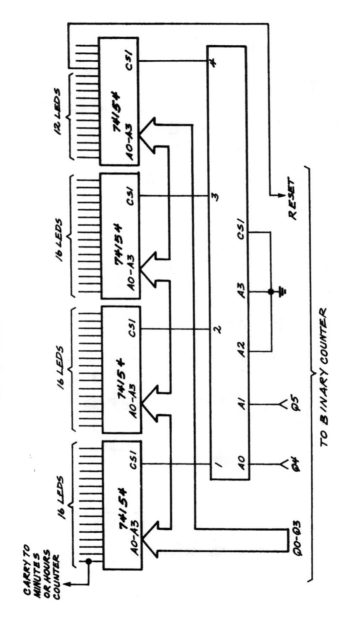

60-LED SEQUENTIAL DRIVER

Fig. 33-7

ELECTRONICS NOW

This circuit is useful as the minute hand for an LED clock or as a general-purpose sequential driver.

MOSFET GATE DRIVER

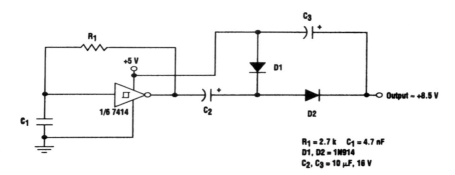

R₁ = 2.7 k   C₁ = 4.7 nF
D1, D2 = 1N914
C₂, C₃ = 10 µF, 16 V

ELECTRONIC DESIGN

Fig. 33-8

This ratiometric 20-kHz voltage-to-frequency converter (VFC) provides superior performance with strain gauges and other ratio-responding transducers, even with noisy, unregulated excitation voltages. Feedback isn't used to achieve the excellent 4-Hz linearity, so there is very low frequency jitter—period measurements can be used to get several digits of resolution even when operating at a fraction of full scale. An operational synchronizing transistor starts the VFC with zero charge at the beginning of each count cycle, eliminating the characteristic digit jumping often encountered with VFC designs. Good linearity is attained by making the comparator's reference voltage vary with the input voltage, which precisely compensates for the finite capacitor reset time:

$$Period = t_1 + t_2$$
$$= t_1 + (V_{CC} - V_{ref})/AV_{in}$$
$$= (t_1 AV_{in} + V_{CC} - V_{ref})/AV_{in}$$

where $AV_{in} = \Delta V/\Delta t$. If V_{ref} is made to include the amount $t_1 AV_{in}$, then the effect of t_1 is eliminated:

$$Period = [t_1 AV_{in} + V_{CC} - (t_1 AV_{in} + V_{ref})]/AV_{in}$$

The MPSA-18 is a remarkably high-gain transistor, even at low currents, giving good current-source linearity down to 0 Hz. In addition, bipolar transistors work well with the low collector voltages encountered in this single-supply, 10-V design. Moreover, most single-supply op amps will work in place of the LM10. But the LM10 also has a reference amplifier that could be used to construct a 10-V excitation regulator. The LM311 propagation delay gives a reset pulse width near 400 ns, which gives the transistor time to discharge the capacitor. Also, the 311's bias current gives a small negative offset that ensures a 0-Hz output for 0 V in.

PIEZOELECTRIC DRIVER CIRCUIT

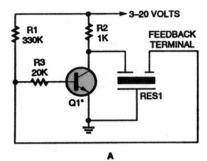

A

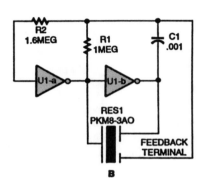

B

Three-terminal piezoelectric elements are typically driven by transistor circuits (A) or logic gates (B).

Fig. 33-9

34

Field-Strength-Measuring Circuits

The sources of the following circuits are contained in the Sources section, which begins on page 1043. The figure number in the box of each circuit correlates to the entry in the Sources section.

DIGITAL FIELD-STRENGTH METER

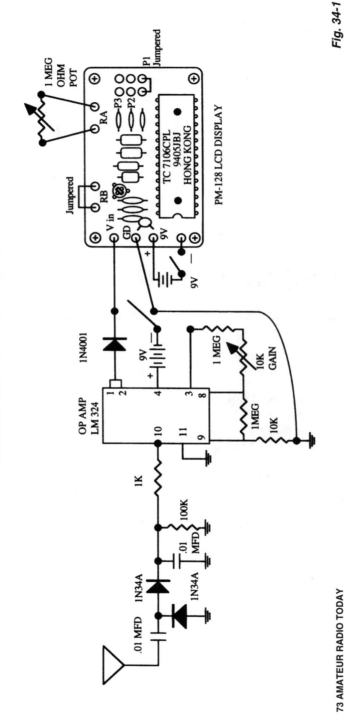

Fig. 34-1

73 AMATEUR RADIO TODAY

This field-strength meter uses a surplus LCD panel meter display, but any suitable unit (1 V full scale, etc.) can be used. RF is detected by a pair of 1N34 diodes and fed to an LM324 op amp acting as a dc amplifier whose gain is set via a 1-MΩ pot. The op-amp output drives the LCD panel meter.

MICROPOWER FIELD DETECTOR FOR 470 MHz

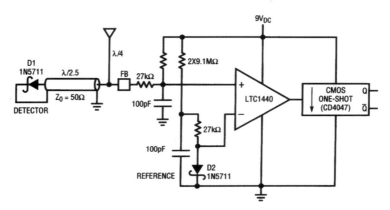

LINEAR TECHNOLOGY

Fig. 34-2

This circuit, which was tested at 470 MHz, contains a couple of improvements over the standard L/C-with-whip field-strength meter. The 0.4-wavelength section presents an efficient, low-impedance match to the base of the quarter-wave whip, but transforms the received energy to a relatively high voltage at the diode for good sensitivity. Biasing the detector diode improves the sensitivity by an additional 10 dB. The detector diode's bias point is monitored by an LTC1440 ultra-low-power comparator and by a second diode, which serves as a reference. Schottky diode D1 rectifies the incoming carrier and creates a negative-going bias shift at the noninverting input of the comparator. Note that the bias shift is sensed at the base of the antenna, where the impedance is low, rather than at the Schottky, where the impedance is high. This introduces less disturbance into the tuned antenna and transmission-line system. The falling edge of the comparator triggers a one-shot, which temporarily enables answer-back and other pulsed functions. Total current consumption is approximately 5 μA. Alternatively, a discrete one-shot constructed from a quad NAND gate draws negligible power. Sensitivity is excellent. The finished circuit can detect 200 mW radiated from a reference dipole at 100 ft. Range, of course, depends on operating frequency, antenna orientation, and surrounding obstacles; in the clear, a more reasonable distance, such as 10 ft, can be covered at 470 MHz with only a few milliwatts. All selectivity is provided by the antenna itself. Add a quarter-wave stub (shorted with a capacitor) to the base of the antenna for better selectivity and improved rejection of low-frequency signals.

FIELD-STRENGTH METER

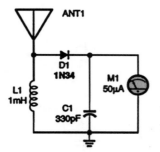

This simple field-strength meter is great for tuning transmitters or antennas. It's a tool no ham should be without. The antenna is a 24-in whip.

Fig. 34-3

EARTH'S FIELD DETECTOR

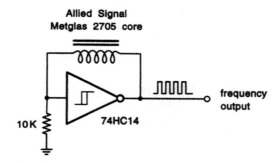

Fig. 34-4

This elegantly simple earth's field detector uses a special variable permeability cored coil. The output frequency varies with orientation.

MAGNETIC FIELD DETECTOR

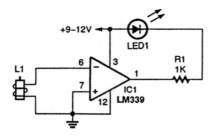

A relay coil makes a great magnetic-field detector.

Fig. 34-5

AMPLIFIED FIELD-STRENGTH METER

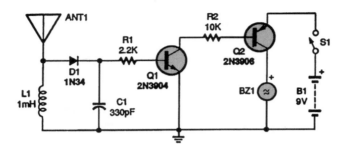

POPULAR ELECTRONICS

Fig. 34-6

This field-strength meter uses Q1 and Q2 to amplify the dc voltage produced by detector D1. A piezo sound BZ1 is used as an audible indicator, rather than using a meter. This circuit could be of use for the visually handicapped. The antenna is a 24-in whip.

ANALOG FLUXGATE MAGNETOMETER ASSEMBLY

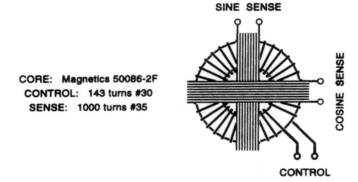

CORE: Magnetics 50086-2F
CONTROL: 143 turns #30
SENSE: 1000 turns #35

ELECTRONICS NOW

Fig. 34-7

An audio input drives the sine-wave control input, switching (or "gating") the core in and out of saturation and drawing in or releasing an external magnetic field. Weak signals at the sense outputs are proportional to field strength.

MAGNETIC FIELD METER

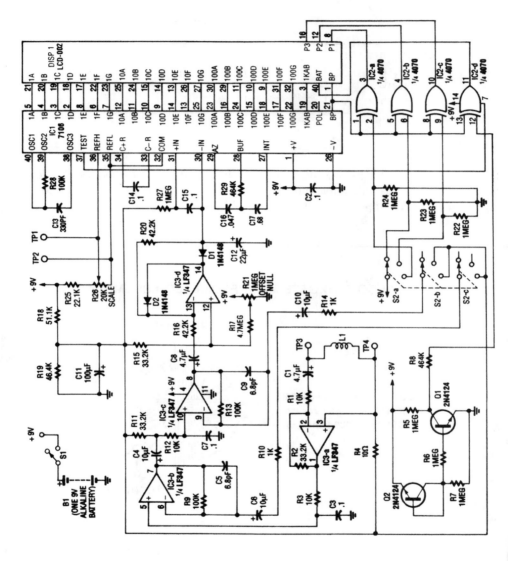

Fig. 34-8

ELECTRONIC EXPERIMENTERS HANDBOOK

MAGNETIC FIELD METER *(Cont.)*

The meter's 12-turn field pickup is integrated into the unit's circuit board. For remote sensing, an external field coil probe can be used. The magnetic field picked up by the coil appears as a voltage, which is proportional to field strength and frequency at the input of a cascaded amplifier IC3-a, IC3-b, and IC3-c. With a first-stage amplifier gain of 3.3 set by R12 to R10, the overall sensitivity is 100 μV/μT, or 100 mV/mT. The meter sensitivity is nominally 2 V full scale, leading to the lowest-level sensitivity of 20 mT full scale. Op amp IC3-a amplifies the signal to a normalized level of 100 μV/μT. The voltage is further amplified by 1, 100, or 10,000 by IC3-b and IC3-c. The three amplifier stages provide the ranges of 2 mT, 200 μT, and 2 μT (full scale). Components R3 and C3 and R12 and C7 establish a frequency rolloff characteristic that compensates for the frequency-proportional sensitivity of the pickup coil, and set the 20-kHz cutoff point. IC3-d is a precision rectifier and peak detector. Its output drives IC1, a combination analog-to-digital (A/D) converter and LCD driver. Components R25 to R29 and C13 to C17 are used by IC1 to set display-update times, clock generation, and reference voltages. The decimal points are driven by IC2, as determined by range-select switch S2. Transistors Q1 and Q2 serve as a low-battery detector, and turn on the battery annunciator in the LCD when the battery voltage drops below 7 V.

RESONANT FLUXGATE MAGNETOMETER

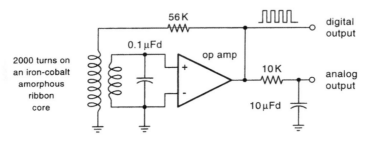

Fig. 34-9

The core will get switched in and out of saturation. The output duty cycle varies in proportion to the single-axis field strength and direction. The high-level output square wave is easy to interface to a PIC or other microcontroller.

35

Filter Circuits

The sources of the following circuits are contained in the Sources section, which begins on page 1043. The figure number in the box of each circuit correlates to the entry in the Sources section.

Peaking Filter Circuit
Amateur Transmitter Absorptive Filter
High-Pass/Low-Pass Filter
SW Receiver CW Filter
1-MHz or 500-kHz Elliptic LP Filter
Active Filter
Notch Filter Circuit
Capacitorless Notch Filter
IF Bandpass Filter Switch
Simple All-Pass Filter
Subsonic Filter
600- to 3000-Hz Tunable Notch-Filter Circuit
Low-Pass Filter Phase Corrector
Ceramic Filter Interfacing
Active HP Filter
Multiple-Feedback Bandpass Active Filter

PEAKING FILTER CIRCUIT

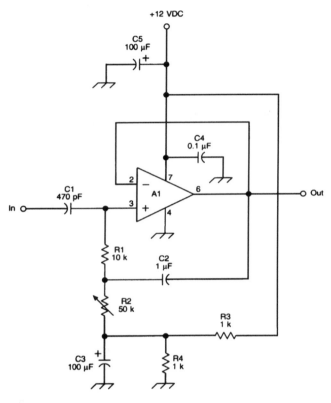

Fig. 35-1

This circuit is basically a noninverting amplifier with positive feedback. The negative feedback is to unity by a direct connection between the output terminal of the operational amplifier and the inverting (−) input. A frequency-selective RC network connects the output to the noninverting (+) input, providing some positive feedback. The center frequency is

$$F_c = \tfrac{1}{2}\pi R_1 (C_1 C_2)^{1/2}$$

where F_c is the center frequency in hertz, R_1 is in ohms, and C_1 and C_2 are in farads.

AMATEUR TRANSMITTER ABSORPTIVE FILTER

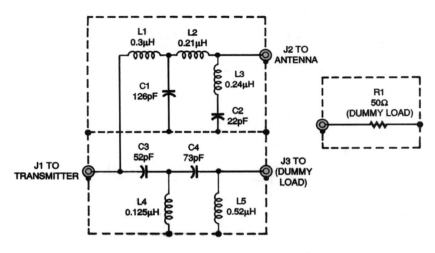

POPULAR ELECTRONICS

Fig. 35-2

This filter diverts harmonic energy above 30 MHz to a dummy load and dissipates it as heat instead of reflecting it back to the transmitter. The absorptive filter is based on the classic design by Weinreich and Carroll from 1968. The high-pass and low-pass sections are shielded from each other (even though they are in the same shielded box) to prevent interaction. If you want to try building this filter, wind the coils on high-powered toroids. Alternatively, you could wind an air-core coil using the nomograms in *The ARRL Handbook for Radio Amateurs*.

HIGH-PASS / LOW-PASS FILTER

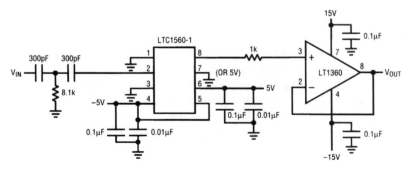

LINEAR TECHNOLOGY

Fig. 35-3

This filter uses a Linear Technology P/N LTC1560-1 and an LT1360 to implement a high-pass / low-pass filter.

SW RECEIVER CW FILTER

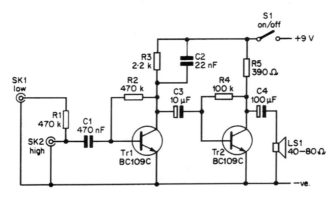

ELECTRONIC EXPERIMENTERS HANDBOOK

Fig. 35-4

The CW filter is connected between the headphone or loudspeaker socket or terminal strip of your receiver and either the headphones or an external loudspeaker. The output of the receiver should have an impedance of 8 Ω or more. As the circuit provides unity gain at pass frequencies and a low-impedance output, there should be no problems with mismatching when the filter is in use. The frequency response of the circuit peaks at approximately 800 Hz, and the −6-dB bandwidth is about 300 Hz or so. The 0-dB points occur at about 350 Hz and 2 kHz. This is sufficient to normally give a substantial reduction in adjacent-channel interference, but the response is not so narrow and peaky that using the receiver with the filter in the circuit becomes difficult, as the wanted signal tends to drift out of the passband and become lost.

1-MHz OR 500-kHz ELLIPTIC LP FILTER

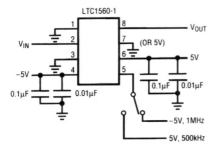

This filter uses a Linear Technology P/N LTC1560-1 to implement a switchable LP filter.

LINEAR TECHNOLOGY

Fig. 35-5

ACTIVE FILTER

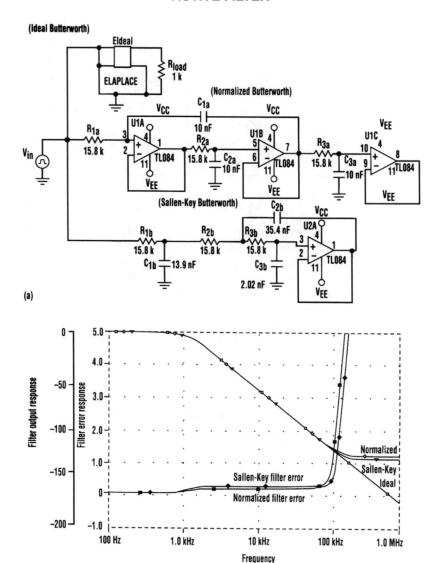

(a)

(b)

Fig. 35-6

ACTIVE FILTER *(Cont.)*

Equal-value components can be quite an advantage in filter designs when the total costs associated with the procurement, stocking, and assembly of the filter are considered. For instance, the Butterworth active third-order low-pass filter (middle) uses equal-value resistors and capacitors. This feature normalizes the filter's 3-dB corner frequency to $1/RC$ (in radians) for both low-pass and high-pass designs. The graphs in (*b*) are plots of the ideal, normalized, and Sallen-Key low-pass filter frequency-domain magnitude and error responses. Note how both the normalized and Sallen-Key filters follow the ideal response well into the stopband. The error plots were created by plotting the difference between the real and ideal filter responses. The plots indicate that the normalized filter achieves performance results that are equal to those of the Sallen-Key low-pass filter.

NOTCH FILTER CIRCUIT

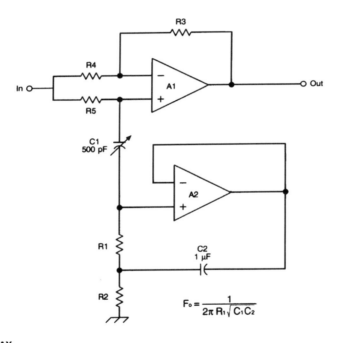

$$F_0 = \frac{1}{2\pi R_1 \sqrt{C_1 C_2}}$$

Fig. 35-7

This circuit is basically a noninverting amplifier with positive feedback. The negative feedback path is set to unity by a direct connection between the output terminal of the operational amplifier and the inverting (−) input. A frequency-selective RC network connects the output to the noninverting (+) input, providing some positive feedback. The center frequency is

$$F_c = \tfrac{1}{2} \pi R_1 (C_1 C_2)^{1/2}$$

where F_c is the center frequency in hertz, R_1 is in ohms, and C_1 and C_2 are in farads.

CAPACITORLESS NOTCH FILTER

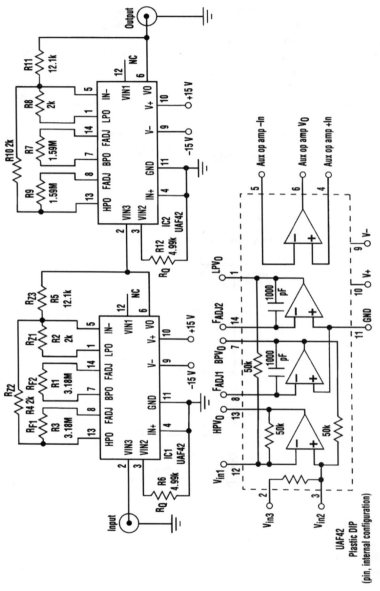

Fig. 35-8

ELECTRONIC DESIGN

CAPACITORLESS NOTCH FILTER *(Cont.)*

The notch frequency for the filter is set by

$$f_{notch} + [(A_{LP}/A_{HP}) \times (R_{Z2}/R_{Z1})]^{1/2} \times f_0$$

where A_{LP} and A_{HP} are the gain from input to low-pass output at $f=0$ Hz, and the gain from input to high-pass output at $f \ll f_0$, respectively. Typically $(A_{LP}/A_{HP}) \times (R_{Z2}/R_{Z1}) = 1$; therefore, $f_{notch} = f_0$, and is given by

$$f_0 = 1/(2\pi)(R_F C)$$

where $R_F = R_{F1} = R_{F2}$, and $C = C_1 = C_2$. The -3-dB bandwidth is determined by the following relation: $BW_{-3dB} = f_{notch}/Q$, where $BW_{-3dB} = f_H - f_L$. The Q of the filter affects the passband gain (which should be adjusted to unity) and is related to the ratio of the resistances R_{Z3} to R_{Z1} and R_{Z3} to R_{Z2}. In other words, $Q = (R_{Z3}/R_{Z1}) = (R_{Z3}/R_{Z2})$. Q also is related to R_Q by the following relation: $R_Q = (25 \text{ k}\Omega/Q - 1)$.

IF BANDPASS FILTER SWITCH

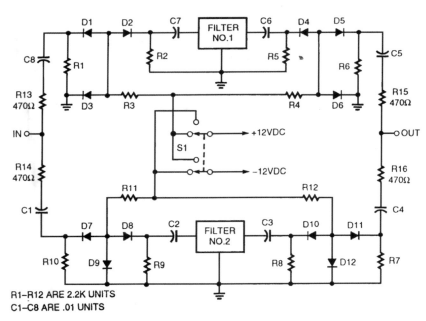

R1–R12 ARE 2.2K UNITS
C1–C8 ARE .01 UNITS

POPULAR ELECTRONICS

Fig. 35-9

Selecting IF bandpass filters via series/shunt PIN-diode switching can be accomplished with this circuit. Diodes can be MV3404 or similar types.

SIMPLE ALL-PASS FILTER

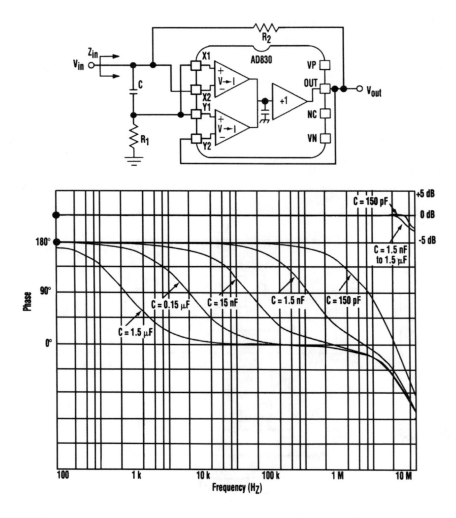

Fig. 35-10

A very simple all-pass implementation can be realized with an active-feedback amplifier like the AD830 or the LTC1193. R1 and C set the filter's actual transfer function, while R2 is needed to provide a purely real input impedance over the frequency range of the AD830 (necessary for measurement reasons). The filter's two basic equations are

$$Z_{\text{in}}(s) = \frac{R_2(1+sCR_1)}{2\left(1+\dfrac{sCR_2}{2}\right)}$$

and

SIMPLE ALL-PASS FILTER *(Cont.)*

$$\frac{V_{out}(s)}{V_{in}} = \frac{-1 - sCR_1}{1 + sCR_1}$$

from which we can see that the magnitude is constant:

$$\frac{V_{out}(\omega)}{V_{in}} = 1$$

and the phase of V_{out}/V_{in} as a function of ω is

$$180° - 2\tan^{-1}(\omega CR_1)$$

From the first equation, it's clear that for Z_{in} to be purely real, R_1 has to be equal to $R_2/2$, which implies $Z_{in}(\omega) = R_1$. Once C is chosen, R_1 and R_2 can be picked according to the termination and required phase shift. The graph shows the circuit's performance for $V_s = \pm 15$ V, $R_1 = 100$, $R_2 = 200$, and values of C from 1.5 µF to 150 pF with 90° phase shifts at one-decade increments up to 10 MHz.

SUBSONIC FILTER

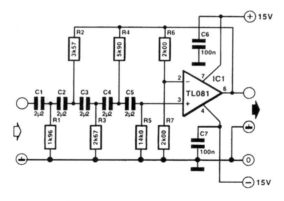

ELEKTOR ELECTRONICS

Fig. 35-11

This filter is a fifth-order high-pass section that provides 1 dB attenuation at 20 Hz. Below that, however, the response drops off very steeply; the −3-dB point is at 17.3 Hz, and at 13.6 Hz, the attenuation is 10 dB. Note that it is important that C1 to C5 are within 1 percent of one another. Their individual tolerance is not so important because that merely affects the cutoff point. However, mutual deviation adversely affects the shape of the response, which should be a Butterworth characteristic as specified. All resistors are 1-percent types.

600- TO 3000-Hz TUNABLE NOTCH-FILTER CIRCUIT

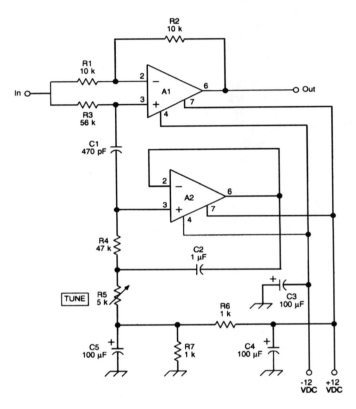

Fig. 35-12

This figure shows a notch filter that will tune roughly from 600 Hz to 3 kHz; it has been used by ham and SWL builders for a number of years. It is used for notching out unwanted CW stations or for notching out heterodynes in receiver outputs. Insert this filter between the headphone output of the receiver and a power amplifier stage.

LOW-PASS FILTER PHASE CORRECTOR

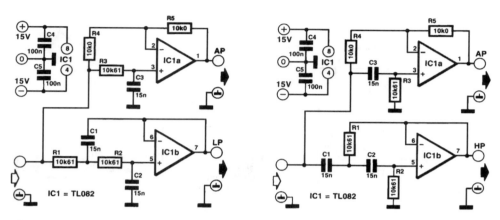

ELEKTOR ELECTRONICS

Fig. 35-13

In some applications, it might be desired, or even be essential, that the bandwidth of the audio signal be limited, but that the phase relationship with the original signal be retained. A surround-sound encoder is a good example of this. The requirement can be met by combining the low-pass filter with an all-pass section and having the filtered signal compared with the signal corrected by the all-pass network. As it happens, the phase transfer of a first-order all-pass filter is exactly the same as that of a second-order critically damped network. The design of such a combination is shown in Figure 1. In this, the all-pass network is based on IC1a and the low-pass section on IC1b. The −6-dB cutoff point is at exactly 1 kHz, and the −3-dB rolloff is at 642 Hz.

CERAMIC FILTER INTERFACING

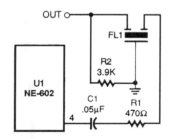

Ceramic or mechanical filters can be used to provide a frequency-selective output.

POPULAR ELECTRONICS

Fig. 35-14

ACTIVE HP FILTER

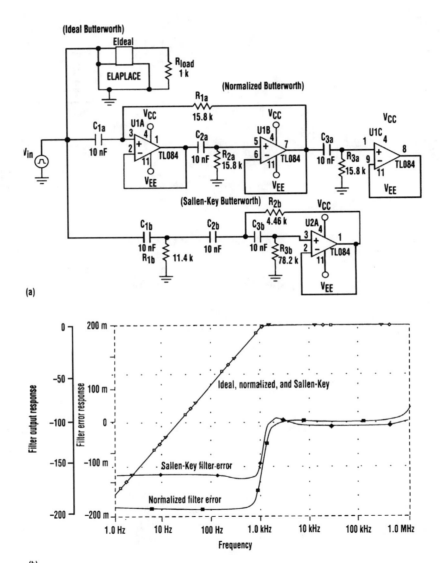

(a)

(b)

Fig. 35-15

ACTIVE HP FILTER *(Cont.)*

Interchanging the resistors and capacitors transforms the normalized low-pass filter into a high-pass filter with the same corner frequency (*a*). Notice that the Sallen-Key filter must be modified according to impedance levels at each node. This yields a filter with equal-value capacitors and unequal-value resistors, an improvement over the traditional low-pass design of equal-value resistors and unequal-value capacitors. The graphs in (*b*) indicate that the normalized high-pass filter compares favorably with the Sallen-Key filter in high-pass applications.

MULTIPLE-FEEDBACK BANDPASS ACTIVE FILTER

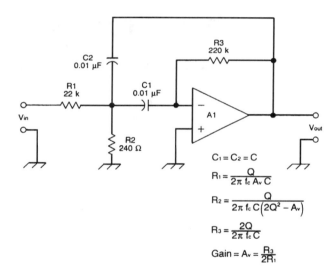

$$C_1 = C_2 = C$$

$$R_1 = \frac{Q}{2\pi\, f_c\, A_v\, C}$$

$$R_2 = \frac{Q}{2\pi\, f_c\, C\left(2Q^2 - A_v\right)}$$

$$R_3 = \frac{2Q}{2\pi\, f_c\, C}$$

$$\text{Gain} = A_v = \frac{R_3}{2R_1}$$

Fig. 35-16

This is a multiple-feedback-path (MFP) bandpass filter.

36

Flasher Circuits

The sources of the following circuits are contained in the Sources section, which begins on page 1043. The figure number in the box of each circuit correlates to the entry in the Sources section.

AC LIGHT FLASHER

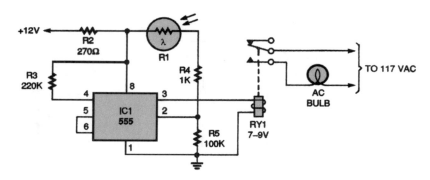

POPULAR ELECTRONICS

Fig. 36-1

This circuit uses a 555 timer that is configured in an unusual way in that trigger pin 2 is activated by the voltage from a photocell to sense when the lamp is lit. The relay time constants affect the flash rate.

XENON FLASHTUBE CIRCUIT

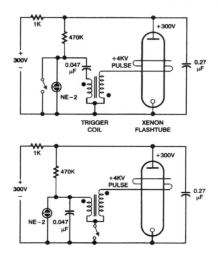

These are basic circuits for firing Xenon flashtubes. Which circuit is applicable depends on trigger coil polarity.

ELECTRONICS NOW

Fig. 36-2

NOVELTY LED FLASHER

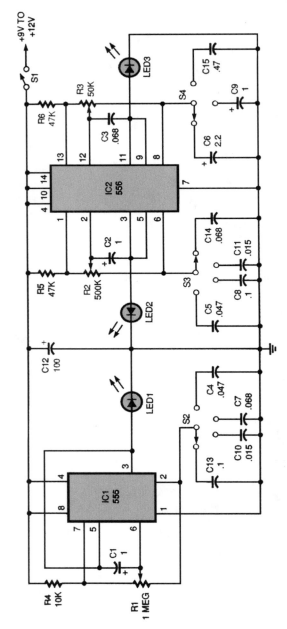

Fig. 36-3

POPULAR ELECTRONICS

The circuit contains three oscillators, one based on a 555 (IC1) and the others centered around a 556 (IC2). When S1 is on, each oscillator's LED (LED1, LED2, or LED3) will flash rhythmically, according to the setting of its potentiometer (R1, R2, or R3), and the capacitor selected by its switch (S2, S3, or S4). Each oscillator output is capable of driving up to 20 jumbo LEDs connected in parallel. Arrange the LEDs to suit your taste. The circuit will operate on a battery or any 9- to 12-Vdc source.

MODEL FIRE ENGINE FLASHER CIRCUIT

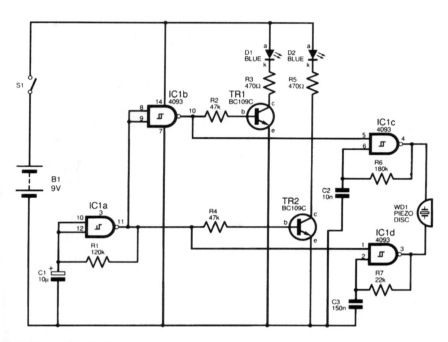

EVERYDAY PRACTICAL ELECTRONICS

Fig. 36-4

This circuit was designed to add both a blue flashing light sequence and a two-tone siren effect to a model fire engine. IC1 is a quad NAND Schmitt trigger, and IC1a provides a low-frequency waveform, which is inverted by IC1b. Complementary outputs are obtained, which drive transistors TR1 and TR2. These driver transistors operate D1 and D2, which are two blue LEDs. The model had a blue translucent molding on its roof and yielded a surprisingly realistic effect. The outputs from IC1a and IC1b are also used to control gated astables IC1c and IC1d, which are set to oscillate at different frequencies to simulate a typical British two-tone siren, sounded by WD1, a piezo disk. Battery B1 should be an alkaline or rechargeable 9-V type because the current consumption is relatively high. S1 was a slide switch fitted on the rear of the vehicle. The sounder is optional.

RAINBOW FLASHER

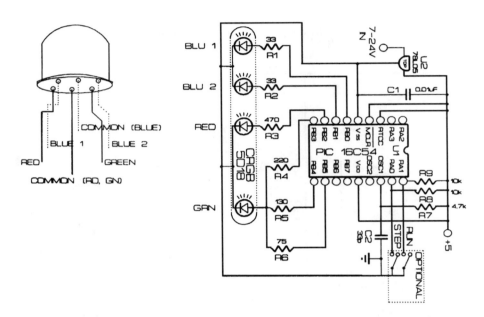

NUTS AND VOLTS

Fig. 36-5

This device flashes a multicolor RGB LED in various colors, in a sequence and speed determined by the PIC 16C54 software. The LED is by Cree Electronics.

SINGLE-LED FLASHER

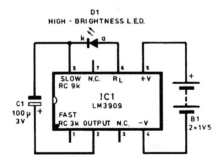

This is a circuit diagram of an LED flasher with few components.

EVERYDAY PRACTICAL ELECTRONICS

Fig. 36-6

12-V LED FLASHER

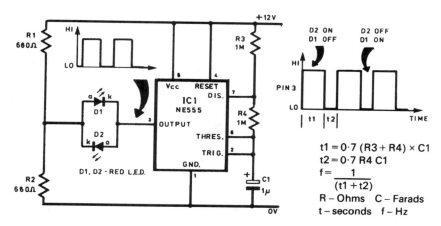

Fig. 36-7

This is the circuit diagram of an LED flasher suitable for 12-V operation. The timing formulas are also shown.

MODEL AIRCRAFT LED FLASHER

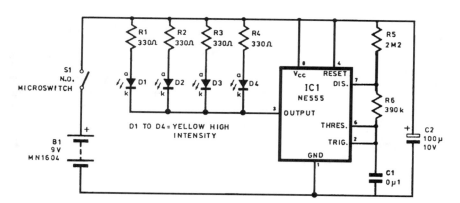

Fig. 36-8

This circuit was used to simulate machine gun firing on a model aircraft, but it can also be used for other applications.

AUDIO-DRIVEN XENON TUBE FLASH CIRCUIT

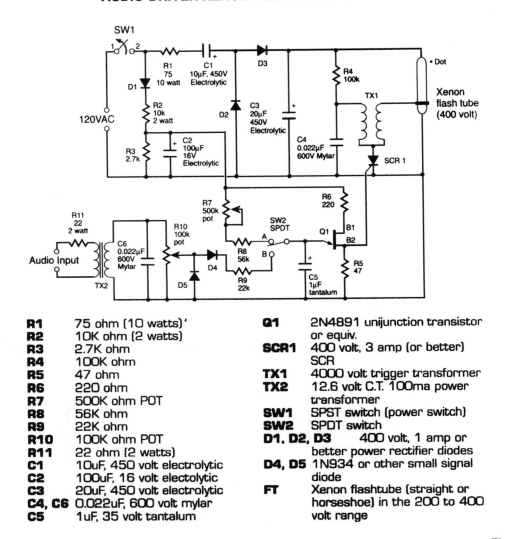

R1	75 ohm (10 watts)*	**Q1**	2N4891 unijunction transistor or equiv.
R2	10K ohm (2 watts)		
R3	2.7K ohm	**SCR1**	400 volt, 3 amp (or better) SCR
R4	100K ohm		
R5	47 ohm	**TX1**	4000 volt trigger transformer
R6	220 ohm	**TX2**	12.6 volt C.T. 100ma power transformer
R7	500K ohm POT		
R8	56K ohm	**SW1**	SPST switch (power switch)
R9	22K ohm	**SW2**	SPDT switch
R10	100K ohm POT	**D1, D2, D3**	400 volt, 1 amp or better power rectifier diodes
R11	22 ohm (2 watts)		
C1	10uF, 450 volt electrolytic	**D4, D5**	1N934 or other small signal diode
C2	100uF, 16 volt electolytic		
C3	20uF, 450 volt electrolytic	**FT**	Xenon flashtube (straight or horseshoe) in the 200 to 400 volt range
C4, C6	0.022uF, 600 volt mylar		
C5	1uF, 35 volt tantalum		

Fig. 36-9

This circuit uses a Xenon flashtube to create a very real lightning effect. It can be driven in one of two ways: by sound input or by straight repeated flashes, depending on where switch 2 is positioned. When producing straight flashes, the rate is determined by pot R7. If the switch is set to be triggered by sound, the sensitivity is determined by the level of the sound driving the transformer

AUDIO-DRIVEN XENON TUBE FLASH CIRCUIT *(Cont.)*

TX2 and the setting of pot R10. This circuit uses high voltages (over 4000 V for the trigger), and extreme care should be taken by anyone that builds it. The high-voltage capacitors can hold dangerous voltages for some time (even after the circuit has been unplugged). Strong caution concerning part polarity is highly urged.

12-V LED FLASHER

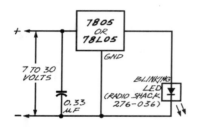

The figure shows a 12-V LED flasher configuration.

ELECTRONICS NOW　　　　*Fig. 36-10*

MULTIPLE-LED FLASHER

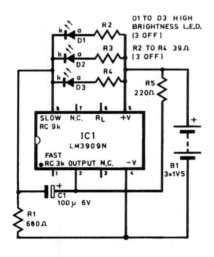

This is a circuit diagram for driving multiple LEDs from a single LM3909 flasher chip.

EVERYDAY PRACTICAL ELECTRONICS　　　　*Fig. 36-11*

MULTIPLE FLASHING LED LIGHT STRING

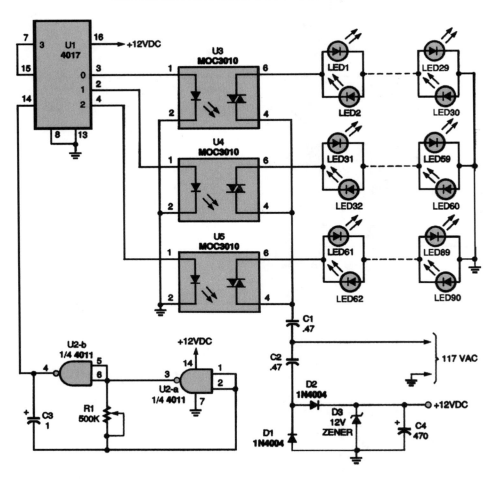

POPULAR ELECTRONICS

Fig. 36-12

This circuit offers up to 10 strings of LEDs that turn on one row at a time in a sequential manner; for the sake of space and simplicity, only three strings are shown. The LED pairs could be all one color, mixed red and green, or any combination of colors. C2, C4, and D1 to D3 make up a 12-Vdc power supply for the five ICs. Two gates of a 4011 quad two-input NAND gate, U2-a and U2-b, are connected in a low-frequency oscillator circuit. Resistor R1 controls the oscillator's frequency. The oscillator supplies a clock output to U1, a 4017 decade counter/divider, which operates as a counter that can be programmed to count from 0 to 9. With each clock pulse, the 4017 makes a single step up in count from 0. The first output, at pin 3, supplies voltage to the LED in U3, a 3010 op-

MULTIPLE FLASHING LED LIGHT STRING *(Cont.)*

tocoupler, turning on the IC's triac and lighting the first string of LEDs. The next clock pulse steps the 4017 to the next output, at pin 2, and repeats the sequence for that string. As each new light string turns on, the preceding string turns off, giving the effect of a climbing string of lights. Changing the setting of R1 will change the rate at which the strings turn on and off. If you want to add additional light strings beyond what is shown in the figure, simply duplicate the circuitry for the light strings as shown and connect to the appropriate output of the 4017, U1. You will also have to connect the first unused output to pin 15. If you use the maximum of 10 strings, pin 15 should be connected to ground.

LIGHT FLASHER

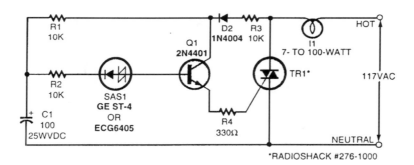

*RADIOSHACK #276-1000

POPULAR ELECTRONICS   *Fig. 36-13*

Shown here is a simple flasher circuit. With the component values shown, the flash rate is approximately once per second. The incandescent-lamp load glows at half brightness for about one-third of the total flasher period and is off for the remaining two-thirds. Electrolytic capacitor C1 charges during the positive half-cycle of the ac waveform through R1, R3, and D2. When the voltage across the capacitor reaches the break-over voltage of the silicon asymmetrical switch (SAS1), the capacitor starts to discharge through R2, SAS1, Q1, R4, and the triac. Emitter follower Q1 is driven by the discharge current from C1, and it, in turn, provides gate drive for the triac. Thus, the triac conducts and the light glows while C1 discharges. The lamp goes dark when C1 is depleted of charge and remains dark until the ac power waveform goes positive again and charges the capacitor sufficiently. The triac should be triggered into conduction by a gate current of no more than 5 mA. The flash rate can be varied by changing the value of capacitor C1. Using more capacitance results in a slower flash rate, and less capacitance results in a faster flash rate.

1.5-V LED FLASHER

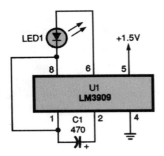

The LM3909 is one of the easiest-to-use LED flashers around. It can run for years using a standard D cell.

POPULAR ELECTRONICS

Fig. 36-14

THRIFTY LED FLASHER

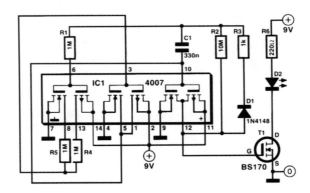

ELEKTOR ELECTRONICS

Fig. 36-15

The dual complementary pair of switching FETs and inverter contained in a CD4007 CMOS IC enables an LED flasher to be made that uses very little energy. The IC is arranged as a three-inverter oscillator. Resistors R4 and R5 in series with the drains of one pair of FETs ensure that the drive current for the following pair of FETs is tiny. The high time of the oscillator is determined by network R3-C1, and its low time by R2-C1 (D1 is then cut off so that R3 is inactive). The LED is provided with current during the high time of the oscillator by T1. The level of this current is determined by R6. The values of R_2, R_3, and C_1 cause an LED OFF time of 1 s and an ON time of 1 ms. Because the high-efficiency diode draws a current of 30 mA, its lighting will be clearly visible. A standard 9-V battery will give continuous operation for about three years.

NEON LAMP FLASHER

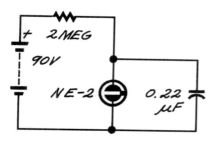

The neon lamp has negative dynamic resistance—the voltage across it falls while conduction is increasing. As a result, it flashes on and off.

ELECTRONICS NOW

Fig. 36-16

1-s FLASHER

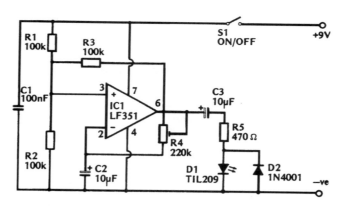

ELECTRONIC EXPERIMENTERS HANDBOOK

Fig. 36-17

The circuit diagram for the 1-s flasher is based on an operational amplifier that is biased by R1, R2, and R3 to act as a form of Schmitt trigger. The output goes to the low state if the inverting input is taken above $\frac{2}{3}$V+, and high if it is taken below $\frac{1}{3}$V+. The output, therefore, goes high initially, but C2 soon charges to $\frac{2}{3}$V+ via R4, and then the output, goes low. Capacitor C2 then discharges to $\frac{1}{3}$V+ via R4, sending the output high again, and producing continuous oscillation. Resistor R4 is adjusted to give an operating frequency of 1 Hz. The 1-s flasher can be calibrated against a watch or clock with a second hand by empirical means. The output of IC1 is coupled to the LED indicator, D1, by way of dc blocking capacitor C3 and current-limiting resistor R5, and the LED is briefly pulsed on as the output voltage swings positive. Diode D2 ensures that there is both a charge and a discharge path for C3 so that the output signal is properly coupled to D1. The current consumption of the unit is about 2 mA.

VARIABLE-FREQUENCY LED FLASHER

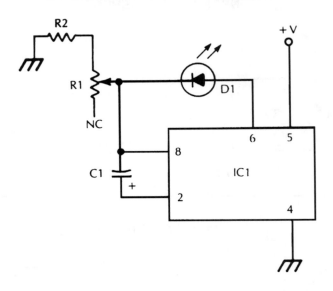

IC1	LM3909 LED flasher/oscillator IC
D1	LED
C1	330 µF 5 V electrolytic capacitor
R1	10 kΩ potentiometer
R2	330 Ω ¼ W 5% resistor

TAB BOOKS   *Fig. 36-18*

R1 varies the flash rate, and R2 limits the minimum resistance in the potentiometer circuit to prevent damage to D1 or IC1.

37

Fluorescent Lamp Circuits

The sources of the following circuits are contained in the Sources section, which begins on page 1043. The figure number in the box of each circuit correlates to the entry in the Sources section.

Cold-Cathode Fluorescent Lamp Driver
Battery-Operated Fluorescent Light

COLD-CATHODE FLUORESCENT LAMP DRIVER

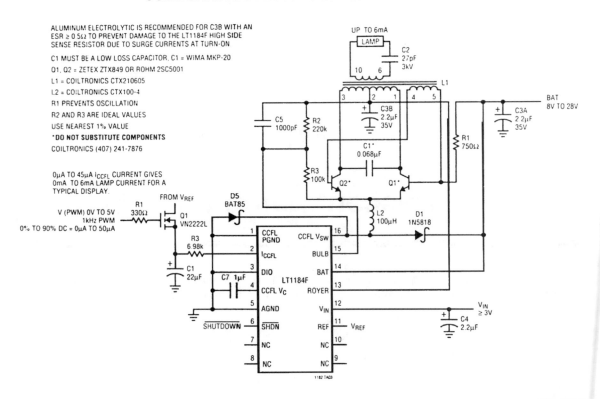

ALUMINUM ELECTROLYTIC IS RECOMMENDED FOR C3B WITH AN
ESR ≥ 0.5Ω TO PREVENT DAMAGE TO THE LT1184F HIGH SIDE
SENSE RESISTOR DUE TO SURGE CURRENTS AT TURN-ON

C1 MUST BE A LOW LOSS CAPACITOR. C1 = WIMA MKP-20

Q1, Q2 = ZETEX ZTX849 OR ROHM 2SC5001

L1 = COILTRONICS CTX210605

L2 = COILTRONICS CTX100-4

R1 PREVENTS OSCILLATION

R2 AND R3 ARE IDEAL VALUES

USE NEAREST 1% VALUE

*DO NOT SUBSTITUTE COMPONENTS

COILTRONICS (407) 241-7876

0µA TO 45µA I_{CCFL} CURRENT GIVES
0mA TO 6mA LAMP CURRENT FOR A
TYPICAL DISPLAY.

LINEAR TECHNOLOGY POWER SOLUTIONS

Fig. 37-1

In this circuit, the lamp is driven sinusoidally, minimizing RF emissions in sensitive portable applications. Lamp intensity is controlled smoothly from zero to full brightness with no hysteresis or "pop on." This floating bulb circuit configuration extends the illumination range for the bulb because parasitic bulb-to-display-frame capacitive losses are minimized. The feedback signal is generated by monitoring the primary-side Royer converter current between the BAT and Royer pins. The LT1184F current-mode switching regulator and L2 provide an average current to Q1 and Q2, which form a Royer-class converter along with L1 and C1. The lamp is driven by L1's secondary. Feedback to the LT1184F is provided on the primary side of L1 for floating bulb configurations, whereas feedback in the grounded configuration is provided by sensing one-half of the average bulb current. The oscillator frequency is 200 kHz, which minimizes the size of the required magnetics.

BATTERY-OPERATED FLUORESCENT LIGHT

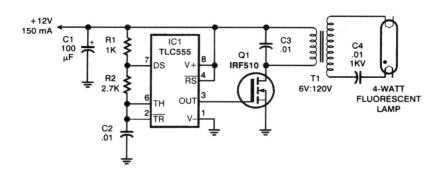

ELECTRONICS NOW

Fig. 37-2

The circuit consists of a 20-kHz oscillator, a switching transistor to amplify its output, and a step-up transformer. A 120- to 6-V power transformer is connected backward, using half of the 12-V side for 6 V. In the circuit, the transformer is working at considerably more than its rated voltage, but the high frequency keeps it from saturating. Although the lamp does not glow at full brightness, the circuit is energy-efficient, requiring only 150 mA.

38

Function-Generator Circuits

The sources of the following circuits are contained in the Sources section, which begins on page 1043. The figure number in the box of each circuit correlates to the entry in the Sources section.

MAX038 HIGH-SPEED FUNCTION-GENERATOR CIRCUIT

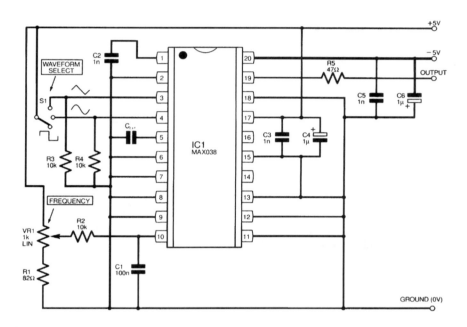

EVERYDAY PRACTICAL ELECTRONICS

Fig. 38-1

For enthusiasts who would like to experiment with this device for themselves, a simple circuit to get it up and running is shown. The only real problem with the MAX038 is inherent in its sheer speed. Maxim suggests careful layout on a double-sided ground-plane PC board for best results. In practice, a single-sided PC board seems to work well, provided it has plenty of copper areas at ground potential, good decoupling, and guard tracks around signal paths carrying rectangular waveforms, especially that from the sync output. The following component recommendations are offered:

Resistors—0.6 W, 1-percent metal film
C1, C2, C3, and C5—ceramic disk
C4 and C6—tantalum bead, 35 V
C_{ext}—as required for frequency, between 47pF and 47μF, polyester or polystyrene

FM DEMODULATOR

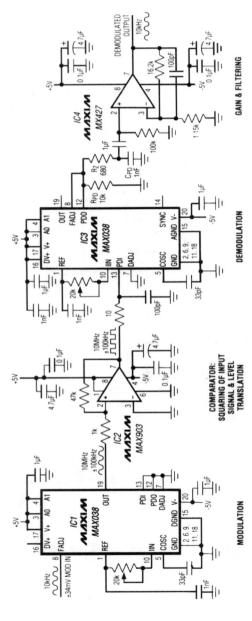

Fig. 38-2

MAXIM ENGINEERING JOURNAL

The frequencies at IC3's phase-detector output are the sum and difference of the frequencies at PDI and OUT. Thus, with appropriate cutoff frequency and gain, the low-pass filter (IC4) passes only the original 10-kHz signal to the demodulated output. The pole for this filter is set by the 16.2-kΩ and 100-pF components. The frequency response for IC3's PLL is set by R_{PD}, C_{PD}, and R_Z. When the loop is in lock, PDI is in approximate phase quadrature with the output signal. Also, when in lock, the duty cycle at PDO is 50 percent, and PDO's average output current is 250 μA. The current sink at FADJ demands a constant 250 μA, so PDO outputs above and below that level develop a bipolar error voltage across R_{PD} that drives the FADJ voltage input. *Note:* The MAX038's internal phase detector is a phase-only detector, producing a PLL whose frequency-capture range is limited by the bandwidth of its loop filter. For wider-range applications, consider an external phase-frequency detector.

SYNCHRONIZED MAX038 FUNCTION GENERATORS

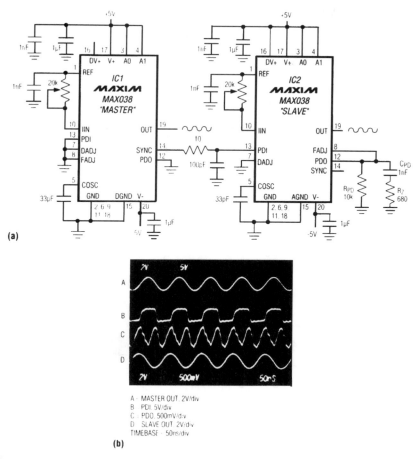

(a)

A - MASTER OUT, 2V/div
B - PDI, 5V/div
C - PDO, 500mV/div
D - SLAVE OUT, 2V/div
TIMEBASE - 50ns/div

(b)

Fig. 38-3

The MAX038's internal phase detector is intended primarily for use in phase-locked-loop (PLL) configurations. In (a), for example, the phase detector in IC2 enables that device to synchronize its operation with that of IC1. You connect the applied reference signal to IC2's TTL/CMOS-compatible phase-detector input (PDI) and connect the phase-detector output (PDO) to the input (FADJ) of the internal voltage-controlled oscillator. PDO is the output of an exclusive-OR gate—a mixer—which produces rectangular current waveforms at frequencies equal to the sum and difference of the PDI frequency and the MAX038 output frequency. These waveforms are integrated by C_{PD} to form a triangle-wave voltage output at PDO (b). The 10-Ω/100-pF pair at PDI limits that pin's rate of rise to 10 ns.

BENCHTOP FUNCTION GENERATOR WITH BUILT-IN COUNTER

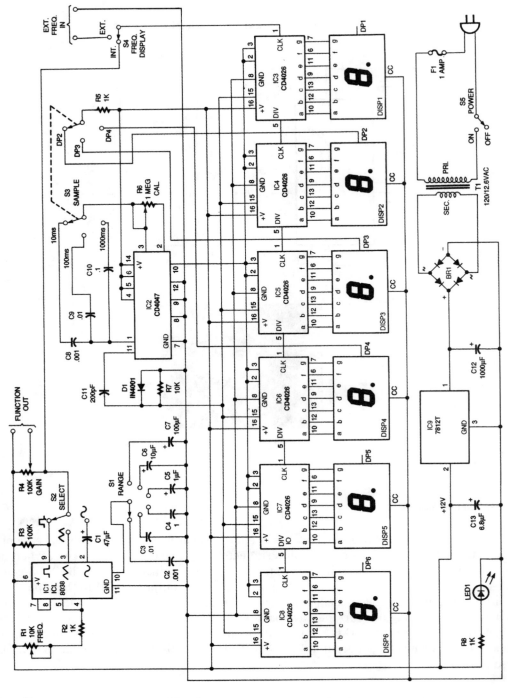

Fig. 38-4

ELECTRONICS NOW

BENCHTOP FUNCTION GENERATOR WITH BUILT-IN COUNTER *(Cont.)*

This circuit will produce sine, square, and triangle waves from 0.1 Hz to 1 MHz and has a counter which will read the frequency of the function generator or an external signal of a few volts peak-to-peak that will drive the CMOS counter.

PLL WITH DIVIDE-BY-*N* CIRCUIT

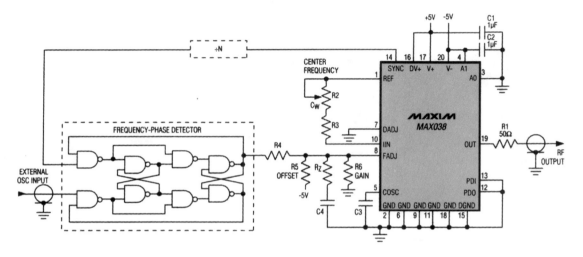

Fig. 38-5

The MAX038 function generator IC can be used in a PLL system. To gain the advantages of a wider capture range and an optional divide-by-*N* circuit (which allows the PLL to lock onto arbitrary multiples of the applied frequency), you can introduce an external frequency-phase detector, such as the 74HC4046 or the discrete-gate version shown. Unlike phase detectors, which can lock to harmonics of the applied signal, the frequency-phase detector locks only to the fundamental. In the absence of an applied frequency, its output assumes a positive dc voltage (logic 1) that drives the RF output to the lower end of its range as determined by resistors R4 to R6. These resistors also determine the frequency range over which the PLL can achieve lock. Again, R4 to R6, C4, and R_z determine the PLL's dynamic performance.

MAX038 FUNCTION GENERATOR

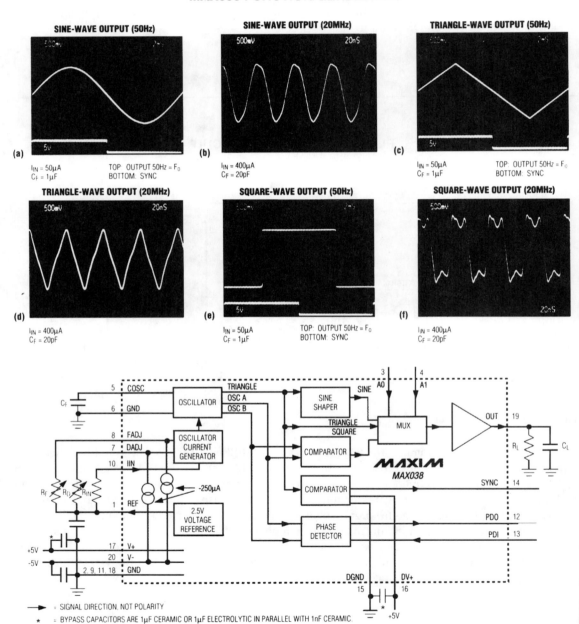

SINE-WAVE OUTPUT (50Hz)

(a)

$I_{IN} = 50\mu A$
$C_F = 1\mu F$

TOP: OUTPUT 50Hz = F_0
BOTTOM: SYNC

SINE-WAVE OUTPUT (20MHz)

(b)

$I_{IN} = 400\mu A$
$C_F = 20pF$

TRIANGLE-WAVE OUTPUT (50Hz)

(c)

$I_{IN} = 50\mu A$
$C_F = 1\mu F$

TOP: OUTPUT 50Hz = F_0
BOTTOM: SYNC

TRIANGLE-WAVE OUTPUT (20MHz)

(d)

$I_{IN} = 400\mu A$
$C_F = 20pF$

SQUARE-WAVE OUTPUT (50Hz)

(e)

$I_{IN} = 50\mu A$
$C_F = 1\mu F$

TOP: OUTPUT 50Hz = F_0
BOTTOM: SYNC

SQUARE-WAVE OUTPUT (20MHz)

(f)

$I_{IN} = 400\mu A$
$C_F = 20pF$

⟶ = SIGNAL DIRECTION, NOT POLARITY

* = BYPASS CAPACITORS ARE 1µF CERAMIC OR 1µF ELECTROLYTIC IN PARALLEL WITH 1nF CERAMIC.

MAX038 FUNCTION GENERATOR *(Cont.)*

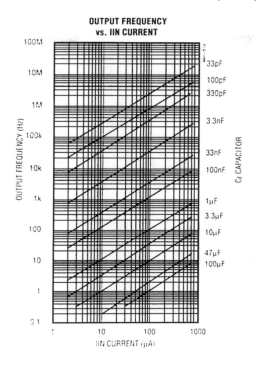

OUTPUT FREQUENCY
vs. IIN CURRENT

Fig. 38-6

The MAX038 is a precision, high-frequency function generator that produces accurate sine, square, triangle, sawtooth, and pulse waveforms with a minimum of external components. The internal 2.5-V reference (plus an external capacitor and potentiometer) lets you vary the signal frequency from 0.1 Hz to 20 MHz. An applied ±2.3-V control signal varies the duty cycle between 10 and 90 percent, enabling the generation of sawtooth waveforms and pulse-width modulation. A second frequency-control input—used primarily as a VCO input in phase-locked-loop applications—provides ±70 percent of fine control. This capability also enables the generation of frequency sweeps and frequency modulation. The frequency and duty-cycle controls have minimal interaction with each other. All output amplitudes are 2 V_{p-p}, symmetrical about ground. The low-impedance output terminal delivers as much as ±20 mA, and a two-bit code applied to the TTL-compatible A0 and A1 inputs selects the sine, square, or triangle output waveform:

A0	A1	WAVEFORM
X	1	Sine wave
0	0	Square wave
1	0	Triangle wave

X=Don't Care.

FREQUENCY SYNTHESIZER

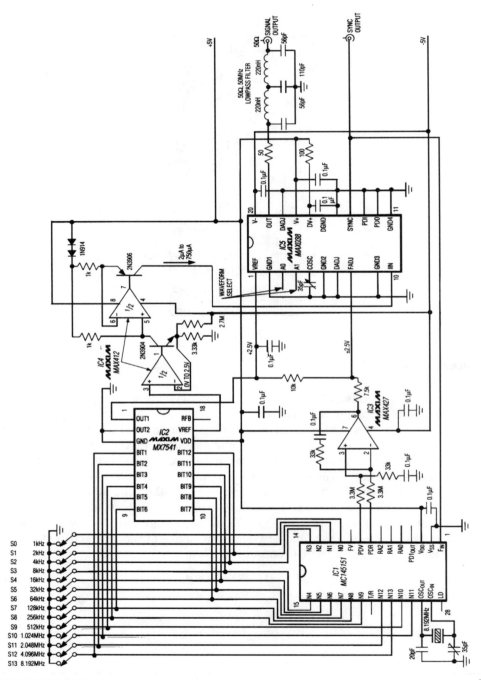

FREQUENCY SYNTHESIZER *(Cont.)*

The MAX038 and four other ICs can form a crystal-controlled, digitally programmed frequency synthesizer that produces accurate sine, square, or triangle waves in 1-kHz increments over the range 8 kHz to 16.383 MHz. Each of the 14 manual switches (when open) makes the listed contribution to output frequency: Opening only S0, S1, and S8, for example, produces an output of 259 kHz. The switches generate a 14-bit digital word that is applied in parallel to the D/A converter (IC2) and a $\div N$ circuit in IC1. IC1 also includes a crystal-controlled oscillator and high-speed phase detector, which form a phase-locked loop with the voltage-controlled oscillator in IC5. The DAC and dual op amp (IC4) produce a 2- to 750-μA current that forces a coarse setting of the IC5 output frequency—sufficient to bring it within capture range of the PLL. This loop, in which the phase detector in IC1 compares IC5's SYNC output with the crystal-oscillator frequency divided by N, produces differential-phase information at PDV and PDR. IC3 then filters and converts this information to a ±2.5-V single-ended signal, which, when summed with an offset and applied to FADJ, forces the signal output frequency to the exact value set by the switches.

MAX038 FUNCTION GENERATOR DIGITAL CONTROL

MAX038 FUNCTION GENERATOR DIGITAL CONTROL *(Cont.)*

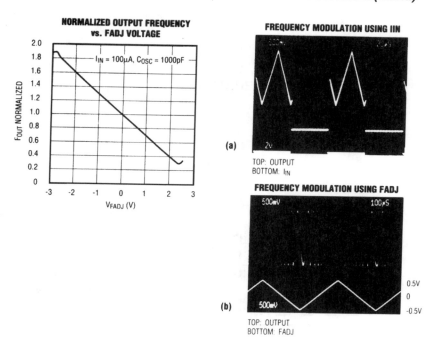

NORMALIZED OUTPUT FREQUENCY vs. FADJ VOLTAGE

I_{IN} = 100µA, C_{OSC} = 1000pF

F_{OUT} NORMALIZED

V_{FADJ} (V)

(a)

FREQUENCY MODULATION USING IIN

TOP: OUTPUT
BOTTOM: I_{IN}

(b)

FREQUENCY MODULATION USING FADJ

TOP: OUTPUT
BOTTOM: FADJ

Fig. 38-8

To adjust the frequency digitally, connect a voltage-output DAC to IIN via a series resistor, as shown. The converter output ranges from 0 V at zero to 2.5(255/256) V at full scale. Current injected by the converter into IIN, therefore, ranges from 0 to 748 µA. The 2.5-V reference and 1.2-Ω resistor inject a constant 2 µA, so (by superposition) the net current into IIN ranges from 2 µA (at a code of 0000 0000) to 750 µA (at 1111 1111). The quad DAC IC operates from 5 V or ±5 V. As described below, it can also provide digital control of FADJ and DADJ. For fine adjustments (±70 percent), apply a control voltage in the range ±2.3 V to the frequency adjust (FADJ) terminal. Both FADJ and IIN have wide bandwidths that allow the output frequency to be modulated at a maximum rate of about 2 MHz. As the more linear input, IIN is preferred for open-loop frequency use in a phase-locked loop. For digital control of FADJ, configure a DAC and external op amp (as shown in the figure) to produce an output ranging from −2.3 V (0000 0000) to 2.3 V (1111 1111).

FUNCTION GENERATOR POWER BUFFER

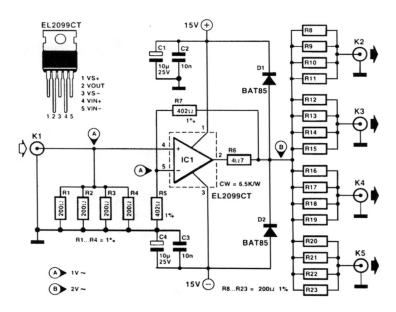

ELEKTOR ELECTRONICS

Fig. 38-9

This buffer circuit can be used as an output booster for any function generator that has to be extended in order to drive several loads. The heart of the circuit is a video distribution amplifier IC from Elantec, the EL2099CT (listed by RS Components). This interesting device has a 3-dB power bandwidth of no less than 65 MHz at a gain of ×2. Here, it is used to drive up to four 50-Ω loads at a maximum signal level in excess of 10 V_{peak}. When used for video applications, the EL2099CT can drive up to six 75-Ω loads. The gain of the amplifier is ×2; unity gain is not possible because of instability problems. The bandwidth of the circuit shown here is >10 MHz, while the output achieves a drive margin of >10 V_{peak}. Current consumption will be of the order of 200 mA.

39

Game Circuits

The sources of the following circuits are contained in the Sources section, which begins on page 1043. The figure number in the box of each circuit correlates to the entry in the Sources section.

Lockout Circuit
Heads or Tails Game Circuit
Mini Roulette
Coin-Toss Circuit

LOCKOUT CIRCUIT

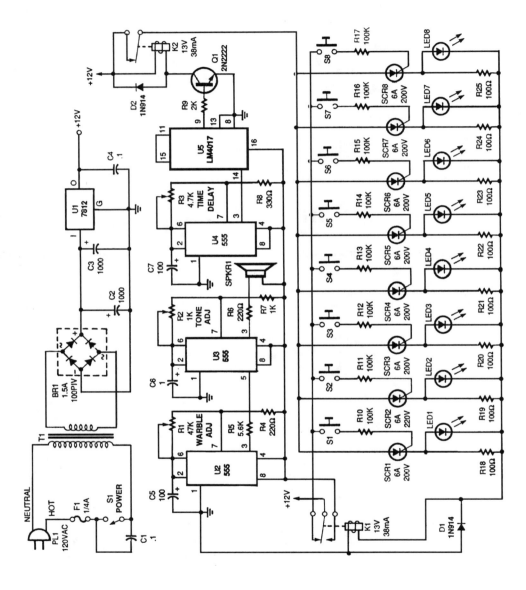

POPULAR ELECTRONICS

Fig. 39-1

LOCKOUT CIRCUIT *(Cont.)*

This circuit is used in contests, games, etc. When a contestant presses one of the buttons, an SCR triggers, lighting the associated LED and preventing anyone else from actuating the system. U2 and U3 form a warble tone generator, while U4 and U5 provide time delay.

HEADS OR TAILS GAME CIRCUIT

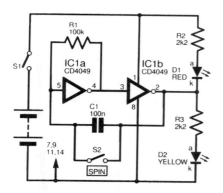

EVERYDAY PRACTICAL ELECTRONICS   *Fig. 39-2*

Designed to simulate by electronic means the tossing of a coin, the circuit is based upon a 4049 hex inverter IC, two of which are used. IC1a and IC1b are wired as an astable oscillator, which causes two LEDs (D1 and D2) to alternate rapidly, at a frequency too high to be distinguished by the naked eye. Both LEDs, therefore, appear to be constantly illuminated. When the SPIN switch S2 is closed, this has the effect of freezing the display, and the LED, which was illuminated at the instant that the switch was closed, will now be continuously alight. Opening the switch enables the oscillator once more. There is an equal chance of either LED lighting, and the circuit can be used in board games, for example, to choose which player will move first.

MINI ROULETTE

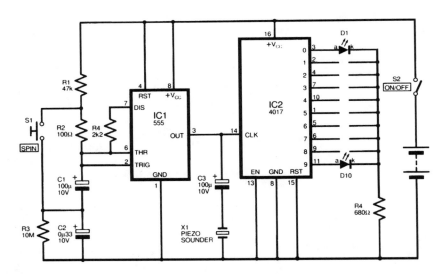

Fig. 39-3

A circuit diagram for a mini battery-powered version of roulette is shown. This circuit uses a 4017 decade counter (IC2) driving 10 LEDs. Because only one LED is ever illuminated at any one time, a common limiting resistor R4 is used. They can also be placed in alternating red/green order for added effect. The counter IC2 is clocked by IC1, a classic 555 timer connected as an astable. When the SPIN button is pressed and then released, full speed is achieved, and then the display gradually slows down until it stops on a single number. Capacitor C2 governs the oscillator speed, and resistor R3 prevents instability when the LED rotation stops. The piezo disk transducer, X1, is placed on the output of the oscillator to provide a sound effect.

COIN-TOSS CIRCUIT

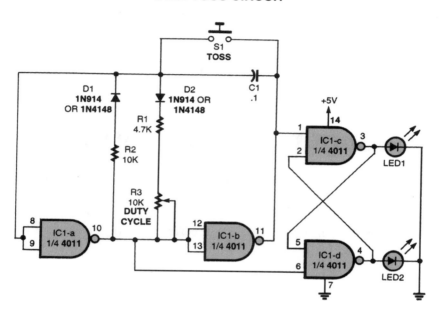

POPULAR ELECTRONICS

Fig. 39-4

Shown is a CMOS coin-toss circuit that works well and can be built with a single 4011 or 4001 CMOS IC. (Note that the IC pin numbers in the diagram apply for both the 4001 and the 4011.) Two gates form a clock, and the others make up a bistable multivibrator. With this circuit, it is necessary to adjust the 10,000-Ω potentiometer (R3) for a 50-percent duty cycle. If you don't have a scope to do this, simply measure the direct current flowing through each LED while adjusting R3. When the same current level flows through each LED, the clock will be adjusted for a 50-percent duty cycle.

40

Geiger Counter Circuits

The sources of the following circuits are contained in the Sources section, which begins on page 1043. The figure number in the box of each circuit correlates to the entry in the Sources section.

GEIGER COUNTER TO IBM INTERFACE

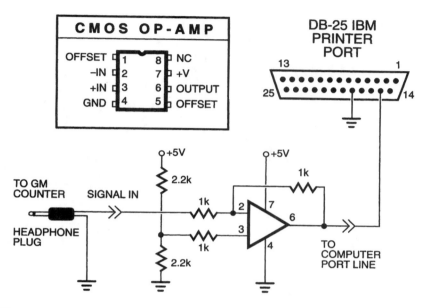

NUTS AND VOLTS

Fig. 40-1

The simplest way to connect a Geiger counter to the IBM computer is to use a headphone jack to the parallel port. The headphone jack provides a +5- to 0-V pulse for each radioactive particle detected. The signal from the Geiger counter is +5 V. When radiation is detected, the signal pulses down to 0 V for a few milliseconds before returning to +5 V. Even though the signal is in the 0- to +5-V range, it is not safe to connect wires from an external circuit or instrument directly to the parallel-port lines without buffering. The op amp operates from a single-pole +5-V power supply. In this particular case, invert the signal from the Geiger counter. Because the pulse signals are +5 to 0 V, we do not require any amplification from our op amp. Therefore, the op amp is configured as an inverting unity-gain follower. Remember that the Geiger counter outputs a steady +5 V through the headphone jack. When it detects radiation, the output pulses down to 0 V for a few milliseconds. By inverting the Geiger counter signal with the op amp, the computer reads a steady 0 V on its line and a +5-V pulse when radiation is detected.

363

GEIGER COUNTER CIRCUIT

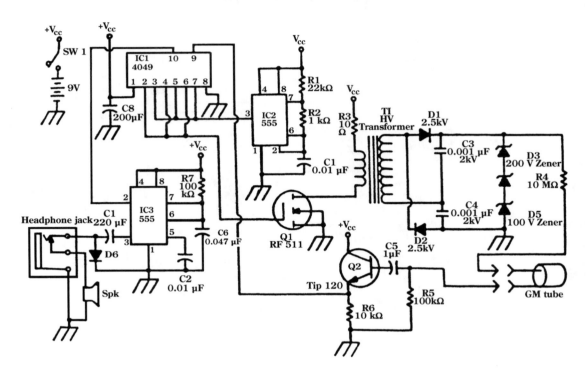

Fig. 40-2

IC2 is a 555 timer set in astable mode. The signal from IC2 is presented to three gates on the 4049 IC1. The 4049 inverts the signal to give an optimum pulse width that switches Q1 on and off. The MOSFET (Q1), in turn, switches the current to a step-up transformer (T1). The stepped-up voltage from T1 first passes through a voltage doubler; the output voltage from this section is approximately 600 to 700 V. Three zener diodes (D3, D4, and D5) are placed across the output of the voltage doubler to regulate the voltage to 500 V. This voltage is connected to the anode on the GM tube through a 10-MΩ resistor. When the GM tube detects a particle, a voltage pulse from the 100-kΩ resistor is amplified and clamped to V_{cc} via Q2, an NPN Darlington transistor. The signal from Q2 is inverted by IC1, where it acts as a trigger signal to IC3. IC3 is another 555 timer. The output of IC3 via pin 3 flashes on LED and provides a click into either the speaker or the headphones. The circuit is powered by a 9-V alkaline battery and draws about 28 mA when not detecting.

GEIGER-MUELLER TUBE CIRCUIT

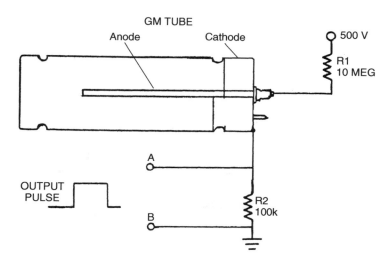

Fig. 40-3

A GM tube is useful for detecting radioactivity. The tube is constructed with a cylindrical electrode (cathode) surrounding a center electrode (anode). The tube is evacuated and filled with a neon and halogen gas mixture. A voltage potential of 500 V is applied across the tube, through a 10-MΩ current-limiting resistor (R1). The detection of radiation relies upon its ability to ionize the gas in the GM tube. The tube has an extremely high resistance when it is not in the process of detecting radioactivity. When an atom of the gas is ionized by the passage of radiation, the free electron and positive ionized atom that are created move rapidly toward the two electrodes in the GM tube. In doing so, they collide with and ionize other gas atoms, which creates a small avalanche effect. This ionization drops the resistance of the tube, allowing a sudden surge of electric current that creates a voltage across the resistor R2, which can be seen as a pulse.

41

Generator Circuits

The sources of the following circuits are contained in the Sources section, which begins on page 1043. The figure number in the box of each circuit correlates to the entry in the Sources section.

Reset Generator
Simple Wideband Noise Generator
Three-Tone Generator
Simple Frequency Synthesizer
Digital Burst Generator
Random-Number Generator
Tone-Burst Circuit
Simple Frequency Generator
Ultra-Fast Monocycle Generator
Sweep Generator
Self-Starting Data Generator
Pink-Noise Generator

RESET GENERATOR

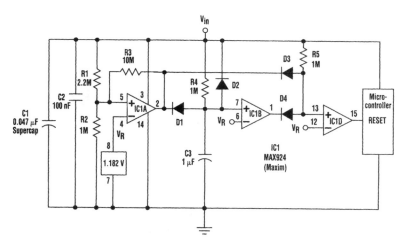

ELECTRONIC DESIGN

Fig. 41-1

This V_{cc} monitor circuit draws less than 10 µA, yet it can generate reliable RESET signals.

Very low quiescent current makes the reset generator shown suitable for use with microcontrollers that spend most of their time in the "sleep" mode. The system's power source, a charged 0.047-µF "Supercap" capacitor (C1), can go without a recharge for intervals exceeding 24 hours. The allowable supply-voltage range for the reset circuit is 2 to 11 V. R1 and R2 make it possible to set the switching threshold as low as 1.2 V (the values shown provide a 4.10-V threshold). Hysteresis introduced by $R3$ minimizes the effect of reference noise and external EMI. Using the values given for R_2 and R_3 (1 MΩ and 10 MΩ), the switching threshold (V_t) is obtained by adjusting the value of R_1 alone: R_1 (in megohms)$=0.770V_t-0.910$. External connections tie the internal reference voltage to the inverting input of each comparator. Comparator A monitors the supply voltage. When this voltage dips below the threshold established by R1 and R2, the comparator output initiates a RESET pulse via D3 and discharges the timing capacitor (C3) via diode D1. Comparator B determines the pulse duration with help from the time-delay components R4 and C3. Because the duration (approximately 250 ms) is well above the 50-ms minimum required in this case, the effect of supply voltage on pulse width can be ignored. D2 allows the circuit to respond to short outages by providing another quick-discharge path for the timing capacitor. Comparator C is a spare that supplies a complementary output or low-line warning. Comparator D acts as an active AND gate for logic-high signals from comparators A and B. The pull-up resistor in many such circuits is 100 kΩ or less (to avoid errors caused by the comparator's input leakage). In this particular case, it would provide a path for excessive battery discharge during reset.

SIMPLE WIDEBAND NOISE GENERATOR

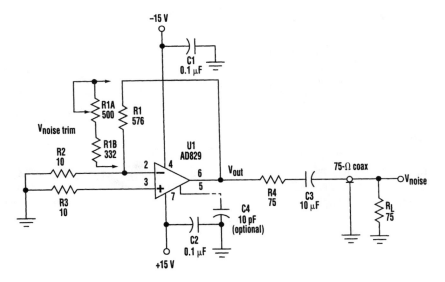

ELECTRONIC DESIGN

Fig. 41-2

This relatively simple op-amp noise generator amplifies the input voltage noise of a wideband, decompensated op amp. If a device with a single-gain stage is selected, the output noise will be spectrally flat up to the closed-loop bandwidth. The op amp is manufactured by Analog Devices Inc.

THREE-TONE GENERATOR

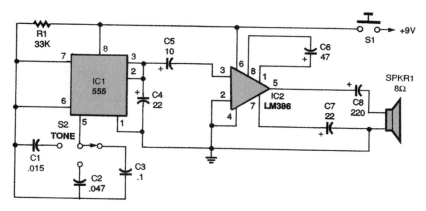

POPULAR ELECTRONICS

Fig. 41-3

The three-tone generator makes a great warning device. It has a lot of uses; for example, if the appropriate switch is used for S1, the circuit can be used as a burglar alarm. When S1 is pressed, it turns on the circuit. The tone frequencies depend on resistor R1 and capacitors C1 through C3. Switch S2 lets you select between those capacitors. Capacitor C1 produces the highest frequency because it has the lowest capacitance, while C3 generates the lowest frequency

SIMPLE FREQUENCY SYNTHESIZER

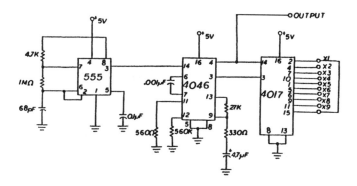

NUTS AND VOLTS

Fig. 41-4

The 555 timer circuit is configured as a 10-kHz astable multivibrator that feeds this signal to the 4046 phase-locked-loop IC. The signal that is fed from this chip is then coupled to a divide-by-N counter. This counter will take the 10-kHz signal locked on by the 4046 and will produce multiples of the fundamental frequency (10 kHz). Therefore, it is possible to generate frequencies as high as 100 kHz with this circuit. When building this circuit, remember to use safety precautions when handling CMOS chips.

DIGITAL BURST GENERATOR

DIGITAL BURST GENERATOR *(Cont.)*

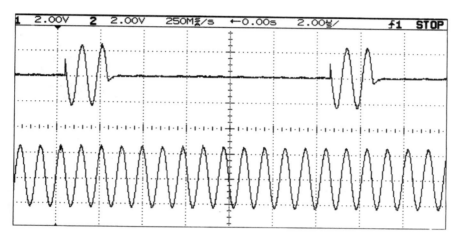

Fig. 41-5

Figure 1 illustrates the transmitted signal and the continuous sine wave from which it is derived. Because the HA4600 buffer has an enable/disable feature, it will pass or reject the input waveform, depending on the state of the enable pin (Fig. 2.). The input signal also is present at the input to the HFA3046 transistor array, which has been configured as a high-speed, high-gain comparator. The comparator squares up the input signal and applies it to the inputs of the two counters, X and Y. The X counter controls the buffer enable, and it determines how many cycles of the input waveform get passed to the output. The four switches S_{X0} through S_{X3} are binary-coded. Consequently, if two switches (S_{X0}=1 cycle and S_{X1}=2 cycles) are closed, three cycles of the input sine wave will be passed to the output. Furthermore, the input signal is connected to the Y counter, which controls the repetition rate by determining the OFF period between pulse bursts. The four switches S_{Y0} through S_{Y3} are binary-coded. When all of these switches are closed, the OFF period will be 16 times the period of the input waveform. With the X and Y counters set as described earlier, the repetition rate is the reciprocal of (16+3) times the period of the incoming waveform. If a longer repetition rate is desired, a flip-flop or another counter can be added in series with the output of the Y counter to extend the OFF time. R6, R7, and R8 bias the long-tailed transistor at 10 mA, which is the optimum point for speed. R5 and R6 are small enough to discharge quickly, thus preventing saturation. Configured as shown, the circuit will handle 10-MHz input signals with little degradation. The limit on frequency response is the speed of the logic and the comparator delay time. The comparator delay time can be eliminated by one-shotting out the delay.

RANDOM-NUMBER GENERATOR

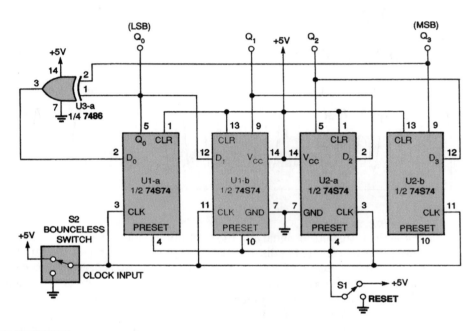

POPULAR ELECTRONICS

Fig. 41-6

The 74S74 flip-flops shown in the circuit are arranged to form a 4-bit shift register. Binary data enter the D0 input on U1-a and are sequentially shifted to each output (Q0, Q1, Q2, Q3) with each clock pulse. The data input to the shift register come from the output of U3-a, one gate of a 7486. That exclusive-OR gate compares two of the output bits from the shift register. If the two bits are the same, then the output of U3-a is 0 V (or low). If the two bits are different, then the output of U3-a is +5 V (or high). Therefore, U3-a acts as a type of logical-feedback network that changes the data at D0, which, in turn, changes the outputs of the flip-flops. The effect of that network is to create a pseudo-random sequence of bits at the outputs of U1 and U2.

TONE-BURST CIRCUIT

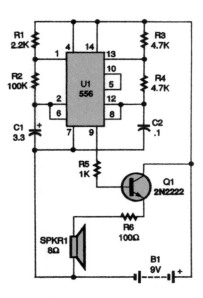

The tone-burst circuit puts out a 500-Hz tone, at a rate of 1 Hz, through an 8-Ω speaker, SPKR1. Besides the 9-V battery and its connecting snap, SPKR1 is the only non-surface-mount component in the circuit. The circuit can also be built from standard-size components.

POPULAR ELECTRONICS

Fig. 41-7

SIMPLE FREQUENCY GENERATOR

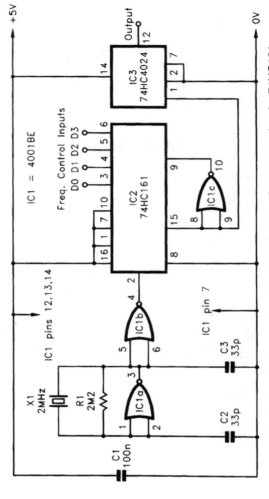

Fig. 2. Circuit diagram for a Simple Frequency Generator using the 74HC161.

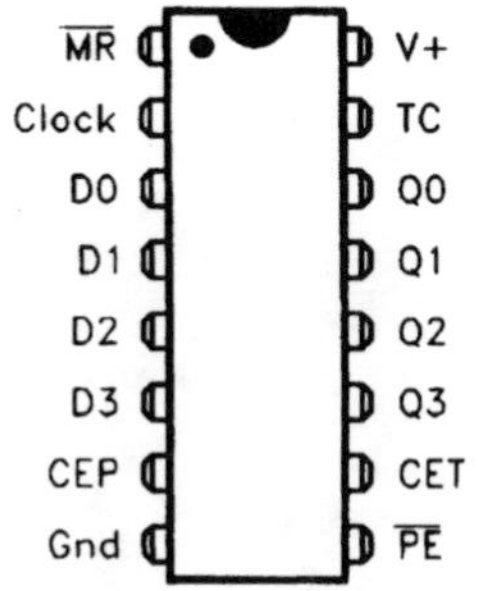

Fig. 1. Pinout details for the 74HC161 binary counter.

SIMPLE FREQUENCY GENERATOR *(Cont.)*

Table 1. Division Rates and Output Frequencies

D0	D1	D2	D3	Div.	Frequency (kHz)
L	L	L	L	16	125
H	L	L	L	15	133·333
L	H	L	L	14	142·857
H	H	L	L	13	153·846
L	L	H	L	12	166·55
H	L	H	L	11	181·181
L	H	H	L	10	200
H	H	H	L	9	222·222
L	L	L	H	8	250
H	L	L	H	7	285·714
L	H	L	H	6	333·333
H	H	L	H	5	400
L	L	H	H	4	500
H	L	H	H	3	666·666
L	H	H	H	2	1000 (1MHz)
H	H	H	H	-	-

Fig. 41-8

The circuit diagram for a simple frequency generator that uses a divide-by-N counter, based on a single 74HC161, is shown. Although IC1a and IC1b are NOR gates, in this circuit, they are used as inverters in a crystal clock generator. This provides an accurate 2-MHz output signal that is fed to the input of the divide-by-N circuit based on IC2 and IC1c. An inversion is needed between the comparator output and the negative active preset enable input, and this inversion is provided by IC1c. Unfortunately, setting the division rate is more convoluted than simply writing the required value for N to the data inputs (D0 to D3). Table 1 shows the division rates and output frequencies for the 16 input codes for IC2. If the data inputs of IC2 are controlled via inverters, the division rate is one more than the value written to the port. Without a hardware inversion, a software inversion is required. This is actually quite easy, and it is just a matter of deducting the required division rate from 16 (e.g., for a division rate of 4, a value of 12 is written to IC2). Notice that the minimum division rate is 2, and that writing a value of 15 to IC2 will not give a division by 1.

ULTRA-FAST MONOCYCLE GENERATOR

0.5 V/div   5 ns/div

(b)

ULTRA-FAST MONOCYCLE GENERATOR *(Cont.)*

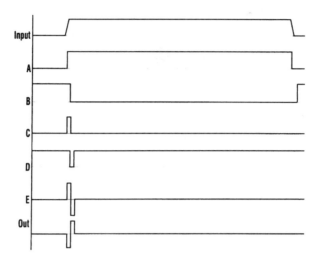

THE NONINVERTING OUTPUT feeds a 2-ns co-axial delay line, which causes the signal to be delayed with respect to the inverting output (A and B). The AND gate's inverting output drives a second 2-ns delay line. This delays the negative-going pulse with respect to the positive-going pulse (C and D).

ELECTRONIC DESIGN

Fig. 41-9

This circuit (Fig. 1*a*) can test an ultra-fast sample-and-hold amplifier, producing the monocycle shown (Fig. 1*b*) when triggered with a TTL input pulse. The comparator (U1) squares up the input pulse and drives the AND gate (U2) with complementary outputs. The hysteresis resistors are chosen to provide reliable triggering despite the 50-Ω loading of the TTL input. The noninverting comparator output feeds a 2-ns coaxial delay line, causing the signal to be delayed with respect to the inverting output (Fig. 2). This forces the outputs of the AND gate to change state for a period of time equal to the delay. The inverting output of the AND gate drives a second 2-ns delay line, delaying the negative-going pulse, with respect to the positive-going pulse (Fig. 2). The power combiner (U3) sums the signals, producing a monocycle output. U4 amplifies the signal. Shorter pulses could be produced by shortening the delay lines and replacing U2 with an AND gate from a faster ECL family.

SWEEP GENERATOR

378

SWEEP GENERATOR *(Cont.)*

TABLE 1
FUNCTION GENERATOR CHARACTERISTICS

Waveform output	Maximum P-to-P	Frequency	Conditions
Sine (1)	5V	10 Hz-100 kHz	1 V@800 kHz
Triangle (1)	8 V	10 Hz- 50 kHz	1 V>500 kHz
Square (2)	5 V		Positive output DC-coupled, ground ref: rise/fall >50 ns
Ramp (3)			Descending, 6 rates

(1) Output level variable frim min. to max.
(2) Output level not adjustable.
(3) X and Y amplitude internally adjustable.

TABLE 2
SWEEP RANGES OF THE FUNCTION GENERATOR

Switch	Condition	Frequency range
1	Preset	20Hz to >2kHz
2	Preset	<400Hz to >10kHz
3	Preset	<1kHz to >25kHz
4	Preset	5kHz to >100kHz
5*	Resistance tuned	2kHz to 100kHz
	Resistance & VCO tuned	<10Hz to >100kHz
6*	Resistance tuned	<40kHz to >800kHz
	Resistance & VCO tuned	<100Hz to >800kHz

* Ranges show for positions 5 and 6 represent the total tuning range of the function generator and do not imply one continuous sweep.

ELECTRONICS NOW   *Fig. 41-10*

Both IC2 and IC4 are Exar XR2206 monolithic function generators; IC4 functions as a ramp generator, and IC2 functions as a generator of sine, triangular, and square waveforms. Dual operational-amplifier IC1 produces a scaled, level-shifted ramp output that is capable of deflecting an oscilloscope's horizontal sweep. This ensures that the sweep generator and the oscilloscope's sweep circuit are always properly synchronized. Any frequency of interest along the horizontal axis of an oscilloscope that is coupled to this function generator can be measured with an external frequency counter by manually tuning the function generator's VCO instead of sweeping it. The performance characteristics of the sweep/function generator are summarized in Table 1. The generator's sweep rate and frequency can be set by front-panel rotary six-position switches, SWEEP RATE switch S5 and FREQUENCY switch S2. The VCO control R30 manually tunes the VCO. Table 2 lists the sweep ranges of the function generator. Sweep ranges other than those covered in ranges 1 to 4 can be set up as required on positions 5 and 6. Selecting the VCO setting on the front-panel toggle switch S4 permits tuning any fixed frequency within the total frequency range of the instrument with both the FREQUENCY switch S2 and VCO control R30.

SELF-STARTING DATA GENERATOR

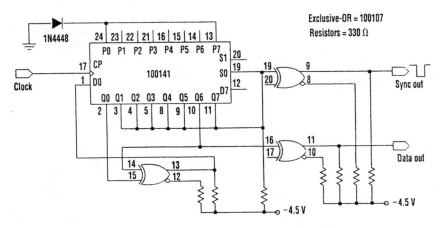

ELECTRONIC DESIGN

Fig. 41-11

Pseudo-random sequence generators built from shift registers and exclusive-OR gates often are used to supply binary test data. If constructed from 100-kΩ ECL parts, such generators can run at up to 200 Mbits/s. Although an N-bit shift register can be connected to generate a sequence that repeats every 2^{N-1} bits, if it should start up in its all-zeros state, no output will be generated. What is needed is a counter to inject a 1 or to preload the shift register whenever N consecutive zeros are detected. The illustrated circuit generates a 127-bit sequence. It provides a synchronization pulse once per repetition without using additional parts, and it is guaranteed to start. It uses the "wired-OR" property of ECL to generate a 1-bit period negative sync pulse when six zeros are present (only five of these zeros are consecutive; a different set of $N-1$ bits might be needed if a longer shift register is used). The sync pulse parallel-loads the shift register with the next state in the sequence. As a result, no seven-zero detector is needed to start the generator.

PINK-NOISE GENERATOR

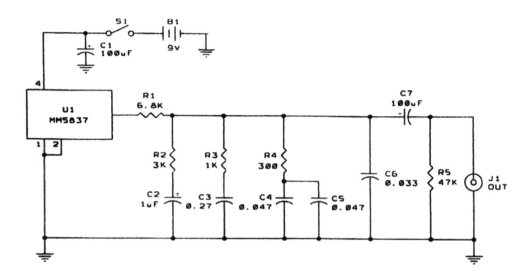

Fig. 41-12

The MM5837 is a digital white-noise generator IC. It produces a clean white noise signal with only a power source. White noise appears at the output of this IC when power is applied. The white-noise signal is then fed through a -3-dB/octave filter to give pink noise. Because the minimum rolloff with a single-stage RC (resistor/capacitor) filter is -6 dB/octave (because of capacitive reactance), an unconventional filter design is needed. The technique involves cascading several stages of lag compensation so that the zeros of one stage partially cancel the poles of the next stage. The result is shown in the schematic of the figure. The response of this circuit is accurate to within $\pm1/2$ dB.

42

Impedance Converter Circuits

The sources of the following circuits are contained in the Sources section, which begins on page 1043. The figure number in the box of each circuit correlates to the entry in the Sources section.

Negative Impedance Converter
Impedance Converter

NEGATIVE IMPEDANCE CONVERTER

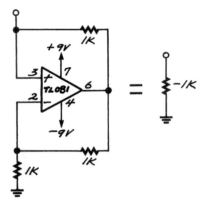

Fig. 42-1

Recall that a resistor is a voltage-dependent current consumer; this circuit does the opposite. If you connect a battery to its input, the battery will get charged, with exactly the current that the battery would have delivered if you had connected it to a 1000-Ω resistor. If you put a 1000-Ω resistor in parallel with a 1000-Ω NIC, you'll get what looks like an infinite resistance. No matter what voltage you apply (within limits), the NIC will match it, and all the current flowing in the resistor will come from the NIC, not from the externally applied voltage.

IMPEDANCE CONVERTER

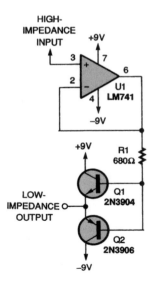

Fig. 42-2

This circuit is a high-input-impedance–to–low-output-impedance converter circuit with unity voltage gain. In the circuit, an LM741 op amp, U1, is connected in a voltage-follower circuit that drives a complementary transistor-emitter-follower circuit. The output of the circuit can be used to drive low-current lamps, relays, speakers, etc. The feedback can be taken from the output to pin 2 of the LM741 if it is desired to include Q1 and Q2 in the feedback loop.

43

Infrared Circuits

The sources of the following circuits are contained in the Sources section, which begins on page 1043. The figure number in the box of each circuit correlates to the entry in the Sources section.

DTMF Infrared Transmitter
Warble Tone Infrared Transmitter
IR Remote Tester
Low-Cost IR Filter
Infrared Alarm Transmitter
Mono Cordless Headphone IR Transmitter
Infrared Alarm Receiver
Mono Cordless Headphone IR Receiver
IR Remote-Control Repeater
Infrared Data Receiver
IR Local Talk Link Receiver
IR Remote-Control Receiver

IR Remote-Control Dc Interface
IR Remote-Control Test Set
IR Remote-Control Relay Interface
Phototransistor IR Receiver
Infrared Data Transmitter
IR Remote-Control Triac Interface
See-Through Sensor
IR Remote-Control Model Railroad
   Application
IR Remote-Control Receiver
IR Relay Circuit
Cordless Headphones

DTMF INFRARED TRANSMITTER

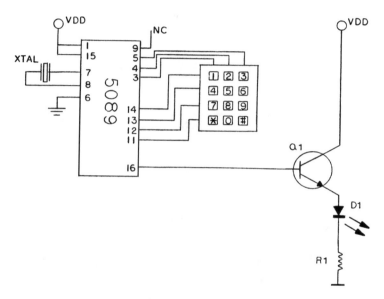

Fig. 43-1

This transmitter uses a 5089 DTMF generator chip and a keypad to generate DTMF signals and modulate them on an IR light beam from an IR LED. Xtal is a 3.579-MHz TV burst crystal.

WARBLE TONE INFRARED TRANSMITTER

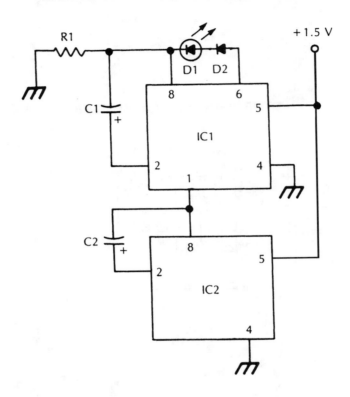

IC1, IC2	LM3909 LED flasher/oscillator IC
D1	infrared LED
D2	diode (1N4148, 1N914, or similar)
C1	1 µF 5 V electrolytic capacitor
C2	47 µF 5 V electrolytic capacitor
R1	470 Ω ¼ W 5% resistor

TAB BOOKS

Fig. 43-2

IC1 and IC2 are LM3909 LED flasher-oscillator devices. IC2 generates a low-frequency square wave that modulates the frequency of IC1. This circuit is useful for IR testing, communications link signal source, etc.

IR REMOTE TESTER

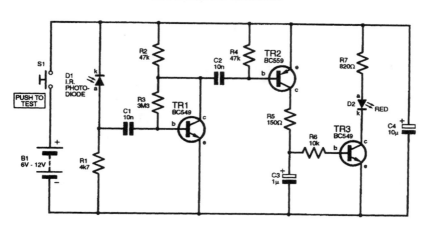

EVERYDAY PRACTICAL ELECTRONICS

Fig. 43-3

This remote handset tester is very useful and convenient, and could quickly pay for itself by helping to recover faulty remote handsets. Pulsed infrared light generated by the handset falls upon D1, which creates a pulse waveform across R1. This is capacitively coupled to the base of TR1, and an amplified signal is coupled to C2 to a pulse-stretching circuit based around TR2, C3, and associated components. Hence, driver transistor TR3 conducts and illuminates the LED D2 whenever infrared light is received by D1. A functional remote controller with fresh batteries will operate the tester from approximately 500 mm, while one with nearly exhausted batteries may work over only a few centimeters. Coincidentally, the design also self-tests its own battery by giving an initial flash of D1 when the switch S1 is first closed. The IR photodiode should be roughly a 940-nm type, while the types of the transistors themselves are not crucial. The circuit will operate from a 6- to 12-V rail (e.g., a 12-V battery), and C4 decouples the power supply rails.

LOW-COST IR FILTER

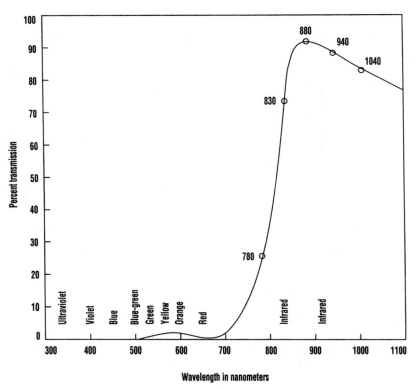

Fig. 43-4

When exposed to "cool white" fluorescent light for 5 s, the color negative (using Kodacolor 100 ASA film) produced after the developing process exhibits a sharp cutoff at about 830 mm. This is perfect for many IR LEDs and other IR devices.

INFRARED ALARM TRANSMITTER

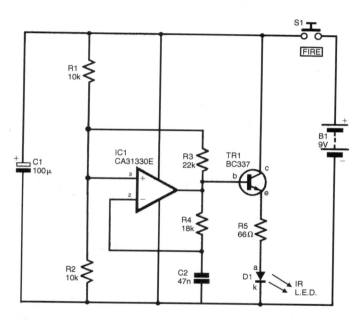

EVERYDAY PRACTICAL ELECTRONICS

Fig. 43-5

The circuit diagram for the infrared transmitter appears in this figure. The oscillator uses IC1 as a relaxation oscillator. Capacitor C2 and resistor R4 are the timing components, and they are connected between the output of IC1 and its inverting input (pin 2). A roughly square-wave signal is generated at pin 6, the output of IC1. This signal is used to drive infrared LED D1, via emitter-follower buffer-stage transistor TR1 and current-limiter resistor R5. The specified value for R_5 sets the LED current at nearly 100 mA, but as the LED is switched off for about 50 percent of the time, the average LED current is a little under 50 mA. This is the maximum acceptable drive current for most normal infrared LEDs.

MONO CORDLESS HEADPHONE IR TRANSMITTER

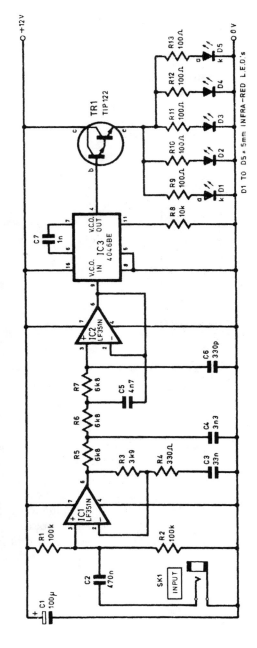

EVERYDAY PRACTICAL ELECTRONICS

Fig. 43-6

The preemphasis is applied by IC1, which is basically just an op amp used in the noninverting mode. At low and middle frequencies, the impedance of capacitor C3 is high in relation to the resistance of R3. Consequently, there is virtually 100 percent negative feedback through resistor R3, and IC1 has a closed-loop gain of little more than unity. At higher frequencies, the impedance of C3 is relatively low, and a significant proportion of the feedback through R3 is decoupled. This gives a closed-loop voltage gain that steadily rises with increased input frequency, with almost 20 dB of gain being provided at the highest audio frequencies. Resistor R4 limits the closed-loop voltage gain of IC1 at frequencies above the audio range. The low-pass filter is a conventional third-order (18 dB/octave) type based on IC2. This filter gives fractionally less than the full 20-kHz audio bandwidth, but does not significantly impair the quality of the input signal. A CMOS "micropower" PLL is used for IC3, but, in this circuit, only the VCO stage is utilized. No connections are made to any of the other sections of IC3. The output of IC2 is direct-coupled to the control input of the VCO (IC3 pin 9). Capacitor C7 and resistor R8 are the VCO timing components, and they set the center frequency at roughly 100 kHz. Transistor TR1 is a high-gain power Darlington device, which is used here as an emitter-follower buffer stage. This can easily source the 500-mA ON current of the LEDs. A bank of five LEDs is used, and each one has a separate current-limiting resistor (R9 to R13).

390

INFRARED ALARM RECEIVER

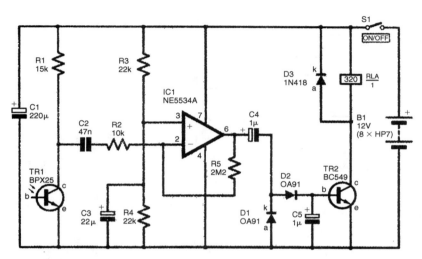

Fig. 43-7

The full circuit diagram for the infrared receiver is given here. TR1 is a phototransistor, and it is used in a common-emitter amplifier. However, no bias current is fed to the base (b) terminal of TR1. The collector (c) current is governed by the light level received by TR1. The higher the received light level, the higher the current flow. The pulses of infrared from the transmitter therefore produce pulses of leakage current through TR1, which give small negative pulses at TR1's collector. The output signal for TR1's collector is coupled by capacitor C2 to a high-gain inverting amplifier based on IC1. Resistors R2 and R5 are the negative-feedback network, and these set the closed-loop voltage gain of IC1 at 220 times. Capacitor C4 couples the amplified output signal from IC1 pin 6 to a conventional half-wave rectifier and smoothing circuit (D1, D2, and C5). Germanium, rather than silicon, diodes are used in the rectifier, because germanium types have a lower forward voltage drop. This gives slightly improved sensitivity. The positive dc signal developed across capacitor C5 drives the base of TR2, which is a simple common-emitter switch that controls the relay RLA coil.

MONO CORDLESS HEADPHONE IR RECEIVER

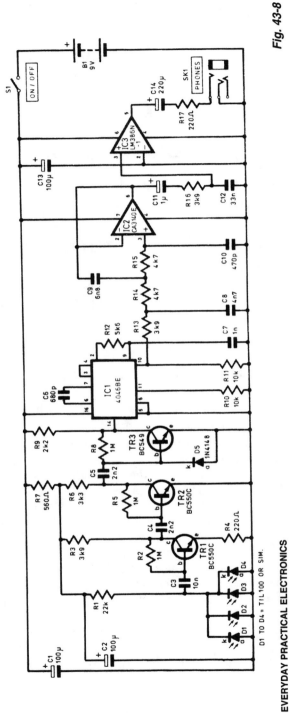

EVERYDAY PRACTICAL ELECTRONICS

Fig. 43-8

The full receiver circuit diagram appears in the figure. As many as four infrared detector diodes can be used (D1 to D4). They are wired in parallel and used in the reverse-biased mode, and under dark conditions only very small leakage currents flow. The pulses of infrared from the transmitter result in increased leakage currents, which generate small negative pulses at the cathodes (k) of the diodes. Transistor TR1 is used as a low-noise preamplifier and buffer stage. The majority of the voltage gain is provided by transistors TR2 and TR3, which are also common-emitter stages. These provide a combined voltage gain of over 80 dB. The signal from TR3 is connected to one of the phase comparator inputs of IC1 (pin 14). This IC provides demodulation and is a CMOS 4046BE p.l.d. Resistor R10 and capacitor C6 are the timing components for the VCO, and they set the center frequency at approximately 100 kHz. Resistor R12 and capacitor C7 form a simple low-pass filter between the phase comparator's output at pin 2 and the input of the VCO at pin 9. The audio output signal is obtained from the low-pass filter via an integral source-follower stage. The output from IC2 is fed to resistor R16 and capacitor C12, which form a simple low-pass filter that provides the deem-phasis. The signal is then fed to the input of IC3, which is a small class-B audio power amplifier. The current consumption of the receiver is about 12 mA.

IR REMOTE-CONTROL REPEATER

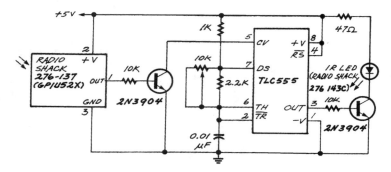

ELECTRONICS NOW

Fig. 43-9

This circuit receives TV remote-control signals and retransmits them. The 555 oscillator provides 40-kHz modulation. The IR LED can be placed in another room.

INFRARED DATA RECEIVER

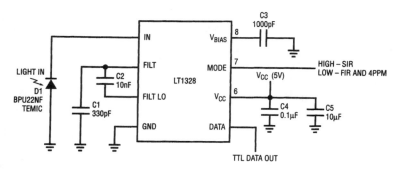

LINEAR TECHNOLOGY

Fig. 43-10

The LT1328 circuit operates over the full 1-cm to 1-m range of the IrDA standard at the stipulated light levels. For IrDA data rates of 115 kb/s and below, a 1.6-μs pulse width is used for a 0 and no pulse for a 1. Light levels are 40 to 500 mW/sr (milliwatts per steradian). The figure shows a scope photo for a transmitter input (top trace) and the LT1328 output (bottom trace). Note that the input to the transmitter is inverted; that is, transmitted light produces a high at the input, which results in a zero at the output of the transmitter. The MODE pin (pin 7) should be high for these data rates.

IR LOCAL TALK LINK RECEIVER

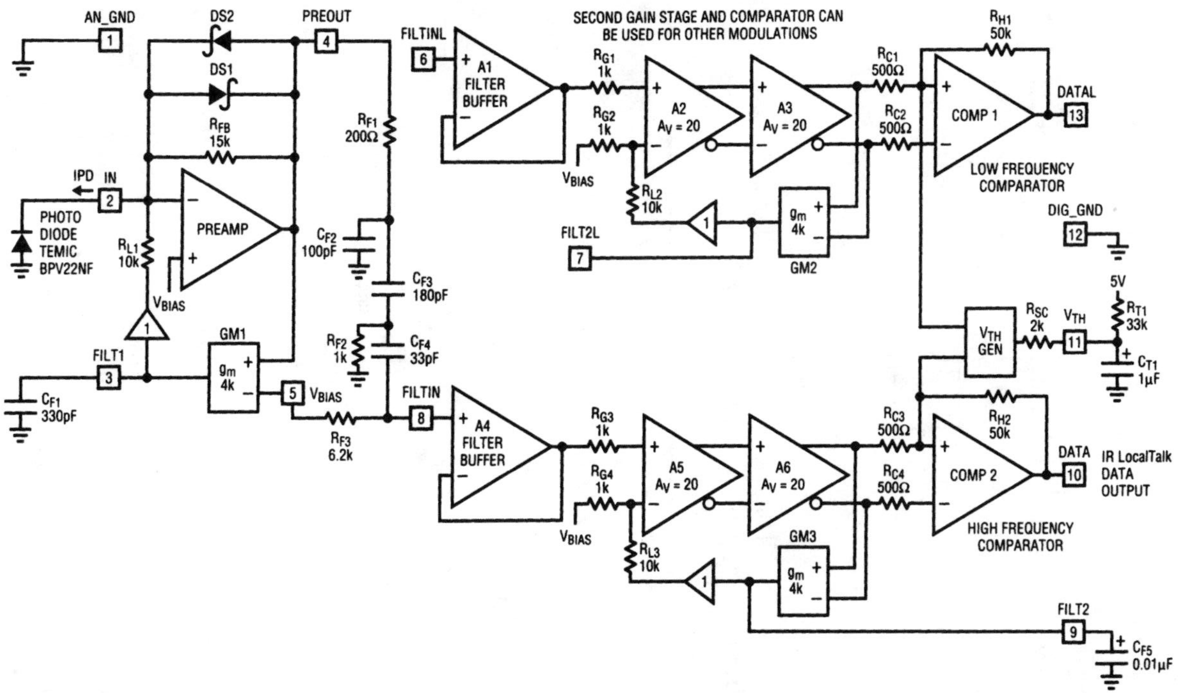

LINEAR TECHNOLOGY

Fig. 43-11

A low-noise (2 pA / $\sqrt{Hz}$), high-bandwidth (7 MHz) current-to-voltage converter formed by the preamplifier and its associated components transforms the reverse current from an external photodiode to a voltage. Although the 7-MHz bandwidth of the preamp supports 4-Mbit data rates, a low-pass filter on the preamp output is used to reduce the bandwidth to just the required amount in order to reduce noise. As shown, capacitance C_{F1} sets the break frequency of an ac high-pass loop around the preamp to 180 kHz. This loop rejects unwanted ambient light, including sunlight and light from incandescent and fluorescent lamps. The preamp stage is followed by two separate channels, each containing a high-impedance filter buffer, two gain stages, high-pass

IR LOCAL TALK LINK RECEIVER *(Cont.)*

loops, and a comparator. The only difference between the two channels is the response time of the comparators: 25 and 60 ns. For the 125-ns pulses of IR LocalTalk, the 25-ns comparator with its active pull-up output is used. The low-frequency comparator with its open collector output (with 5 kΩ internal pull-up) is suitable for more modest speeds, such as the 1.6-μs pulses or the IrDA-SIR. Each gain path has an ac coupling loop similar to the one on the preamp. Capacitance C_{F5} sets the high-pass corner at 140 kHz for IR LocalTalk. The loops serve the additional purpose here of maintaining an accurate threshold for the comparators by forcing the dc level of the differential gain stages to zero. As the preamp output is brought out and the inputs to the two comparator channels are buffered, the user is free to construct the exact filter required for the application by the careful selection of external components. R_{F1}, C_{F2}, C_{F3}, R_{F2}, C_{F4}, and R_{F3} form a bandpass filter with a center frequency of 3.5 MHz and 3-dB points of 1 MHz and 12 MHz. Together with the high-pass ac loop in the preamp and the 7-MHz response of the preamp, this forms an optimal filter response for IR LocalTalk.

IR REMOTE-CONTROL RECEIVER

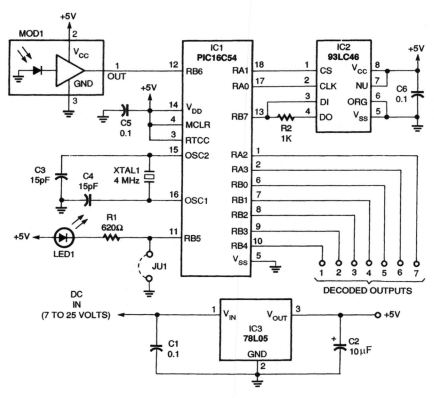

ELECTRONICS NOW

Fig. 43-12

A schematic diagram of the remote-control receiver is shown. The heart of the circuit is IC1, a PIC16C54 8-bit CMOS manufactured by Microchip. The microcontroller stores its data in IC2, a 93LC46 1-kbit serial EEPROM (electrically erasable programmable read-only memory), also manufactured by Microchip. In this application, the 93LC46 has a three-line interface with the microcontroller. The three lines are CHIP SELECT, CLOCK, and DATA IN/OUT. Because DATA IN and DATA OUT share the same line, a resistor (R2) limits the current flow during transitions between writing and reading when there are conflicting logic levels. The microcontroller communicates with the 93LC46 by placing a logic high on the CHIP SELECT pin. Data are then transferred serially to and from the 93LC46 on the positive transition of the CLOCK line. Each read or write function is preceded by a start bit, an opcode identifying the function to be performed (read, write, etc.), then a 7-bit address; this is followed by the 8 bits of data which are being written to or read from that address. Immediately preceding and following all write operations, the microcontroller sends instructions to the 93LC46, which enables or disables the write function, thereby protecting the data that have been stored. In the programming mode, IC1 reads an IR data stream from MOD1 and converts it to data

IR REMOTE-CONTROL RECEIVER *(Cont.)*

patterns that can be stored in IC2. These data patterns are held for comparison while the unit is in normal operation. Power for the circuit is conditioned by IC3, a 78L05 low-current, 5-V regulator, which will accept any dc input voltage between 7 and 25 V. Capacitors C1 and C2 stabilize the operation of the regulator. Crystal XTAL1 sets the internal oscillator of IC1 to 4 MHz. Jumper JU1 consists of two closely spaced pads on the PC board that, when momentarily jumpered with a screwdriver or other piece of metal, places IC1 in the programming mode and lights LED1. The source and object code are available on the Gernsback BBS (516-293-2283, v.32, v.42bis) as a file called IREC.ZIP for those who wish to program their own PICs and have the proper equipment to do so.

IR REMOTE-CONTROL DC INTERFACE

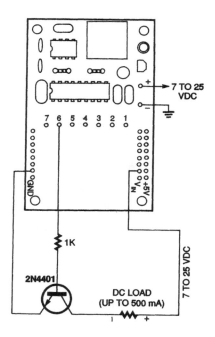

Fig. 43-13

This circuit can be used to interface dc loads up to 500 mA to the IR remote control.

IR REMOTE-CONTROL TEST SET

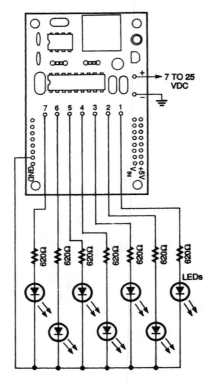

Fig. 43-14

This circuit can be used to check out operation of the IR remote control.

IR REMOTE-CONTROL RELAY INTERFACE

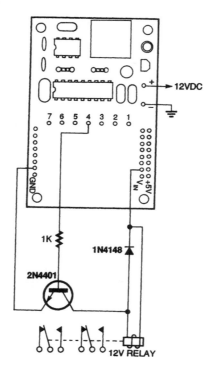

Fig. 43-15

This circuit can be used to interface a relay to the IR remote control in order to control large ac or dc loads.

PHOTOTRANSISTOR IR RECEIVER

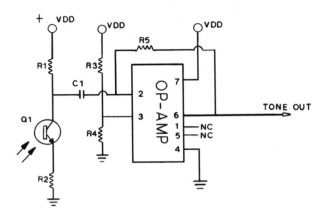

NUTS AND VOLTS

Fig. 43-16

This receiver uses an op amp and a phototransistor as a receiver for IR signals. Q1 is a phototransistor. The op amp is a CMOS or FET, such as a TL081, etc. R1 and R2 depend on the phototransistor, but are typically 2.2 kΩ and 330 Ω, respectively. R3=R4=10 kΩ, R5=100 kΩ, and C1 is 0.1 μF (or larger).

INFRARED DATA TRANSMITTER

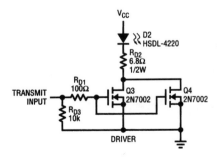

LINEAR TECHNOLOGY

Fig. 43-17

This figure shows an IrDA transmitter.

IR REMOTE-CONTROL TRIAC INTERFACE

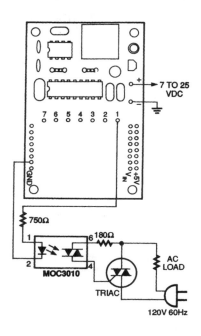

ELECTRONICS NOW

Fig. 43-18

This circuit can be used to interface a triac to the IR remote control in order to control ac loads.

SEE-THROUGH SENSOR

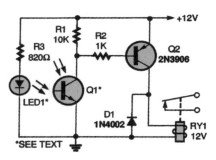

POPULAR ELECTRONICS

Fig. 43-19

In this circuit, an infrared emitter LED (LED1) is aimed at an infrared phototransistor (Q1). As long as the IR light path between the two remains uninterrupted, transistor Q2 will keep relay RY1 closed. Any opaque object blocking the light path will cause RY1 to open. This circuit is a see-through-type light sensor. Such units are often used as part-in-place detectors and parts-counter sensors.

IR REMOTE-CONTROL MODEL RAILROAD APPLICATION

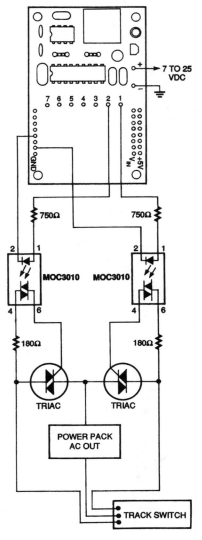

Fig. 43-20

This circuit can be used to interface two triacs to the IR remote control in order to control model railroad track switches.

IR REMOTE-CONTROL RECEIVER

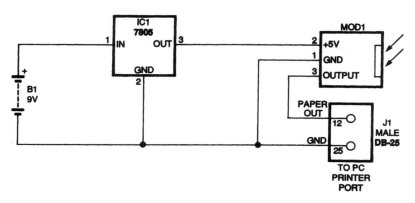

Fig. 43-21

The figure shows a schematic of the IR receiver circuit. The heart of the circuit is MOD1, an infrared detector module that removes the IR carrier frequency and transmits only the data that are encoded in the received IR signal. A suitable IR module is available at Radio Shack (No. 276-137). The IR module needs a clean 5-V power supply, which is provided by IC1, a 7805 regulator. Power is supplied to the regulator by 9-V battery B1. The output of the module is wired to a male DB-25 multipin connector. The infrared detector module receives a signal, filters it, and removes the 40-kHz carrier. The output of the module is a TTL-level signal consisting of long and short pulses. The PC records those voltage levels over time, while the signal is being sent, and stores the data in a file. The line normally used by the PC's printer port to indicate that the printer is out of paper (pin 12) is used in this project to accept data from the IR module. The I/O port is located at the address ox379. Bit 5 corresponds to input pin 12. Various software programs are required to let a PC store information input to its printer port. (All of the software is available on the Gernsback BBS—516-293-2283, v.32, v.42bis—contained in a file called IR-TEST.ZIP.) The source code of the first program, IRLOG.EXE, is written in C. The program stores the value it reads from the PC's printer port into an array. When the input line is logic high, the ASCII character 1 is stored in the array. When the input line is logic low, ASCII character 0 is stored.

IR RELAY CIRCUIT

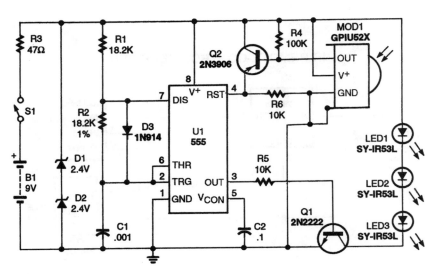

Fig. 43-22

All IR is received by the module via an IR photodiode. The diode is operated in its reverse-bias current it permits depends on the intensity of the received infrared light. The fluctuations in the reverse-bias current are then amplified by a high-gain stage. The output of the amplifier is then "limited" by the next stage. The limiter chops the extreme highs and lows off the amplified signal, and the result is a quasi-digital pulse train. The simplified wave then passes through a bandwidth filter that has its center frequency at 40 kHz. At that point, the circuit has effectively retrieved the 40-kHz remote carrier. The reproduced carrier is then integrated. The next stage is an "inverting Schmitt trigger." It will not go low unless the filter's output signal surpasses a certain amplitude, and it will not go high again until the signal drops below a certain minimum. Thus, the Schmitt trigger responds only to large changes in the filter's output (caused by bursts and pauses) and ignores small changes (caused by the 40-kHz carrier, to which the filter can't respond quickly). The Schmitt trigger's output is thus low when a 40-kHz burst is received, and high during pauses between bursts. The resulting waveform is an inverted version of the pulses that were modulated and transmitted by the remote. The oscillator circuit is based on a 555 timer. With D3, R2 is bypassed (or shorted out) while C1 is charging, but is in the current path during discharge. Therefore, with R1 and R2 made equal, the charge time equals the discharge time, yielding an output with a 50-percent duty cycle. Components R1, R2, and C1 (all precision, drift-free units) have been chosen to provide a 40-kHz output in this configuration. That output strobes the IR LEDs via Q1. The oscillator functions only when pin 4 of the 555 is high. Because that pin is connected to the inverter circuit, the oscillator functions when the inverter's input is low. The inverter's input is connected to MOD1's output, which goes low with each remote burst received. So overall, the circuit produces a 40-kHz IR burst when it receives one.

CORDLESS HEADPHONES

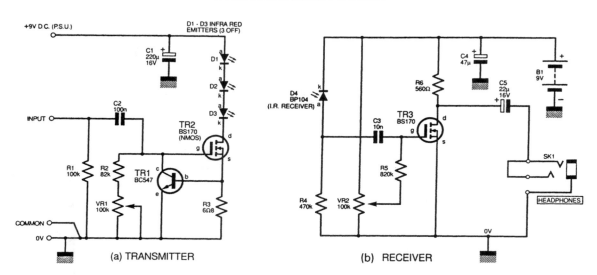

+9V D.C. (P.S.U.)

D1 - D3 INFRA RED EMITTERS (3 OFF)

(a) TRANSMITTER

(b) RECEIVER

EVERYDAY PRACTICAL ELECTRONICS

Fig. 43-23

It is possible to use headphones with no direct connection to the audio source. This figure illustrates a simple transmitter circuit suitable for connecting to an audio source. Diodes D1 to D3 are three infrared LEDs in series, driven by TR2, an n-channel MOSFET with a drain-current rating of 500 mA maximum. The illumination of the LEDs is modulated by the audio signal applied, with TR1 current-limiting and shunting the bias of TR2 to ground when TR2 source current exceeds roughly 100 mA. Potentiometer VR1 should be adjusted for best results. A 9-V power supply is best used in this circuit because of the sustained level of current consumption. The transmitter range is about 1 to 2 m, but this can be extended by using reflectors behind the diodes.

A suggested receiver circuit, using an infrared photodiode, D4, to detect the IR light, is shown in (*b*). The received signal is ac-coupled to TR3, another MOSFET. The output is taken from the drain terminal via capacitor C5. Bias can be adjusted with potentiometer VR2, and this circuit will run from a PP3-size battery. The circuit could be adapted for other applications.

44

Inverter Circuits

The sources of the following circuits are contained in the Sources section, which begins on page 1043. The figure number in the box of each circuit correlates to the entry in the Sources section.

Gate Inverters
Regulated Low-Noise Voltage Inverter
D-Bistable

GATE INVERTERS

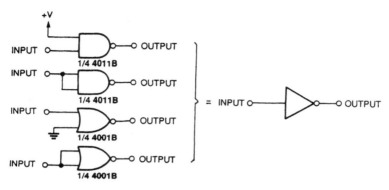

WILLIAM SHEETS

Fig. 44-1

Logic gates can be configured as buffers or inverters as shown.

REGULATED LOW-NOISE VOLTAGE INVERTER

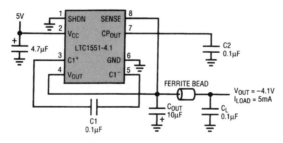

LINEAR TECHNOLOGY POWER SOLUTIONS

Fig. 44-2

Analog cell phones rely on a quiet GaAsFET bias supply to maximize the signal-to-noise ratio, providing a high-quality transmission. The LTC1551CS8-4.1 switched-capacitor voltage inverter circuit shown reduces output noise to less than 1 mV on the −4.1-V regulated output voltage. The 900-kHz operating frequency allows the charge pump to use small 0.1-μF capacitors. The 10-μF and 0.1-μF output capacitors effectively reduce the 900-kHz ripple noise to relatively insignificant levels. The LTZ1550CS8-4.1 has an active low shutdown input (SHDN), whereas its counterpart, the LTC1551CS8-4.1, has an active high shutdown input. This choice minimizes component count. The LTC1550 and 1551CS8-4.1 both operate in doubler mode, meaning that they only invert the input voltage and regulate it down to −4.1 V. The input voltage range is 4.5 to 7 V.

D-BISTABLE

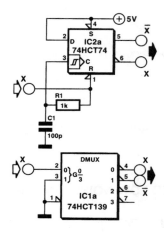

Fig. 44-3

The interesting point of this circuit is that a D-type bistable is used as an inverter. When the level at the input changes from high to low, the bistable is reset and its Q output goes high. When the input becomes low, the reset is removed and the Q output goes low. The delay introduced by network R1-C1 between the RESET input and the CLOCK input makes it possible to trigger the bistable at the leading edge of the input signal. As an example, in the case of a dual D-bistable Type 74HCT74, the time needed for a clock pulse to be accepted after the reset has been removed is 5 ns. Therefore, an RC introducing a delay of 7.5 ns gives a reasonable safety margin. The reduced edge gradient of the clock pulse does not create any problems because the maximum allowed rise time of the clock input is 500 ns. To obviate asymmetrical output signals, it is advisable to limit the input frequency to about 1 MHz with component values as specified.

45

Laser Circuits

The sources of the following circuits are contained in the Sources section, which begins on page 1043. The figure number in the box of each circuit correlates to the entry in the Sources section.

LASER-BEAM-ACTIVATED RELAY

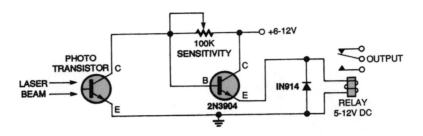

Fig. 45-1

This two-transistor relay circuit will activate whenever the phototransistor is illuminated by the laser beam.

LASER-ACTIVATED RELAY WITH AMPLIFIER

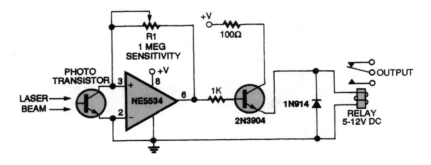

Fig. 45-2

A low-noise amplifier is used in conjunction with a phototransistor to allow higher sensitivity than the detector alone would provide.

LASER-DIODE DRIVER

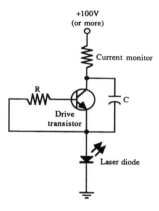

TAB BOOKS

Fig. 45-3

One way to drive a single heterostructure laser diode. The transistor is driven in avalanche mode, producing short-duration pulses of current.

LASER-DIODE DRIVER

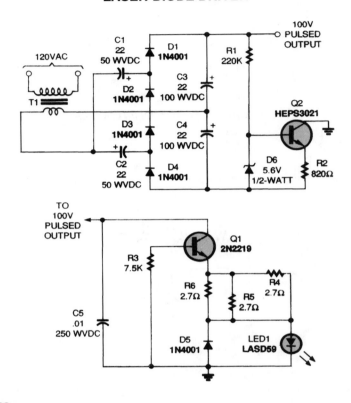

POPULAR ELECTRONICS

Fig. 45-4

T1, a 24-Vac wall transformer, drives a full-wave quadrupling rectifier, which produces an output of 120 Vdc. Components R1, R2, D6, and Q2 form a 5.7-mA constant-current sink. Use a heatsink on Q2, which can be any 1-W or higher transistor with a V_{CE} of 150 V or more. Components C5, R3 to R6, D5, LED1, and Q1 form an avalanche oscillator that runs at 4 to 5 kHz. Transistor Q1 must be selected for a breakdown voltage of between 90 and 110 Vdc. Build the power-supply section on a separate circuit board using standard methods. Resistors R4 to R6 must be carbon, ½-W units, or can all be replaced by a single 1-W, 1-Ω carbon unit. Use short leads throughout on this assembly to avoid current undershoot. Use a socket to mount Q1, and a female pin for the anode of LED1.

120-VAC SINGLE HETEROSTRUCTURE LASER DRIVER SUPPLY

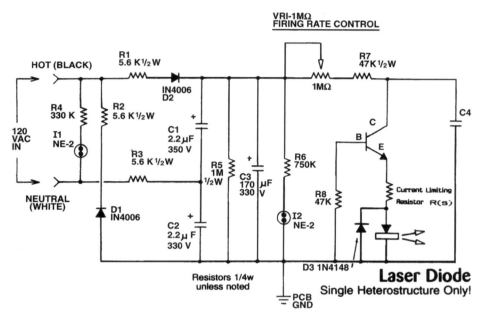

VRI-1MΩ
FIRING RATE CONTROL

Laser Diode
Single Heterostructure Only!

NUTS AND VOLTS

Fig. 45-5

This figure shows the complete schematic of the 120-Vac SH power supply. The circuit consists of a polarized plug, a voltage doubler, a large-value storage capacitor, two neon indicator lamps to show that ac and dc power is present, a 1-MΩ firing-rate control, and the "business end" of the supply to drive the laser. I2 is also a safety light. When it is on, the laser should be considered operating. The current to the voltage doubler is limited by resistors R1 to R3, and by the capacitive reactance of capacitors C1 and C2. R7—a 47-kΩ, ½-W resistor—limits the input power to the avalanche circuit to avoid overheating the diode or Q1. The capacitors and resistors serve to isolate the circuit from the ac line sufficiently to prevent a shock hazard; however, you can still get shocked. Be sure you put this device in a nonconductive case, and obey the polarity of the polarized plug. Don't touch any part of the circuit while it's plugged in, either. Be sure you use a plastic potentiometer for VR1, or a separate control-type potentiometer with a plastic knob.

LOW-COST LASER-DIODE DRIVER

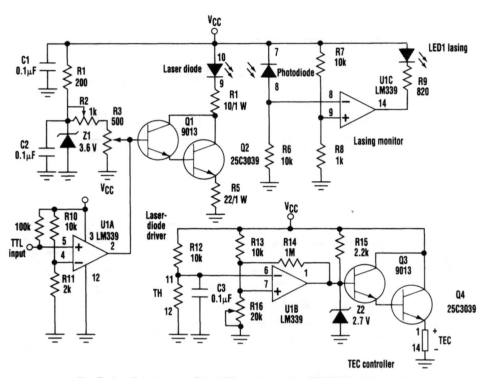

Note: The laser diode, photodiode, TH, and TEC come in one package (QLM5S876 laser-diode module).

ELECTRONIC DESIGN

Fig. 45-6

The circuit presented is a low-cost laser-diode driver with current limiting and a lasing monitor for safe operation of the laser. The driver provides temperature stabilization using a built-in thermo-electric-cooler (TEC) controller. This circuit is designed around a QLM5S876 1.55-m laser-diode module with a built-in monitor photodiode, TEC, and thermistor. The laser is driven by the constant-current source Q2 and Q1. The drive current is stabilized against supply changes by Z1, limited by R2, and adjusted by R3 to the desired operating point. When the current is small, the laser doesn't lase and the monitor photodiode detects no optical power. As a result, comparator U1C goes below its trip point and LED1 turns off. When the current is adjusted and passes the laser threshold, sufficient optical power is generated by the laser, causing photo current to flow through the photodiode and the comparator to drive the lasing monitor indicator LED1 to on. The TEC controller circuit is a feedback system. If the temperature is higher than the set point, the comparator output will be high and will drive the TEC element through R15, Q3, and Q4. The TEC drive current is limited by Z2. When the TEC is driven on, it cools the laser. This increases the voltage of the inverting output until it passes the comparator's upper trip point and turns off the TEC's drive current. The temperature set point can be adjusted by R16.

SINGLE HETEROSTRUCTURE LASER DRIVER

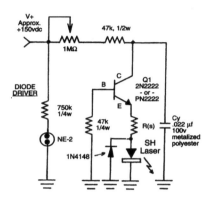

Fig. 45-7

One way to reliably achieve the fast rise-time pulse of short duration to power SH lasers is to use the phenomenon called *avalanching* of a transistor. Transistors have a breakdown voltage, abbreviated as V_{ceo}; if this voltage is exceeded, the transistor will spontaneously conduct with its collector-to-emitter circuit open and the voltage applied to its collector. This produces a pulse whose rise time is extremely fast—on the order of 1 or 2 ns—and whose duration is related to the capacitance at the collector. The operation is as follows: Current rises at the base of transistor Q1 when the voltage on Cy has risen enough to jump over and break down the collector-to-emitter junction. For the 2N2222, this is typically 75 to 80 V. When the transistor breaks down, it sends a high-current pulse of energy through a low-value carbon film resistor R(s). With the 0.022-μF capacitor shown, the duration of the pulse to the laser is only about 75 ns. This capacitor must be specifically designed for low inductance. Note the high-speed switching diode D1. Its purpose is to damp out any reverse voltage that might occur in the circuit due to inductance after the laser turns off because a reverse voltage of only 3 V can destroy the SH laser. The rise time of this diode is comparable to that of the avalanche circuit, being only 6 ns before it begins to switch. The values of the capacitor and the current-limiting resistor R(s) depend upon the type of SH laser diode used.

LASER RECEIVER DETECTOR AND AUDIO CIRCUIT

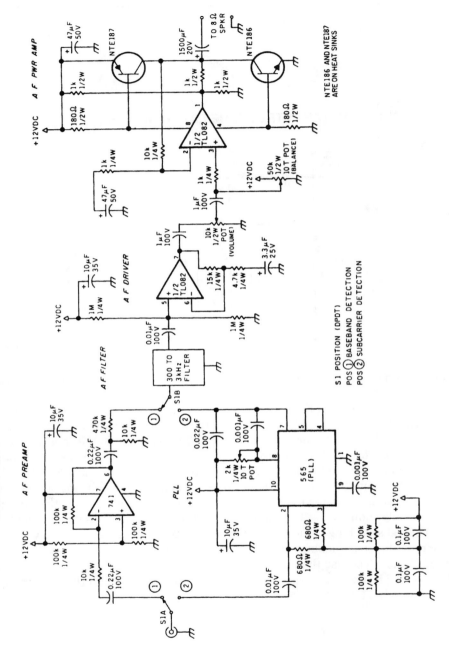

Fig. 45-8

LASER RECEIVER DETECTOR AND AUDIO CIRCUIT *(Cont.)*

Video preamplifier output can be switched between an audio amplifier that establishes the correct level of the demodulated baseband audio when used for short distance test links and a phase-locked loop (PLL) that demodulates an FM subcarrier when used for longer distances. The PLL VCO is set to a 100-kHz subcarrier frequency. The error signal generated as the VCO tracks the FM signal produces the demodulated audio. Either of these outputs is sent to a 300- to 3000-Hz bandpass filter and on to an audio power amplifier capable of driving a low-impedance speaker or headphones. The dc power supply for these circuits is provided by a filtered 12 Vdc input.

LASER CURRENT MODULATION

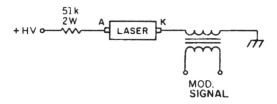

COMMUNICATIONS QUARTERLY   *Fig. 45-9*

Depending on laser characteristics, only a small percentage of modulation (around 10 percent) is possible with this scheme.

LASER MODULATION

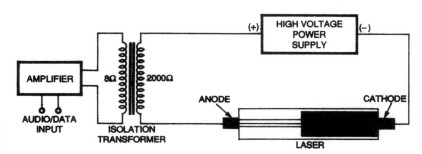

ELECTRONICS NOW   *Fig. 45-10*

Electronic modulation of a laser beam can be handled in this fashion. Obviously, the transformer you use must be able to handle the high voltage from the laser power supply without breaking down.

LASER TRANSMITTER

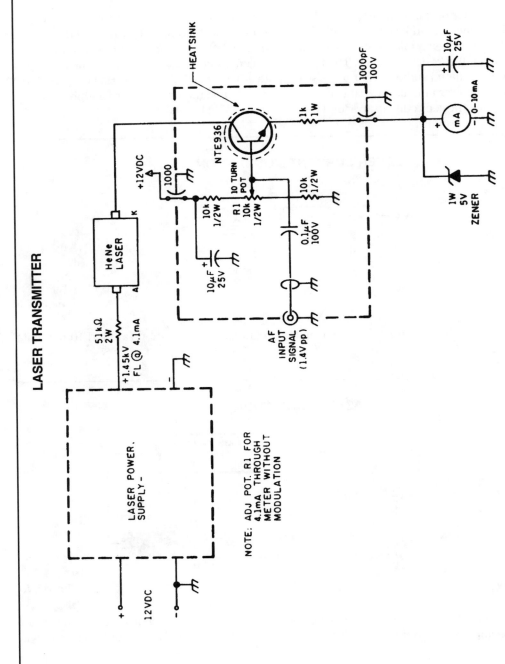

NOTE: ADJ POT. R1 FOR
4.1mA THROUGH
METER WITHOUT
MODULATION

Fig. 45-11

LASER TRANSMITTER *(Cont.)*

A current modulator is connected to the cathode of the laser. The quiescent current is set at 4.1 mA, and the modulating signal is set to vary the current only by ±0.7 mA. A well-protected and by-passed 0- to 10-mA meter is used in the ground leg to set and monitor the laser's quiescent operating current. A 51-kΩ, 2-W ballast resistor close to the anode lead of the laser stabilizes the discharge. The voltage across the sustaining supply should be 1.45 kV dc at full load. Remember that the sub-carrier frequency must be no greater than 200 kHz, and that the peak-to-peak signal voltage into the modulator must not exceed 1.4 V p-p. (There are any number of ICs that operate as voltage controlled oscillators to produce the FM subcarrier.)

LASER MODULATION WITH CURRENT SOURCE

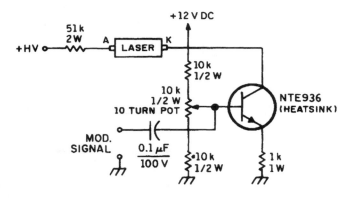

Fig. 45-12

Because atmospheric turbulence can become bothersome, any low-frequency baseband intensity modulation will become corrupted over long distances. To overcome this limitation, the information must be frequency-modulated up to 200 kHz in this manner. To obtain a more desirable current modulator for subcarrier modulation, use a transistor as a variable-current source to vary the current through the laser. Here the quiescent operating current through the laser is set at 4.1 mA by a potentiometer in the base bias circuit. Assuming a 0.7-Vdc drop across the transistor base-emitter junction and a 1000-Ω resistor in the emitter lead, the quiescent bias voltage from base to ground is 4.8 Vdc. The input voltage must vary 1.4 V p-p to cause a ±0.7-mA current variation through the laser. To effect this, the modulation signal source must provide up to a 1.4-V p-p signal into the laser current modulator circuit.

PHOTOMULTIPLIER LASER RECEIVER WITH VIDEO AMPLIFIER

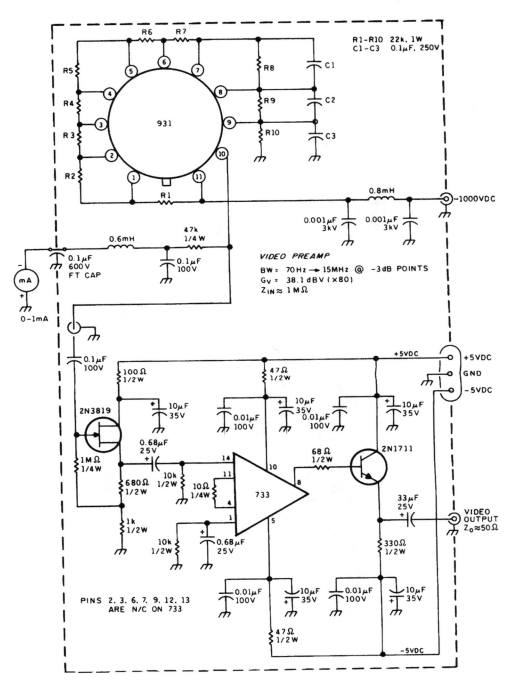

Fig. 45-13

PHOTOMULTIPLIER LASER RECEIVER WITH VIDEO AMPLIFIER *(Cont.)*

This is a circuit diagram for the inexpensive 931 side-looking PMT and its resistor divider network. This circuit maintains the proper voltages for the dynodes. The capacitors across the last three network resistors improve the frequency response of the PMT to modulated signals. A milliammeter in the PMT anode circuit registers the average PMT current under the medium to high illumination encountered during diagnostic tests. A PMT preamplifier uses a 733 (or equivalent) wideband amplifier. Because the demodulated signal can be either a direct baseband signal or a baseband signal modulated onto a subcarrier, the preamplifier must be capable of amplifying all the demodulated signal frequencies. The preamplifier is designed for a 70-Hz to 15-MHz bandwidth at the -3-dB points and for a gain of 80. Because the signal detection circuits will probably be placed in a separate housing, the video preamplifier is designed to drive a 50-Ω cable. With the high-current multiplication inherent in the PMT, the PMT noise will predominate over the video preamplifier IC noise.

46

Latch Circuits

The sources of the following circuits are contained in the Sources section, which begins on page 1043. The figure number in the box of each circuit correlates to the entry in the Sources section.

Transparent Latch
PB Switch-Activated Latch
Latching Switch Circuit
Digital Latch with Safety Reset Feature
555 Latch
One-Button Latching Switch

TRANSPARENT LATCH

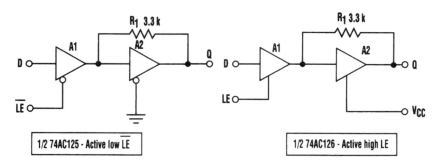

R₁ labels as R_1 3.3 k

1/2 74AC125 - Active low $\overline{LE}$

1/2 74AC126 - Active high LE

ELECTRONIC DESIGN

Fig. 46-1

The circuit depicted here, which yields two independent latches per chip, is rather simple and inexpensive to build (see the figure). The second buffer A2 is wired with resistor R1 in the feedback path (ignore A1 for the moment). A2 with feedback resistor R1 is a stable latch. Because of the very low input current requirements of A2, there is hardly any voltage drop across R1. As a result, the input is the same as the output, and that is fed forward through the buffer, maintaining a stable level. When buffer A1 is enabled, the input of A2 is driven to the same level as D. Even if A2's output (Q) is at an opposite logic level, which can happen momentarily (for a gate delay) when D is opposite to Q, A1 is required to sink or source current through R1. After a gate delay, this signal propagates to Q. Both sides of R1 are now at the same logic level. When the LE signal goes inactive, A1 tristates and the latch will hold the level present at D one setup time prior to LE's going inactive.

PB SWITCH-ACTIVATED LATCH

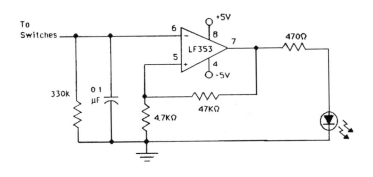

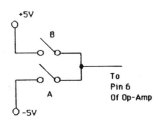

A	B	LED
ON	OFF	ON
OFF	ON	OFF

Last Input

TRUTH TABLE

NUTS AND VOLTS

Fig. 46-2

LATCHING SWITCH CIRCUIT

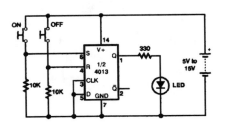

One button turns the LED on, the other turns it off. If you apply a positive pulse to the SET input, the output turns on. It remains on until a positive pulse is applied to the RESET input to turn it off. (TTL flip-flops such as the 74LS74 are actuated by negative, rather than positive pulses.)

ELECTRONICS NOW

Fig. 46-3

DIGITAL LATCH WITH SAFETY RESET FEATURE

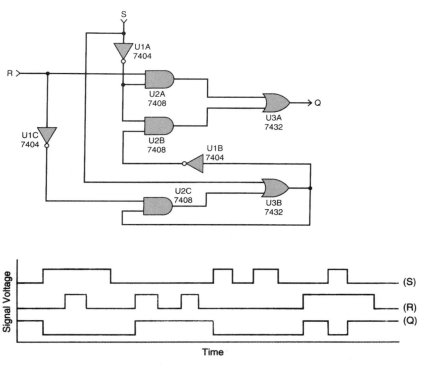

Fig. 46-4

The time diagram illustrates the various modes of operation of the circuit. A high S input causes the output (Q) to go low. Thereafter, a high R input can reset Q to high, but only so long as S remains low.

The asynchronous digital latching circuit is designed for use in a safety-related application, like turning off power in response to an alarm signal. During normal operation in the absence of an alarm, the SET (S) and RESET input voltages are low or off, while the output voltage (Q) is high or on. The SET input constitutes the alarm signal: Whenever "S" goes high (on), Q goes low (off), and thereafter remains low, even when S goes low. Thus, for example, the circuit keeps a power supply turned off even when the alarm has been shut off. If a safe condition has been restored, then the circuit can be reset to Q high by applying a high (on) signal to the RESET (R) input terminal. However, regardless of the R input level, Q cannot be driven high as long as S remains high; that is, the circuit cannot be reset if the alarm signal is still on. Thus, the RESET signal cannot override the alarm signal and thereby provide a false indication of safety. Also, this does not go into oscillation when the SET and RESET inputs change simultaneously.

555 LATCH

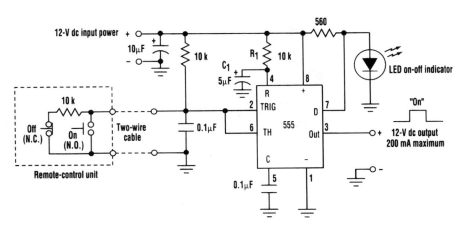

ELECTRONIC DESIGN

Fig. 46-5

Using a 555 chip in the memory mode, this push-button-controlled latch switch can source up to 200 mA of load current. Only one pair of wires is required to interface the ON and OFF push buttons to the control circuitry. The memory-mode feature of the 555 chip is implemented by connecting the trigger (pin 2) and threshold (pin 6) inputs together and applying one-half the supply voltage via a resistor network. Momentarily forcing the input low causes the output to go high, while forcing the input high causes the output to go low. To facilitate remote operation of the latch switch using one pair of wires, one resistor in the voltage-divider network is installed in the remote-control unit. Shorting this resistor out with the ON push button causes the output to go high. Conversely, opening this resistor with the OFF push button induces the output to go low. The R1-C1 network connected to the RESET input (pin 4) forces the latch to come up in the OFF state when power is first applied. The LED on-off indicator is kept off whenever the discharge output (pin 7) is conducting. When the output (pin 3) goes high.

ONE-BUTTON LATCHING SWITCH

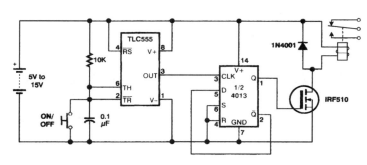

ELECTRONICS NOW

Fig. 46-6

The SET and RESET inputs are grounded, the inverted (-Q) output is fed back to the D input, and the pulses go into the CLOCK input. Each positive pulse makes the flip-flop toggle from one state to the other. The TLC555 chip in the figure serves two purposes. It inverts the pulses so that you can get a positive pulse from a switch that is connected to ground. More importantly, it also "debounces" the switch. When you press a button, it doesn't just make contact once—the contacts "bounce," opening and closing three or four times. The 4013 would toggle once on each bounce, leading to unpredictable results. The TLC555 uses a resistor and capacitor to smooth out these fluctuations so that each press of the button produces only one pulse.

47

Light-Control Circuits

The sources of the following circuits are contained in the Sources section, which begins on page 1043. The figure number in the box of each circuit correlates to the entry in the Sources section.

Dc Lamp Dimmer
Incandescent Lamp Touch Control
Christmas Light Dimmer
Incandescent Lamp-Life Extender
Timed Night Light
Twinkle Tree

DC LAMP DIMMER

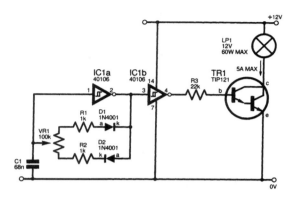

EVERYDAY PRACTICAL ELECTRONICS

Fig. 47-1

The dc lamp dimmer circuit will adjust the brightness of a dc lamp or the speed of a suitable dc motor. An oscillator is formed around Schmitt inverter IC1a, and is buffered by IC1b. This, in turn, drives TR1, which is a Darlington power transistor. The lamp LP1 forms the collector (c) load for the transistor, and this is driven at a nominal 330 Hz. The duty cycle is, however, fully variable using potentiometer VR1. Consequently, the power delivered to the load can be controlled between 5 and 95 percent, approximately. No flicker will be apparent at this frequency. The circuit consumes little power itself and is ideal for controlling car instrument panel lights.

INCANDESCENT LAMP TOUCH CONTROL

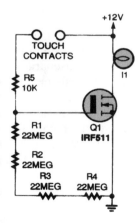

An IRF511 power MOSFET controls an incandescent lamp. This circuit is useful where low-voltage dc is used.

POPULAR ELECTRONICS

Fig. 47-2

CHRISTMAS LIGHT DIMMER

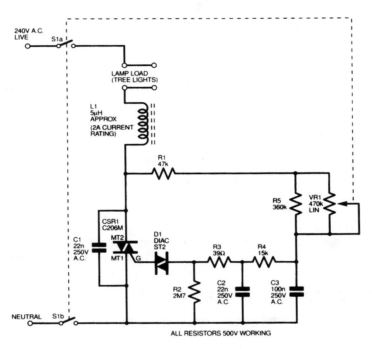

EVERYDAY PRACTICAL ELECTRONICS

Fig. 47-3

This is based on a standard diac/triac dimmer configuration, which has been optimized for this application. Resistor R2 is, however, optional and was included only to help discharge the capacitors quickly. All capacitors are 260-Vac X2 rated (*this is very important*), all resistors are 500-V working voltage, and triac CSR1 was chosen for its low holding current so that it would function with small loads. Note that VR1 is fitted with a double-pole mains switch for complete isolation. Even with only one 20-lamp string plugged in (22 W), control is very smooth with barely any trace of hysteresis. This circuit can be built safely on a piece of matrix board, wired point-to-point, provided it is mounted in a *completely insulated plastic case* and that no external components (i.e., S1 and VR1) are metal. (*Do not build this circuit if you do not understand the safety requirements.*)

INCANDESCENT LAMP-LIFE EXTENDER

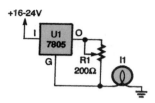

Fig. 47-4

The cold resistance of an incandescent lamp is normally very low compared to its operating resistance. Each time such a lamp is turned on, the initial current is several times greater than its rated operating current. The life-extender circuit will work with any incandescent lamp that operates at a voltage of 1.5 to 12 V and a current of 1 A or less. Look up the lamp's normal operating current and, using an ammeter, set R1 so that the normal operating current flows to the lamp. Now, each time the lamp is switched on, the initial current will be limited to its preset value.

TIMED NIGHT LIGHT

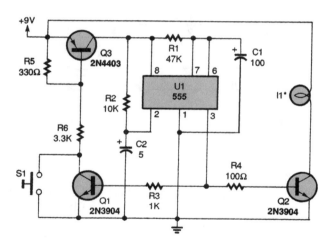

Fig. 47-5

A small lamp is turned on via switch Q2 for a predetermined time. S1 initiates the 555 timer cycle, holding on Q1 and switch Q3, supplying power to the circuit, and Q2, turning on the lamp. At the end of the cycle, power is removed from the circuit because Q3 is cut off. Therefore, no current is drawn from the battery during standby.

TWINKLE TREE

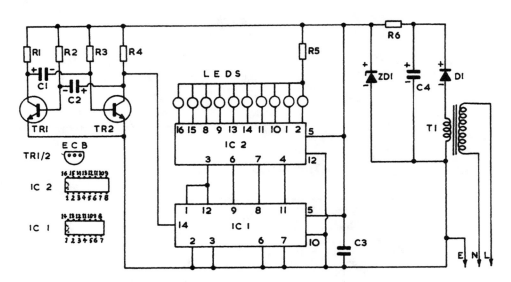

Fig. 47-6

Twinkle Tree is an easy project for beginners to build, and its basic circuit has a number of useful applications. The circuit's visible action appears as a string of 10 LEDs (light-emitting diodes) flashing on, one at a time, in sequence, this being repeated so long as the circuit is powered. The LEDs can be used on a small table decoration, in the form of a Christmas tree, or around a picture frame; or they can be placed at various points on a hanging decoration, such as a bunch of mistletoe, or can be incorporated in other decorations or modes. The LED light display provides an interesting and novel twinkling effect.

48

Light-Controlled Circuits

The sources of the following circuits are contained in the Sources section, which begins on page 1043. The figure number in the box of each circuit correlates to the entry in the Sources section.

HYSTERESIS-STABILIZING LIGHT SENSOR

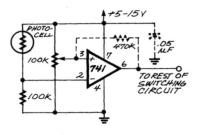

ELECTRONICS NOW

Fig. 48-1

This light turn-on circuit originally had oscillation problems. The 470-kΩ resistor introduces the hysteresis to prevent the oscillation. You can experiment with the value of the resistor for best results in your application.

AUDIO-CONTROLLED LAMP DIMMER

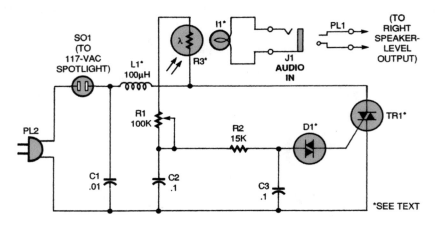

POPULAR ELECTRONICS

Fig. 48-2

The audio input to J1 lights low-voltage lamp I1. This light illuminates light-dependent resistor R3 (100 kΩ to 1 MΩ dark resistance) and modulates the intensity of the lamp plugged into SO1. L1 and C1 suppress RFI caused by the triac phase-control circuit. D1 and TR1 are a diac and a triac, respectively, and are the types used in common lamp dimmers.

DARK ALARM

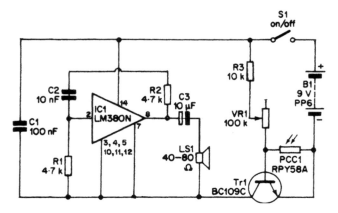

ELECTRONIC EXPERIMENTERS HANDBOOK

Fig. 48-3

This alarm senses darkness. Photocell PCC1 is normally irradiated with light and has a low resistance. In darkness, the resistance increases, and enough bias is available to turn on Tr1, activating the oscillator circuit consisting of IC1 and associated components. This produces a tone in speaker LS1.

MAGIC WAND

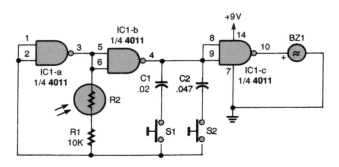

POPULAR ELECTRONICS

Fig. 48-4

This device is easy to use: Just aim the wand at a light source and listen to the tone. Two gates, IC1-a and IC1-b, of a quad two-input NAND gate are connected in an audio-oscillator circuit with R1, the photocell (R2), C1, and C2 setting the tone of BZ1. When the photocell is aimed at a light source, its resistance drops and the oscillator's output tone increases in frequency. When either of tone switches (S1 or S2) is pressed, the frequency range goes up. Pressing S2 shifts the oscillator into its highest-frequency range. Only one tone switch can be pressed at a time; if both switches are pressed simultaneously, the output stops.

LOSS-OF-LIGHT DETECTOR

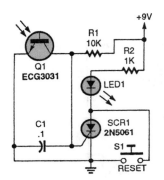

Light falling on Q1 causes Q1 to conduct. When the light fails, C1 charges toward +9 V, triggering SCR1, lighting LED1. S1 resets the SCR.

POPULAR ELECTRONICS   *Fig. 48-5*

LIGHT-ACTIVATED SWITCH

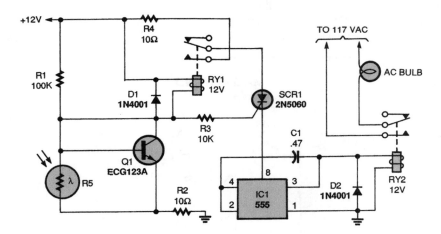

POPULAR ELECTRONICS   *Fig. 48-6*

When even a little light hits light-dependent resistor R5, transistor Q1 is turned off because the base has less resistance to the ground than to the positive rail. In that situation, the base is at a negative potential. When the sun sets, R5 is no longer illuminated, giving the transistor's base a high resistance to ground—higher than 100,000 Ω. With less resistance to positive potential, the base is biased, turning on Q1. Relay RY1 is then energized and pulls in, connecting the SCR1's anode to positive potential. The 555 timer, IC1, powers up, and its output goes high to approximately 10.67 V, which is sufficient to energize RY2. Relay RY2 then pulls in, keeping the ac bulb on the whole night, and turning it off again at sunrise.

LIGHT BLOCK SENSOR

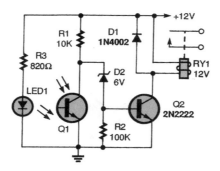

POPULAR ELECTRONICS

Fig. 48-7

In this circuit, relay RY1 remains open until the light source is blocked. As long as the IR light source is uninterrupted, phototransistor Q1's collector voltage is near zero. The voltage at zener diode D2's cathode is too low for conduction, keeping transistor Q2 off and the RY1 open.

PROGRAMMABLE CONTROLLER LIGHT SENSOR INTERFACE

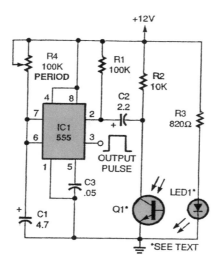

POPULAR ELECTRONICS

Fig. 48-8

As long as an object remains between LED1 and Q1, no output occurs. But as soon as the object moves out from between the LED and the phototransistor, IC1 (a 555 timer) is triggered, producing a timed output pulse. This circuit can be used to indicate when a part has moved away from a location.

ELECTRONIC TURBIDIMETER

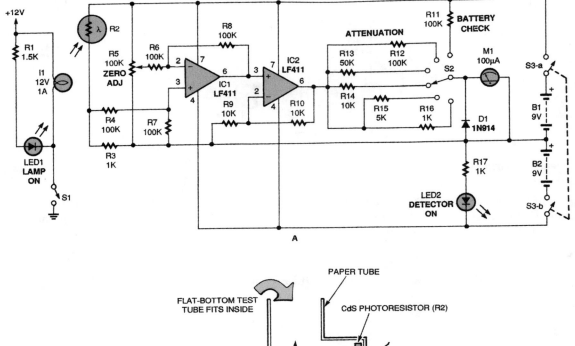

POPULAR ELECTRONICS

Fig. 48-9

A turbidimeter is a scientific instrument used to measure the cloudiness of solutions, such as water. If you measure the amount of scattered light, you can measure the turbidity. Paint the inside of the test-tube holder flat black. Glue a cadmium sulfide photoresistor (R2) to a cardboard disk, and glue the assembly to the open end of the 90° tube. Next, cement a biconvex lens (one that curves outward on both sides) to the bottom of the test-tube holder. Power up I1, a 12-V, 1-A automotive

ELECTRONIC TURBIDIMETER *(Cont.)*

light bulb, and hold it at various distances below the biconvex lens. Find the distance at which a beam of light comes straight up the tube or converges slightly, but does not diverge. Mount the lamp and play it on the 100-μA meter; the resistors on the rotary switch set the sensitivity or attenuation. Power is supplied to the circuit by two 9-V batteries in a split supply; I run the lamp off a separate 12-V supply. If you want truly calibrated readings, you'll need to buy a turbidity standard solution, available in JTU (Jackson Turbidity Units) and NTU (Nephelos Turbidity Units), from a science-supply company. In operation, zero the meter to a clean-water blank, note the reading from a known standard, note the reading from the unknown sample, and calculate the sample's turbidity, based on the needle's position.

LIGHT ALARM

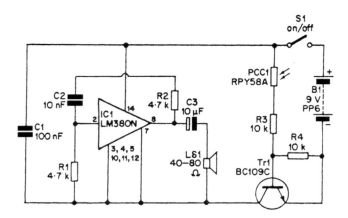

ELECTRONIC EXPERIMENTER'S HANDBOOK

Fig. 48-10

The circuit is activated if Tr1 is switched on by a suitable base current and voltage. The voltage and current available at the base of Tr1 are dependent on two main factors: the resistance provided by R4 and the setting of control VR1. If VR1 is set at maximum value, photocell PCC1 needs to have a resistance of about 10,000 Ω to bias Tr1 into conduction and activate the audio alarm circuit, of which IC1 is a primary part. Fixed resistor R4 has been used across the base-emitter terminals of the switching transistor so that the sensitivity of the circuit is preset. R4 can be raised somewhat in value if increased sensitivity is required. The audio-alarm generator uses an LM380N (IC1) in a simple audio-oscillator circuit, and drives high-impedance loudspeaker LS1 via coupling capacitor C3. Provided the losses through this coupling are less than the voltage gain provided by the amplifier, this will give sufficient positive feedback to sustain oscillation. The values for R1, R2, and C2 shown in the circuit diagram give considerably more feedback than is needed to just sustain oscillations, and the circuit oscillates strongly, producing a square-wave output at a frequency in the region of 1 kHz (1000 Hz).

PROGRAMMABLE CONTROLLER PHOTOELECTRIC INTERFACE

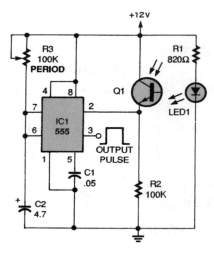

POPULAR ELECTRONICS

Fig. 48-11

This circuit is an IR sensor with a timed positive output pulse that will operate with older and slower PLCs. Some of the early PLCs have scan times of 15 ms or more. The sensor's extended output pulse can be set, by R3, to a time period longer than the controller's scan time. The sensor can also be used as a stand-alone circuit to operate a counter, a valve, an indicator, or any other electrically controlled device. As long as nothing is blocking the IR light source, the emitter of phototransistor Q1 is high and 555 timer IC1 is set in the READY state. When an object blocks the light source, Q1 turns off, sending a negative pulse to the trigger input at pin 2 of IC1, producing a timed output pulse. R_3 and C_2 determine the length of the output pulse. Larger values produce longer output pulses.

REFLECTIVE SENSOR

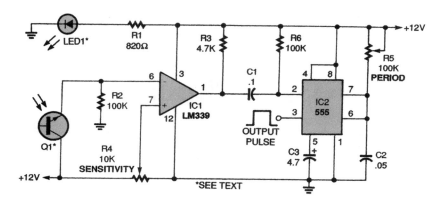

POPULAR ELECTRONICS

Fig. 48-12

The IR sensor circuit shown produces an output when LED1's light is reflected from an object back to phototransistor Q1. The LED and phototransistor should be mounted parallel to each other and aimed in the same direction. Without a reflective object, the voltage at the input of the LM339 comparator (IC1) is at or near ground level and the output at pin 1 of IC1 is high. When the phototransistor detects a reflected light signal, the voltage at Q1's emitter goes high, causing the comparator's output to go low. The 555 timer (IC2) is then triggered, and produces a timed output pulse at pin 3. The circuit's sensitivity is set by R4 and its output time period by R5. Note that in this circuit, all of the unused input and output pins of the LM339 must be tied to circuit ground.

49

Load Circuits

The sources of the following circuits are contained in the Sources section, which begins on page 1043. The figure number in the box of each circuit correlates to the entry in the Sources section.

Bidirectional Active Load
Active Load Resistor
Mixer Load Differential Amplifier

BIDIRECTIONAL ACTIVE LOAD

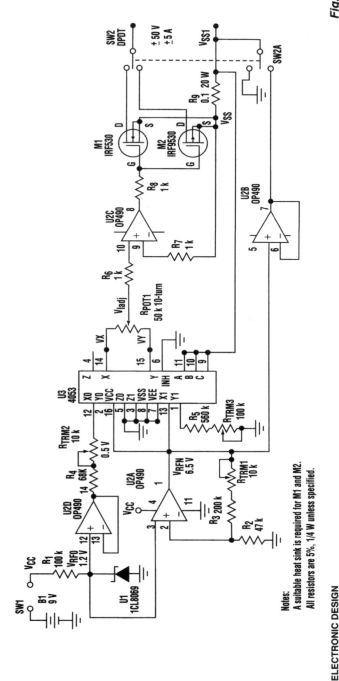

ELECTRONIC DESIGN

Fig. 49-1

The design presented here is a single +9-V, battery-operated, bidirectional active load that can sink and source current. This is a low-power design, consuming only about 140 μA. The power MOSFET selected (IRF530 N-channel and IRF9530 P-channel), after derating, accepts a maximum load of ±50 V at ±5 A. The key to the design is having two different sets of voltage levels at V_X, V_Y, and V_{SS1}. One set is for current-sinking test, while the other is for current-sourcing test.

443

ACTIVE LOAD RESISTOR

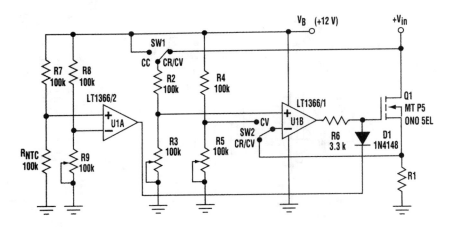

Fig. 49-2

The design idea presented here concerns an active power resistor. It can be used as a load resistor when probing or servicing power supplies. The circuit can work in three different modes. It can act as a constant resistor (mode CR), or as a constant current from a supply of any positive voltage. Finally, it can be in constant-voltage mode (CV), in which the circuit loads the voltage across supply terminals to a constant value adjusted by the user. The power MOSFET transistor Q1 works as a resistive component. The transistor gate is controlled by an op amp (U1B). The feedback voltage, which can be selected by switch SW2, is connected to the amplifier's inverting input. In CC and CR modes, the feedback voltage is the voltage between the source resistor (R1) terminals, which is proportional to the amplifier supply voltage (V_B) through a voltage-divider circuit (R4-R5). The amplifier noninverting input is controlled by a control voltage. The control-voltage input can be selected by switch SW1. In CR and CV modes, that input is the resistor input voltage (V_{in}), and in CC mode, it is the amplifier supply voltage (V_B). The control voltage is set in voltage divider R2-R3 to a proper value to control the amplifier. The other op amp (U1A) protects the MOSFET transistor. It is controlled by a resistor bridge circuit consisting of three resistors (R7 through R9) and an NTC resistor. The NTC is in contact with the transistor cooling element. With moderate element temperatures, the output voltage of U1A is high, and thus has no effect on the transistor gate because of the reverse-biased diode (D1). If the element temperature becomes high, the amplifier output has a zero value, which takes the transistor gate voltage to zero through the diode. The amplifiers also can be powered directly from the resistor input voltage. This circuit can work in CR mode without any control voltage. In CC and CV modes, however, it should be controlled by the external voltage CC/CV input. The circuit acts in CR mode as a resistor that has a resistance of

$$R = [(R_2 + R_3)/R_3] \times R_1$$

The required resistance value can be adjusted by potentiometer R3. In CC mode, the circuit sinks a current

$$I = R_3/[(R_2 + R_3)R_1] \times V_B$$

ACTIVE LOAD RESISTOR *(Cont.)*

The sink current also can be adjusted by potentiometer R3. In CV mode, the voltage between the terminals of the resistor is:

$$V=[(R_2+R_3)/R_3][R_5/(R_4+R_5)]\times V_B$$

The required constant voltage can be adjusted by both R3 and R5.

MIXER LOAD DIFFERENTIAL AMPLIFIER

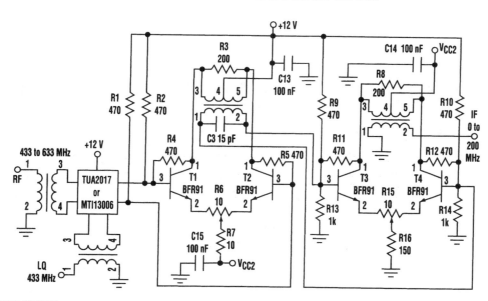

ELECTRONIC DESIGN

Fig. 49-3

Two different amplifier stages are used as the load of the mixer TUA2017 or MTI13006 (products developed by Siemens Co.). The LQ frequency is fixed at 433 MHz, because a commercial SAW oscillator at 433 MHz can be used as a local oscillator. If the RF frequency is swept from 433 to 633 MHz, then the IF frequency also is swept from 0 to 200 MHz. Compared to a passively loaded mixer, this combination of mixer and differential amplifiers exhibits excellent performance. The relative amplitude (amplitude difference between the fundamental frequency and the largest distortion frequency) is typically at least 50 dB. But it is only 30 dB (or less) in the case of a passive load mixer. Resistors R4, R5, R11, and R12 are used as the shunt feedback to obtain the flat output amplitude when the RF frequency is swept. The amplitude ripple with the IF bandwidth (0 to 200 MHz) is less than 0.5 dB.

50

Measuring and Test Circuits—Cable

The sources of the following circuits are contained in the Sources section, which begins on page 1043. The figure number in the box of each circuit correlates to the entry in the Sources section.

Cable Tester Transmitter
Cable Tester Receiver
Cable Short Detector
Multiconductor Short/Open Cable Tester
Cable Reflection Tester

CABLE TESTER TRANSMITTER

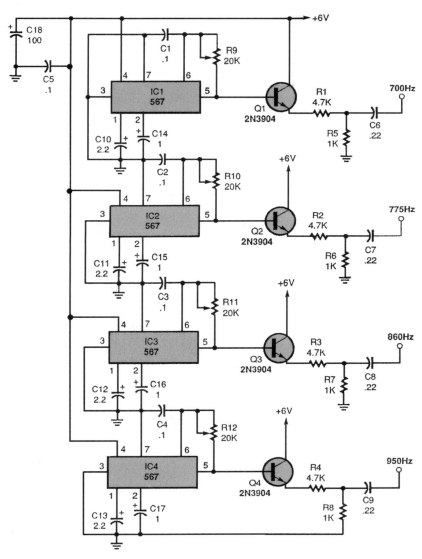

POPULAR ELECTRONICS

Fig. 50-1

Four 567 tone-decoder ICs, IC1 to IC4, are used to generate four different audio frequencies between 700 and 950 Hz. The capacitor from pin 6 of each 567 to circuit ground, along with the resistor between pins 5 and 6, sets the operating frequency. A 2N3904 NPN transmitter is connected in an emitter-follower circuit to isolate the ICs from output loading. Each output supplies a different audio tone to four wires in a cable.

447

CABLE TESTER RECEIVER

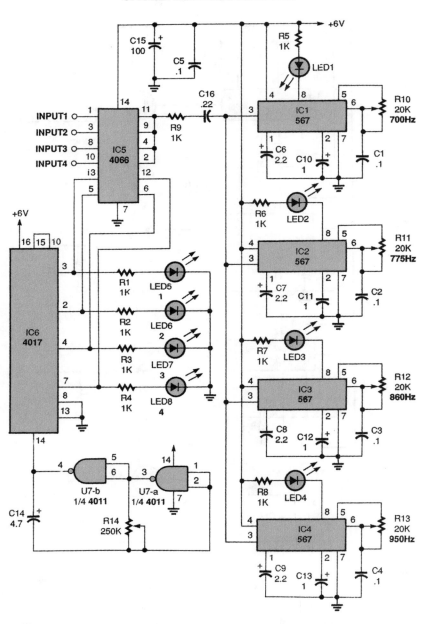

CABLE TESTER RECEIVER *(Cont.)*

The four inputs are connected to four input contacts of a quad bilateral switch, IC5, with the output contacts tied together and connected to the inputs of the 567 decoders through R9 and C16. The bilateral switch's four control inputs are operated by IC6, a 4017 decade counter IC, which is driven by a low-frequency oscillator made up of two gates of IC7, a quad two-input NAND gate. Four LEDs, LED 5 through LED8, are connected to IC6's four outputs and indicate which input is being checked, just as in manual mode. When LED5 is on, input 1 is switched through to the four decoder circuits. The decoder that responds, by lighting its LED, indicates which output—700, 775, 880, or 950 Hz— is connected to that input. Resistor R14 sets the input-stepping rate, which should be slow enough for recording.

CABLE SHORT DETECTOR

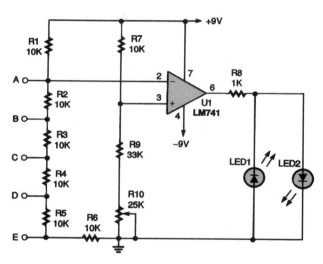

POPULAR ELECTRONICS

Fig. 50-3

A five-input short-detector circuit can be used to check long runs of a two- to five-wire cable for shorts between any of the wires. An LM741 op-amp IC, U1, is connected in a comparator circuit. The inverting input (pin 2) of U1 is connected to a +9-V source and a series of six resistors, while the noninverting input (pin 3) of the op amp is connected to an adjustable voltage-divider circuit. To use the circuit, potentiometer R10 should be used to set the voltage at pin 3 to a slightly more negative voltage than that at pin 2; U1's output will swing low, lighting LED1. Make that adjustment slowly until LED1 just barely turns on. Then connect one end of a cable to points A to E. When any of resistors R2 to R6 are shorted out, the voltage at pin 2 will go negative with respect to pin 3. That will make the op amp's output go positive, which will light LED2, indicating a short in the cable.

MULTICONDUCTOR SHORT/OPEN CABLE TESTER

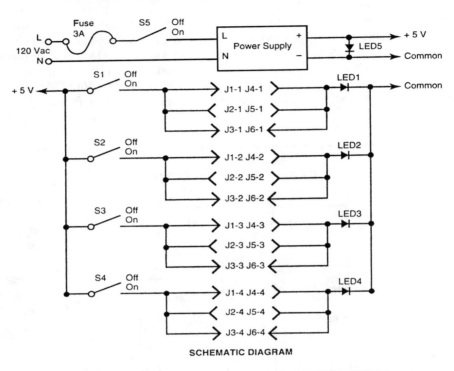

SCHEMATIC DIAGRAM

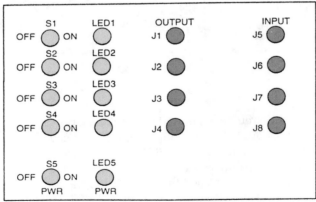

PANEL LAYOUT

MULTICONDUCTOR SHORT/OPEN CABLE TESTER *(Cont.)*

Frequent or regular testing of multiconductor cables terminated in multipin conductors can be a tedious, if not impossible, task. This is especially true if the cables must be tested for both open and short circuits for all of the various combinations of conductors. The inexpensive circuit shown in the figure simplifies the task. In operation, a pair of connectors is selected to match the pair of connectors installed on each of the cables to be tested. A single-pole/single-throw (SPST) toggle switch is provided for each conductor of whichever cable in the system has the greatest number of conductors. One conductor of each of the output connectors is connected in parallel to the anode of an LED. One LED is required for each conductor of the cable that has the greatest number of conductors. The cathodes of the LEDs are tied together and connected to the common side of the 5-V power supply. At the beginning of a test, all of the switches are turned off. First, one turns on power switch S5 and verifies that LED5 illuminates. Then one turns on switches S1 through S4, one at a time. Only the LED corresponding to the switch that is on should illuminate. An open conductor in the test cable is indicated if the corresponding LED is not illuminated. A short circuit is indicated if any other than the corresponding LED is illuminated.

CABLE REFLECTION TESTER

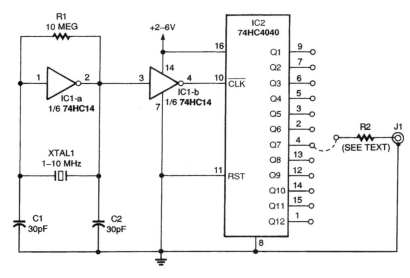

ELECTRONICS NOW

Fig. 50-5

This device is a square-wave generator and frequency divider. A square-wave generator supplies a signal to the cable through a resistance (R_2) equal to the characteristic impedance of the cable. Any reflections caused by mistermination will show up on a scope hooked to J1. IC2 acts as a frequency divider to produce various pulse rates for testing long and short cable.

51

Measuring and Test Circuits— Capacitance

The sources of the following circuits are contained in the Sources section, which begins on page 1043. The figure number in the box of each circuit correlates to the entry in the Sources section.

High-Resolution Bridge
Simple Capacity Meter
Auto-Ranging Capacitance Meter
DVM Capacitance Meter Adapter
Capacitor Checker
Capacitive Sensor System
Capacitor Leakage Tester

HIGH-RESOLUTION BRIDGE

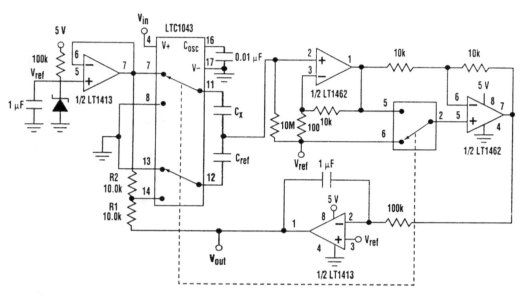

ELECTRONIC DESIGN

Fig. 51-1

A JFET-input op amp (LT1462) amplifies the signal before demodulation for good noise performance, and the integrator's output is attenuated by R1 and R2 to increase the sensitivity of the circuit. If $\Delta C_X << C_X$ and $C_{ref} \approx C_X$, then

$$V_{out} - V_{ref} \approx V_{ref} (\Delta C_X / C_{ref})[(R_1 + R_2)/R_2]$$

With $C_{ref} = 50$ pF, the circuit has a gain of 5 V/pF and can resolve 2 fF. Supply current is 1 mA. The synchronous detection makes this circuit insensitive to external noise sources; in this respect, shielding isn't terribly important. However, to achieve high resolution and stability, care should be taken to shield the capacitors being measured. Bridge circuits are particularly suited for differential measurements. When C_X and C_{ref} are replaced with two sensing capacitors, these circuits measure differential capacitance changes, but reject common-mode changes. CMRR for the circuit in this figure exceeds 70 dB. In this case, however, the output is linear only for small relative capacitance changes.

SIMPLE CAPACITY METER

Fig. 51-2

Parts List

BT1 9V alkaline battery
C1, C3, C4, C6   0.01 µF disc ceramic capacitors
C2 10 µF 16V electrolytic capacitor
C5 220 pF polystyrene, mylar or dip mica capacitor
D1 Germanium diode: 1N34, 1N60, 1N90, 1N270
J1, J2 Binding post
M1 Meter, 100-300 µA full scale (see text)
R1, R2 560 ohm 1/4W 5% resistor
R3 Carbon potentiometer, 200 to 300 Ohms
R4 1K 1/4W 5% resistor
S1 SPST toggle or slide switch
U1 74LSOO
U2 78LO5
Y1 3.57955 MHz crystal

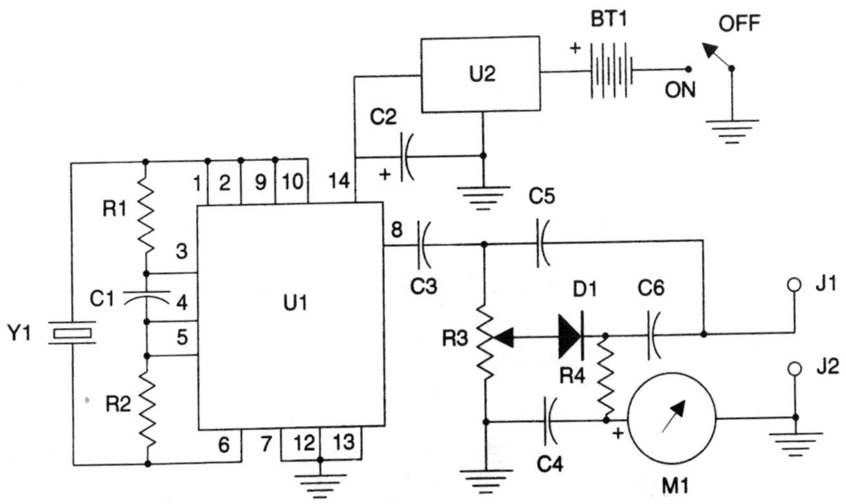

73 AMATEUR RADIO TODAY

SIMPLE CAPACITY METER *(Cont.)*

The figure shows the schematic diagram of the simple capacity meter. U1 (a 74LSOO logic chip), two resistors, a capacitor, and a crystal form a crystal oscillator operating near the marked frequency of the crystal. The RF voltage is taken from pin 8 through isolation capacitor C3 and applied to the modified Wheatstone bridge circuit. R3 forms two arms of the bridge, with the arms ratio variable through the position of the wiper of R2. C5, which should be the stable capacitor specified in the parts list, and the unknown capacitor to be measured form the remaining bridge arms. R3 is adjusted for bridge balance, indicated by a dip minimum shown on microammeter M1, and the value of the unknown capacitor is indicated on the calibrated dial attached to R3. The instrument is powered by BT1, a 9-V battery, controlled by on/off switch S1. This 9 V is reduced and regulated by U2, a 78LO5, to the +5 V required by U1.

AUTO-RANGING CAPACITANCE METER

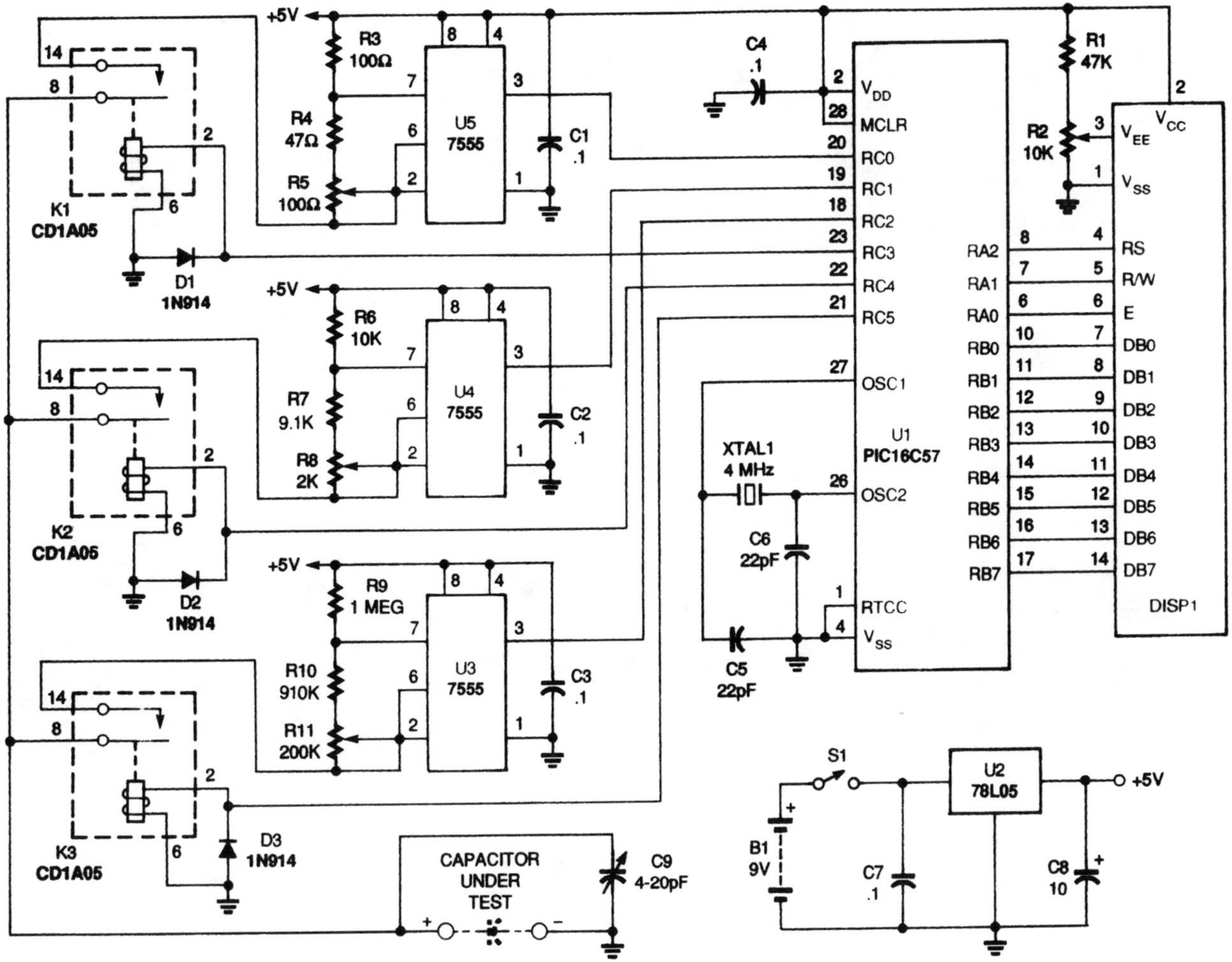

Fig. 51-3

POPULAR ELECTRONICS

456

AUTO-RANGING CAPACITANCE METER *(Cont.)*

The schematic for the meter is shown. When switch S1 is closed, a 9-V battery, B1, supplies power to a 78L05 regulator, U2, and to filter capacitors C7 and C8. The result is a 5-Vdc supply for the rest of the circuit. Because the current draw is low, the 9-V battery will provide close to 40 $\frac{1}{2}$ hours of usage. A microchip PIC16C57, U1, is at the heart of the circuit. The internal EPROM is programmed with the meter's program. The circuit uses one of three 7555 CMOS timers, U3 to U5, to generate a pulse train. Microcontroller U1 turns on one relay to connect the capacitor to one of the three timing circuits. The timer circuit, made up of U5, R3, R4, R5, and C1, has the lowest resistance and impedance; U4, with its surrounding components, has a medium resistance and impedance; and U3's circuit has the highest. All the timer circuits have two nominally equal resistors; one of them is the series combination of a fixed resistor and a trimmer potentiometer. That variable combination is for calibration. The high-impedance (1-MΩ) circuit is the default and is used for small-valued capacitors, the medium-impedance (10,000-Ω) circuit is used for medium-valued capacitors, and the low-impedance (100-Ω) circuit is used for high-valued capacitors. When the capacitor under test is connected to the appropriate circuit, the microcontroller measures the frequency. A 4-MHz crystal, XTAL1, gives U1 a 1-MHz internal clock frequency. The microcontroller can, therefore, count the number of pulses from the appropriate timer circuit in a known period of time. It then goes on to convert the frequency to capacitance using 32-bit, floating-point arithmetic, and displays the results on a 16 × 1 LLCD, DISP1, with the correct unit and decimal point. Capacitor C9 is connected in parallel to the capacitor under test to aid in final calibration and for stability. Diodes D1 to D3 protect the microprocessor from the back EMF of the relays' coils. Before you can use the PIC microprocessor in the meter, you have to program it. If you have the equipment to do that, you can obtain the software from the Gernsback BBS (516-293-2238).

DVM CAPACITANCE METER ADAPTER

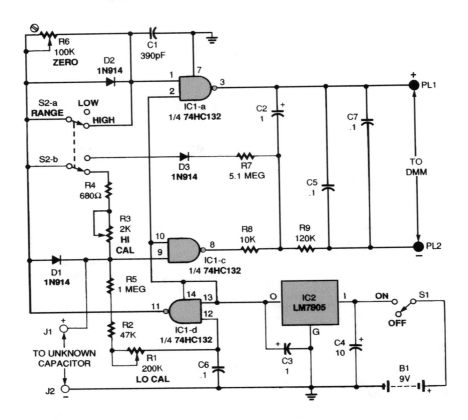

POPULAR ELECTRONICS

Fig. 51-4

The figure shows the schematic for the capacitance meter adapter. When power switch S1 is in the ON position, an LM7805 regulator (IC2) provides a 5-V source to the circuit. Switch S2 is used to select between the high- and low-capacitance ranges. The circuit measures from 0 to 2200 pF in its low range and from 0 to 2.2 μF in its high range. The adapter outputs 1 mV/pf in the lower range and 1 V/μF in the higher range. Those voltages would be displayed on a DMM, and would have to be read as capacitance values. Gate IC1-d is configured as a free-running oscillator whose frequency is determined by the setting of potentiometer R1. The output of IC1-d, pin 11, is a square-wave voltage that is fed to the other two gates configured as inverters (IC1-a and IC1-c). A capacitor with an unknown value (C_x) is connected across input terminals J1 and J2. Note that the positive lead of a polarized capacitor must be connected to J1. The capacitor under test charges through diode D1 on positive half cycles and discharges through R5 on negative half cycles in the low range, or through R3, R4, and R5 in the high range. In the high range, the square-wave output from IC1-d is fed directly to the inputs of the other two gates. With no capacitor connected to J1 and J2, the outputs of IC1-a and IC1-c are identical inverted versions of the input signal. The average voltage across the

DVM CAPACITANCE METER ADAPTER *(Cont.)*

outputs of the gates at pins 3 and 8, in this case, is zero. With C_x in place, and depending on its value, the input voltage to IC1-c, pin 9, stays higher longer than the input to IC1-a, pin 2. The output pulses from IC1-a and IC1-c are filtered and smoothed by R8, R9, C2, and C5. The average dc voltage across the outputs is then proportional to the value of C_x. The circuit is a bit different in the low range because compensation must be made for stray capacitance. In the low range the output from IC1-d is fed to pin 1 of IC1-a through diode D2, which also charges C1. Capacitor C1 charges quickly and discharges slowly through potentiometer R6. The charge on C1 holds the input to IC1-a higher for a slightly longer time than would normally be the case. That causes its inverted output to be low for a slightly longer time, and conversely high for a slightly shorter time. Stray capacitance across input terminals J1 and J2 does the same thing to IC1-c, holding its input higher for a slightly longer time than would be the case if there were none. Potentiometer R6 is adjusted so that the discharge time for C1 matches that of the stray capacitance and so cancels it out. In the low range, an offset voltage is applied to the negative output terminal via D3, R7, and R8 to prevent the gates from locking together with such close trigger points.

CAPACITOR CHECKER

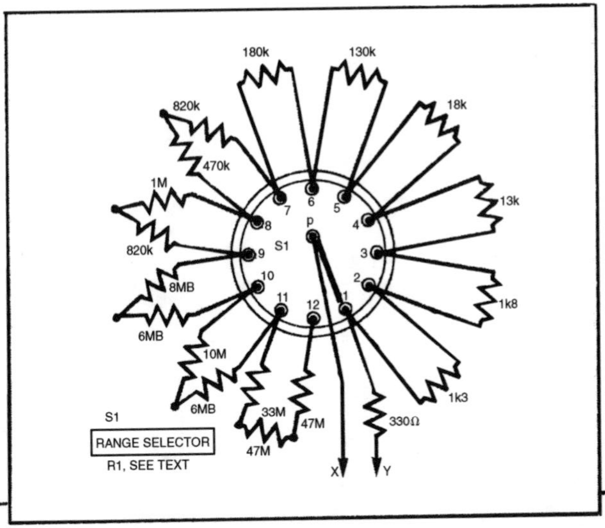

Table. 1

Range	Total resistance needed	Resistors used cumulatively
		330Ω
5000μF	333Ω	1k3
1000μF	1k67	1k8
500μF	3k33	13k
100μF	16k7	18k
50μF	33k3	130k
10μF	167k	180k
5μF	333k	820k + 470k
1μF	1M67	1M + 820k
500nF	3M33	6M8 + 6M8
100nF	16M7	10M + 6M8
50nF	33M33	47M + 47M + 33M (or 100M
10nF	167M	+ 33M)

S1
RANGE SELECTOR
R1, SEE TEXT

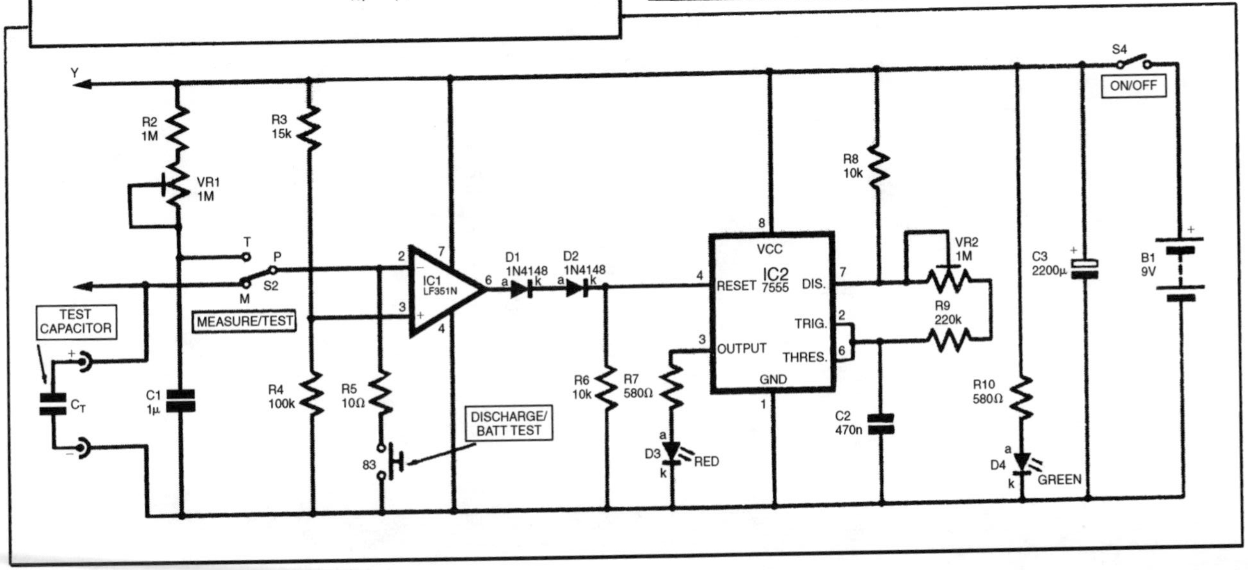

Fig. 51-5

EVERYDAY PRACTICAL ELECTRONICS

460

CAPACITOR CHECKER *(Cont.)*

The circuit for the capacitor checker is shown in the figure. Switch S2 is used to select the Measure (M) or Test (T) mode, as required. Assume for the moment that it is in the M position, as shown. Operational amplifier IC1 is configured as a voltage comparator. The values of the resistors have been chosen to provide 86 percent (actually, it is nearer 87 percent) of the supply voltage to the noninverting input, pin 3. Meanwhile, the capacitor being measured, C_t, charges through resistor R_1, so the voltage across it rises and follows the form shown in Fig. 2. R_1 is selected from a chain of resistors using the rotary range switch, S1 (inset). The value of R_1 in the chain will be such that with a capacitor at the top of the range selected (e.g., a 100-nF capacitor on the 100-nF range), the time constant will be 1.66 s. The values of R_1 required to provide this time constant on the various ranges are shown in Table 1. Assuming that the test capacitor C_t is discharged to begin with, after two time constants have elapsed, the voltage across the capacitor will rise above 86 percent of the supply. The op amp switches off, with pin 6 becoming low. While IC1 pin 6 is high (during the initial stages of C_t charging), IC2 pin 4 is maintained in a high state, which enables the IC. When the capacitor charges through two time constants, IC1 pin 6 becomes low, IC2 is disabled, and the flashing stops. The nominally low state of IC1 output would still be high enough to enable IC2 if connected directly, and the two diodes (D1 and D2) correct this. Resistor R6 maintains IC2 pin 4 *low* in the absence of a *high* state from IC1 output. The rate at which pulses are produced depends on the values of fixed resistors R8 and R9 and preset potentiometer VR2, in conjunction with capacitor C2. With the values chosen, they will be delivered at between 1 and 10 per second approximately, depending on the adjustment of VR2. VR2 will be adjusted so that exactly *three* flashes per second are provided. Ten flashes, therefore, correspond to 3.33 s, which is equal to two time constants for a capacitor at the limit of the range selected.

CAPACITIVE SENSOR SYSTEM

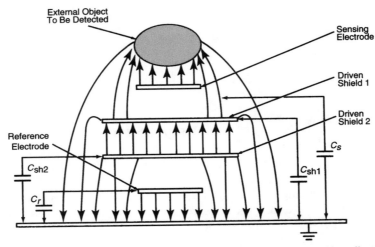

Figure 1. Driven shield 2 isolates the reference electrode from the capacitive effects of an object in the vicinity of the sensing electrode and of driven shield 1.

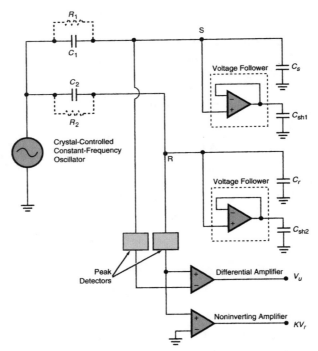

Figure 2. Unbalance and the bridge circuit give rise to nonzero V_u, which serves as a measure of proximity of an external object to the sensing electrode

Fig. 51-6

CAPACITIVE SENSOR SYSTEM *(Cont.)*

Figure 1 illustrates the electric-field configuration of a capacitive proximity sensor of the "capaciflector" type. This one includes a sensing electrode driven by an alternating voltage, which gives rise to an electric field in the vicinity of the electrode; an object that enters the electric field can be detected by its effect on the capacitance between the sensing electrode and electrical ground. Also, it includes a shielding electrode (in this case, driven shield 1), which is excited via a voltage follower at the same voltage as that applied to the sensing electrode to concentrate more of the electric field outward from the sensing electrode, increasing the sensitivity and range of the sensor. Because the shielding electrode is driven via a voltage follower, it does not present a significant electrical load to the source of the alternating voltage. In this case, the layered electrode structure also includes a reference electrode adjacent to ground, plus a second shielding electrode (driven shield 2), which is excited via a voltage follower at the same voltage as that applied to the reference electrode. Driven shield 2 isolates the reference electrode from the electric field generated by driven shield 1 and the sensing electrode, so a nearby object exerts no capacitive effect on the reference electrode. The excitation is supplied by a crystal-controlled oscillator and applied to the sensing and reference electrodes via a bridge circuit, as shown in Fig. 2. Fixed capacitors C1 and C2 (or, alternatively, fixed resistors R1 and R2) are chosen to balance the bridge—that is, to make the magnitude of the voltage at the sensing-electrode node S equal the magnitude of the voltage at the reference-electrode node R. The voltages at S and R are peak-detected and fed to a differential amplifier, which puts out voltage $V_u = 0$. When an object intrudes, it changes C_s, unbalancing the bridge and causing V_u to differ from 0. The closer the object comes to the sensing electrode, the larger $|V_u|$ becomes. An additional output voltage KV_r is also available, where K is the amplification factor of a noninverting amplifier and V_r is the voltage on the reference electrode.

CAPACITOR LEAKAGE TESTER

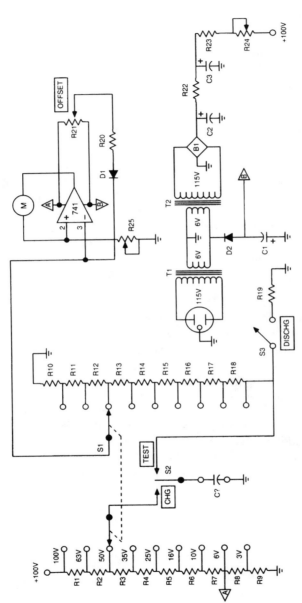

Fig. 51-7

In this circuit, a capacitor is charged to a voltage near its working voltage, as selected by S1. Then S2 is placed in the TEST position, and high-resistance divider R10 to R18 is used to discharge it. The total resistance is 100 MΩ. Taps selected by S1 are used to reduce the voltage "seen" by the op-amp metering circuit to an appropriate value (around 3 V). The discharge curve is watched on the meter. If it is too rapid, the capacitor might be leaky. This technique is useful for values of C that produce a sufficiently large time constant, 10 s or more ($C>0.1$ μF).

52

Measuring and Test Circuits—
Continuity/Resistance

The sources of the following circuits are contained in the Sources section, which begins on page 1043. The figure number in the box of each circuit correlates to the entry in the Sources section.

Milliohm DMM Adapter
Kelvin Connection
Wheatstone Resistance Bridge
Continuity Checker
Simple Continuity Tester
Milliohm Tester
Low-Range Ohmmeter

MILLIOHM DMM ADAPTER

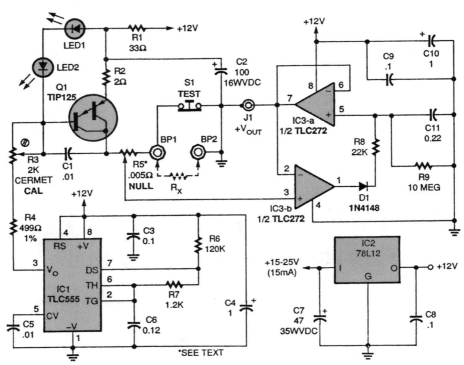

POPULAR ELECTRONICS

Fig. 52-1

The circuit provides a 1.000-A constant-current pulse of 100-μs duration, at a 100-Hz repetition rate, to the unknown (R_x) resistance, and provides an output voltage to your DMM of 1.000 V per ohm of input resistance. Accuracy is better than ±1 percent up to 5.00 Ω, and the resolution using a 3½-digit DMM is 100 μΩ. Extremely low resistance readings are limited only by the quality of the mechanical connections to R_x and the accuracy of the NULL setting. IC1 (a TLC555) provides the pulse timing to TIP125 transistor Q1, which is configured as a capacitive discharge current source. Capacitor C2 provides the current through R2, a stable metal-film or wire-wound resistor. Two red LEDs, LED1 and LED2, are selected to have a forward voltage drop of about 1.75 V each (3.50 V total) at a current range of 4 to 15 mA, as provided in the circuit. Transistor Q1 has a dynamic V_{be} of about 1.50 V, so the LEDs provide a 2.00-V reference across R2 with some temperature compensation. Integrated circuit IC3 (a TLC272) is configured as a peak detector. The 78L12 provides a regulated 12 V to the circuit. The R_x terminals, BP1 and BP2, must be heavy-duty binding posts; do not use test leads because that will introduce errors greater than the value of R_x. Resistor R5, the 0.005-Ω trimmer potentiometer, is simply a 2½-in length of tinned, 24-gauge busbar wire, with the lead from pin 3 of IC3 terminated in a clip that will slide on the busbar. To calibrate the circuit, first use a heavy shorting

466

MILLIOHM DMM ADAPTER *(Cont.)*

bar between the input terminals, push S1 (TEST), and adjust R5 (NULL) for an output of 1.00 mV, or whatever value is convenient. You will have to subtract that NULL value from the actual readings. Next, replace the shorting bar with an accurate resistance, such as two 10.00-Ω, 1-percent resistors in parallel, and adjust R3 for the proper output (5.000 V).

KELVIN CONNECTION

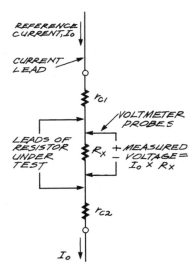

ELECTRONICS NOW

Fig. 52-2

For low-resistance measurements, contact and test lead resistance introduce errors. This figure shows the way to avoid these. Current is applied through test leads, but separate leads are used to read voltage drop.

WHEATSTONE RESISTANCE BRIDGE

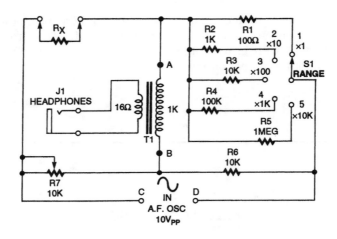

POPULAR ELECTRONICS

Fig. 52-3

R7 is a calibrated potentiometer. At null, R_x equals $R_n(R_7/R_6)$, where $R_n = R_1, R_2, R_3, R_4,$ or R_5, as selected by switch S1.

CONTINUITY CHECKER

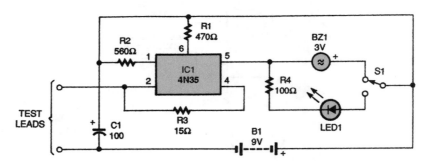

POPULAR ELECTRONICS

Fig. 52-4

The checker is built around a 4N35 optocoupler. When the test leads are shorted by a resistance, it turns IC1 on by pulling pins 2 and 4 low, allowing current to flow through pins 6 and 5. The current to pin 5 flows through either the buzzer or the LED, depending on the position of S1. To test continuity of 50 Ω or less, use the buzzer as the indicator. For continuity of up to 1000 Ω, use the LED. The circuit can also be used to check PN junctions of transistors and diodes, with less possibility of damaging such sensitive components. For that, use the LED, and connect the lead from pin 2 of IC1 to the P side. The supply can be from 7.5 to 9 V. To catch fast pulses, trigger a monostable with pin 5, and use it to drive the LED and the buzzer. The monostable will act like a pulse stretcher.

SIMPLE CONTINUITY TESTER

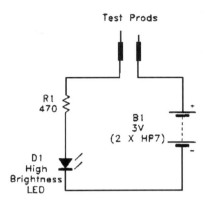

EVERYDAY PRACTICAL ELECTRONICS

Fig. 52-5

This figure shows the circuit for a simple, but safe, continuity tester.

MILLIOHM TESTER

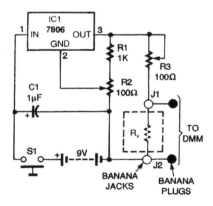

ELECTRONIC EXPERIMENTERS HANDBOOK

Fig. 52-6

The milliohm adapter circuit is powered from a 9-V battery. A resistor to be tested (R_x) is connected across banana jacks J1 and J2, and a pair of banana plugs, connected directly to J1 and J2, is plugged into the voltage input jacks of a DMM. Switch S1 applies battery power to 7806 voltage regulator IC1. Capacitor C1 removes voltage transients. Resistors R1 and R2 form a voltage divider for the GROUND pin of IC1. Potentiometer R2 trims IC1's output voltage to exactly 6 Vdc. Potentiometer R3 sets the output current through R_x to 100 mA. Because R3 is a relatively large resistance compared to R_x, the error introduced by different values of R_x (1 mΩ to 1 Ω), the effect it will have on the 100-mA current source, is below 2 percent.

LOW-RANGE OHMMETER

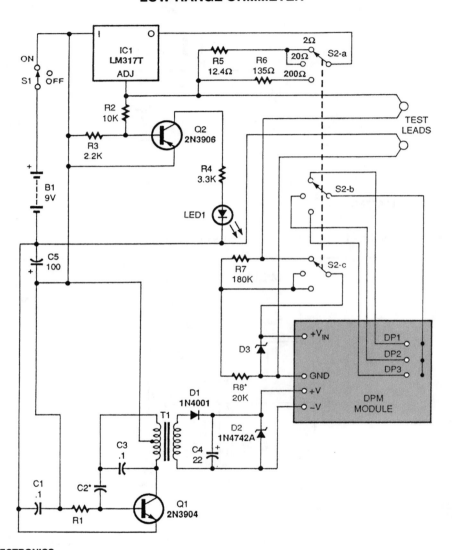

Fig. 52-7

This direct-reading self-contained low-ohms meter uses the four-wire, constant-current method of resistance measurement. That eliminates any effect of test-lead resistance on measurements.

T1 is a 1:1 audio transformer, 600:600 Ω. D3 is an optional Zener diode to protect the digital panel meter module DPM from excessive input voltage. The lowest reading for this circuit is 1 mΩ.

53

Measuring and Test Circuits—Current

The sources of the following circuits are contained in the Sources section, which begins on page 1043. The figure number in the box of each circuit correlates to the entry in the Sources section.

PICOAMPERE MEASURER

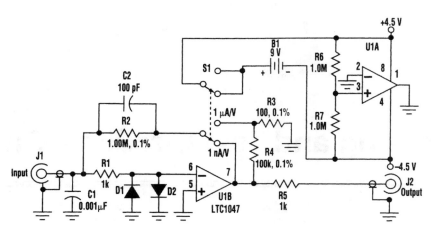

ELECTRONIC DESIGN

Fig. 53-1

Many times, the need arises to measure current below 1 μA. The circuit shown turns any voltmeter into a picoammeter with scales of 1 nA/V and 1 μ1A/V. With a 3½- digit voltmeter with a resolution of 1 mV, the readout will be in picoamperes or nanoamperes. In addition, it can be attached to an oscilloscope. The frequency response is about 1 kHz for the 1 μA/V setting and 150 Hz for 1 nA/V. U1B forms a transimpedance amplifier. With S1 in the position shown, the transimpedance is 1 MΩ. In the other position, a gain of 1000 is added, to make the total transimpedance 1GΩ. R1, C1, D1, and D2 protect the input from high voltages, and R5 isolates the op amp's output from any load capacitance. The op amp's input current and voltage offset must be low for this circuit to work. In this case, a Linear Technology LTC1047 was used. It has a nominal input bias current of ±5 pA and a VOS of ±3 μV at room temperature. U1A is used to split the 9-V battery into positive and negative supplies. The total current is essentially the supply current of the op amps. For the prototype, the total current measured for the LTC1047 was <100 μA, so a standard 9-V battery should last six months if you forget to turn it off.

INTEGRATING CURRENT METER CIRCUIT

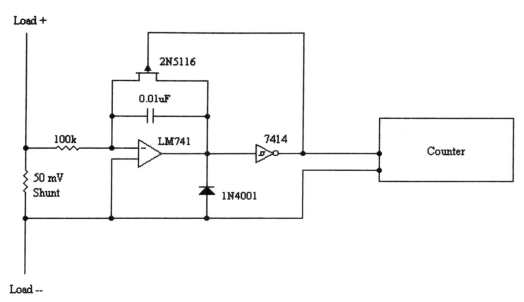

NUTS AND VOLTS

Fig. 53-2

The current in the line produces a dc drop across the shunt, which feeds an integrator. When the integrator output reaches the level required to activate the Schmitt trigger, the 0.01-μF capacitor that is used in the integrator is discharged and a pulse is produced at the output of the Schmitt trigger (7414). A digital counter counts these pulses, and the counter reading is proportional to the product of current flow and time.

CURRENT SOURCE

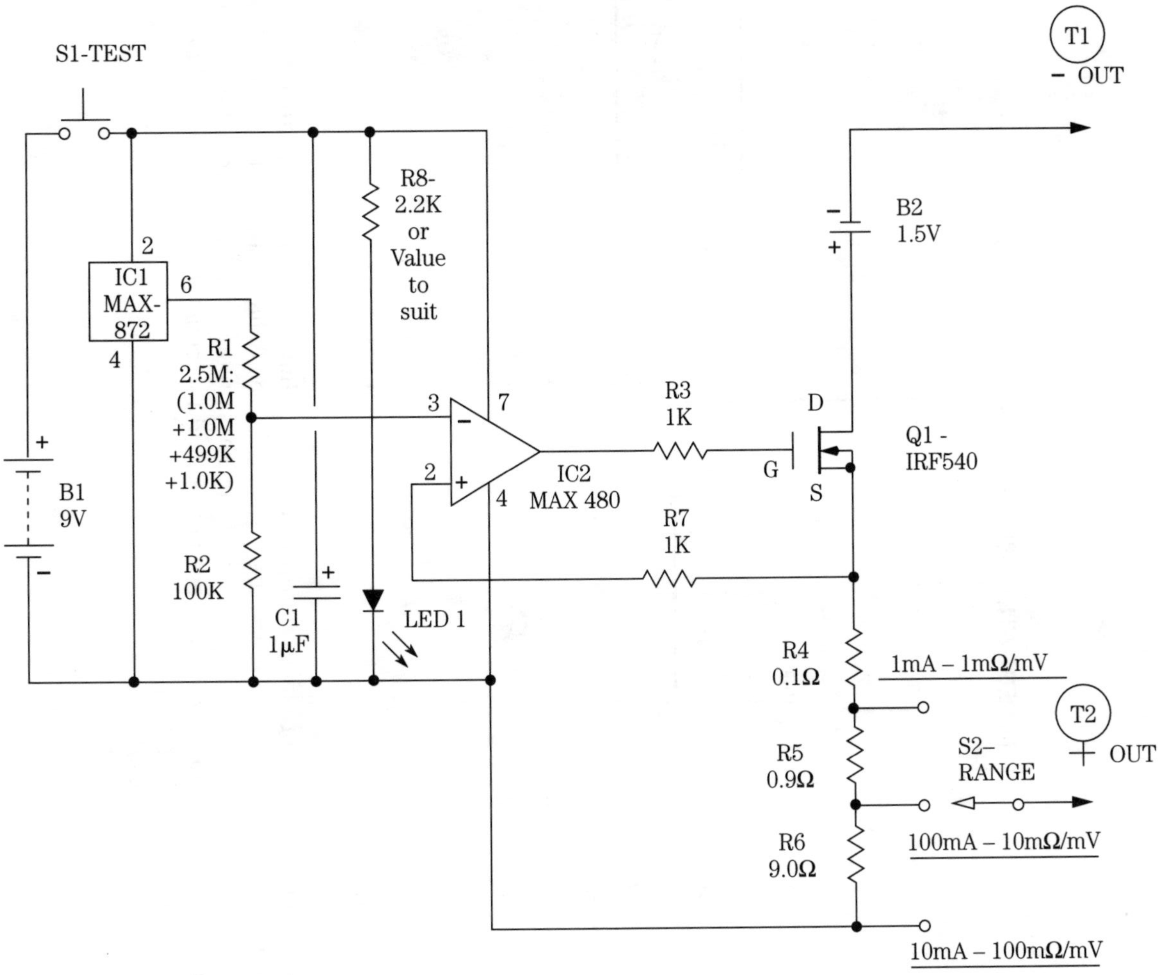

Fig. 53-3

CURRENT SOURCE *(Cont.)*

The figure shows the circuit of the milliohm adapter. It can apply 1 A, 100 mA, or 10 mA to the resistance under test. The current application will indicate the resistance directly in millivolts if your voltmeter is set on the 2-V range. B1 is the high-current source, an alkaline battery capable of generating the 1 A used in the circuit's highest-current application mode. This *must* be an alkaline battery. A three-position switch enables user selection of the desired applied current to the DUT. S1 performs as a dual-function on/off switch. When a measurement is to be made, pressing S1 turns on the circuit, which consists of a micropower voltage reference chip IC1 and an op amp, IC2. IC1's output is 2.500 V, and that voltage is divided by R1 and R2, producing 0.1 V at the input of IC2. This turns on Q1, creating a 0.1-V potential at its source terminal. The resistive divider (R4, R5, and R6) forms a current source along with Q1, the 1.5-V battery and the DUT's resistance. That is applied to the DUT through the two output terminals, T1 and T2. When you connect your voltmeter across these terminals, you can directly read the resistance of the DUT. An optional LED shows when the circuit is in test mode. The 1-A range of the instrument will drain the 1.5-V alkaline battery faster than the 9-V battery, so it is advisable for reasons of economy not to keep S1 depressed for protracted periods on that range, or else the battery life will suffer. Releasing S1 will halt the current source through Q1 and the associated circuit path, and thus will shut off the current flow from the 1.5-volt battery. The same effect will occur if you disconnect the test leads.

LED CURRENT INDICATOR CIRCUITS

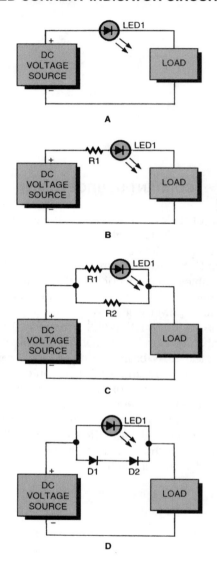

POPULAR ELECTRONICS

Fig. 53-4

Here are some ways to connect an LED as an indicator that a load is receiving current: with the LED alone (A), with a resistor in series (B), with a resistor in series and one in parallel (C), and with two diodes in parallel (D).

DC AMMETER I

Resistance Values for DC Ammeter

I FULL SCALE	R_A [Ω]	R_B [Ω]	R_f [Ω]
1 mA	3.0	3k	300k
10 mA	.3	3k	300k
100 mA	.3	30k	300k
1A	.03	30k	300k
10A	.03	30k	30k

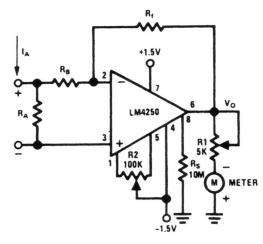

NATIONAL SEMICONDUCTOR

Fig. 53-5

For dc readings higher than 100 μA, the inverting amplifier configuration shown in the figure provides the required gain. Resistor R_A develops a voltage drop in response to input current I_A. This voltage is amplified by a factor equal to the ratio of R_f/R_B. R_B must be sufficiently larger than R_A so as not to load the input signal. The figure also shows the proper values of R_A, R_B, and R_f for full-scale meter deflections from 1 mA to 10 A.

DC AMMETER II

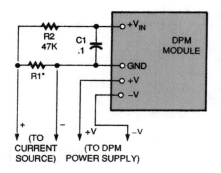

POPULAR ELECTRONICS

Fig. 53-6

As shown here, a DPM module can be used as a dc ammeter.

R1 is chosen to provide a voltage drop so that the meter reads full scale at the maximum desired current reading. This drop should be kept as small as possible. Typically, it will be 100 to 200 mV dc.

54

Measuring and Test Circuits— Frequency

The sources of the following circuits are contained in the Sources section, which begins on page 1043. The figure number in the box of each circuit correlates to the entry in the Sources section.

555 ASTABLE FREQUENCY METER

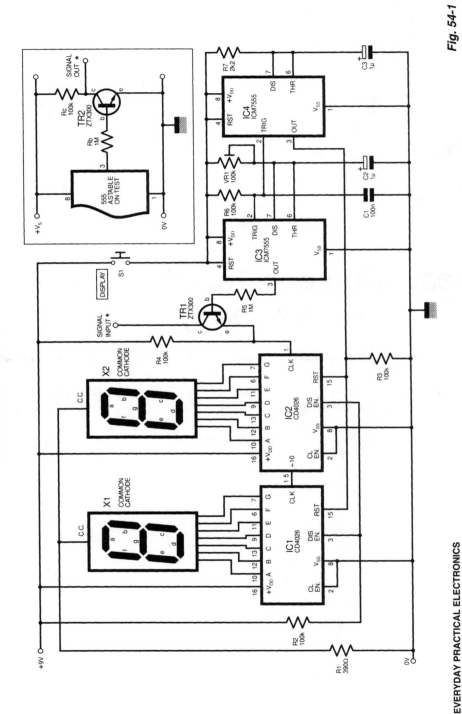

Fig. 54-1

EVERYDAY PRACTICAL ELECTRONICS

The circuit shown is a simple digital frequency meter, which indicates the frequency in hertz of an astable 555 timer. It could, perhaps, be adapted for other uses. An NPN transistor, TR2, is connected to the 555 astable to be measured, as shown in the inset, and the output signal obtained is "hooked" to the main circuit via transistor TR1 collector (c). Two CMOS 7555 timers, IC3

555 ASTABLE FREQUENCY METER *(Cont.)*

and IC4, are wired as monostables, and these will trigger via R6 and C1 when the DISPLAY switch S1 is closed. The period of IC3 is adjustable using VR1, while IC4 times for a fraction of a second. When S1 is closed, IC4 sends a RESET signal to IC1 and IC2, which are two decade counters, each driving a seven-segment common-cathode LED display directly. IC3 then drives transistor TR1, which allows pulses generated by the 555 under test to be fed into pin 1 (CLOCK) of IC2, the first decade counter. Monostable IC3 needs to be trimmed so that it triggers for approximately 100 ms. Pressing switch S1 then allows "100 ms' worth" of pulses to clock up on the display, and simply multiplying the result by 10 will yield the value in hertz. It would be feasible to cascade several counters to show a higher range of frequencies.

FREQUENCY METER

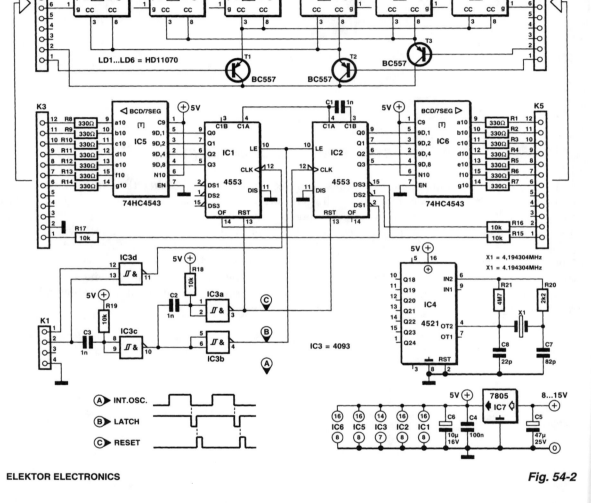

Fig. 54-2

The counter makes use of two 3-digit DCD counters that have a multiplexed output with buffer memory and a RESET input. The multiplexed output signal is converted by an integrated BCD-to-seven-segment converter and then applied to an LED display. The necessary clock signal is generated by a crystal of 4.194304 MHz and a counter that divides the signal by 2^{23}. The result is a stable

FREQUENCY METER *(Cont.)*

digital signal with a frequency of 0.5 Hz. The time during which the clock signal is high, and measurements can occur, is exactly 1 s. Owing to this arrangement, a new measuring value is displayed every 2 s. If the meter is used to measure frequency, the internal clock signal at pin 3 of K1 must be linked to pin 2 of K1. Gate IC3d then operates as a lock; every time the clock is high, the measurand at pin 1 of K1 is applied to the clock input of IC1. This circuit is connected in series with IC2 to form a six-digit counter. At the end of the measurement, a latch pulse is generated at the output of IC3c with the aid of R19 and C3. After this signal has been inverted in IC3b, it is applied to the LATCH input of IC1 and IC2. At the command of this signal, the current counter state is stored in the buffer memory. At the end of the LATCH pulse, network R18-C2-IC3a generates a RESET pulse with which the counter is reset. In the circuit as shown, the counter can be used for signals with frequencies from 1 Hz to 1 MHz. If higher frequencies are envisaged, the measurement time must be adapted accordingly. There are limits to this, though: With a supply voltage of 5 V, the 4553 can be used for frequencies up to 1.5 MHz; with 7 V, up to 5 MHz; and with 15 V, up to 7 MHz. Still higher frequencies require the use of a prescaler. The current drawn by the circuit (as shown) does not exceed 50 mA, which makes a battery supply feasible.

55

Measuring and Test Circuits—
Miscellaneous

The sources of the following circuits are contained in the Sources section, which begins on page 1043. The figure number in the box of each circuit correlates to the entry in the Sources section.

Component Checker
Chronometer/Counter
Simple SWR Bridge
Simple Inductance Meter
Multimeter Conductance Adapter
LED SWR Indicator Circuit
Simple Polarity Indicator
Bench Power Supply Meter
RF Signal-Strength Indicator
Tilt Sensor
LC Self-Calibrating Meter
Maximum Acceleration Recording Circuit
Voice Stress Analyzer
Sound-Level Meter
Electroscope Circuit

Decibel Meter
LED Peak Indicator
Audible Carpenter's Level Circuit
Frequency Response Tester
Strip-Chart Recorder from Old
    Computer Printer
Ripple Adder
Bike Speedometer Circuit
Chart Recorder Range Extender
Time and Period Adapter
Pulse Counting Circuit for Shaft Encoders
Electronic Stethoscope
Signal Tracer
Broadband RF Tester
Voltage Divider

COMPONENT CHECKER

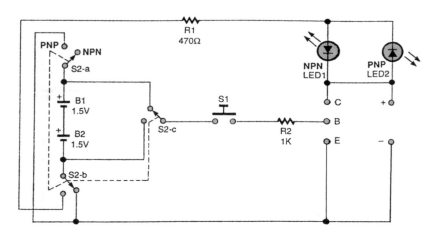

POPULAR ELECTRONICS

Fig. 55-1

The circuit is a component checker that works by lighting corresponding LEDs. If you insert a transistor's B and E leads (or a shorted transistor) into the C and E sockets, an LED will light, but you will know that the transistor is placed in the wrong position because the proper LED should light only when you press S1. You might want to use five sockets placed "C-E-B-C-E" to avoid bending the transistor leads. A bicolor LED can be used instead of the two plain LEDs shown. With the transistor-lead sockets you can also check SCRs, LEDs, etc. For SCRs, place the polarity switch in the NPN position, and insert the A, K, and G leads in the C, E, and B sockets, respectively. Momentarily press S1 and the "NPN" LED should light and stay on when you release the switch. To check LEDs, place the polarity switch in the NPN position, and insert the LED's anode and cathode in the B and E sockets, respectively. Pressing S1 should cause a good LED under test to light. Two more sockets, labeled "+" and "−," can be used to check diodes, continuity, capacitors, etc. To check a diode, place the polarity switch in the NPN position and place the anode and cathode of the diode into the + and − sockets. Press S1, and the "NPN" LED should light. To check continuity, place the polarity switch in any position, insert test leads into the + and − sockets, press S1, and a polarity LED should light if the path under test conducts. To check capacitors, insert the capacitor leads with the correct polarity in the + and − sockets. Then rapidly change the polarity to PNP, then to NPN, then back to PNP. The LEDs should light alternately. This test checks only charge and discharge, not capacitance.

CHRONOMETER/COUNTER

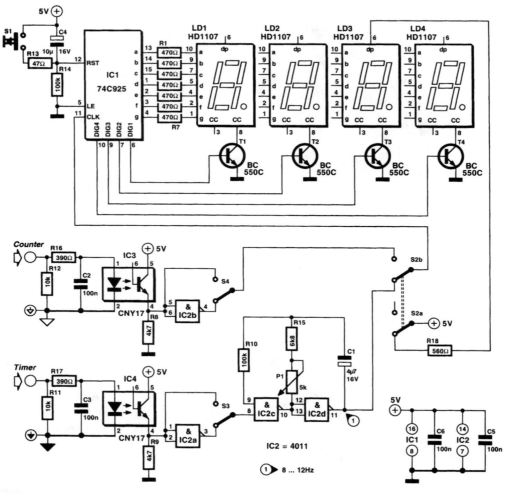

Fig. 55-2

The chronometer has four 7-segment displays, which can show a time lapse of 000.0 s to 999.9 s with a resolution of 0.1 s or a count from 0000 up to 9999. The chronometer is based on the type 74C925 counter IC with integrated display driver from National Semiconductor. The device draws a current of about 40 mA from a +5-V supply. On power-up, the counter is set to 0000 by network R14-C4 or by S1. The IC can derive a clock from two different sources: the internal oscillator or an external one via the COUNT input. The oscillator is formed by IC2c and IC2d and is enabled via the timer input. The enabling is effected manually by S3 or by inputting a given level, high or low, depending on

CHRONOMETER/COUNTER *(Cont.)*

the position of S3, into CHRON. The timer and counter inputs are identical, but are separated electrically from one another and from the signal source by optoisolators. This allows input potentials of up to 25 V_{p-p} to be applied to either of them. As with the timer input, the level at the COUNT input can be either high or low and is selected by S4. The position of S2b determines whether the time or count function is active. Switch section S2a inserts the decimal point between the third and fourth display digits when the timer is selected. Whereas a simple level is needed to start the oscillator, the signal at the COUNT input needs more if the module is to work error-free. The signal must have steep edges; it must be free of interference; at low level, it must be well below 1 V; and at high level, it must well above 2 V. Moreover, if a switch is used at the COUNT input, it must be debounced adequately.

SIMPLE SWR BRIDGE

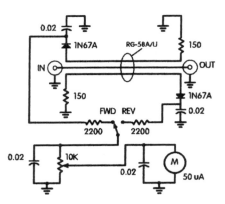

Fig. 55-3

This circuit can be used to measure SWR. It uses a coupler made from a 6.5-inch piece of RG-58A/U coax. First remove the outer plastic jacket, then open up the braid a little bit. Next, thread two lengths of #22 tinned bus wire into Teflon spaghetti. Then thread the insulated wires inside the braid, trying to keep them on opposite sides of the center conductor. Be careful to keep the leads to the 1N67A diodes and the 150-Ω resistors very short. If you can't find 1N67A diodes, use 1N34. The RG-58A/U braid should, of course, be grounded at both ends. If the device is carefully constructed, the result should be a perfectly balanced VSWR meter. It should need no adjustment.

SIMPLE INDUCTANCE METER

Parts List

BT1 9V alkaline battery
C1,C3,C4   0.01 µF ceramic disc
                      capacitor
C2   10 µF 16V electrolytic capacitor
C5   0.001 µF ceramic disc capacitor
C6   365 pF variable capacitor
C7   560 pF NPO, COG, Mylar™ or poly
          capacitor
D1   Germanium diode: 1N34, 1N60,
          1N90, 1N270, etc.
J1,J2 Binding post
M1   0-1 mA DC meter (see text)
R1,R2 560 ohm 5% 1/4W resistor
R3,R4 100 ohm 5% 1/4W resistor
R5   1k ohm 5% 1/4W resistor
R6   10k ohm linear potentiometer
S1   SPST toggle or slide switch
U1   74LS00 two-input quad NAND
          gate

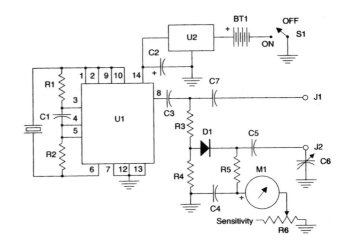

Fig. 55-4

This figure shows the schematic diagram of the simple inductance meter. U1, a 74LS00 two-input quad NAND gate logic integrated circuit, two resistors, a capacitor, and a surplus microprocessor crystal form a stable crystal oscillator near the marked frequency of the crystal. The RF voltage is taken from pin 8 through isolation capacitor C3 to the measuring circuit. RF voltage is applied through capacitor C7 to J1, a binding post. This same RF voltage is applied to a resistive voltage divider consisting of R3 and R4. Germanium diode D1 has its anode connected to the junction between R3 and R4. RF across the variable tuning capacitor C6 is applied back through C5 to the cathode of D1 and load resistor R5, the lower end of which is bypassed to ground through C4 and applied to the positive terminal of meter M1. R6 is a sensitivity control connected between the negative terminal of meter M1 and ground. This instrument operates by measuring the RF voltage developed across C6, which will be the highest when the series circuit made up of C6 and the unknown inductance is at resonance at the crystal frequency. In other words, the value of the unknown is indicated on the dial attached to C6 when the voltage indicated by M1 peaks, just the opposite of bridge operation.

MULTIMETER CONDUCTANCE ADAPTER

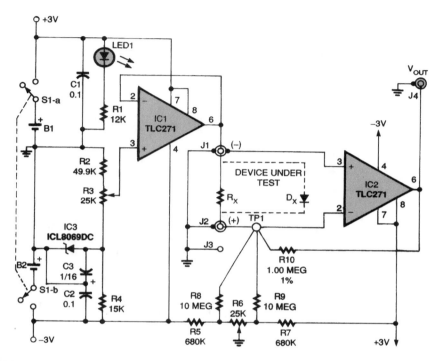

Fig. 55-5

A 1.000-V reference is buffered by IC1, a TLC271, and is connected to J1 for use when measuring conductance. The voltage reference is derived from IC3, a 1.25-V temperature-compensated band-gap reference. The current drop developed by the component under test at J1 and J2 is converted by IC2 to an output voltage that can be displayed on a digital voltmeter. The output voltage is limited to a maximum of 2.0 V and a minimum of 1.0 V. With the inverting input of IC2 at "virtual ground," the maximum input current that can be measured is -2.0 μA. In reading the conductance of an unknown resistor (labeled R_x in the figure), -1.0 V is applied to J1, with J2 being at virtual ground. Based on the maximum input current that can be converted by IC2, the smallest resistance that can be measured is 500,000 Ω. The values chosen for the circuit generate a conductance reading of 1.0 μmho/V, or 1.0 nmho/mV output. A 1.0-MΩ resistor will have a conductance of 1.0 μmho, or 1000 nmho. A 1.0-GΩ resistor would have a conductance of 1.0 nmho. The current-to-voltage conversion is set by R10, which gives an output of 1 V/μA, or 1 mV/nA. The lowest tolerance you should use for R10 is 1 percent.

LED SWR INDICATOR CIRCUIT

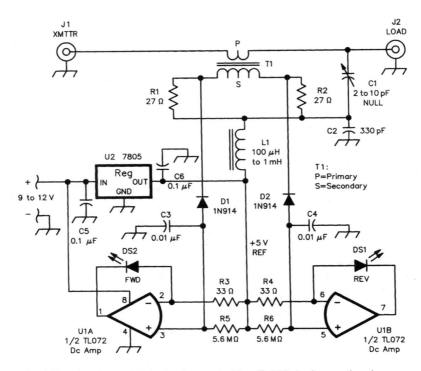

A standard directional coupler design teamed with a TL072 dual operational amplifier work together in this simple SWR indicator.

C1—Ceramic, plastic or air-dielectric trimmer
C2-C6—Ceramic
D1, D2—1N914 (or equivalent) silicon switching diode
J1,J2—Coaxial jacks
L1—RFC choke, 100 μH to 1 mH.
DS1, DS2—Light-emitting diodes (LEDs); red for **REV** and green for **FWD** suggested
R1, R2—5% or 10%-tolerance carbon film or composition

R3, R4—5% or 10%-tolerance carbon film or composition. The value of these resistors sets the bridge's power sensitivity; see text
T1—Toroidal transformer wound on a T-50-2 powdered-iron core. Secondary: 30 turns of #30 enameled wire; primary: 1 turn (just pass the wire through the core)
U1—TL072 dual JFET-input operational amplifier IC
U2—7805 or 78L05 5-V regulator IC

QST

Fig. 55-6

D1 and D2 act as half-wave diode detectors for the forward and reverse voltages. In traditional SWR-meter circuits, the output of these detectors feeds a panel meter via a forward/reflected switch and a scaling resistor, which typically provides the detector diodes with a load on the order of 25 kΩ. This relatively low load impedance limits the diodes' performance at low power levels because it causes their forward voltage drop to be higher than the sampled RF voltages to be detected. To use

LED SWR INDICATOR CIRCUIT *(Cont.)*

LEDs (DS1 and DS2) as indicators, some sort of dc amplifier was needed. Operational amplifiers were used because their high input impedance lets us load the diodes so lightly that their detection range can extend down to about 50 mV. U1A and U1B, halves of a TL072 dual JFET-input op amp, do the dc amplification. The dc voltage developed by each detector is presented to its op amp across a 5.6-MΩ resistor. Each op amp operates at a voltage gain of 1 (unity), but provides enough current gain to drive an LED and to cause a matching voltage to appear across R3. To provide each half of U1 with split (positive and negative) power supplies in environments in which only single-polarity, negative-ground supplies are typical, it was necessary to provide an artificial ground. U2 provides a stiffly regulated dc voltage between ground and the positive supply. (Note that the bridge circuit's dc reference is also set to +5 V because the RF-grounded end of L1, which is connected to dc ground in traditional directional-wattmeter circuits, is also connected to the +5-V supply provided by U2.) The overall sensitivity of the circuit can be adjusted to match the power level expected by changing the values of R_3 and R_4 to drive the LEDs with an appropriate current level. If you use different LEDs or want to use the bridge at a different power level, you may want to adjust these resistor values according to the formula R_3 and R_4 (ohms) $= \sqrt{P_0} \times 50/30 \times I_{f(LED)}$, where P_0 is the transmitter power in watts, 50 is the system impedance in ohms, 30 is T1's turns ratio, and $I_{f(LED)}$ is the current level (in amperes) that produces the LED brightness you need. With the resistor values chosen to light the FED LED (DS2) brightly, the REV LED (DS1) will be easily visible at SWRs around 2:1, allowing no-guess SWR dipping with tuner adjustment.

SIMPLE POLARITY INDICATOR

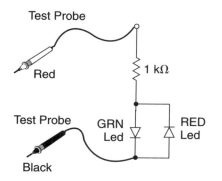

Values shown
for dc circuits
to 24V

For higher voltages
increase value of 1k resistor

WILLIAM SHEETS

Fig. 55-7

Use this circuit to determine the polarity of low-voltage dc circuits. If the green LED lights, the red probe is positive. If the red lights, the red probe is negative.

BENCH POWER SUPPLY METER

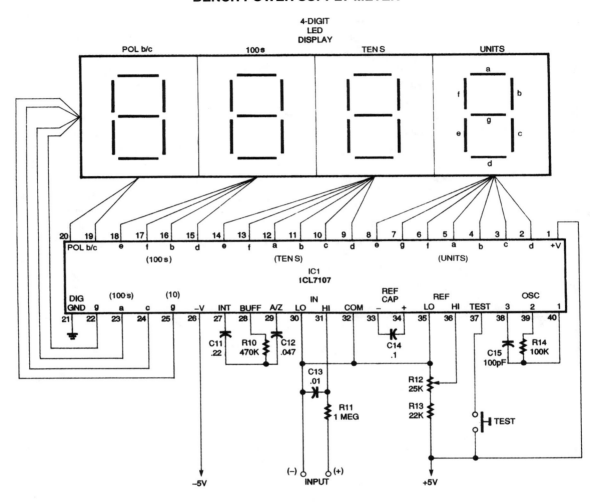

Fig. 55-8

The heart of the digital meter is the Harris ICL7107 A/D converter. The 7107 contains all the active devices needed to construct the voltmeter. These include the segment decoders and LED display drivers. The optional push-button switch connected to pin 37 of the ICL7107 permits the display to be tested. When that pin is connected to +5 V, the display will indicate "-1888," testing all the segments. If a liquid-crystal display is preferred, substitute an ICL7106CPL for the 7107. All pin connections will be the same, except that pin 21 will become the backplane instead of the digital ground.

RF SIGNAL-STRENGTH INDICATOR

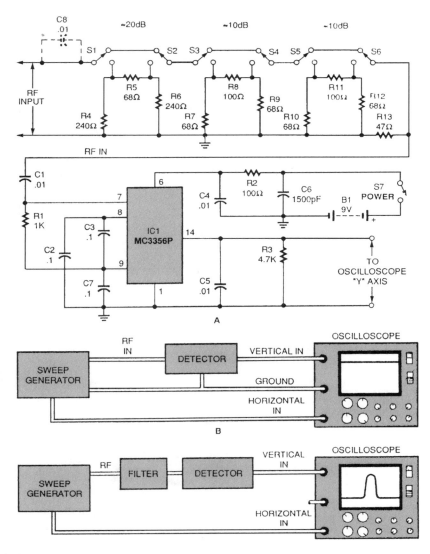

POPULAR ELECTRONICS

Fig. 55-9

This circuit is useful for test purposes because it produces an output that varies logarithmically with RF input level. A Motorola MC3356 IC is used for its received signal strength output (pin 14), the other functions remain unused. An attenuator is placed at the input of the circuit to increase usable range. For best results, the circuit should be enclosed in a metal shielded box.

TILT SENSOR

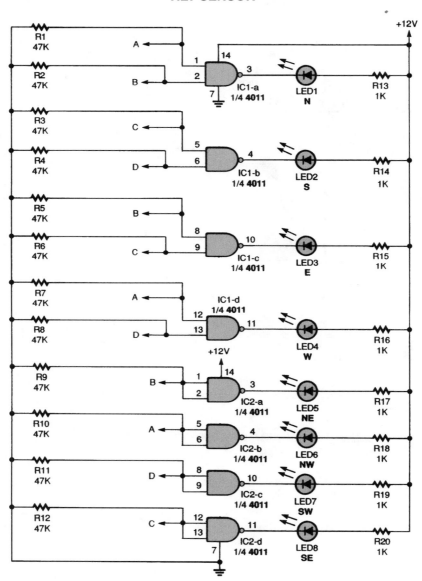

Fig. 55-10

TILT SENSOR *(Cont.)*

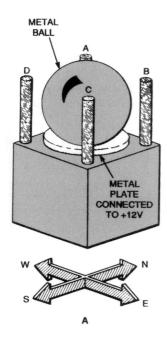

METAL BALL

METAL PLATE CONNECTED TO +12V

A

DIRECTION	BALL POSITION
N	A,B
S	C,D
E	B,C
W	A,D
NE	B
NW	A
SE	C
SW	D

B

When the sensor in A tilts, a current is applied through the metal ball to one or two of the pins, indicating one of the directions shown in B.

A metal ball sits on top of a metal plate that is mounted on an insulated stand. A metal pin is located at each corner. The metal plate is connected to +12 V, and the four insulated pins go to the sensor's input circuitry. When the ball is centered (in a nontilt position), it does not contact any of the pins. No electrical connection is made between the ball and the pins. If the sensor tilts toward the north (as shown in the diagram) the ball will roll between pins A and B. That will complete an electric circuit, placing +12 V on both those pins. The table shows which pins will be contacted when the sensor tilts in each direction. Pins A through D are connected to the inputs of several 4011 NAND gates (IC1 and IC2). When both inputs of one of the NAND gates go high, its output goes low, lighting the LED connected to it. Note that each of the IC2 gates has both its input pins tied together and to one directional pin to accommodate tilting in directions between the four cardinal points. For example, if the sensor tilts to the northwest position, the ball touches only the A pin. That means that IC2-b's output will go low, lighting LED6 (the "northwest" indicator). All other directions operate in a similar manner.

LC SELF-CALIBRATING METER

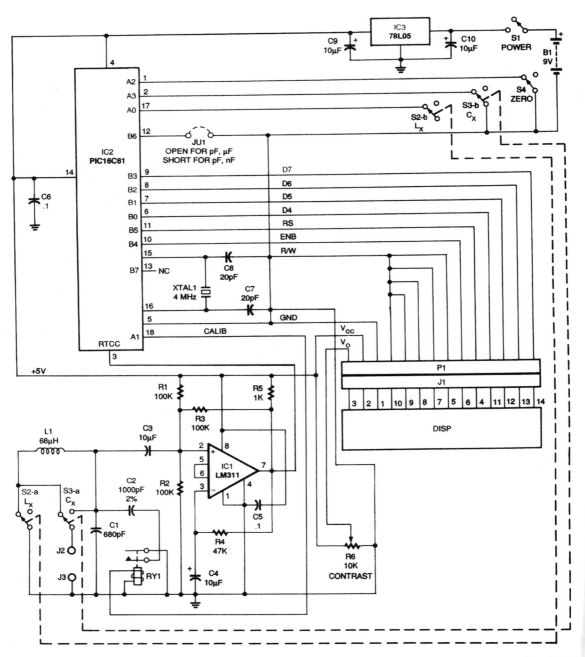

Fig. 55-11

LC SELF-CALIBRATING METER *(Cont.)*

LISTING 1

```
INITIALIZE THE CPU AND I/O PORTS
INITIALIZE THE LCD DISPLAY
WHILE Lx OR Cx are ON
      DISPLAY "SWITCH ERROR"
WEND
(The computer cannot calibrate itself if Lx or Cx are on.  The unit
waits for the operator to clear the switches.)
DISPLAY "WAIT" (wait 10 seconds  for the oscillator to stabilize.)
CALIBRATE:
      DISPLAY "CALIBRATING"
      MEASURE F1
      SWITCH IN THE CALIBRATION CAPACITOR
      MEASURE F2
      SWITCH OUT THE CALIBRATION CAPACITOR
      COMPUTE C1=F2^2 / (F1^2 - F2^2) C2
      COMPUTE L1=1 / (4 p^2 F1^2 C1)
DO (loop continuously)
      IF Lx and Cx are OFF
            IF ZERO
                  GOTO CALIBRATE (re-calibrate the unit)
            ELSE
                  DISPLAY "READY" (ready to measure Lx,Cx,or be ZEROed)
                  MEASURE and STORE F1
            END IF
      ELSEIF Lx ON AND Cx OFF
            MEASURE F2
            IF ZERO ON
                  MEASURE and STORE F1
                  DISPLAY "0.000"
            ELSE (ZERO OFF)
                  COMPUTE Lx=(F1^2 / F2^2 -1) L1
                  DISPLAY "Lx="
                  DISPLAY VALUE in engineering units
            END IF
      ELSEIF Cx ON AND Lx OFF
            MEASURE F2
            IF ZERO ON
                  MEASURE and STORE F1
                  DISPLAY "0.000"
            ELSE (ZERO OFF)
                  COMPUTE Cx=(F1^2 / F2^2 -1) C1
                  DISPLAY "Cx="
                  DISPLAY VALUE in engineering units
            END IF
      ELSE (Lx and Cx both ON)
            DISPLAY "SWITCH ERROR"
      END IF
LOOP
```

497

LC SELF-CALIBRATING METER *(Cont.)*

This circuit uses a microcontroller and a LED display. IC1 is an oscillator in which the unknown component is placed for measurement, and IC2 is a PIC16C61 microcontroller that calculates the unknown value and drives the LCD. The microcontroller must be programmed; the program is given in Listing 1.

MAXIMUM ACCELERATION RECORDING CIRCUIT

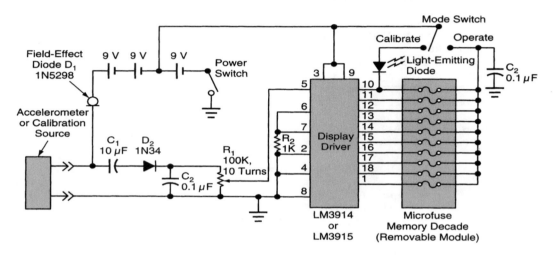

Fig. 55-12

The circuit (see figure) includes three 9-V batteries, one of which supplies the 9 V needed by the circuit, and all of which supply the 27 V needed by the accelerometer. Power is supplied to the accelerometer through field-effect diode D1, which regulates the accelerometer current to keep it in the range of 2 to 4 mA. The accelerometer puts out an ac signal that peaks at a full-scale value of 5 V when the ac component of acceleration reaches $50g$ (where g denotes normal Earth gravitation). The acceleration signal is coupled through C1 and D2 into C2, which retains the peak value for a short time. The signal is fed through potentiometer R1 to the input terminal (pin 5) of a 10-level display driver, LM3914, that has equally spaced levels, each representing $5g$, or of a similar circuit, LM3915, that has logarithmically spaced levels, with each succeeding level representing division of the next higher level by a factor of $\sqrt{2}$ (3 dB/step). Depending on the level of the input signal, the display driver energizes one of its ten output lines, each of which is connected to one of ten 2-mA transparent-cap microfuses plugged into a module. If the fuse on a line is still intact, then when that line is energized, the driver delivers a current of 10 mA, blowing the fuse. The fuses can be inspected visually or electrically at any convenient time thereafter to determine which (if any) has blown, thereby determining what level of acceleration was reached.

VOICE STRESS ANALYZER

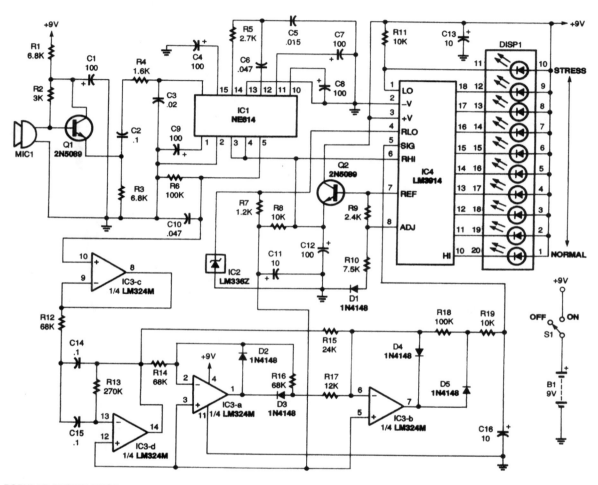

POPULAR ELECTRONICS

Fig. 55-13

This circuit analyzes the 7- to 15-Hz voice-frequency components that are associated with the stressed voice that occurs if a person is lying or under duress of some kind. These components appear as an amplitude modulation on the voice and are thought to be inversely proportional to psychological stress. Their relative amplitudes are read out on a bar-graph display.

SOUND-LEVEL METER

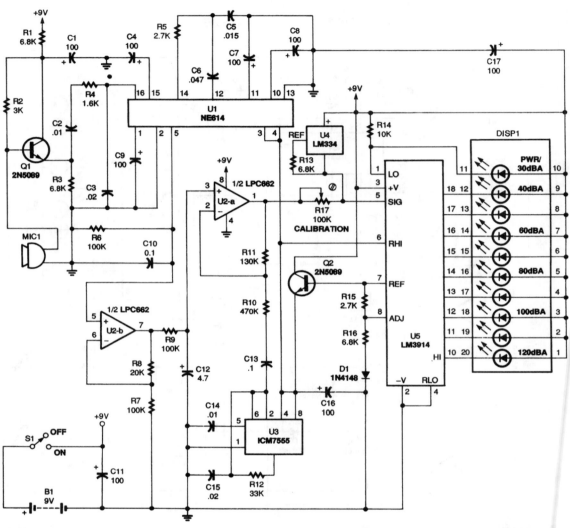

Fig. 55-14

Power for the circuit is provided by B1, a 9-V battery. Total battery current is 14.5 mA, so an alkaline battery should last about 250 hours. Audio signals are picked up by microphone MIC1. The output of MIC1 is buffered by Q1 to maintain the 3000-Ω load of resistor R2. The input impedance at pin 16 of the NE614 (U1) is 1600 Ω to reduce the gain on the high end and preserve the RSSI linearity. The RSSI signal is a current source that flows through R6 to establish a voltage at pin 5 of U1;

SOUND-LEVEL METER *(Cont.)*

the RSSI output voltage is a function of the input SPL. Capacitor C10 removes high-frequency components. The slope of the RSSI line is nominally $0.084V_{cc}/10$ dB. Op amp U2-b is configured as a non-inverting voltage buffer with a gain of 1.2; multiplied by the RSSI slope, this produces a slope of $0.1V_{cc}/10$ dB. The display, DISP1, is a 10-bar LED that is used to indicate sound level in 10-dB increments from 30 to 120 dB(A). The display is driven by an LM3914 linear bar-graph driver, U5. The internal voltage regulator of U5 functions by establishing a constant current through R15 because the voltage between pins 7 and 8 is maintained at 1.25 V. The resulting voltage at the emitter of Q2 is about 5 V, which powers U1, U3, and the internal voltage-divider string of U5 at pin 6. Each segment of the display will increase by 10 dB, independent of the exact magnitude of V_{cc}. While the slopes of the display driver and the RSSI signal are the same, they might not have the same offset. For that reason, an LM334, U4, is used to source a constant 10-μA current through R17 to generate a fixed offset voltage on top of the composite RSSI voltage at the output of U2-a. By adjusting the value of potentiometer R17, differences in microphone sensitivity can be accommodated so that the display reads properly from 30 to 120 dB(A). The proper readout of the display is also ensured by an ICM7555 timer, U3, which is configured as a self-excited square-wave oscillator; the resulting peak-to-peak, 1-kHz signal of the timer is almost one step size of DISP1.

ELECTROSCOPE CIRCUIT

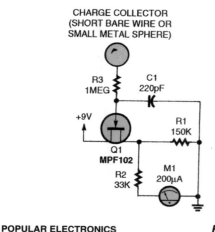

CHARGE COLLECTOR
(SHORT BARE WIRE OR
SMALL METAL SPHERE)

R3 1MEG
C1 220pF
+9V
R1 150K
Q1 MPF102
R2 33K
M1 200μA

POPULAR ELECTRONICS

Fig. 55-15

This circuit detects a charged object at a respectable distance and displays the polarity (positive, negative, or earth-grounded) and relative intensity. In operation, C1 reduces ac noise but lowers the sensitivity a bit. The MPF102 and R1 form a voltage divider. When the FET's gate is earth-grounded, the divider's output will be about 4.5 V, giving a half-scale reading on M1, a 200-μA meter. A positively charged object (like cotton-rubbed glass) will give a positive deflection from half scale, and a negatively charged object (a plastic comb, for example) will give a negative meter deflection. The whole circuit (including the 9-V supply) should be in a metal enclosure.

DECIBEL METER

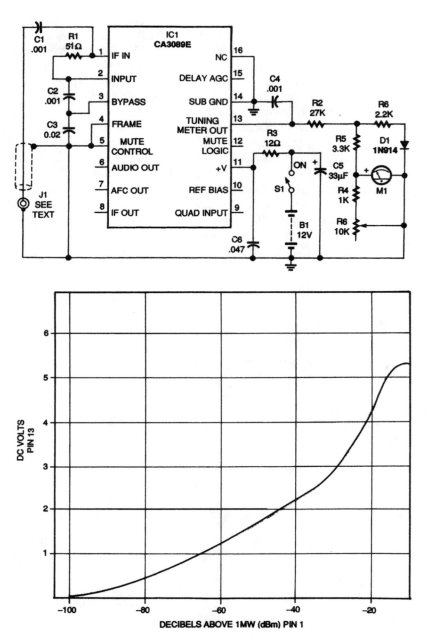

Fig. 55-16

DECIBEL METER *(Cont.)*

The circuit consists essentially of IC1, the CA3089E, a 0- to 100-μA meter M1, on-off switch S1, a few resistors and capacitors, and a diode. The graph shows the plot of dc volts out of METER OUT, pin 13 of IC1 (33 kΩ to ground), vs. the input signal in microvolts on IF IN 1, pin converted to decibels above 1 mW. IC1 detects the signal level and generates a nearly logarithmic dc output. To linearize the high end of this curve, diode D1 and resistor R6 shunt some of the dc as the voltage on pin 13 rises above 3 V. The battery pack, consisting of eight AA alkaline cells, makes the instrument portable and eliminates the possibility of 50/60-Hz hum interfering with the readings. The meter draws only about 16 mA, so battery life will be long. The meter will provide accurate readings as long as the output of the battery pack remains above 8 V. No regulation of this dc is required.

LED PEAK INDICATOR

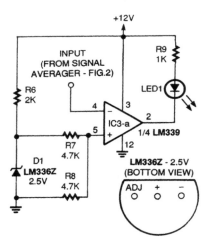

ELECTRONICS NOW

Fig. 55-17

The voltage divider consisting of resistor R6 and 2.5-V precision voltage reference IC2 forms a precision voltage reference whose voltage is divided in half at the junction of resistors R7 and R8. This places a reference voltage of 1.25 V at the comparator noninverting input. Whenever the inverting input voltage exceeds the reference voltage, the output will be connected to ground internally, and LED1 will light. With the input of this circuit connected to the output of the averager circuit, the LED will flash on whenever the audio input exceeds the level set with gain potentiometer R2. Adjust R2 by applying a maximum level signal to the input of the averager, and then reduce the level slightly to allow for headroom. With the input level now set, adjust R2 until the peak LED just turns on.

AUDIBLE CARPENTER'S LEVEL CIRCUIT

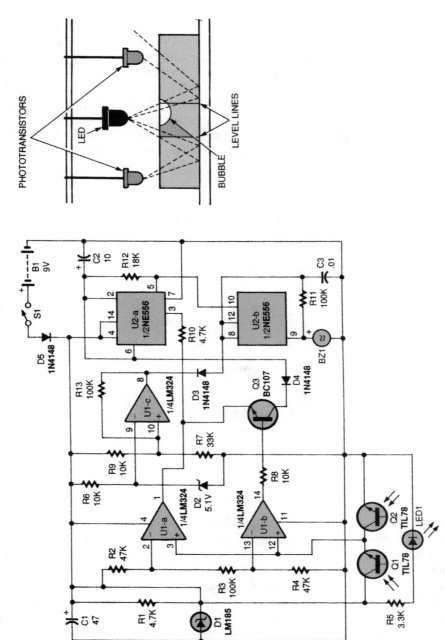

Fig. 55-18

POPULAR ELECTRONICS

AUDIBLE CARPENTER'S LEVEL CIRCUIT *(Cont.)*

This is the optical geometry for the sensor used in the audible level. Because of the refraction caused by the bubble in the alcohol-filled vial, less infrared light reaches a phototransistor as the bubble moves closer to it.

The schematic for the audible level is shown. In this circuit, the amounts of infrared radiation received by phototransistors Q1 and Q2 is translated by op amp U1 and dual timer U2 into either a steady tone or a fast- or slow-pulsing one. The circuit is powered by a 9-V battery, B1. Unfortunately, there is a drawback to using a battery for power: The ratio of the phototransistors' resistance will change as the supply voltage goes down. To solve that problem, D1, an LM185, 2.5-V, voltage-reference diode, is used. That way, when the battery voltage declines to a point at which the level would become inaccurate, the tones abruptly cease, indicating the need for a new battery.

Resistors R2, R3, and R4 set two voltage levels on the inverting inputs of U1-a and U1-b. With the bubble off center in the direction of Q1, Q2 will receive the most radiation. Under that condition, the voltage at the Q1/Q2 junction is near zero. That is below the voltage thresholds on the inverting inputs of both op amps, putting both outputs low. Both halves of a 556 dual timer, U2, are oscillators. The output of U2-a (which runs at a low frequency) is fed to the RESET input of U2-b at pin 10. With U2-a active, the second oscillator (responsible for the tone) is switched on and off, producing the required pulsed sound. The output of U1-a is connected via R10 to the control-voltage input (pin 3) of U2-a. That expands or contracts the limits within which the voltage on timing capacitor C2 is held, altering its charge/discharge period and frequency. With the output of U1-a low, a high-speed pulse is generated. The output of op amp U1-b is also low, which keeps transistor Q3 switched off. As the bubble moves to a central position, the voltage at the Q1/Q2 junction increases. When the voltage at pin 12 of U1-b exceeds that applied to pin 13 (the inverting input), Q3 turns on. The output of U1-a is still low and sinks current from U2-a, the slow oscillator. Because the trigger input of U2-a (pin 6) is kept low, its output is forced high, producing the constant tone. Finally, with the bubble off center in the direction of Q2, Q1 receives the most radiation. The output of U1-a then goes high and blocks the current through Q3. Simultaneously, the voltage at the control-voltage input (pin 3) of U2-a increases, and the pulsing tone reappears, but at a lower frequency.

The circuit centered around U1-c is responsible for the low-battery alert. Op amp U1-c normally has its output high, but as the battery voltage drops to about 6 V, the noninverting input goes below the inverting input. That switches the output low with the help of the positive feedback introduced by R13 and turns off of U2-b.

FREQUENCY RESPONSE TESTER

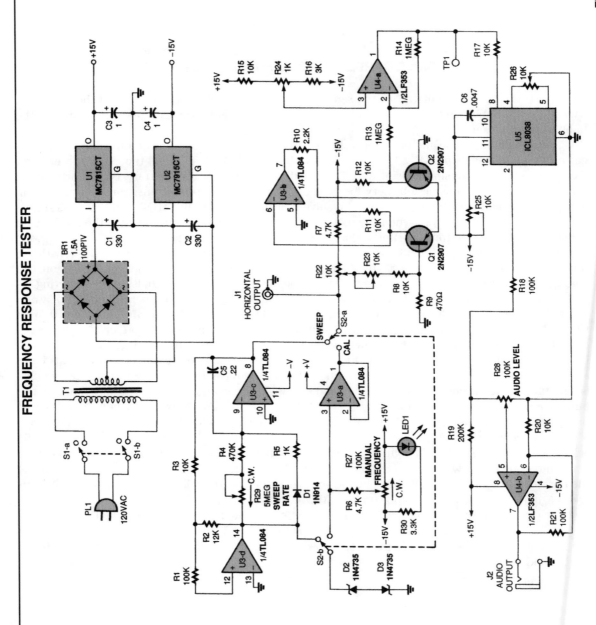

ELECTRONICS HOBBYISTS HANDBOOK

Fig. 55-19

FREQUENCY RESPONSE TESTER *(Cont.)*

The two quad op-amp sections, U3-c and U3-d, are configured as a linear ramp generator. With switch S2 in the SWEEP position, as shown, the output of U3-d is low. Zener diodes D2 and D3 limit that output to about -7 V, which is the 6.2-V reverse drop across D3 plus the forward drop of D2. The internal short-circuit protection of the op amp limits the zener current to several milliamperes. Because the output of U3-d is negative, integrator U3-c generates a linear ramp in a positive direction, at a rate determined by the resistance of R4 and R29 and the capacitance of C5. A portion of that output is fed back to the noninverting input of U3-d. Because U3-d is basically operating as a comparator with its inverting input grounded, its output will switch positive as soon as its noninverting input crosses zero volts. The overall effect is to produce a sawtooth waveform with an amplitude of over 10-V p-p. That is more than sufficient for just about any scope's horizontal input. Next, the linear ramp undergoes a series of level and offset adjustments via resistors R7, R8, R9, R22, and R23, and is then applied to the base of Q1. The emitter-coupled transistors, Q1 and Q2, in combination with U3-b, produce an antilog transfer function. The output at the collector of Q2 will be logarithmic with respect to the input, yielding a one-decade voltage differential for every few volts of input. Next, U4-a inverts the polarity of the signal, so that it starts high and ends low. That is necessary because the output frequency of the 8038 function generator, U5, is inversely proportional to its input voltage. The ICL8038 function generator produces constant-amplitude sine waves, as well as triangle- and square-wave outputs, from 20 Hz through 20 kHz. Also, its output frequency is nearly a perfect inverse proportional to its input voltage, so driving its input with a logarithmic sweep circuit will produce the balanced, three-decade frequency sweep. The sine-wave output from U5 is fed to the voltage divider made up of R18 and R19, which restores the dc offset to zero. The voltage-divider output is then applied to potentiometer R28 for output-level control, and is subsequently routed to U4-b for a boost in both amplitude- and current-drive capability. The output can range from 0 to over 10 V p-p.

STRIP-CHART RECORDER FROM OLD COMPUTER PRINTER

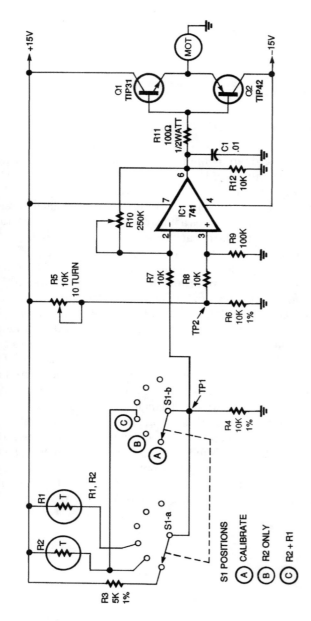

ELECTRONICS NOW

Fig. 55-20

Converting an obsolete computer printer into a strip-chart recorder is both easy and inexpensive. The chart recorder works by comparing the voltage on a potentiometer to the voltage on a temperature-sensing thermistor. The potentiometer is mechanically connected to the printer carriage. A marking pen is also connected to the printer carriage. Any difference in the voltages on the potentiometer and the temperature sensor causes an op amp to attempt to bring the voltages back into balance. In the course of this, the print-head carriage is moved left or right, thereby causing a mark on the paper. A standard 741 op amp is used as a comparator. The comparator monitors the voltages at the junctions, marked TP1 and TP2. The op amp compares the voltages from the two networks at its inputs (pins 2 and 3). When the voltages are identical, the op amp's output (pin 6) is zero, so neither driver transistor (Q1 and Q2) turns on; therefore, the motor does not turn. But if thermistor R2's resistance changes because of a change in air temperature, a new voltage occurs at the op amp's inverting input (pin 2). For example, assume that R2 decreases.

STRIP-CHART RECORDER FROM OLD COMPUTER PRINTER *(Cont.)*

The pin 2 voltage then increases, so the op amp's output swings negative, thereby biasing Q2 on. That in turn starts the motor, which moves the carriage left or right. Potentiometer R5 is also coupled to the carriage, so, as the carriage moves, R5's wiper moves, thereby varying the voltage on the noninverting input of IC1. When the op amp's input voltages become equal, the output goes to zero, both transistors turn off, and so does the motor. Now the circuit remains quiescent until the temperature changes again.

RIPPLE ADDER

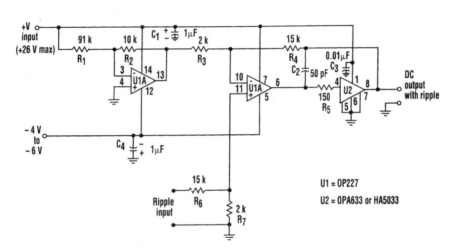

Fig. 55-21

When a device under test needs a dc voltage applied with ripple superimposed, this circuit can handle the task. The power supply for the $+V$ input will need to be set about 20 percent higher than the desired output. The supply should be well regulated, because it directly affects the output voltage. A function generator set to the desired waveform, amplitude, and frequency must be connected to the ripple input. Because of resistor tolerances, the final voltage adjustments will have to be made by monitoring the output. U2 will drive capacitive loads as long as the dc + peak ac doesn't exceed 200 mA. The frequency response of the ripple adder with a 5-V p-p output ranges from dc to past 100 kHz. R5 and C3 should be located as close as possible to U2. R_1 can be altered to change the ratio between the input and output voltages.

BIKE SPEEDOMETER CIRCUIT

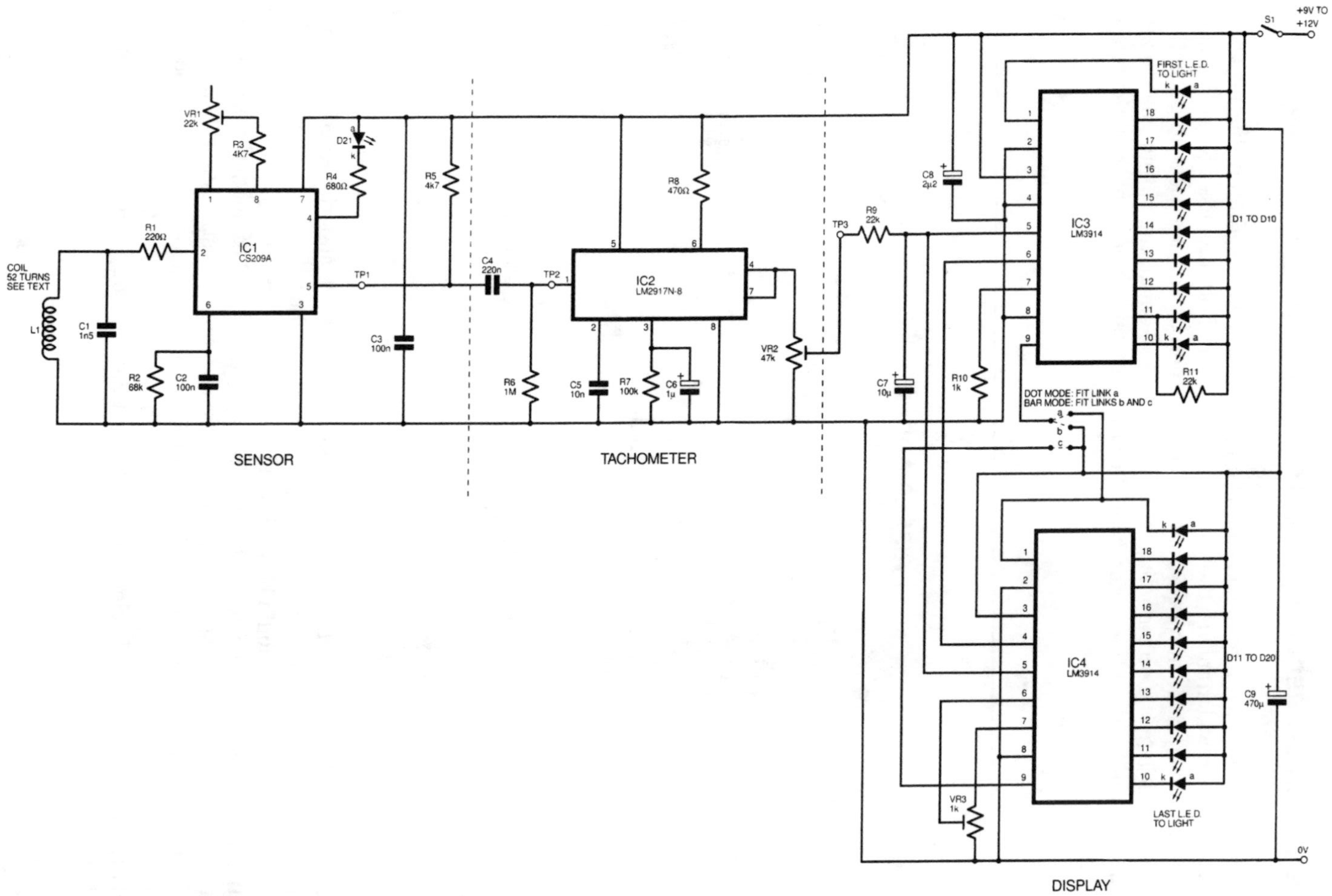

Fig. 55-22

EVERYDAY PRACTICAL ELECTRONICS

510

BIKE SPEEDOMETER CIRCUIT *(Cont.)*

The proximity detector senses the passing of the PC board pieces as the wheel rotates, providing an output signal, which has a clean transition between high and low voltage levels, and is ideal for triggering counting or processing circuits. Next is the frequency-to-voltage converter, and, because a reasonable degree of accuracy is required, the LM2917N-8 IC has been chosen. The output from this IC is fed to two bar-graph ICs, which drive a 20-LED display.

CHART RECORDER RANGE EXTENDER

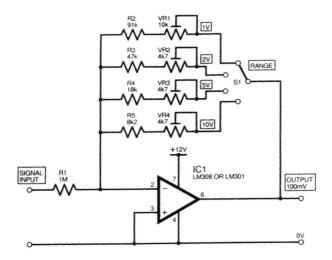

EVERYDAY PRACTICAL ELECTRONICS

Fig. 55-23

The circuit diagram shown was designed to adapt a laboratory chart recorder with a 100-mV input for full-scale deflection so that it will accept input voltages of 1 V, 2 V, 5 V, or 10 V at the flick of a switch. The op amp (IC1) is configured as an inverting amplifier whose feedback resistance is much less than the input resistance (R1); therefore, the output voltage will be a fraction of the input signal. Range switch S1 selects among four different feedback resistors, which consist of fixed metal-film resistors and a 20-turn cermet trimmer, which is adjusted to produce 100 mV output at the selected input voltage range. In practice, VR2 should be adjusted to 100 mV output at 2 V input. The other ranges are set up in the same manner.

TIME AND PERIOD ADAPTER

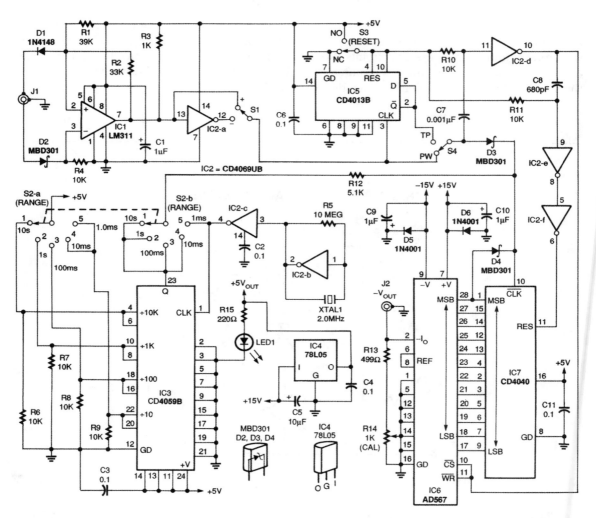

Fig. 55-24

This circuit enables time period measurements to be made on a DMM. Integrated circuit IC1, an LM311 comparator, conditions the input waveform for the 5-V logic required by the circuit. The gate signal is applied to IC7, a CD4040B CMOS 12-stage binary counter, to measure pulse width, or through IC5, a CD4013B CMOS dual flip-flop, to measure time period. Integrated circuit IC2, a CD4069UB CMOS hex inverter, provides the precision 2.000-MHz crystal-controlled pulse train for IC3, a CMOS divide-by-*N* counter. It divides the pulse train by 1, 10, 100, 1000, or 10,000 to cover the five ranges selected by rotary selector switch S2. The divider's output is gated by the input pulse at

TIME AND PERIOD ADAPTER *(Cont.)*

IC7 through a resistor-diode NOR gate. Integrated circuit IC4, a 78L05 voltage regulator, supplies +5 V to the circuit; IC2-d, IC2-e, and IC2-f provide the properly timed RESET signal for IC7 and the LATCH ENABLE signal for IC6, an AD567 digital-to-analog converter (DAC), the key component in this adapter. DAC IC6 converts the 12-bit count stored in IC7 to a negative output current. That current is converted to a negative voltage by the combination of R13 and R14. The voltage is sent to a DMM at jack J2 over a twisted-wire pair terminated with a phono plug at one end and a suitable matching plug for the DMM at the other end.

PULSE COUNTING CIRCUIT FOR SHAFT ENCODERS

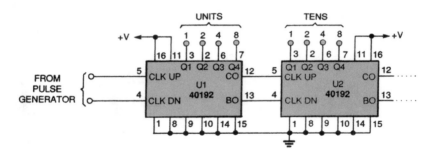

POPULAR ELECTRONICS

Fig. 55-25

The output pulses of an encoding circuit can be counted with a BCD up/down counter such as the 40192. A binary-coded-decimal (BCD) number is produced. The counters have separate "up" and "down" clock inputs. For example, as long as the COUNT input (pin 11) is high, the counter will increment one count for every positive edge at the CLOCK-UP input (pin 5), and decrement one count for every positive edge at the CLOCK-DOWN input (pin 4). Counters can be cascaded by connecting the BORROW output (pin 13) to the following CLOCK-DOWN input and the CARRY output (pin 12) to the following CLOCK-UP input.

ELECTRONIC STETHOSCOPE

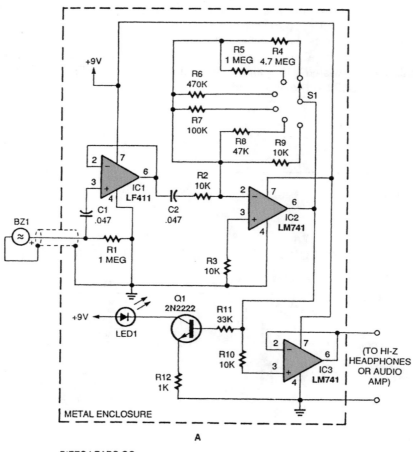

A

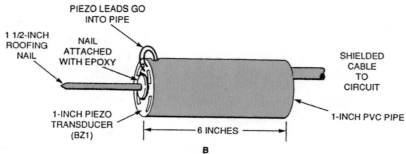

PIEZO LEADS GO
INTO PIPE

1 1/2-INCH
ROOFING
NAIL

NAIL
ATTACHED
WITH EPOXY

SHIELDED
CABLE
TO
CIRCUIT

1-INCH PIEZO
TRANSDUCER
(BZ1)

1-INCH PVC PIPE

6 INCHES

B

ELECTRONIC STETHOSCOPE *(Cont.)*

Most electronic stethoscope circuits feature a microphone assembly. Here's one based on a piezo disk used as the transducer. Using 5-min epoxy, glue a roofing nail to the center of the top side of a piezo disk. Be generous with the epoxy, and lay a bead over the top of the nail head. Later, epoxy the outer rim of the backside of the disk to a 1-in diameter PVC pipe roughly 6 in in length, which will serve as a handle. Attach a shielded cable from the piezo-disk leads and secure them in place. The piezo disk's signal is coupled by the cable to IC1, an LF411 FET op amp configured as a buffer. The LF411's output is coupled to IC2, an LM741 op amp configured for high gain. Resistors R4 to R9 set the gain via a rotary switch, but a potentiometer could be used instead. The amp's output goes to both IC3 (a final buffer stage) and Q1 (an MPS7965 general-purpose NPN transistor). The output of the final buffer IC3, an LM741, is connected to high-impedance headphones or a small LM386-based audio amplifier. Power is supplied by two 9-V batteries configured as a split supply.

SIGNAL TRACER

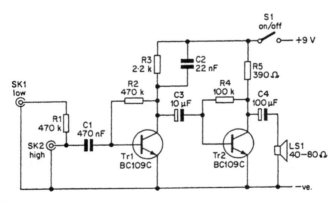

Fig. 55-27

The signal tracer has good sensitivity, so it can be used for any normal type of audio signal tracing. The circuit provides AM demodulation, and the signal tracer can therefore be used for IF and RF testing on AM radios, provided that a suitably strong antenna signal is available. The unit has an integral loudspeaker, and although the maximum output power is only a few tens of milliwatts, the volume level obtained is more than adequate for this application. If preferred, though, the loudspeaker can be replaced by an earpiece or headphones. Low-, medium-, and high-impedance headphones are all suitable for use with the unit.

BROADBAND RF TESTER

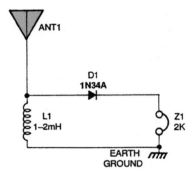

POPULAR ELECTRONICS

Fig. 55-28

This circuit is useful for crystal radio experimenters for evaluating an antenna and ground setup. If you hear some signals, a crystal set will probably work with the antenna and ground.

VOLTAGE DIVIDER

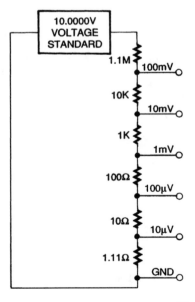

This divider will produce low dc voltages from a 10-V supply. The resistor values can be scaled proportionally for other requirements.

ELECTRONICS NOW

Fig. 55-29

56

Measuring and Test Circuits—Power

The sources of the following circuits are contained in the Sources section, which begins on page 1043. The figure number in the box of each circuit correlates to the entry in the Sources section.

Power Check
Ac Wattmeter Circuit
Simple Wattmeter

POWER CHECK

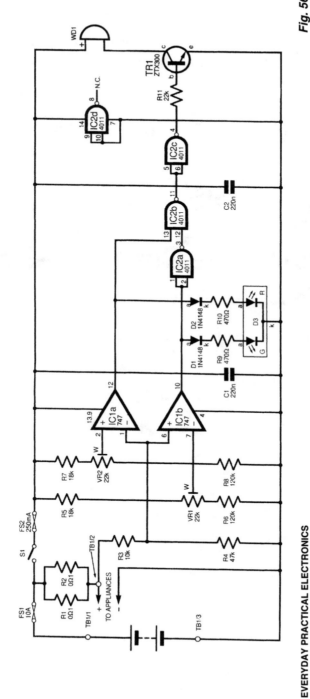

Fig. 56-1

EVERYDAY PRACTICAL ELECTRONICS

This system is basically an ammeter with a tricolor LED readout in place of a conventional meter. In this system, the tricolor LED shows green for a low demand, yellow for a medium one (acceptable for a temporary heavy load, such as a water pump), and red to indicate excessive use and the need to switch appliances off right away. When red shows, an audible signal is also given. The circuit is built in a small metal box, with the LED mounted on the front panel. A piece of terminal block inside is used to make the connections to the external system.

This circuit must not be used in installations using more than 10 A. Having said that, though, the values at which the colors operate are freely adjustable. In the prototype unit, a current below 2 A was judged to be *low*; one between 2 and 4 A, *medium*; and one more than 4 A, *high*. With these criteria, by keeping green, at least 20 hours of operation would be obtained from a standard 60-Ah battery.

AC WATTMETER CIRCUIT

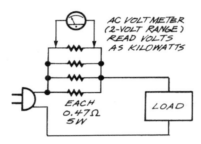

AC VOLTMETER
(2-VOLT RANGE)
READ VOLTS
AS KILOWATTS

EACH
0.47Ω
5W

LOAD

ELECTRONICS NOW

Fig. 56-2

A true wattmeter is a complicated instrument. But if all the wattages that you want to measure are at the same voltage (120 Vac) and the loads are resistive (such as lights, heaters, or motors under load), you can just pass the current through an 0.12-Ω resistance (four 0.47-Ω, 5-W resistors in parallel), connect an ac voltmeter across the resistance, and read volts as kilowatts. That is, every volt across the resistance corresponds to 1000 W (equal to 8 A) drawn by the load. Remember that the voltmeter terminals are "hot;" use insulated pin sockets so that you won't touch the contacts accidentally. On inductive or capacitive loads, such as unloaded motors or transformers and computer power supplies, your wattmeter will read high because the current and voltage are not in phase.

SIMPLE WATTMETER

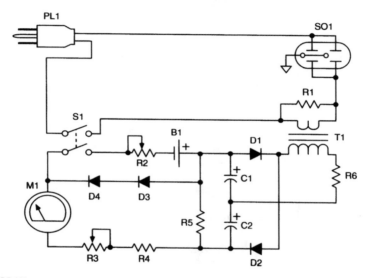

Fig. 56-3

A current transformer (T1) is used to derive a larger ac voltage, which is rectified by D1 and D2 to produce a dc voltage proportional to ac line current. B1, R2, and D3, and D4 are used to linearize the circuit.

57

Measuring and Test Circuits—
Semiconductors

The sources of the following circuits are contained in the Sources section, which begins on page 1043. The figure number in the box of each circuit correlates to the entry in the Sources section.

DIODE MATCHING CIRCUIT I

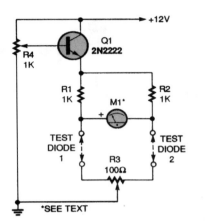

A 2N2222 transistor is connected in an emitter-follower circuit that is powered by a 12-Vdc source. Potentiometer R4 sets the voltage feeding the diodes under test. Two 1000-Ω, 1 percent resistors, R1 and R2, make up two legs of a four-element bridge circuit. The diodes you want to test make up the other two legs. Potentiometer R3 is a fine-balance control. Meter M1 is a 100–0–100-μA, center-zero unit. Connect jumpers in place of the test diodes, set RC to midposition, and set R4 to its maximum-voltage position (minimum resistance). Apply power and adjust R3 for a zero meter reading. Disconnect the power, set R4 to its minimum-voltage position (maximum resistance), and connect two diodes in the test positions. Slowly increase the bridge voltage with R4 and watch the meter for any change. The diodes are perfectly matched if the meter remains at zero as the voltage is increased.

POPULAR ELECTRONICS

Fig. 57-1

DIODE MATCHING CIRCUIT II

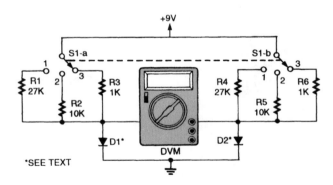

POPULAR ELECTRONICS

Fig. 57-2

To match diode pairs with this circuit, simply insert diodes of equal rating in the D1 and D2 positions. When the DVM reading drops to zero or nearly zero, you have found a set. D1 and D2 are diodes under test. S1-a and S1-b select currents of 8 mA, 800 μA, or 250 μA.

LED TEST CIRCUIT

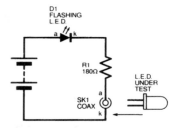

The circuit is intended as a simple one to enable spot testing of bulk surplus LEDs. A flashing LED is used (D1) with a 9-V battery, and is connected to a coaxial socket. When a test LED is placed across the socket (anode to the inner conductor), both the flashing LED and the test LED will flash. No on-off switch is necessary

EVERYDAY PRACTICAL ELECTRONICS   *Fig. 57-3*

SIMPLE TRANSISTOR CHECKER

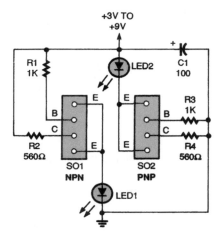

POPULAR ELECTRONICS   *Fig. 57-4*

When a T092-type transistor is placed in the correct socket, if the transistor is good, the circuit turns on, according to the kind of transistor being tested. For example, if a good NPN unit is tested, LED1 turns on. Note that two emitter positions are in each socket. Transistors have different pin configurations. Some are EBC, and some are ECB. If you don't know the configuration, you can test all the possible pin positions until an LED turns on. You'll need to know if the transistor is an NPN or a PNP type, however. The supply voltage can be 3 to 9 V. There is no power switch because, without a socketed transistor, capacitor C1 is the only load on the battery. For the sockets, use an eight-pin, mini-DIP socket.

FET TESTER

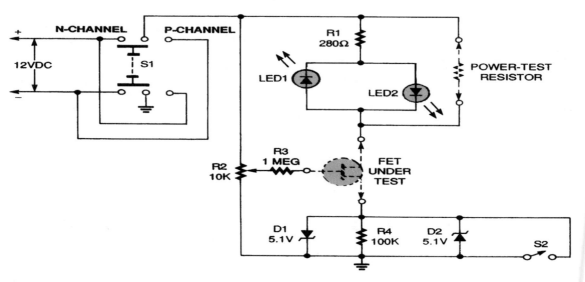

Fig. 57-5

In operation, either a P-channel or an N-channel FET can be plugged in and tested by selecting the proper polarity with the power switch. With the control potentiometer (R2) at ground potential, the FET under test will conduct until the pinchoff voltage is reached. The LED will not turn on at this point because of the low current. This pinchoff voltage can be read at the top of resistor R4 with a high-impedance voltmeter. Increasing the potentiometer will decrease the pinchoff voltage until the FET conducts and an LED turns on. The LEDs give a rough indication of conduction. A more accurate reading is obtained by monitoring the voltage drop across the drain resistor. For a quick go/no-go FET check, the switch (S2) across the zeners can be closed. The associated LED should light and should go out when the switch is opened. If it does not, the FET is bad or is connected wrong, or the selector switch is set for the wrong-polarity FET. The 1-MΩ resistor (R3) at the gate connection is protection for the potentiometer should a unit under test short, it provides current limiting for the protection diodes in the gate of the CMOS FETs.

58

Measuring and Test Circuits—Voltage

The sources of the following circuits are contained in the Sources section, which begins on page 1043. The figure number in the box of each circuit correlates to the entry in the Sources section.

Tonal Voltmeter
Voltage Calibrator
Audible Voltmeter
10.000-V Standard
Power Supply Monitor
Ac Line-Voltage Monitor
Voltage Monitor
Digital Voltmeter Circuit

TONAL VOLTMETER

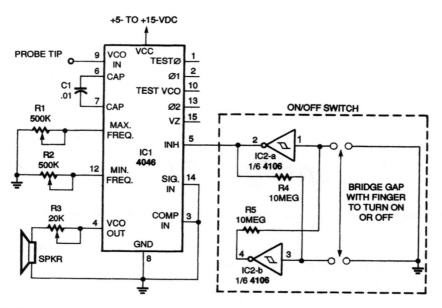

ELECTRONICS NOW

Fig. 58-1

The circuit is based on a CD4046B phase-locked loop (PLL) IC with a built-in voltage-controlled oscillator (VCO). The frequency of the VCO is determined by the voltage at pin 9, capacitor C1 between pins 6 and 7, and potentiometer R2 at pin 12, which sets the tonal voltmeter's minimum frequency. The VCO output, which appears at pin 4, is normally fed back into the comparator input at pin 3. However, in this circuit, the VCO output drives a speaker directly. The VCO operates while the inhibit line (INH) at pin 5 is held logic low, and it turns off when INH is logic high. A touch switch consisting of two Schmitt-trigger inverters (IC2-a and IC2-b) turns the circuit on and off to conserve power when the TVM is not being used. The touch switch can be replaced with a standard SPST switch, if so desired. This is recommended if you don't have a Schmitt-trigger inverter and don't want to purchase one. The tonal voltmeter emits a low tone when reading a logic low. As the voltage input increases, the tone pitch increases until the input voltage reaches a logic high. As the voltage input decreases, the pitch decreases.

VOLTAGE CALIBRATOR

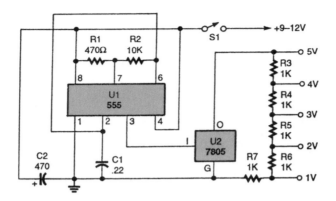

POPULAR ELECTRONICS

Fig. 58-2

A NE555 oscillator drives a regulator to provide a regulated square-wave voltage to a voltage divider. The output is 1 to 5 V p-p in 1-V steps.

AUDIBLE VOLTMETER

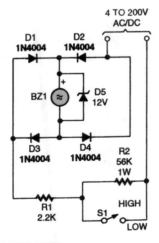

The audible voltmeter can be used to test for ac or dc voltages in a circuit. With S1 closed, the circuit can be used to test for voltages between 4 and 24 V. When S1 is open, it can be used to check for the presence of voltages of up to 200 V

POPULAR ELECTRONICS

Fig. 58-3

10.000-V STANDARD

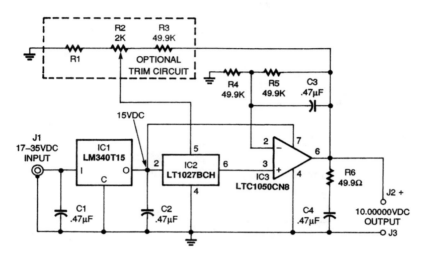

Fig. 58-4

The heart of the voltage standard is the LT1027BCH voltage reference IC, IC2, manufactured by Linear Technology Corp. It sports a temperature coefficient of two parts per million per degree Celsius (2 PPM/°C), and it comes pretrimmed within ±0.05 percent. Provision for optional trim circuitry is included on the board; however, it requires an additional resistor-selection step. To eliminate sensitivity to input-voltage variations, the incoming supply is regulated at 15 Vdc. This is accomplished with a conventional LM340T15 voltage regulator, which powers both the reference IC and the output amplifier. The LT1050 amplifier doubles the 5-V output of the LT1027 to the desired 10 Vdc. It is chopper-stabilized, providing far lower drift than a conventional op amp. Noise filtering is provided by the 0.47-μF capacitor across the feedback resistor. Note that the LT1050 does not have the output drive of a bipolar device—thus, the slightly higher than normal feedback resistors. You should not try to use the voltage standard with loads below 10 kΩ. The 0.47-μF/49.9-Ω output network improves stability when cables and capacitive loads are driven.

POWER SUPPLY MONITOR

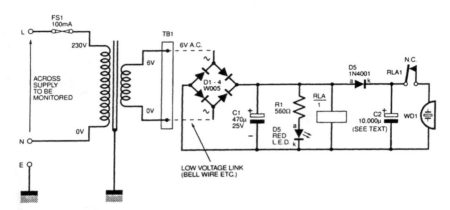

EVERYDAY PRACTICAL ELECTRONICS

Fig. 58-5

The power supply is monitored by T1, a low-voltage step-down transformer. A standard full-wave power supply is used on the secondary winding to illuminate LED D5, and also operate the relay RLA1 and charge a large-value electrolytic capacitor (C2). The relay has normally closed (NC) contacts, which remain open while mains power is present through transformer T1, allowing the output from the bridge rectifier to power the relay coil. Power failure causes the relay to drop out and the normally closed contacts will now connect the piezo buzzer across capacitor C2. The stored voltage across C2 is enough to sound the buzzer for several minutes. The relay contacts could alternatively be used to drive an isolated circuit, such as a buzzer and battery. The circuit might also appeal to tropical fish keepers or possibly deep-freeze owners.

AC LINE-VOLTAGE MONITOR

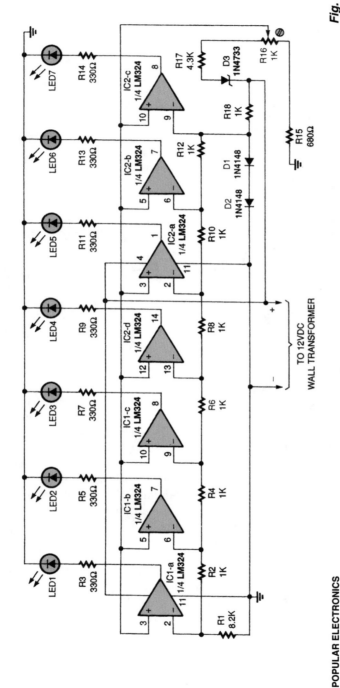

Fig. 58-6

POPULAR ELECTRONICS

This figure shows the schematic of the ac line-voltage monitor circuit. The circuit receives 12 Vdc from a wall transformer. The circuit is centered around two quad LM324 op-amp ICs (IC1 and IC2) that receive regulated operating power from a clamped portion of the dc supply provided by a 5.1-V zener diode, D3. The op amps drive an LED bar graph, consisting of LED1 through LED7. The op amps receive an adjustable reference voltage from the center contact of potentiometer R16 and an input voltage from the voltage divider, consisting of resistors R1, R2, R4, R6, R8, R10, R12, and R18. Those resistor values were chosen so that the op-amp outputs sequentially turn on and light the LEDs as the ac line voltage, or one-tenth of it, varies from 100 to 132 V. Potentiometer R16 sets the midpoint of the LED bar graph—usually 118 V—which can be shifted, if you like.

VOLTAGE MONITOR

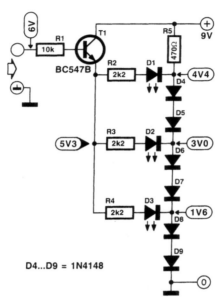

ELEKTOR ELECTRONICS

Fig. 58-7

When voltages are measured, it is not always necessary to find the exact value. Often, it is sufficient to check whether a potential lies within a certain range of voltages. The present circuit indicates by means of three LEDs whether a voltage is greater than 4 V, 5.7 V, or 7.4 V. The reference potential is provided by a series network of diodes D4 to D9. The LEDs are connected to various junctions in this network. The measurand is applied to the LEDs via emitter follower T1 and series resistors. Any one LED will light only if the input voltage exceeds the sum of the base-emitter junction of T1, the drop across the relevant series resistor, the LED voltage, and the drop across the diodes following the LED.

DIGITAL VOLTMETER CIRCUIT

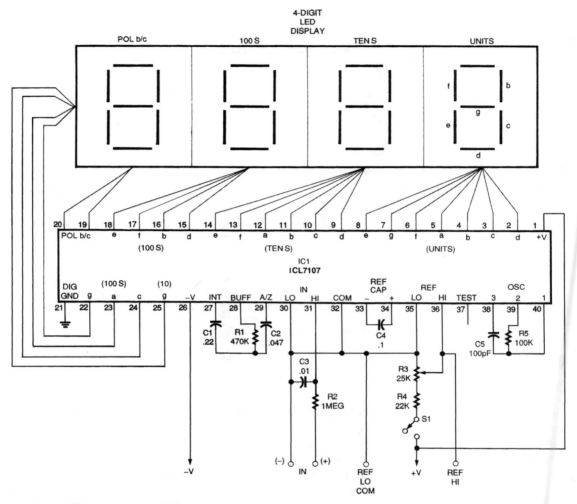

Fig. 58-8

A basic digital voltmeter circuit using the Harris Semiconductor ICL7107 is shown. It has a 2-V range. Calibration consists of applying a known voltage of 1.2 V to the input and adjusting R3 for a correct reading on the display. Supply is ±5 V, and S1 selects either the supply voltage or an external reference.

59

Medical Circuits

The sources of the following circuits are contained in the Sources section, which begins on page 1043. The figure number in the box of each circuit correlates to the entry in the Sources section.

Prosthetic Hand Control Circuit
Drink Aid for the Blind
Biological Signal Waveform Generator
Pathfinder
Heart Sound Simulator
Blood Oxygen Measurement

PROSTHETIC HAND CONTROL CIRCUIT

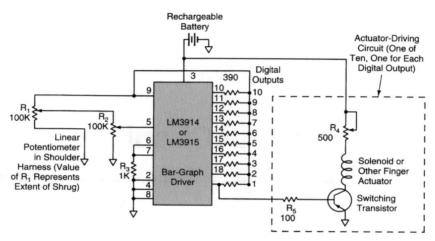

Fig. 59-1

A proposed circuit for the control of an electromechanical prosthetic hand would derive electrical control signals from shoulder movements. The harness would contain a linear potentiometer (R1 in the figure), the resistance of which would be varied by shrugging the shoulder, as in the older mechanical system. The variable output voltage of the potentiometer would be fed to an attenuating potentiometer (R2), which would be set to scale the voltage to the range of shoulder movement. The scaled voltage would be fed to an analog-to-digital converter of the type used to control a bar-graph display. Either a linear or a logarithmic converter could be used, depending on the requirements of the user. Each digital output, in continuous or single-pulse mode, would be fed to a transistor switch, which would supply current to a solenoid or motor to actuate one of the prosthetic fingers. With no shrug, the prosthetic thumb and all of the prosthetic fingers would be extended. As the "shrug" is increased, the digital outputs would turn on in sequence, thereby causing the thumb and fingers to move sequentially to the closed position.

DRINK AID FOR THE BLIND

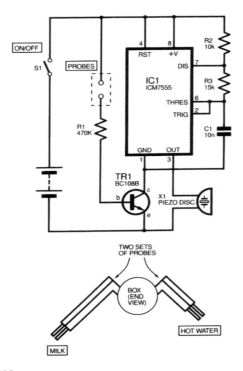

EVERYDAY PRACTICAL ELECTRONICS

Fig. 59-2

The circuit is an audio oscillator based around IC1, a 555 astable that drives a piezo disk directly. Transistor TR1 acts as a switch and will enable the audio oscillator when the two probes detect liquid. By using two sets of probes formed at different lengths (see inset diagram), the audio alarm will activate when enough milk, then enough water, has been added. The circuit is powered by a small 12-V battery (e.g., MN21 type), which should give many years of continuous service because the quiescent current is virtually nil. The device was housed in a small circular plastic box, approximately 4 cm in diameter. The probes were formed from sheathed, thick solid-core copper wires protruding from either side so that the container could clip onto the side of the cup.

BIOLOGICAL SIGNAL WAVEFORM GENERATOR

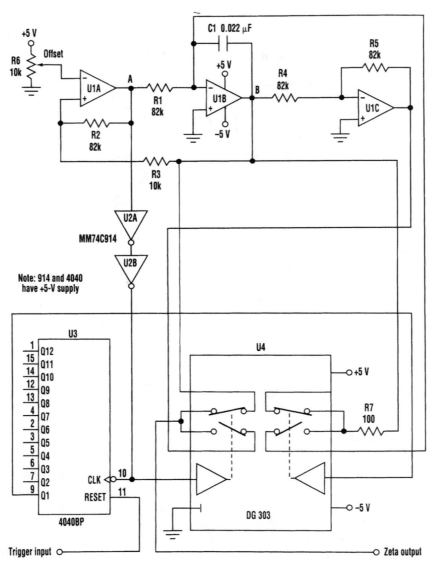

ELECTRONIC DESIGN

Fig. 59-3

The circuit operates as follows: U1A and U1B form a triangle waveform generator, using R6 to offset the output so that it is a positive-going signal. The amplitude is set by R2 and R3, and the ramp's rate of rise is set by R1 and C1. Using the values shown, the output has an amplitude of >1 V with rise and fall times of 0.5 ms, giving a total zeta duration of 1 ms. U1C provides an inverted out-

BIOLOGICAL SIGNAL WAVEFORM GENERATOR *(Cont.)*

put of U1B. When the RESET pin on U3 is triggered, its Q1 output goes low and the short is removed across C1 by one half of the analog switch (U4). At this time, the output of U1A will be negative and the output of U1B will start to ramp up. During this time, the output of the circuit is taken from U1B via the other half of the analog switch. When the ramp reaches its positive voltage threshold, the output of U1A goes positive and U1B will start to ramp down. The output of U1A is taken via U2A and U2B (both inverting Schmitt triggers with special input circuitry that protects the input from negative voltages and allows them to level-shift) to the CLOCK input of U3. It is also taken to the analog switch, which now takes the output of the circuit from U1C (which has an output ramping up from the negative maximum to 0 V). At 0 V, the output of U1A will go negative, resulting in the Q1 output of U3 going high and again shorting C1. This produces 0 V output until the next trigger input resets U3 and the cycle is repeated.

PATHFINDER

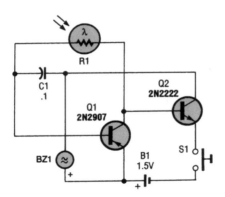

POPULAR ELECTRONICS

Fig. 59-4

This circuit is little more than a simple two-transistor oscillator (a relaxation type) whose frequency is controlled by a light-dependent resistor (LDR). The oscillator can be built using nearly any NPN and PNP transistor combination desired, and its operation is simple, as its parts count is low. The resistance of R1 and the value of C1 determine the oscillating frequency of the circuit, which varies from a slow tick in darkness (when R1 is about 100,000 Ω) to a high-pitched tone in bright light (when R1 is 1000 Ω). The change in frequency is heard through the transducer (BZ1). Power for the circuit is furnished by a single 1.5-V AA battery; because the Pathfinder draws so little power, it will function normally until the battery voltage drops to about 0.9 V.

HEART SOUND SIMULATOR

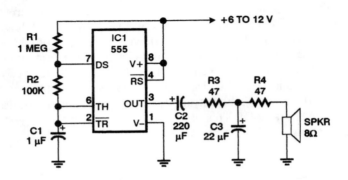

ELECTRONICS NOW

Fig. 59-5

This oscillator circuit imitates the sound of a heartbeat. Use a 4-in or larger speaker to get good bass response.

BLOOD OXYGEN MEASUREMENT

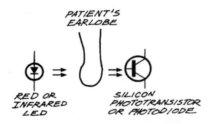

ELECTRONICS NOW

Fig. 59-6

An oximeter measures the amount of oxygen in your blood by shining light through some part of your body—traditionally the earlobe, nowadays often the finger—and comparing the transmission of two different wavelengths, 800 nm (in the near infrared) and 640 nm (bright red). If they're about the same, your oxygen is very low; if your blood is well oxygenated, the 640-nm transmission will be much higher. The light sources are LEDs, and the detector is a silicon photodiode or phototransistor.

60

Microphone Circuits

The sources of the following circuits are contained in the Sources section, which begins on page 1043. The figure number in the box of each circuit correlates to the entry in the Sources section.

Electret Mike Circuit
Dynamic Microphone Preamp

ELECTRET MIKE CIRCUIT

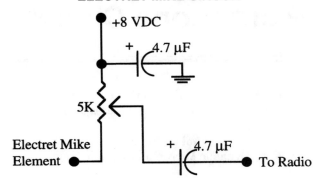

Fig. 60-1

This circuit shows how to power an electret microphone. The 5-kΩ potentiometer acts as a gain control.

DYNAMIC MICROPHONE PREAMP

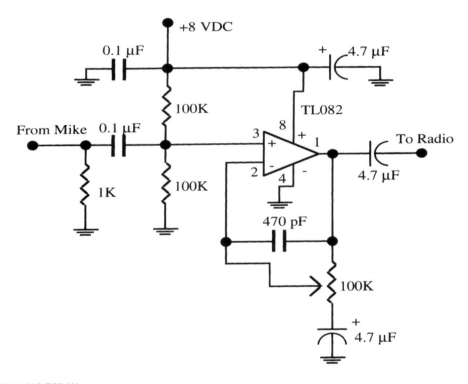

Fig. 60-2

This circuit uses a TL082 op amp to boost the output of a dynamic mike to a level sufficient for a radio transceiver, audio amplifier, or PA system.

61

Miscellaneous Treasures

The sources of the following circuits are contained in the Sources section, which begins on page 1043. The figure number in the box of each circuit correlates to the entry in the Sources section.

BASIC CHARGE PUMP CIRCUIT

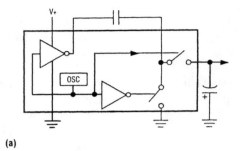

(a)

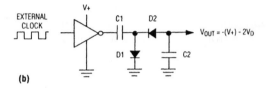

(b)

MAXIM

Fig. 61-1

A basic charge pump provides voltage doubling or inversion. It can be implemented with on-chip switches (*a*) or discrete diodes (*b*).

SCHMIDT TRIGGER MEMORY-CELL CIRCUIT

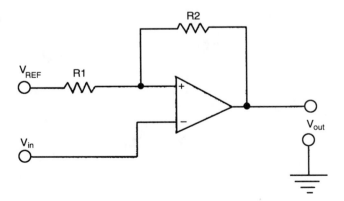

Schmitt Trigger
used in simple
memory cells

NUTS AND VOLTS

Fig. 61-2

IMPROVED CLIPPING CIRCUIT DESIGN

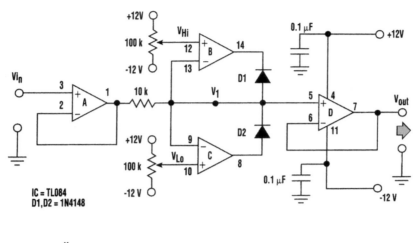

IC = TL084
D1,D2 = 1N4148

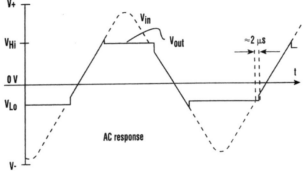

AC response

Fig. 61-3

Op amps A and D form unity-gain buffer amplifiers. Op amps B and C compare the input voltage (V_{in}) with the reference voltages (V_{Hi} and V_{Lo}). These references set the high and low clipping levels of the circuit. When the input voltage lies in the range ($V_{Hi}-V_{Lo}$), the outputs of the comparators make both diodes (D1 and D2) reverse-biased. As a result, the circuit's output V_{out} follows the input voltage. When the input voltage exceeds the value V_{Hi}, op amp B's output goes negative, forward-biasing D1 and thus reducing V_1, the voltage at the inverting input of op amp B. The circuit reaches an equilibrium condition at which $V_{out}=V_1=V_{Hi}$. Similarly, when the input voltage goes below the value V_{Lo}, op amp C's output goes positive, forward-biasing D2. This increases the voltage at V_1. At the equilibrium condition, $V_{out}=V_1=V_{Lo}$. The circuit's output continues to follow the input for about 2 μs, even after the input voltage crosses the reference voltages, because of the slew rates of the op amps. The supplies (5 V≤$V+$≤15 V, −5 V≥$V-$≥−15 V).

FLASH MEMORY PROGRAMMER

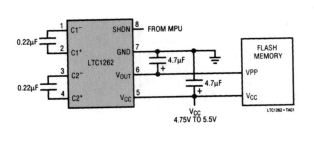

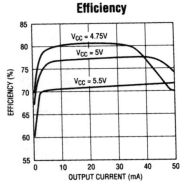

Efficiency

LINEAR TECHNOLOGY

Fig. 61-4

Generating 12 V from a 5-V supply for flash memory usually requires a switching regulator and inductor to produce the necessary 30-mA byte-wide flash. The LTC1262 reduces to just four small capacitors and no inductors the components required for this dc-to-dc conversion function, which generates 30 mA guaranteed output current from a 5-V ±5 percent supply. This surface-mount solution requires less than 0.2 in². The LTC1262 features a charge-pump-gated oscillator/tripler architecture operating at 300 kHz. The gating of the oscillator regulates the output voltage at 12 V. Operating supply current for the LTC1262 is 1 mA maximum, which drops just 10 μA max when the device is shut down. Output voltage ramp-up is well behaved, with absolutely no overshoot. The LTC1262 is specified to operate with input voltages down to 4.75 V and up to 5.5 V.

KEYING CIRCUIT

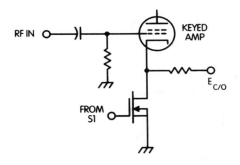

73 AMATEUR RADIO TODAY

Fig. 61-5

A power MOSFET can cathode-key a high-power RF amplifier.

ELECTROMAGNETIC LEVITATOR

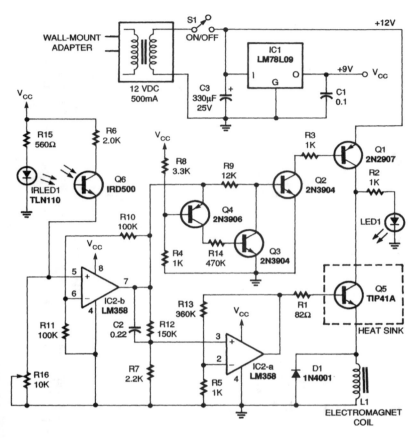

Fig. 61-6

This circuit is used to levitate a steel ball as a demonstration of levitation via an electromagnet. An infrared photodetector is used to sense the position of a steel ball and control the current in the electromagnet via a feedback loop. An infrared emitter and detector mounted across from each other create an invisible beam that passes slightly below the coil. As the object rises toward the electromagnet, it begins to block the beam. When the beam becomes blocked, the output of the detector is reduced, which, in turn, reduces the current in the electromagnet's coil. The reduced current weakens the magnetic field, the object begins to drop, and the detector once again sees more of the beam. This causes the circuit to increase the magnetic field, and the cycle repeats as the object is attracted upward again. The circuit is designed so that eventually an equilibrium is reached where the magnetic attraction exactly balances the force of gravity pulling on the object. The object then remains perfectly suspended in the infrared beam's path with no visible means of support!

SINGLE POT SWINGS AMPLIFIER GAIN POSITIVE OR NEGATIVE

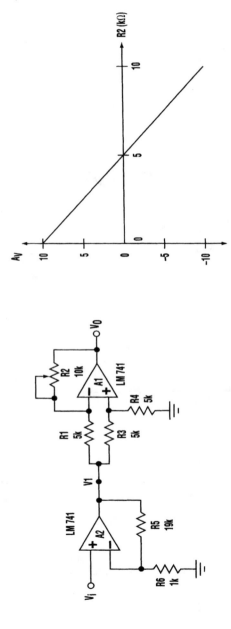

Fig. 61-7

ELECTRONIC DESIGN

It is handy to have a gain block with a gain that can be varied smoothly from positive to negative with a single potentiometer. The circuit shown accomplishes this function with R2. Op amp A1 is configured with both inputs tied together. Op amp A2 functions as a buffer. With $R_1 = R_3 = R_4 = R = 5$ kΩ, the gain of the differential amplifier is given by $A_{V1} = (V_0/V_1) = \frac{1}{2}(1 - R_2/R)$. With a buffer amplifier gain of 20, the overall gain of the amplifier is given by $A_V = (V_0/V_1) = 10(1 - R_2/R)$. By using a 10-turn potentiometer, the gain can be varied from positive to negative.

PIN DIODE SWITCH AND ATTENUATOR CIRCUIT

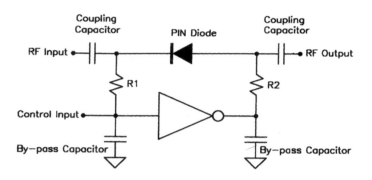

RF DESIGN

\Fig. 61-8

This circuit is a basic PIN-diode switch building block. R1 is chosen from $I_{diode} = (V_{cc} - V_{diode})/(R_1 + R_2)$, where I_{diode} is the specified diode current.

GYRATOR CIRCUIT

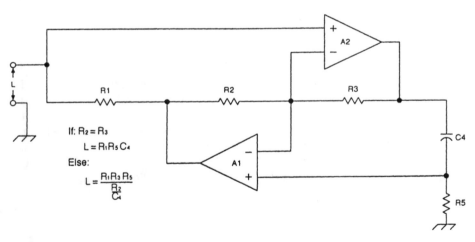

If: $R_2 = R_3$

$L = R_1 R_5 C_4$

Else:

$L = \dfrac{R_1 R_3 R_5}{\dfrac{R_2}{C_4}}$

73 AMATEUR RADIO TODAY

Fig. 61-9

The gyrator circuit has been around for a number of years. This active circuit can be used in place of an inductor. With two operational amplifiers (two sections of a dual or quad op amp are preferred), the value of the inductance simulated by this circuit is $L=R_1R_3R_5/(R_2/C_4)$ or, in the case where $R_2=R_3$, $L=R_1R_5C_4$.

POWER-ON CONTROL WITH POWER-ON RESET

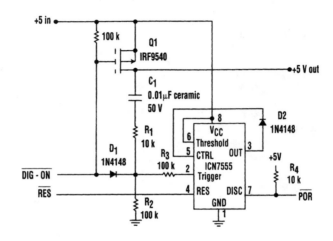

ELECTRONIC DESIGN

Fig. 61-10

A power-management control device for high-power interfaces in battery-operated equipment can incorporate a power-on-reset (POR) function. The power control is provided by a P-channel MOSFET in a conventional switching configuration. The POR function is supplied by the DIS-CHARGE pin of a CMOS version of a 555 timer. The circuit holds the POR line low while power to the device interface is off. However, when the interface device is turned on, its power rises immediately, and a delay to release the POR line is initiated. The threshold voltage on the 555 trigger comparator is about 1.7 V. If the component values for R1, R2, and C1 are applied, the capacitor voltage decays to the threshold value in about 1.6 ms. When the capacitor voltage decays below the threshold value, the POR line is released. Diode D2 provides threshold hysteresis and raises the POR threshold voltage to approximately 2.5 V at the output-state transition.

INEXPENSIVE TWO-TRANSISTOR XOR GATE

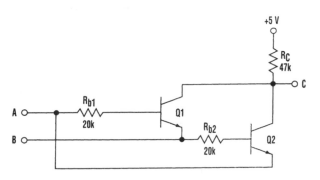

ELECTRONIC DESIGN

Fig. 61-11

For some applications, the cost or size of XOR-gate chips can be prohibitive. One alternative is the two-transistor combination shown, which forms an inverting XOR gate for just pennies. If both inputs A and B are low, both Q1 and Q2 are off, and the output at C is high. Likewise, if both inputs are high, both transistors are turned off, and again the output at C is high. When A is high and B is low, Q1's base-emitter junction is forward-biased, while Q2's base-emitter junction is reverse-biased. This turns on Q1, pulling the output low. The last condition is B high and A low. In this state, Q1's base-emitter junction is reverse-biased, but Q2 is forward-biased. This turns on Q2 and pulls the output low. The sink and source currents driving the two-transistor gate are very low when the values shown are used. Even though the signals are driving the emitters of transistors, when the base and emitter are at the same potential (A and B are both high or both low), no current flows. When one of the transistors is turned on, the emitter drive must be able to sink only the base plus collector currents. This current will be approximately $I_E = I_B + I_C$ where: $I_B \approx (V_H - V_{BE})/R_B$ and $IC \approx (V_{CC} - V_{sat})/R_C$. From this, V_H + TTL high output level = 3 V; V_{BE} = forward-biased base-emitter voltage $\approx$ 0.7 V; V_{CC} = 5 V typical; and V_{sat} = 0.2 V typical. Also, for this example, R_b = 20 kΩ and R_C = 47 kΩ. Thus, $I_C \approx 215$ μA.

HF AMPLIFIER GROUNDING TECHNIQUE

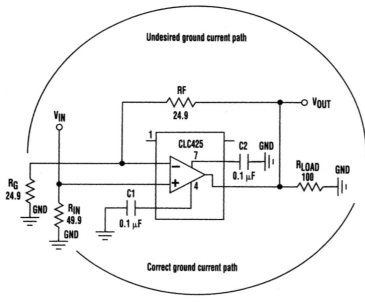

Incorrect grounding

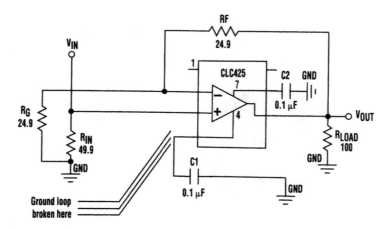

ELECTRONIC DESIGN

Fig. 61-12

Incorrect grounding on a high-frequency amplifier allows the addition of polarized signal back into the input. Improved grounding is provided by common grounds at the input, physical orientation of a bypass capacitor, and breaking a ground loop.

ONE-IC TRANSMIT TONE ENCODER

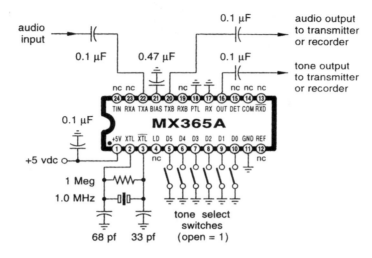

ELECTRONICS NOW

Fig. 61-13

This circuit can be used in a communications transmitter or repeater to place an encoding tone on an audio signal.

BACKUP SUPPLY CAPACITOR

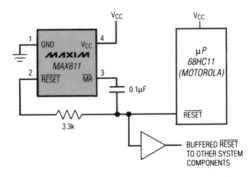

MAXIM

Fig. 61-14

These connections allow dual control of the buffered RESET line and extend the duration of resets issued by the μP. The parts shown are CMOS RAM.

FEEDBACK AMPLIFIER INTEGRATOR

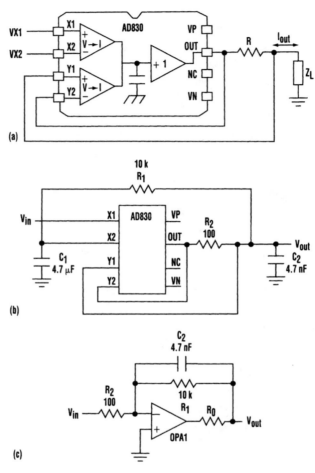

(a)

(b)

(c)

ELECTRONIC DESIGN

Fig. 61-15

The active-feedback-amplifier topology, such as that in the AD830 from Analog Devices and Linear Technology's LTC1193, can be used to produce a precision voltage-to-current converter. This, in turn, makes possible the creation of grounded-capacitor integrators. This circuit can be used as a differential input integrator by making Z_L a capacitor. Figure 1b and 1c compares the grounded-capacitor integrator using the AD830 with a standard op-amp integrator. R1 in both figures and C1 in Fig. 1b define the dc operating point for testing purposes. R2 and C2 determine the integrator time constant. If the op amp in Fig. 1c is ideal in the sense that it is modeled as an ideal integrator with zero output and infinite input resistance, then the only difference between the two circuit topologies is the finite input resistance of the op-amp circuit determined by R2.

FEEDBACK AMPLIFIER INTEGRATOR *(Cont.)*

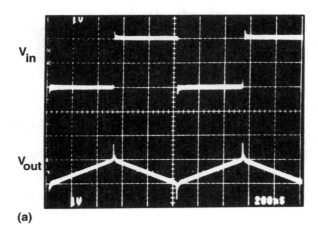

(a)

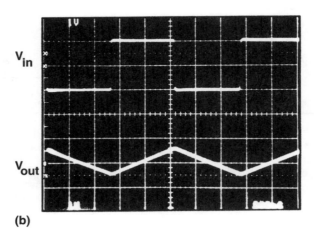

(b)

A large spike is produced on the output voltage waveform whenever the input switches using the circuit with the standard op-amp integrator (a). The integrator using the AD830 has no spike in its output waveform because the capacitor is connected to a true ground (b).

SIMPLE INTERCOM

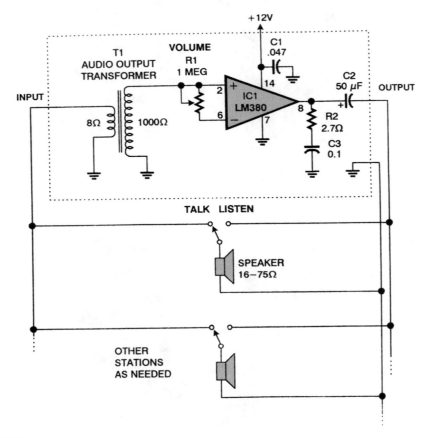

Fig. 61-16

This intercom uses an LM380 IC to act as an amplifier and speaker driver.

SCHMIDT TRIGGER MEMORY-CELL CIRCUIT

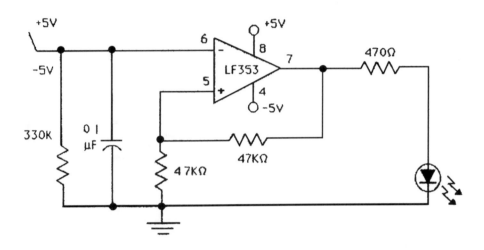

Schematic of a memory cell circuit

-5V	+5V	LED	
ON	OFF	ON	Last Input
OFF	ON	OFF	

TRUTH TABLE

Fig. 61-17

SOLAR-POWERED INSECT CONTROLLER

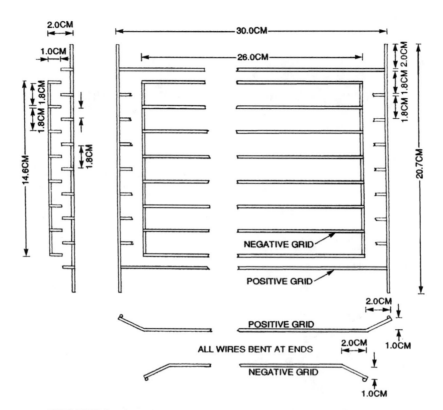

THE GRID for the solar-powered fly controller is fabricated from stiff wire that can be assembled with solder, braising material, or by welding. No. 12 solid copper wire without insulation is a good choice. It forms easily and solders well with a 100-watt iron. (Note: 2.54 cm equals 1.0 inch.)

POPULAR ELECTRONICS

Fig. 61-18

The circuit for the solar-powered controller consists of a switching circuit, a pulsing circuit, and a high-voltage output circuit. The external power components are a 1- to 5-W solar panel and a 12-V motorcycle or camcorder battery. The output of the high-voltage ignition coil connects to a network of paralleled electrodes, called a *grid*, upon which flies land and are destroyed. Voltage input to the LM3909 LED-flasher/oscillator (IC2) is kept at 6 to 9 V by a LM317T voltage regulator (IC1). The exact voltage is not crucial as long as it is regulated. The LM3909 produces a series of pulses that are coupled to a 2N2222A transistor (Q1) to form a switching circuit. An output of positive pulses from IC2 to the 2N2222A transistor boosts the pulse current so that it closes a 5-V relay (RY1) for approximately 0.1 s at intervals of 1 to 2 s. Diode D1 shunts out high-voltage spikes

SOLAR-POWERED INSECT CONTROLLER *(Cont.)*

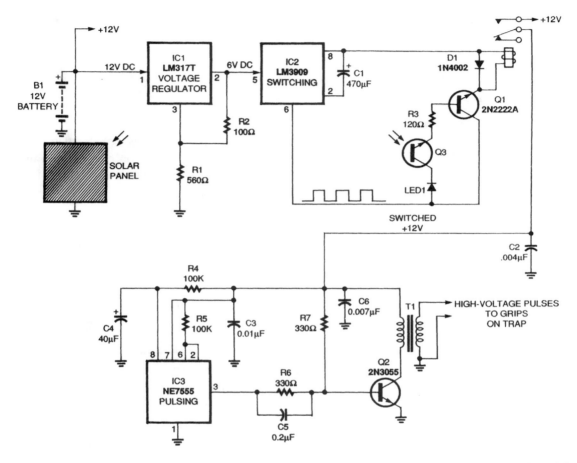

produced by the switching voltage across the relay's coil. The switching circuit is turned off at night or during periods of heavy cloud cover by phototransistor Q3, whose internal resistance increases as the ambient light diminishes. This reduces the positive bias on Q1's base, causing the transistor to cut off. LED1 serves as a voltage-dropping device. Single-pole, single-throw relay RY1 provides brief pulses of the 12-V battery voltage to IC3, an NE755 timer that is wired as a free-running audio-frequency pulse generator. The pulses are amplified by a 2N3055 power amplifier transistor, Q2. The output of Q2 drives an automobile ignition coil, T1, to generate the high-voltage pulses for the external grid. The output voltage at the secondary winding of T1 is approximately 12,000 V p-p. The circuit is powered by a 12-V rechargeable lead-acid or nickel-cadmium battery. A 1-W or better solar panel of the type used to keep automobile batteries charged should be used to eliminate the need to recharge the battery frequently.

LOGIC GATE CURRENT BOOSTER

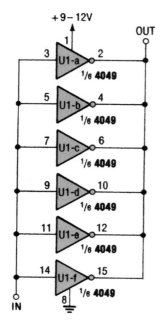

By connecting all six gates of a 4049 hex inverting buffer in parallel, you can obtain a much higher output current than would otherwise be available.

POPULAR ELECTRONICS

Fig. 61-19

PHOTOVOLTAIC ISOLATOR CIRCUIT

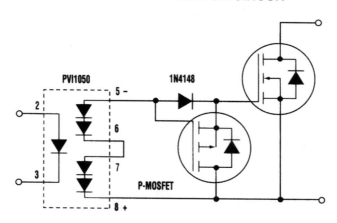

ELECTRONIC DESIGN

Fig. 61-20

The advantage of the programmable junction transistor (PUT) over a P-MOSFET used as a gate pull-down is that the PUT will pull the gate well below the MOSFET threshold voltage much faster at lower cost. The PUT can discharge a gate capacitance of 5 nF at 12 V in 100 ns.

VOLTAGE-CONTROLLED AMPLITUDE LIMITER

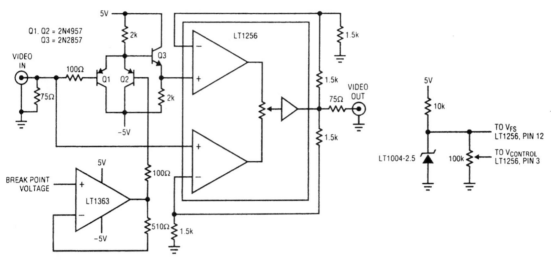

Fig. 61-21

Amplitude-limiting circuits are useful when a signal should not exceed a predetermined maximum amplitude, such as when feeding an A/D or a modulator. A clipper, which completely removes the signal above a certain level, is useful for many applications, but there are times when it is not desirable to lose information. The circuit in this figure is a voltage-controlled breakpoint amplifier. When the input signal reaches a predetermined level (the breakpoint), the amplifier gain is reduced. As both the breakpoint and the gain for signals greater than the breakpoint are voltage programmable, this circuit is useful for systems that adapt to changing signal levels. Adaptive highlight compression finds uses in CCD video cameras, which have a very large dynamic range. Although this circuit was developed for video signals, it can be used to adaptively compress any signal within the 40-MHz bandwidth of the LT1256. The LT1256 video fader is connected to mix proportional amounts of input signal and clipped signal to provide a voltage-controlled variable gain. The clipped signal is provided by a discrete circuit consisting of three transistors. Q1 acts as an emitter follower until the input voltage exceeds the voltage on the base of Q2 (the breakpoint voltage, V_{BP}). When the input voltage is greater than V_{BP}, Q1 is off and Q2 clamps the emitters of the two transistors to V_{BP} plus a V_{BE}. Q3, an NPN emitter follower, buffers the output and drops the voltage (V_{BE}); thus, the dc level of the input signal is preserved. The breakpoint voltage at the base of Q2 must remain constant when this transistor is turning on or the signal will be distorted. The LT1363 maintains a low output impedance well beyond video frequencies and makes an excellent buffer.

561

PHOTOVOLTAIC ISOLATOR WITH FAST RESET

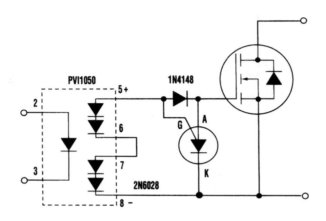

Fig. 61-22

Photovoltaic isolators (PVIs) provide an isolated gate drive for MOSFETs and IGBTs. At turn-off, however, they don't pull the gate down to drain the gate charge. Using a programmable junction transistor (PUT) provides a very fast gate pull-down, pulling as much as 5 A to discharge the gate in less than 0.2 s. The PUT is a four-layer structure, much like a silicon-controlled rectifier, but with a much more sensitive anode gate. In this application, the PUT gate is driven by the PVI to a voltage equal to or greater than that of the anode that holds the PUT in the blocking state so that no current flows through the anode-cathode channel. When the PVI is turned "off", the PUT gate voltage drops below the anode voltage, triggering the PUT and causing the anode-cathode channel to conduct. Conduction continues until the MOSFET gate voltage drops to about 0.5 V, at which time the PUT stops conducting and recovers for another cycle.

ZENER-DIODE SIMULATOR CIRCUIT

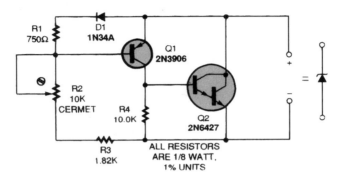

POPULAR ELECTRONICS

Fig. 61-23

Conventional zener diodes, unless specially compensated, all exhibit a similar problem: As the zener bias current increases, the temperature coefficient also increases in a positive direction. Also, the temperature compensation can be negative or positive, depending on the zener voltage and the particular bias current used—even for a single diode over its characteristic curve! Also, zener diodes available for less than approximately 5.0-V values exhibit very weak "knees," where the zener impedance causes a large voltage change over the full bias-current range. This circuit is an attempt to reduce those problems. The bias current range is 1 to 20 mA, and this can be extended by using a higher-power transistor for Q2. This circuit has an adjustable "zener" voltage of from 1.5 to 6.5 V. That voltage (V_z) can be determined using

$$V_z = 1.5 = (5R_2 / 10^4)$$

For any setting of R2, the voltage varies less than 1 percent over the 2- to 20-mA bias current. The base-emitter voltage of Q1 provides the reference voltage that is temperature D1, a 1N34A germanium diode. A stable voltage of about 375 mV is then present across R1, and a constant current of 0.5 mA flows through the resistor divider. The net temperature coefficient is complex, but it remains below −2mV/°C because of the interaction of the diode and transistors. Darlington pair Q2 handles all the bias current, except for the 600 μA needed to bias the reference.

VOLTAGE FOLLOWER STABILITY

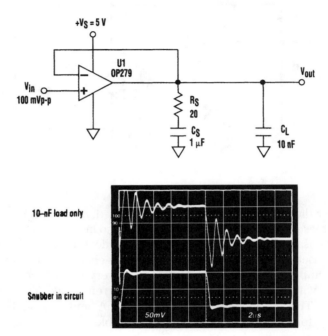

ELECTRONIC DESIGN

Fig. 61-24

A unity-gain voltage follower driving a capacitive load is prone to stability problems. Using an RC snubber network (R_S–C_S) allows capacitive loads, such as C_L to be compensated without loss of load drive.

The comparison of pulse response with and without the snubber network shows a dramatic reduction in overshoot. The top trace is without the snubber and the bottom trace is with the snubber.

OP-AMP CIRCUIT DC POWER SUPPLY CONNECTIONS

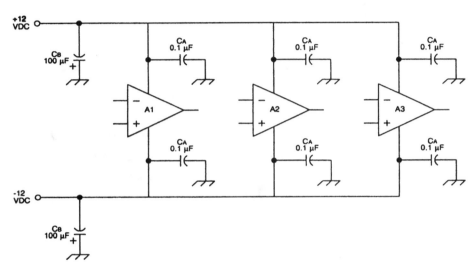

Fig. 61-25

The dc power distribution of multistage, or multi-op-amp, filter circuits needs to be well conditioned to prevent either oscillation in one op amp or, because of poor power supply, decoupling between op amps. The figure shows the proper way to keep these problems under control. The capacitors marked C_A are 0.1 μF to 1 μF, and must be mounted as close to the body as possible. A larger capacitor (marked C_B), 100 μF to 500 μF, is connected between each power-supply line and ground, usually at the point of connection to the power supply. Note that these capacitors are polarity-sensitive; they must be connected into the circuit properly or they might be damaged.

RAIL SPLITTER VIRTUAL GROUND CIRCUIT

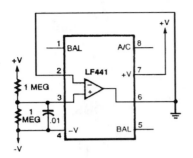

Use a low-power op amp, of which there are many to choose from. The National LF441 is a good, inexpensive choice (don't confuse it with the 411). Pin 6 is the virtual ground "output." This circuit will work correctly down to a battery voltage of about 6 V. The current consumption is about 160 μA.

ELECTRONICS NOW

Fig. 61-26

PHOTOVOLTAIC ISOLATOR WITH FAST RESET

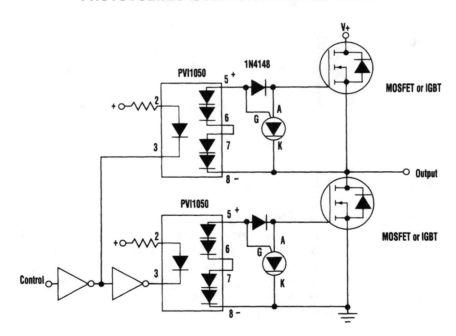

ELECTRONIC DESIGN

Fig. 61-27

The combination of a weak photovoltaic isolator pull-up and a strong programmable junction transistor pull-down leads to a very simple half-bridge FET or IGBT driver, which virtually guarantees that the devices won't crowbar the power supply. The device turning "off" will do so rapidly, but the device turning "on" will do so very slowly. A dead time is inherent in the circuit dynamics, so no additional circuitry is required to prevent a crowbar or "shoot-through" effect.

GENERIC TRIAC SWITCH INTERFACE FOR INDUCTIVE LOAD

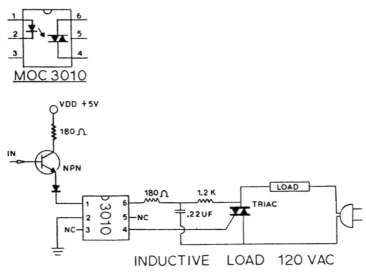

INDUCTIVE   LOAD   120 VAC

NUTS AND VOLTS

Fig. 61-28

A TTL level input turns on the MOC3010 optocoupler and triggers the triac, applying 120 Vac to the load.

HUMAN BODY MODEL

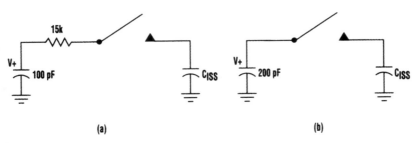

(a)   (b)

ELECTRONIC DESIGN

Fig. 61-29

The human-body model (HBM) is approximated with a series capacitor and resistance (*a*); the machine model (MM) is simply a capacitor (*b*). The model is charged to the required test voltage (*V+*) and switched into the device. If the input capacitance of the device (C_{ISS}) charges to higher-than-rupture voltage, the device will be destroyed.

GUARD DRIVE

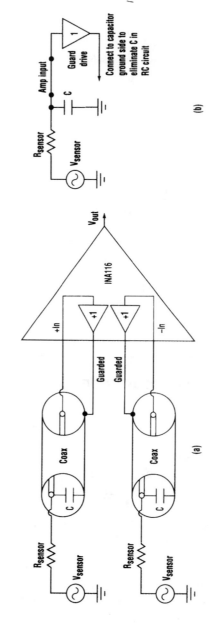

(a)

(b)

Fig. 61-30

ELECTRONIC DESIGN

A guard drive can improve common-mode rejection and ac gain when an IA includes an RC circuit made up of the sensor source resistance and the capacitance associated with a coax cable (*a*). The guard drive prevents a voltage from developing across the capacitance, thus eliminating it from the RC filter's equivalent circuit (*b*).

EXTENDED 68HC11 RESET TIMEOUT

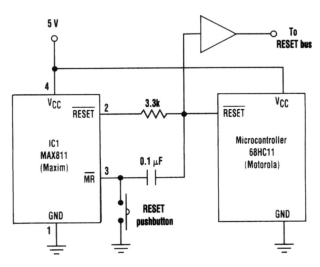

ELECTRONIC DESIGN

Fig. 61-31

Some microcontrollers (such as the Motorola 68HC11) are implemented with a bidirectional RE-SET pin that can create a contention problem with the RESET output of certain supervisory ICs. If the supervisor RESET is high, for example, and the microcontroller tries to pull it low, the result might be an intermediate logic level that fails to reset the microcontroller. In addition, some devices on the RESET bus might require a RESET pulse of longer duration than the microcontroller pro-vides. IC1 in the circuit shown eliminates this concern by producing a RESET output with a mini-mum duration of 140 ms. Its RESET output becomes asserted either when the microcontroller initiates a reset, when the MR pushbutton is depressed, or when V_{CC} dips below a threshold internal to IC1. When the reset line is pulled low by IC1 or the microcontroller, MR is pulled low and then re-turns high (initiating a timeout period of 200 ms typical) as the capacitor is charged via the internal MR pull-up resistor. IC1's RESET deasserts following the timeout, allowing the capacitor to discharge through the internal pull-up resistor and ESD-protection diode.

POSITION-SENSOR POTENTIOMETER

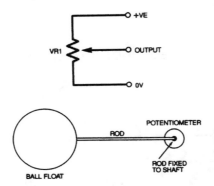

Fig. 61-32

A potentiometer can be used for position-to-voltage sensing. This figure shows the use of a potentiometer to sense water-level changes.

RESTRICTED-RANGE POTENTIOMETER

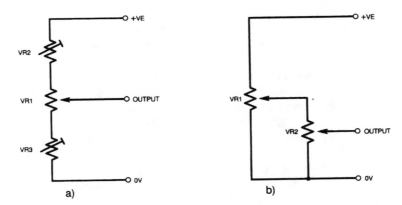

Fig. 61-33

The figure shows two methods for restricting the output voltage range from a potentiometer.

CMOS INTERFACE

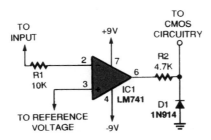

POPULAR ELECTRONICS

Fig. 61-34

This CMOS interface makes it possible for a simple or complex analog circuit to input data into a digital system.

GENERIC TRIAC SWITCH INTERFACE FOR RESISTIVE LOAD

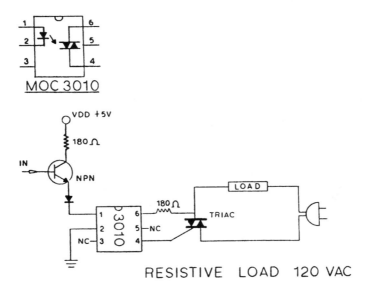

RESISTIVE LOAD 120 VAC

NUTS AND VOLTS

Fig. 61-35

A TTL-level input turns on the MOC3010 optocoupler and triggers the triac, applying 120 Vac to the load.

CONSTANT OFF-TIME REGULATOR WITH FREQUENCY COMPENSATION

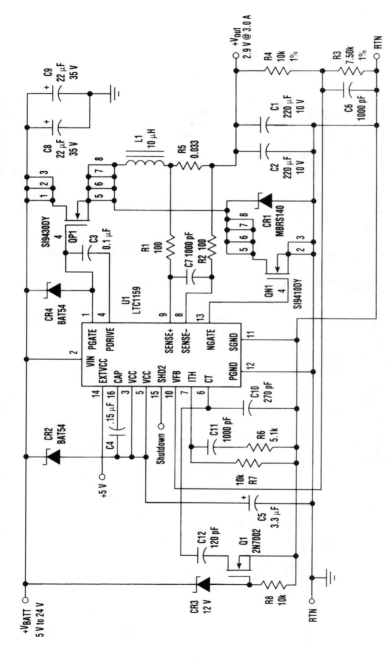

Fig. 61-36

ELECTRONIC DESIGN

Constant off-time switching regulators offer several advantages over constant-frequency designs. The only potential problem is that the switching frequency increases with rising input voltage. In designs that have large ratios of the high-line to low-line supply voltage, this frequency shift can get quite large. As a result, the switching losses can become excessive at high input voltages. To offset this problem, the simple circuit shown detects the high input voltage condition and lowers the switching frequency to keep switching losses under control. The frequency-shift circuit consists of D3, R8, Q1, and C12. When V_{in} exceeds the zener voltage plus the FET threshold, Q1 turns on and adds an extra timing capacitor (C12) in parallel with the timing capacitor (C10). This increases the off-time, lowering the frequency.

INEXPENSIVE EMERGENCY LIGHT

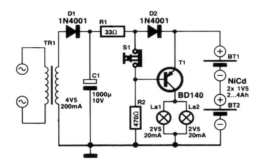

ELEKTOR ELECTRONICS

Fig. 61-37

This low-cost circuit charges two nickel-cadmium cells to provide a standby power supply for a two-lamp emergency light, which is automatically switched on in the event of a power failure. Diode D1 and smoothing capacitor C1 form a conventional single-phase rectifier circuit with an output voltage of about 6 V. This voltage is used to continuously charge two series-connected NiCd batteries with a current of 80 to 100 mA, which flows via R1 and D2. At this rate, a 2-Ah (ampere-hour) NiCd battery can be safely charged for extensive periods. The voltage drop across D2 reverse-biases the base-emitter junction of PNP transistor T1. Consequently, the transistor is turned off, and the lamps connected to its collector do not light. When a power outage occurs, T1 is supplied with base current (from the batteries) via R2, whereupon the miniature lamps are turned on. As soon as the power is restored, T1 is switched off again, and the batteries are charged again via R1 and D2.

573

LOW-RFI AC LIMITER

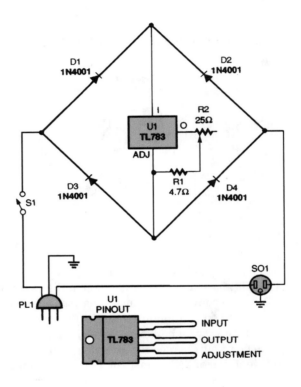

Fig. 61-38

This current limiter will work with any ac device. It does not need a zero-crossing network because it is not based on a triac. Instead, a current-limiting circuit reduces the peaks of the ac waveforms. That is preferable to the action of a triac in a light dimmer, which suddenly turns on and generates significant RF harmonics (even with a zero-crossing network) in all but the fully ON state.

PB SWITCH-ACTIVATED MEMORY

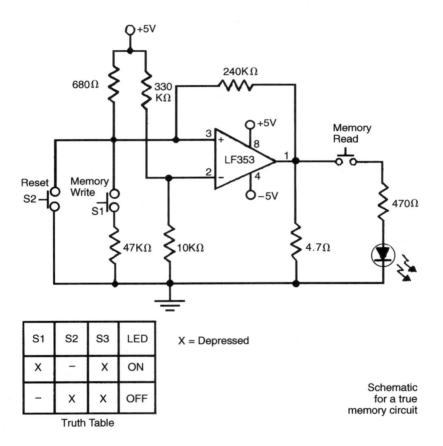

S1	S2	S3	LED
X	–	X	ON
–	X	X	OFF

X = Depressed

Truth Table

Schematic
for a true
memory circuit

Fig. 61-39

BACK-EMF SENSOR

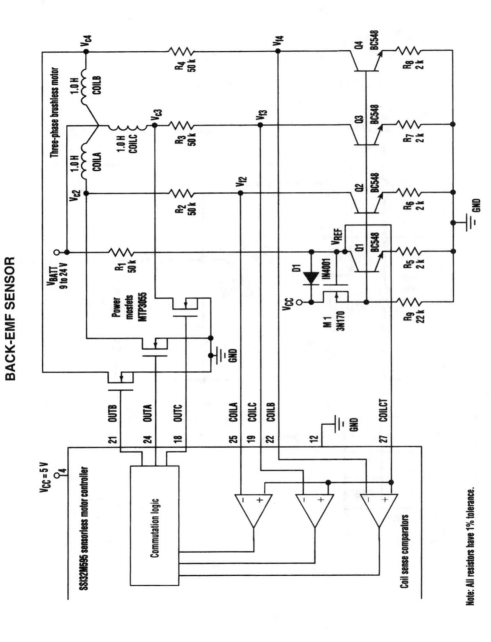

Fig. 61-40

Note: All resistors have 1% tolerance.

ELECTRONIC DESIGN

BACK-EMF LOW SENSOR *(Cont.)*

Many new brushless-motor control schemes use ICs that determine the rotor position by sensing the motor's back-EMF signals. In a unipolar configuration, the control chip synchronizes the three phases of drive currents by detecting the moments when each back-EMF signal rises just above, or falls just below, the motor supply voltage, V_{BATT} (see the figure). These instances are known as *zero crossings*. This circuit shifts the signals down to a reference level, V_{REF}. The MOSFET, M1, along with resistor R1 and NPN transistor Q1, creates a reference current (I_1) that is equal to $(V_{BATT} - V_{REF})/R_1$. This reference current is replicated in Q2, Q3, and Q4. Because $R_2 = R_3 = R_4 = R_1$, the voltage drop across each resistor is the same. That means that V_2, V_3, and V_4 vary above and below V_{REF} the same amount as V_2, V_3, and V_4 vary above and below V_{BATT}. Matching resistors R1 to R4 within 1 percent and matching R5 to R8 within 1 percent will resolve back-EMF signals as small as 200 mV. Diode D1 protects M1's gate from large transients on V_{BATT}. The ESD diodes on the SSI595, along with resistors R2, R3, and R4, protect the IC from damaging transients at V_2, V_3, and V_4.

FAST-PULSE CARRYING MICROSTRIP LINE

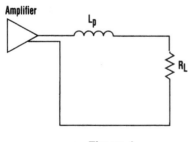

Figure 1

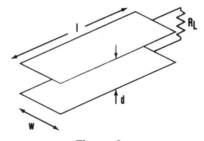

Figure 2

Figure 1. Parasitic external inductance in the interconnect wire may limit the switching speed of this amplifier.

Figure 2. To reduce the inductance evidenced in Fig. 1, a microstrip line can be used in place of the interconnect wire.

CUTTING NOISE PICKUP WITH REMOTE GAIN

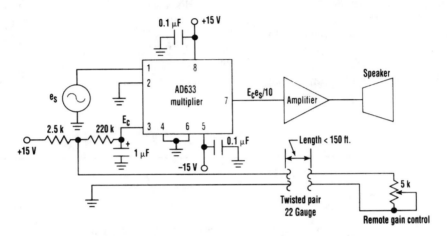

ELECTRONIC DESIGN

Fig. 61-42

At times, gain control that is remote from the amplifier and speaker is required. However, merely extracting the gain-control potentiometer from a typical amplifier and placing it remotely will contaminate the signal with pickup. One simple solution is to use an analog multiplier as a gain-control element. The 2.5-kΩ resistor and the remote gain-control potentiometer form an attenuator of the +15-V supply, whose output is filtered through a low-pass RC filter. The resistance of the twisted pair of leads is negligible compared with the potentiometer's 5-kΩ resistance. When the remote gain-control potentiometer is at maximum resistance, the control voltage, E_c, at pin 3 of the multiplier is at a nominal value of +10 Vdc. This results in a maximum input amplitude to the amplifier, which is equal to the original input signal, e_s. Correspondingly, when the gain-control potentiometer is set to zero resistance, the signal input to the amplifier is nominally zero. Any ac signals that might be picked up by the long run of wire are effectively attenuated by the large time constant of the low-pass filter.

FREQUENCY MULTIPLIER MMIC AMPLIFIER

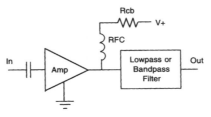

RF DESIGN

Fig. 61-43

With sufficient drive, an MMIC can be an efficient frequency multiplier.

STRAY CAPACITANCE-CANCELLATION TECHNIQUE

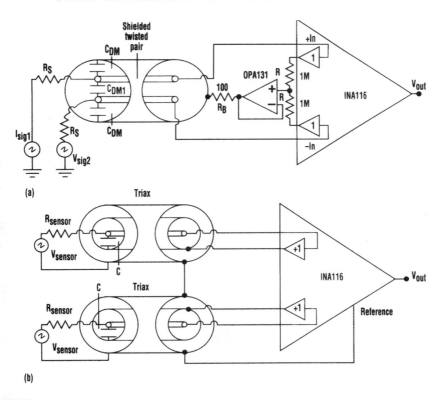

ELECTRONIC DESIGN

Fig. 61-44

A shielded twisted-pair cable eliminates common-mode capacitance, but the differential capacitance remains (*a*). A triax cable eliminates both common-mode and differential capacitances (*b*).

ECHO SYSTEM DYNAMIC RANGE BOOSTER

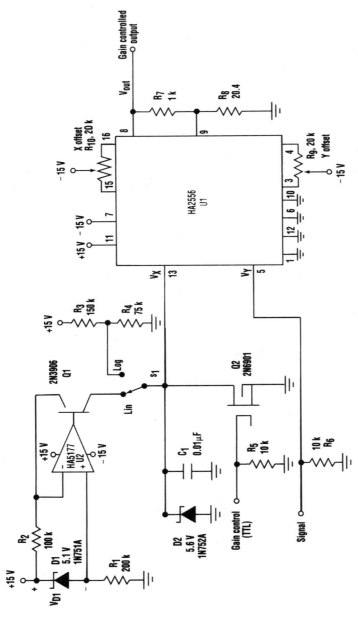

Fig. 61-45

ELECTRONIC DESIGN

Using a variable-gain preamplifier helps improve the dynamic range of echo systems. Here, Harris Semiconductor's HA2556 multiplier is used to implement the variable-gain preamplifier and establish the signal bandwidth and noise figure. An echo system with a fixed-gain preamplifier exhibits poor dynamic range because close targets (long return times) have high signal amplitudes, whereas distant targets (long return times) produce amplitudes that are much lower. One solution involves a preamplifier that has a gain that is proportional to time so that the gain will be small for close targets and large for distant targets. The pre-

ECHO SYSTEM DYNAMIC RANGE BOOSTER *(Cont.)*

amplifier must meet all of the other preamplifier criteria, such as bandwidth and noise performance. Moreover, the added time-dependent gain function must not degrade the signal. This type of variable-gain preamplifier can be built with a multiplier IC, the Harris Semiconductor HA2556. This IC establishes the signal bandwidth and noise figure because it is the only component in the signal path. The HA5177 op amp and its associated circuitry make up a constant-current source with a current (I) of $V_{D1}/R_2 = 51$ µA. If the switch (S1) is in the LINEAR-position with Q2's gate held high, the current source is shorted to ground by Q2 and the multiplier gain is set to zero. When the transmission of the outgoing signal is complete, Q2's gate is brought low, forcing it into a very high drain resistance state (almost an open circuit). Consequently, the HA5177 current can charge C1 in a linear manner. The voltage across C1 then ramps up from 0 to 5 V in 1 ms. During the first portion of the ramp, when the returned signal is very large, the multiplier gain is small because V_x is small. As time increases, so does V_x, providing more gain through the multiplier as the expected echo decreases in amplitude. As a result, the output-voltage swing of the multiplier tends to stay constant for large changes in input signal. In addition, the dynamic range is improved to the amount of the ramp change, which is more than 60 dB with the values shown in the figure.

4013 DIVIDE-BY-2 CIRCUIT

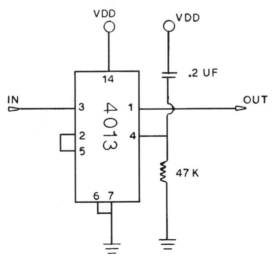

NUTS AND VOLTS

Fig. 61-46

The first pulse into the input drives the output high. The second pulse drives the output low. V_{CC} can be 5 to 15 V.

CHARGE PUMP GENERATING GATE DRIVE

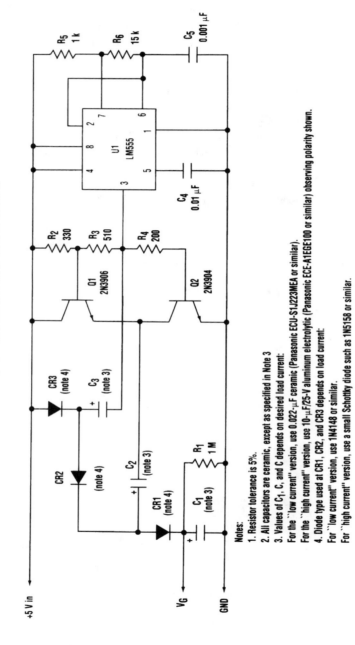

Notes:
1. Resistor tolerance is 5%.
2. All capacitors are ceramic, except as specified in Note 3.
3. Values of C_1, C, and C depends on desired load current:
For the "low current" version, use 0.022-μF ceramic (Panasonic ECU-S1J223MEA or similar).
For the "high current" version, use 10-μF/25-V aluminum electrolytic (Panasonic ECE-A1EGE100 or similar) observing polarity shown.
4. Diode type used at CR1, CR2, and CR3 depends on load current:
For "low current" version, use 1N4148 or similar.
For "high current" version, use a small Schottky diode such as 1N5158 or similar.

ELECTRONIC DESIGN

Fig. 61-47

A charge pump built as a voltage tripler can provide just the right amount of voltage needed for gate drive in a 5-V system.

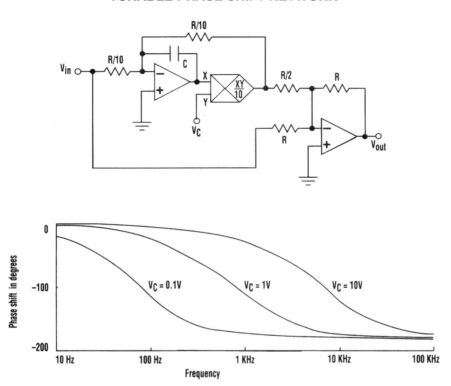

TUNABLE PHASE SHIFT NETWORK

Fig. 61-48

Frequently, the need arises to control the phase shift or delay of an audio signal. The all-pass active filter shown performs this function. The transfer function has the form

$$T(s) = V_{out}(s) / V_{in}(s) = 1 - sRC / V_c / (1 + sRC / V_c)$$

where V_c is the control voltage and $V_c > 0$. As a result, this filter has unity voltage gain for all frequencies. V_c can be used to vary the phase over the theoretical range from $-180° \leq \phi \leq 0$. In the circuit, the left op amp and analog multiplied combination produce an inverting, first-order, low-pass filter whose pole frequency is set by the control voltage, according to $\omega_p = V_c / RC$. An MC1495 analog multiplier has an adjustable scale factor, which, in this case, is set to 0.1 for convenience. Choosing a center frequency of f_0 at the geometric mean of the audio range, or $f_0 = \sqrt{20 \times 20}$ kΩ = 633 Hz, centers the phase shift at $-90°$. For $V_c = 1$ Vdc, using the relationship $f_0 = V_c\omega_0 / 2\pi = V_c / 2\pi RC$ sets the RC time constant to 252 μs. Selecting $R = 10$ kΩ makes $C = 25$ nF.

LONGWAVE TRF TUNER

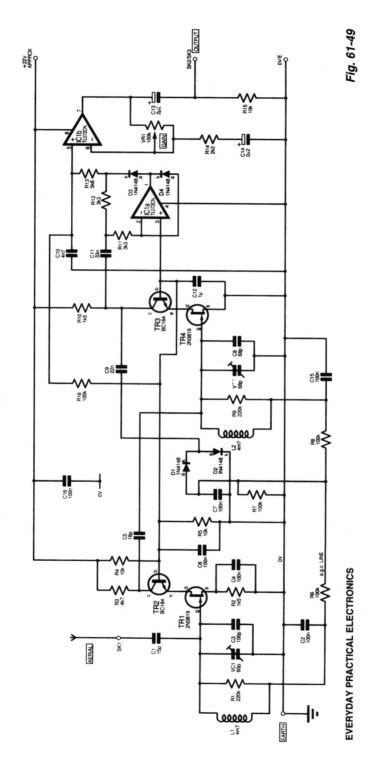

Fig. 61-49

EVERYDAY PRACTICAL ELECTRONICS

This TRF tuner was intended for reception of Radio 4 AM broadcast signals in the United Kingdom on 198 kHz, with the best possible audio fidelity. It can be used as an IF amplifier or for reception of LW or MW AM broadcasts in other areas. The aerial wire is directly connected, via SK1, to the input tuned circuit (coil L1 and capacitors VC1 and C3) via a small-value capacitor C1. The second tuned circuit (L2, VC2, and C8) is similarly connected to the output from transistor TR2. The output from transistor TR3 is taken to the op-amp demodulator (IC1a) and also, via capacitor C9, to the "diode-pump" circuit D1/D2. The "pump" is used to generate a small negative AGC. For AM detection, if a diode is used as the rectifier, the tuner will not work very well on input voltages less than 0.5 V peak. However, if a pair of diodes are hooked around an op amp, IC1a, as shown in the figure, the rectification threshold is reduced to a few millivolts, which increases the demodulator sensitivity by a few hundred times. The other half of the dual op amp, IC1b, is a straight AF amplifying stage, whose gain is adjustable by potentiometer VR1. Resistor R13 and capacitor C10, on its noninverting input (pin 5), act as a simple RF filter to remove any residual 198-kHz signal voltage from the audio output.

TRANSIENT SWITCH SIGNAL SUPPRESSOR

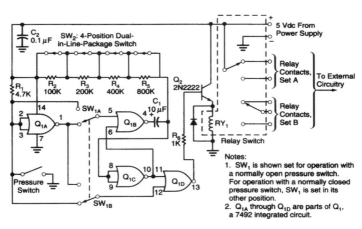

Fig. 61-50

    The figure shows a circuit that delays the transmission of a switch-opening or switch-closing signal until after a preset suppression time. The circuit is used to prevent the transmission of an undesired momentary switch signal. For example, a pressure switch that is meant to be held steadily on or off by a given static pressure in a piping system can also be actuated momentarily by a transient overpressure (sometimes called "water hammer") or underpressure caused by the sudden closure or opening of a valve. The basic mode of operation is simple. The beginning of the switch signal initiates a timing sequence. If the switch signal persists after the preset suppression time, then this circuit transmits the switch signal to the external circuitry. If the switch signal is no longer present after the suppression time, then the switch signal is deemed to be transient, and this circuit does not pass the signal on to the external circuitry. From the perspective of the external circuitry, it is as though there were no transient switch signal. The suppression time is preset at a value large enough to allow for the damping of the underlying pressure wave or other mechanical transient.

MULTIPLIER-SIMULATING VOLTAGE-CONTROLLED RESISTOR

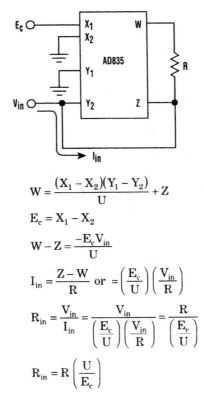

$$W = \frac{(X_1 - X_2)(Y_1 - Y_2)}{U} + Z$$

$$E_c = X_1 - X_2$$

$$W - Z = \frac{-E_c V_{in}}{U}$$

$$I_{in} = \frac{Z - W}{R} \text{ or } = \left(\frac{E_c}{U}\right)\left(\frac{V_{in}}{R}\right)$$

$$R_{in} = \frac{V_{in}}{I_{in}} = \frac{V_{in}}{\left(\dfrac{E_c}{U}\right)\left(\dfrac{V_{in}}{R}\right)} = \frac{R}{\left(\dfrac{E_c}{U}\right)}$$

$$R_{in} = R\left(\frac{U}{E_c}\right)$$

ELECTRONIC DESIGN

Fig. 61-51

A two- or four-quadrant multiplier can be configured to simulate a resistor with a value that is the ratio of a fixed resistor and a control voltage. If the voltage-controlled current source is made to depend on input voltage V_{in} (i.e., by driving Y2), as well as being allowed to "float" on top of V_{in} (i.e., by connecting Y2 to Z), then the input of Y2 will simulate a resistor with the transfer function $R_{in} = R(U / E_c)$. One node of the simulated resistor will always be referred to ground.

NE602 DC BIAS CIRCUIT

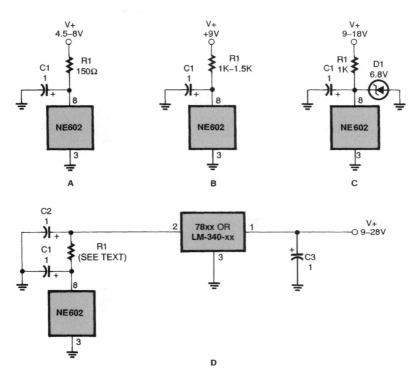

ELECTRONICS NOW

Fig. 61-52

There are several ways to power the NE602. A resistor should be placed in series between the power supply and the NE602 (*a* and *b*). A zener diode (*c*) or a voltage regulator (*d*) can also be used. The resistor R1 should provide 4.5 to 8.0 V at pin 8 of the NE602. The current draw is about 2.4 to 2.8 mA.

NE602 OUTPUT CIRCUITS

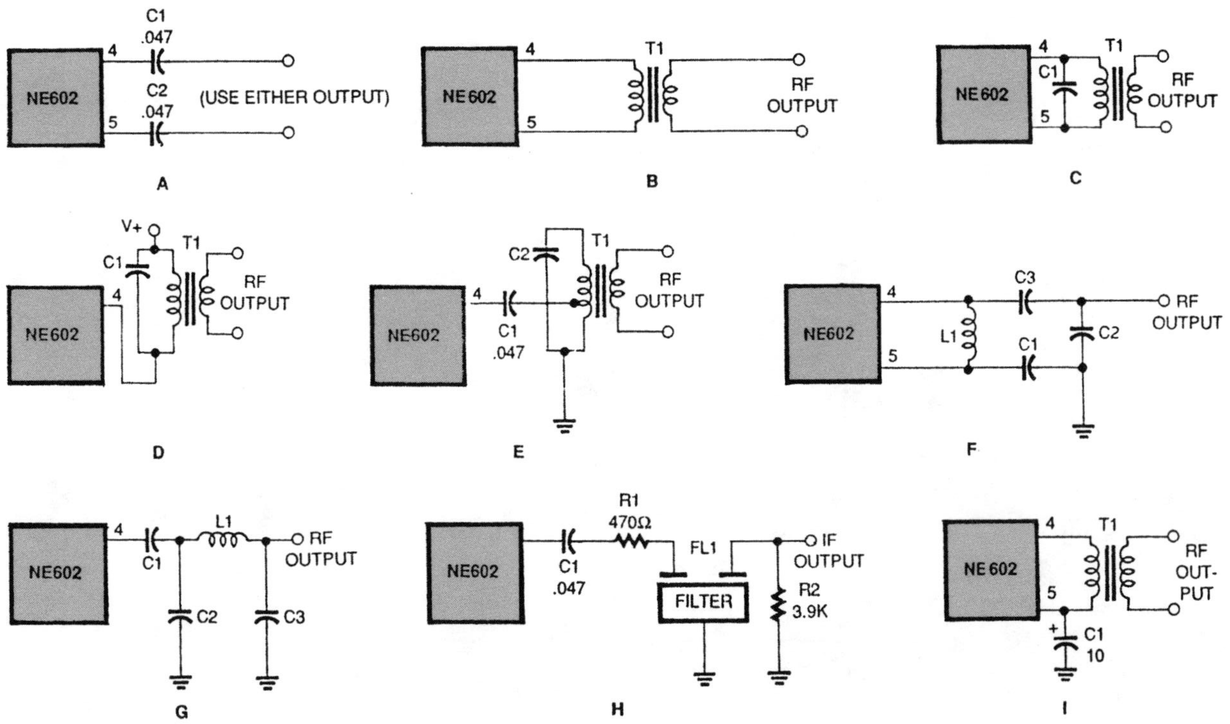

ELECTRONICS NOW

Fig. 61-53

The various output circuits shown here demonstrate how to either pass all the frequencies from the NE602 or allow only the sum or difference frequencies through, depending on which circuit is used.

NE602 IF OUTPUT CIRCUITS

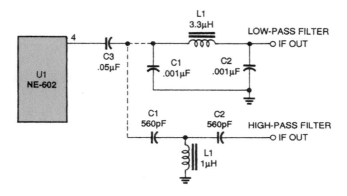

POPULAR ELECTRONICS

Fig. 61-54

Either the low-pass filter or high-pass filter outputs can be used, depending on the situation.

BROADBAND NE602 INPUT CIRCUITS

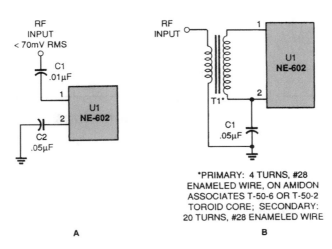

*PRIMARY: 4 TURNS, #28
ENAMELED WIRE, ON AMIDON
ASSOCIATES T-50-6 OR T-50-2
TOROID CORE; SECONDARY:
20 TURNS, #28 ENAMELED WIRE

A   B

POPULAR ELECTRONICS

Fig. 61-55

A variety of single-ended input circuits can be used with the NE602, including a broadband capacitor-coupled input and a broadband transformer-coupled input.

TUNED-OUTPUT NE602 CIRCUITS

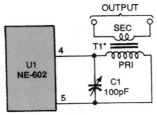

*PRIMARY: 33 TURNS,
#28 ENAMELED WIRE
ON T-50-2 OR
T-50-6 TOROID CORE;
SECONDARY: 5 TURNS,
#28 ENAMELED WIRE

A tuned output like this one will reject all but the desired output signal.

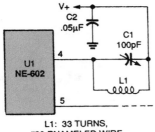

L1: 33 TURNS,
#28 ENAMELED WIRE
ON T-50-2 OR T-50-6
TOROID CORE

Here is a single-ended tuned-output circuit.

POPULAR ELECTRONICS   **Fig. 61-56**

NE602 OUTPUT CONFIGURATIONS

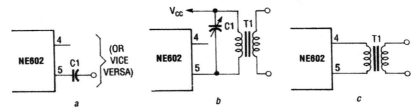

RADIO-ELECTRONICS   **Fig. 61-57**

Here are (*a*) a simplest single-ended approach without impedance matching, (*b*) a single-ended approach for a tuned LC circuit load, and (*c*) a balanced approach for better suppression of input and LO signals.

NE602 INPUT CIRCUITS

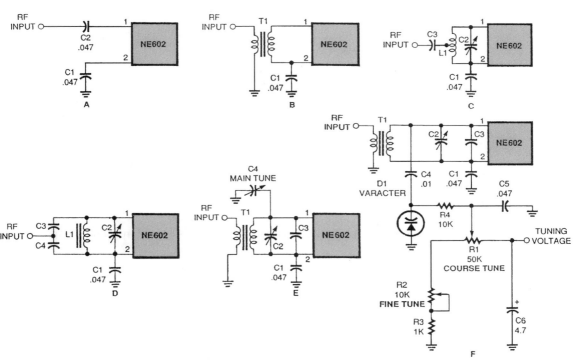

ELECTRONICS NOW

Fig. 61-58

A few of the many ways to input a signal into the NE602. Simple untuned methods (*a* and *b*) are acceptable. If you need to tune to a specific frequency, you can use an LC resonant circuit with ungrounded trimmer capacitors (*c* and *d*) or with grounded variable capacitors (*e*). You can even use a tuning voltage in connection with a varactor (*f*).

BROADBAND NE602 SINGLE-ENDED OUTPUT CIRCUITS

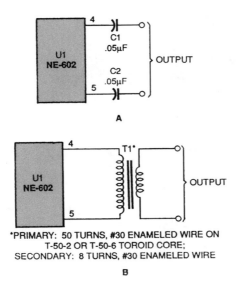

A

*PRIMARY: 50 TURNS, #30 ENAMELED WIRE ON
T-50-2 OR T-50-6 TOROID CORE;
SECONDARY: 8 TURNS, #30 ENAMELED WIRE

B

POPULAR ELECTRONICS

Fig. 61-59

Here are two single-ended output circuits. A capacitor output is shown in A and a single-ended transformer output is shown in B.

NE602 RF INPUT CONFIGURATIONS

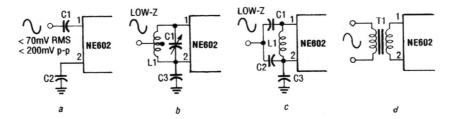

RADIO-ELECTRONICS

Fig. 61-60

Here, (a) to (c) are for single-ended coupling, with (a) being for no impedance matching, (b) for inductive matching, and (c) for capacitive matching. By contrast, (d) is a balanced input with reduced second harmonic.

TUNED INPUT CIRCUITS FOR NE602

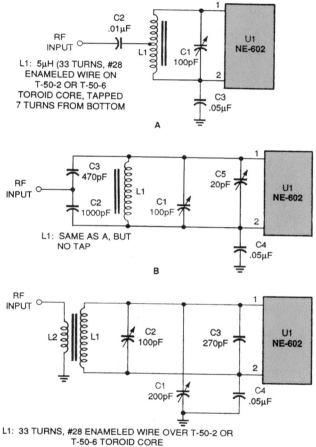

POPULAR ELECTRONICS

Fig. 61-61

Here are three tuned input circuits. A tapped-inductor LC tuned circuit is shown in A and a flexible-tuned input circuit that can be used in a variety of situations is shown in C.

30-MHz IF AND POWER SUPPLY FOR 10-GHz GUNN-DIODE COMMUNICATIONS SYSTEMS

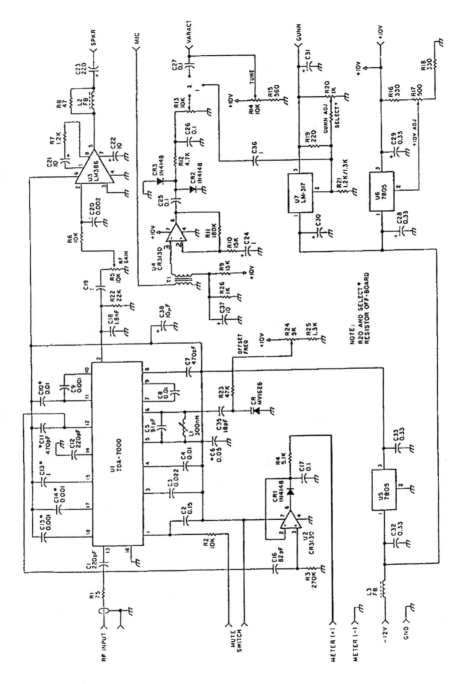

Fig. 61-62

30-MHz IF AND POWER SUPPLY FOR 10-GHz GUNN-DIODE COMMUNICATIONS SYSTEMS *(Cont.)*

This unit consists of a TDA7000 IF amplifier and detector system, an LM386 audio amplifier, a CA3130 modulator circuit, a Gunn-diode supply (U7) that can be modulated with audio for production of wideband FM, and a 10-V regulator.

AUDIO INTERPOLATING PHASE SHIFTER

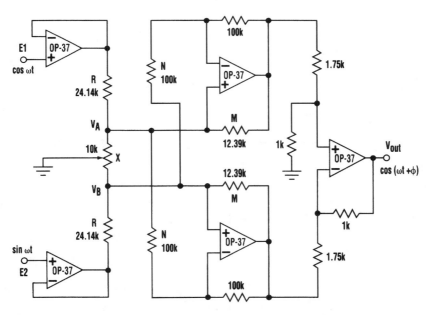

ELECTRONIC DESIGN

Fig. 61-63

A simple interpolating phase shifter places a negative resistance across potentiometer X to obtain constant-amplitude, linear-phase performance.

VARIABLE-CURRENT SOURCE (0 TO 5 A)

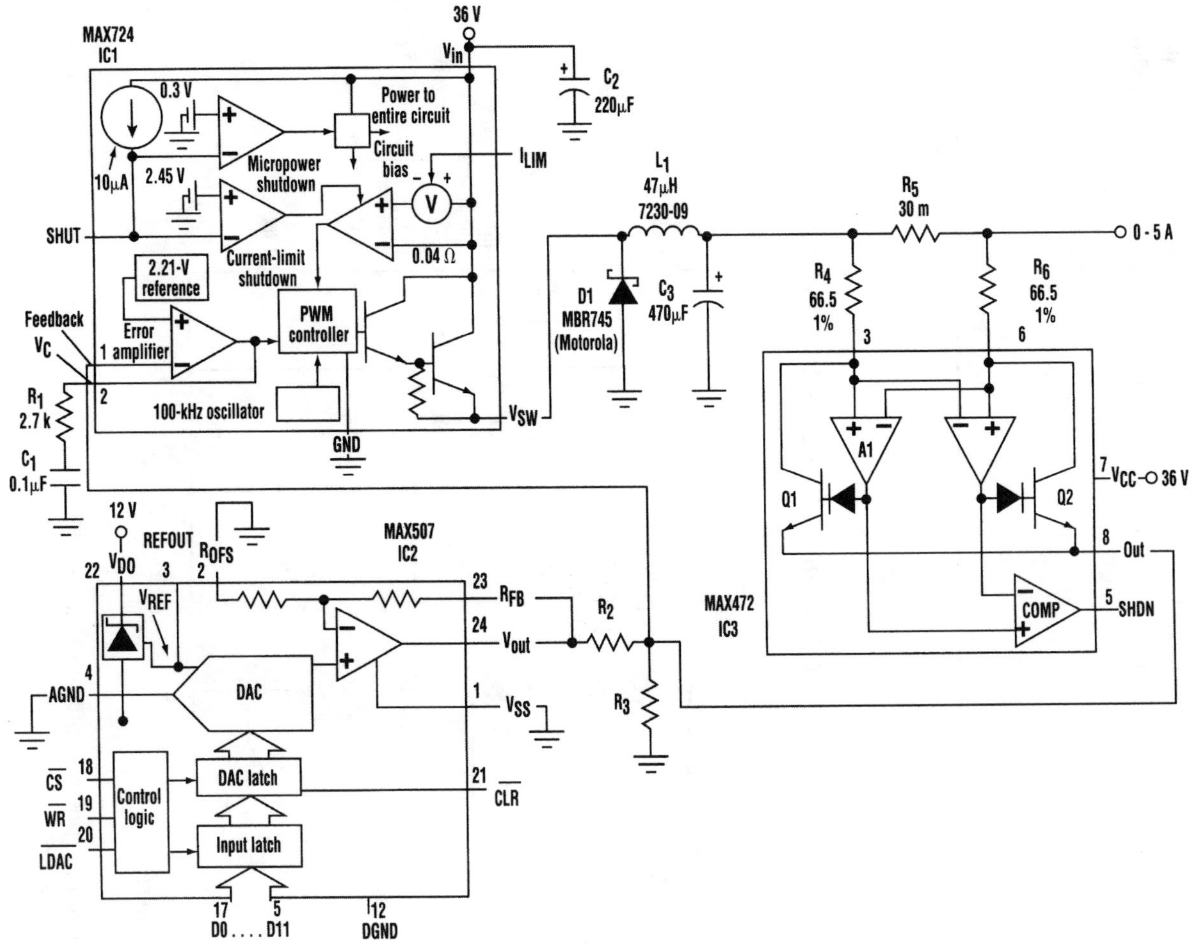

Fig. 61-64

ELECTRONIC DESIGN

VARIABLE-CURRENT SOURCE (0 TO 5 A) *(Cont.)*

The variable-current source generates 0 to 5 A with a compliance range of 4 to 30 V. The 12-bit digital-to-analog converter (DAC), IC2, makes it digitally programmable. The switch-mode step-down regulator, IC1, is more efficient than the alternative current source with linear pass transistor. IC3 is a high-side current-sense amplifier. In this circuit, it senses output current as a voltage drop across R5, and produces a proportional signal current at pin 8. The regulator's feedback voltage (IC1 at pin 1) is set by the DAC and modified by IC3's current feedback, which flows across a parallel combination of R2 and R3. This current feedback opposes any change in load current because of a change in load resistance. The DAC generates 0 to 10 V, producing a source current that varies inversely with code: FFF_{HEX} (10 V from IC2) produces 0 mA, and 000_{HEX} (0 V from IC2) produces 5 A. The circuit can be reconfigured for other ranges of output current (I_{source}) by resizing R_2 and R_3:

$$I_{source} = 2217[V_{FB}(R_2 + R_3) - (R_3 V_{DAC})] / R_2 R_3$$

where $V_{FB} = 2.21$ V and V_{DAC} can range from 0 to 10 V.

The desired range for I_{source} defines values for R_2 and R_3: $V_{DAC} = 10$ V for the low value of I_{source} and $V_{DAC} = 0$ V for the high value of I_{source}. Substituting these two sets of values in the equation yields two equations to be solved simultaneously for the values of R_2 and R_3.

SIGNAL AVERAGER CIRCUIT

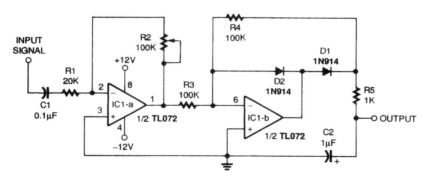

Fig. 61-65

The first operational amplifier, IC1-a, is an inverting amplifier whose gain can be adjusted up to 5. Adjustable feedback resistor R2 sets the circuit gain. The second amplifier, IC1-b, forms a half-wave rectifier and the signal averager clips the brief peaks and displays them over a longer period of time.

FM IF STRIP

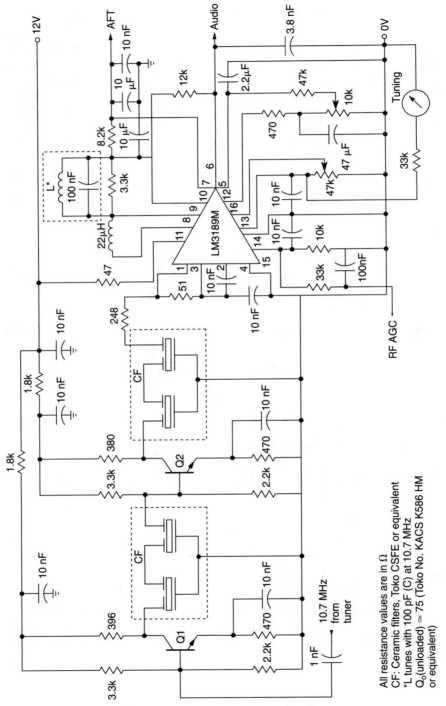

All resistance values are in Ω
CF: Ceramic filters, Toko CSFE or equivalent
*L tunes with 100 pF (C) at 10.7 MHz
Q_0(unloaded) ≈ 75 (Toko No. KACS K586 HM
or equivalent)

NATIONAL SEMICONDUCTOR

Fig. 61-66

62

Mixer Circuits

The sources of the following circuits are contained in the Sources section, which begins on page 1043. The figure number in the box of each circuit correlates to the entry in the Sources section.

JamMix Stereo Mixer
NE602 RF Translator Circuit
MMIC Amplifier Mixer
Audio Mixer
Two-Channel Audio Mixer
NE602 Mixer Circuit

JAMMIX STEREO MIXER

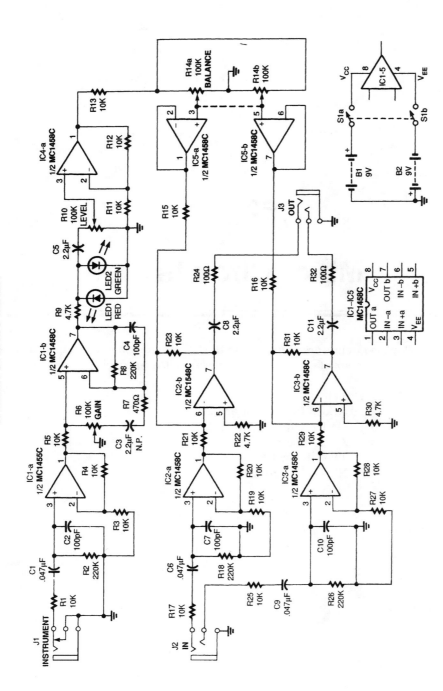

Fig. 62-1

600

JAMMIX STEREO MIXER *(Cont.)*

Here is the schematic diagram for JamMix. Note that nine out of ten sections from five MC1458C dual op amps are used. The INSTRUMENT jack, J1, provides a signal to IC1-a, which is a noninverting buffer amplifier with a gain of 2. The gain is determined by R3 and R4. The amplified instrument signal is fed to audio-gain stage IC1-b, which is controlled by potentiometer R6. The circuit location of control R6 provides a desirable interplay with volume control on the guitar. Whenever the output of IC1-b exceeds about 2 V p-p, light-emitting diodes LED1 and LED2 begin to illuminate and clip the audio signal, providing a high-quality distortion effect. The LEVEL control, R10, sets the level of the output of the high-gain distortion stage IC4-a, which, in turn, drives the ganged potentiometers (R14a and R14b). Unity-gain amplifiers IC5-a and IC5-b feed summing amplifiers IC2-b and IC3-b, which in turn drive the external load connected to stereo OUT jack J3. The load can be either stereo headphones or additional external stages used for audio amplification. Integrated-circuit sections IC2-a and IC3-a are noninverting amplifiers, each with a gain of 2. Their inputs are the line-level stereo input signals, each having a gain of 2. Their outputs mix with the instrument signal in amplifiers IC2-b and IC3-b; both of these amplifiers also have a gain of 2.

NE602 RF TRANSLATOR CIRCUIT

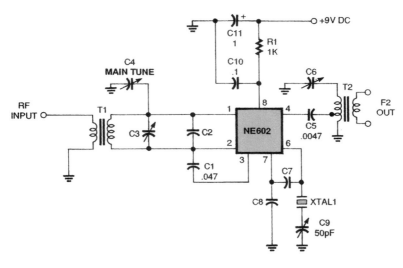

Fig. 62-2

This circuit shows how an NE602 is used as an RF translator or mixer in a typical receiver application. Component values depend on operating frequency.

MMIC AMPLIFIER MIXER

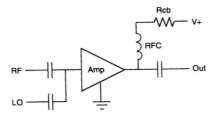

RF DESIGN

Fig. 62-3

This figures shows a simple mixer using an MMIC amplifier.

AUDIO MIXER

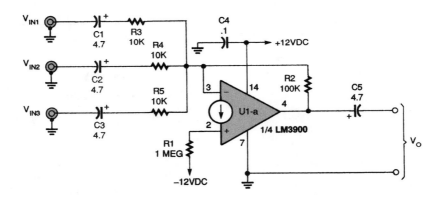

POPULAR ELECTRONICS

Fig. 62-4

An audio mixer is a handy circuit that will combine two or more audio signal sources into one signal channel. The figure shows an audio mixer that is built around the LM3900 CDA. The crux of this circuit is the three input networks, principally R3, R4, and R5. These resistors are connected to different input sources (labeled V_{IN1}, V_{IN2}, and V_{IN3}). Those three resistors all connect to the inverting input of the CDA. The gain is approximately

$$A_V = R_2 / R_X$$

where R_X is the value of any one input resistor. The output voltage is

$$V_O = R_2[(V_{IN1} / R_3) + (V_{IN2} / R_4) + (V_{IN3} / R_5)]$$

If resistor R2 is made variable, then the potentiometer used for R2 will serve as a master gain control for the audio mixer.

TWO-CHANNEL AUDIO MIXER

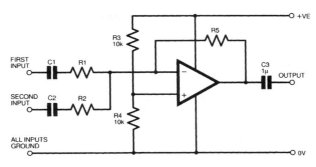

EVERYDAY PRACTICAL ELECTRONICS

Fig. 62-5

The figure shows a basic two-channel op-amp mixer circuit. The gain is equal to R_5/R_1 or R_5/R_2 for channel 1 or 2, respectively. Typically R1 and R2 are 2.2 to 22 kΩ and R5 is 10 to 100 kΩ, and the op amp is any suitable type, such as a 741 or its numerous variations.

NE602 MIXER CIRCUIT

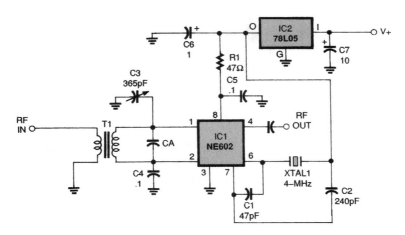

ELECTRONICS NOW

Fig. 62-6

This circuit uses an NE602 as a mixer in a typical application. Component values depend on operating frequency and are typical for a 4-MHz operation. Note that no IF filter is shown in the RF output circuit. A filter is necessary to select the desired output product and reject the others.

63

Model Circuits

The sources of the following circuits are contained in the Sources section, which begins on page 1043. The figure number in the box of each circuit correlates to the entry in the Sources section.

Model Train Sounder
PWM Model Railroad Controller
Model Sports Car Lighting System
Model Railway Crossing Lights
Shuttle Track Controller
Model Train Crossing Incandescent Flasher
Model Railroad Lighthouse Lamp Simulator
Model Train Crossing Flasher with Sensor Switch
Grand Prix Starting Lights
Model Train Track-Control Signal

MODEL TRAIN SOUNDER

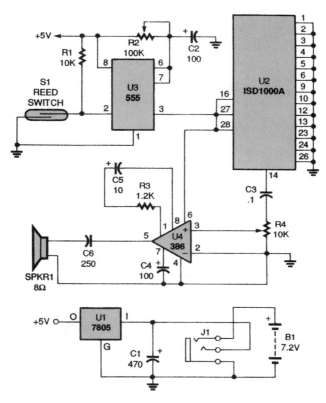

Fig. 63-1

This circuit can fit inside a model-train car and output any sound that you record in the ISD1000A. Imagine a cattle car that moos! The circuit can be used in model railroading to sound an electronic diesel horn when the train approaches a railroad crossing. It is mounted on the bottom of the car. When the "sound car" approaches the crossing, S1 is momentarily closed by a magnet that must be installed under the track, starting the timing cycle and sounding the horn. Before you can use U2, the ISD1000A record and playback IC, you must first program it by breadboarding the circuit found in the applications data included with the IC. To record sound on U2, you'll need a sample of the sound that you want the IC to play. The phone jack, which is mounted on the bottom of the car, serves two purposes. It serves as a means of charging B1, the 7.2-V NiCd battery, and it also provides a way to turn the circuit off.

PWM MODEL RAILROAD CONTROLLER

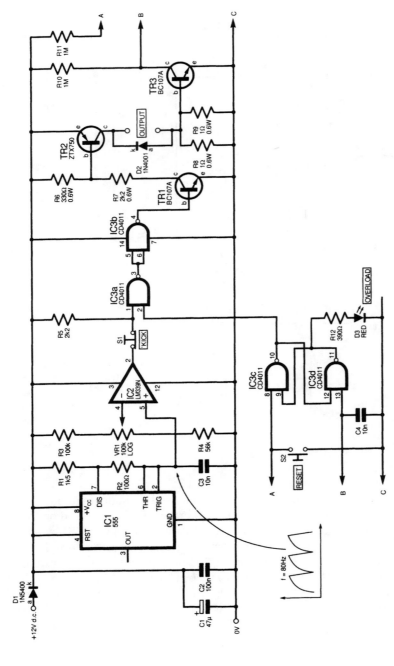

Fig. 63-2

EVERYDAY PRACTICAL ELECTRONICS

The PWM model train controller shown offers results that are far superior to those of a traditional rheostat-type model train controller and provides better fine control of train speeds. A 555 timer chip (IC1) is configured as a free-running oscillator operating at 80 kHz. The exponential ramp waveform generated across capacitor C3 feeds a comparator (IC2), whose threshold reference is set by logarithmic potentiometer VR1. The output of comparator IC2 is inverted by IC3a and IC3b and it drives the

PWM MODEL RAILROAD CONTROLLER *(Cont.)*

output stage, which comprises transistors TR1 and TR2; the latter can handle 2 A and dissipate 1 W, which is more than adequate for a saturated switch operation. Diode D2 protects against back EMF in the train motor. Short-circuit protection is provided with a 0.5-Ω series sense resistor (R8/R9). Transistor TR3 switches on when the current flowing reaches approximately 1.2 A. This sets the output of an S-R flip-flop (IC3c/IC3d) low. LED D3 illuminates to indicate a fault, and TR1 is disabled by the inverters. Capacitor C4 introduces a short time period across TR3 to prevent the cutout from triggering falsely. Push-switch S1 is a "kick" button. Pressing this causes a short, higher-voltage burst to be applied to reluctant motors.

MODEL SPORTS CAR LIGHTING SYSTEM

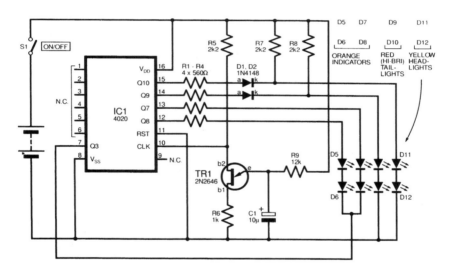

EVERYDAY PRACTICAL ELECTRONICS

Fig. 63-3

In order to add some realism to a model sports car, this simple circuit can be built into virtually any type of toy car, battery-operated or not. The design uses a CD4020 14-stage binary counter to operate LED "turn indicators" (D5 to D8), "taillights" (D9 and D10), and "headlights" (D11 and D12). The counter IC1 is clocked by a unijunction oscillator based around TR1. The red taillights and the yellow headlights draw a small current, supplied constantly by resistors R7 and R8. For added realism, this current is supplemented through resistors R1 and R2 to give an impression of headlights "dipping" and taillights "braking." The model's turn indicators flash whenever they are enabled, and so the cathodes (k) of D5 to D8 are taken to another output (Q3, pin 7) of IC1. With their anodes (a) taken positive via resistors R3/R4 and the relevant counter output, they flash at the rate determined by Q3. The result is that one set of indicator lights flashes, then the other set, followed by both. The circuit is powered by a PP3-type or MN1604 9-V battery, and a small slide switch S1 can be mounted on the vehicle's underside.

MODEL RAILWAY CROSSING LIGHTS

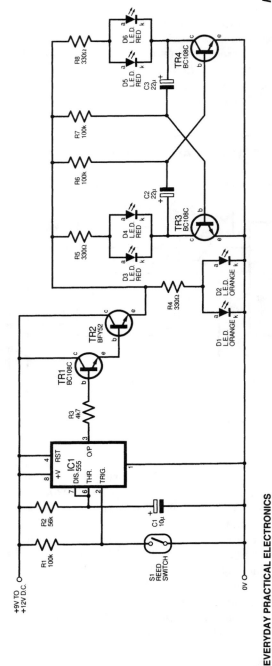

EVERYDAY PRACTICAL ELECTRONICS

Fig. 63-4

The circuit is for use on model railway layouts and approximates the flashing light signals used at level crossings. It is operated by a reed switch (S1) embedded in the track, which is closed by a magnet on the underside of a passing train. This triggers the monostable timer IC1, which powers a Darlington transistor pair (TR1 and TR2). Diodes D1 and D2 (orange) illuminate, and also TR3 and TR4, connected as an astable, cause the red LEDs D3 to D6 to flash alternately for about 6 s, long enough for the train to pass. One set of LEDs is used on each side of the level crossing, for added realism. For "00" scale, 3-mm LEDs are best.

SHUTTLE TRACK CONTROLLER

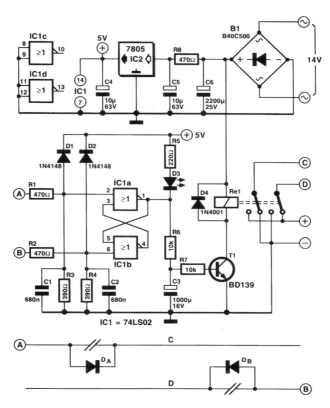

ELEKTOR ELECTRONICS

Fig. 63-5

This circuit enables a model train to shuttle continuously between two tracks. When the train travels from left to right across the track in the lower part of the diagram, the lower rail (D) is connected to the positive supply line. After it has passed diode D_B, the train will stop. Because the locomotive short-circuits the diode when it passes the break in the rail, a short positive pulse is generated on rail section B. The pulse is used to set bistable IC1, whereupon D3 goes out and capacitor C3 is being charged. When the potential across the capacitor rises to a sufficiently high level, transistor T1 switches on, whereupon the relay is energized. This causes the polarity of rails C and D to be reversed. Diode D_B is then on, so the train departs in the direction of A. Rail C is then connected to the positive supply rail; so when the train passes diode D_A, a positive pulse is generated on rail section A. This pulse is used to reset the biastable, whereupon the LED lights and the relay is deenergized. After the relay contacts have reversed the polarity of C and D again, diode D_A comes on and the train departs again in the direction of B.

MODEL TRAIN CROSSING INCANDESCENT FLASHER

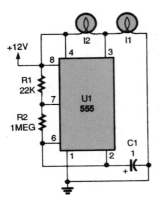

This simple circuit can operate the incandescent bulbs in a railroad-crossing flasher.

POPULAR ELECTRONICS

Fig. 63-6

MODEL RAILROAD LIGHTHOUSE LAMP SIMULATOR

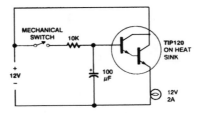

ELECTRONICS NOW

Fig. 63-7

The lamp turns on and off gradually, controlled by the charging and discharging of the capacitor. Here is a circuit that will do the job. It uses a TIP120 power transistor as a voltage-regulating element. The voltage across the lamp is always 1.2 V lower than the voltage across the capacitor. Closing the switch charges the capacitor gradually over a period of a few seconds, causing the voltage to go up. When the switch opens, the capacitor slowly discharges and the voltage goes down again. You might have to experiment with component values to get the effect you want. The capacitor controls both the rise time and the fall time; the resistor controls the rise time only. In both cases, larger values make the action go more slowly. Be sure to mount the TIP120 transistor on a heatsink, and remember that its mounting hole is connected to the collector (and thus to +12 V in the circuit).

MODEL TRAIN CROSSING FLASHER WITH SENSOR SWITCH

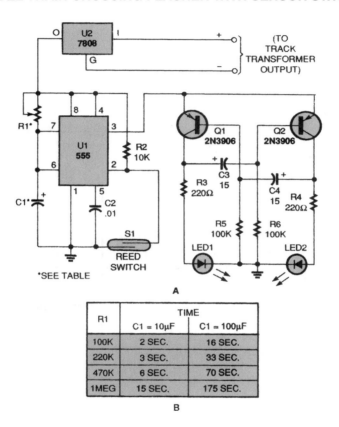

A

R1	TIME	
	C1 = 10µF	C1 = 100µF
100K	2 SEC.	16 SEC.
220K	3 SEC.	33 SEC.
470K	6 SEC.	70 SEC.
1MEG	15 SEC.	175 SEC.

B

POPULAR ELECTRONICS

Fig. 63-8

Featuring a reed switch operated by train proximity, this circuit (A) has everything you need to activate an LED-based crossing signal. To change the time period of U1, use these substitutions (B).

GRAND PRIX STARTING LIGHTS

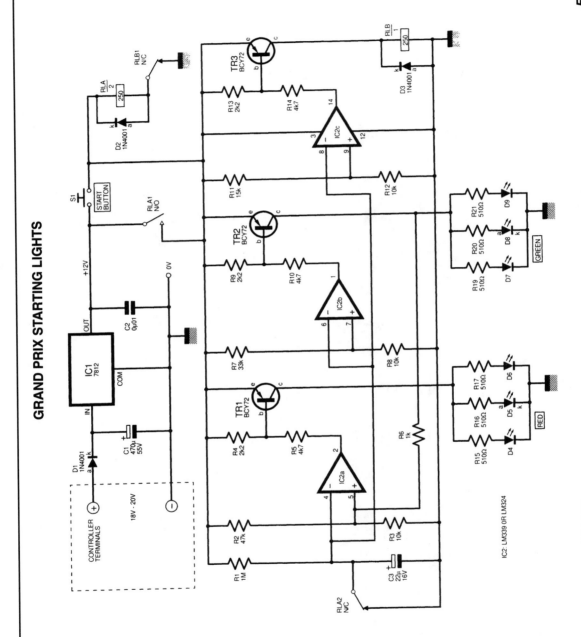

IC2: LM339 OR LM324

EVERYDAY PRACTICAL ELECTRONICS

Fig. 63-9

GRAND PRIX STARTING LIGHTS *(Cont.)*

The figure shows a circuit design for a "starting gantry" with three red and three green starting lights. It can be powered from the racing set's power-supply unit. A 12-V regulator (IC1) provides a stabilized supply for the circuit and diode D1 prevents accidental reverse connection by the young motor enthusiast. The circuit uses a system of relays to start and stop the sequence. When START switch S1 is pressed, relay contacts RLA1 close and this applies +12 V to the sequencer and also latches the relay. Contacts RLA2, across capacitor C3, open. Three comparators are used (IC2a, IC2b, and IC2c), and their noninverting (+) terminals each have a reference voltage set by a potential divider at about +2.10 V, +2.75 V, and +4.75 V, respectively. Either an LM339 quad comparator or an LM324 quad op amp can be used. When capacitor C3 charges in excess of +1.75 V, transistor TR1 conducts and the red LEDs D4 to D6 illuminate. At +2.75 V, TR2 conducts and the green LEDs D7 to D9 light. IC2a also swings high, via resistor R6, and so the red lamps extinguish. When capacitor C1 reaches +4.75 V, transistor TR3 conducts and completes the circuit to relay RLB. Contacts RLB1 (normally closed) now open and disconnect relay RLA. Contacts RLA2 then close, which discharges the timing capacitor C1, ready for next time.

MODEL TRAIN TRACK-CONTROL SIGNAL

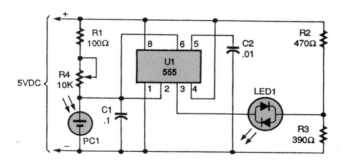

POPULAR ELECTRONICS

Fig. 63-10

The circuit is based on a 555 timer IC, U1, and operates on 5-Vdc to make it usable with digital logic. The operating power of the circuit can be converted to 12 V by increasing the values of R_2 and R_3. Train detection is accomplished by photocell PC1, which should be mounted between the rails, flush with or slightly below the ties. A small light must be mounted directly above PC1 when it is used in a dark area. Resistor R4 is adjusted so that when a train covers PC1, the circuit triggers. Resistor R1 reduces the current to U1 if R4 is set for minimum resistance. Variable resistor R4 can also be adjusted so that each individual car in a train will change the colors of LED1, a bicolor unit, back and forth, indicating train movement. That can be handy if two circuits are used, one at each end of the track block. It will indicate that the train has begun to move. Mount LED1 on the control panel. The circuit is designed to indicate a green when no train is present. When a train covers PC1, U1 switches LED1 to red, indicating that the train is in position.

64

Modulator and Demodulator Circuits

The sources of the following circuits are contained in the Sources section, which begins on page 1043. The figure number in the box of each circuit correlates to the entry in the Sources section.

AM Modulator Limiting Amplifier
ADC Converter Synchronous Demodulation

AM MODULATOR LIMITING AMPLIFIER

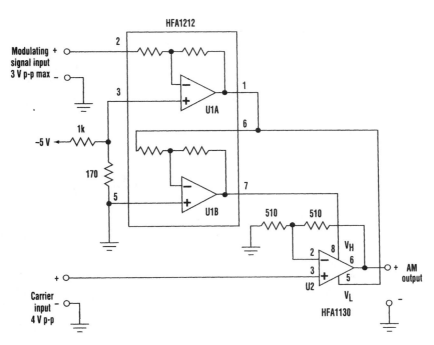

ELECTRONIC DESIGN

Fig. 64-1

Limiting amplifiers also can be put to work in AM modulators. The fast overdrive recovery time and wide bandwidth of the clamp inputs allows these inputs to be driven by high-frequency ac, can as well as dc signals. When driven at the appropriate levels, the clamp inputs can be used to form an amplitude modulator. The HFA1130 limiting amplifier is driven by a 4-V p-p carrier signal. The gain of 2 through the HFA1130 ensures that the carrier amplitude is sufficient to drive the output over its ±3.3-V range. The HFA1212 performs the necessary level shifting and inversion to convert the modulating signal input into a pair of antiphase signals that control the high and low clamp inputs. U1A inverts the signal and level-shifts to −1.5 V. U1B inverts that signal, forming a complementary signal centered at +1.5 V. With a signal input of 0 V, U2 produces a 3-V p-p output at the carrier frequency. As the signal input varies, U2 produces a symmetrically modulated carrier with a maximum amplitude of 6 V p-p. The 2300-V/μs slew rate of the HFA1130 limits 6-V p-p amplitude carrier signals to a frequency of 61 MHz. If adjusted for lower output-signal levels, the carrier and modulating frequencies can be increased to well above 100 MHz.

ADC CONVERTER SYNCHRONOUS DEMODULATION

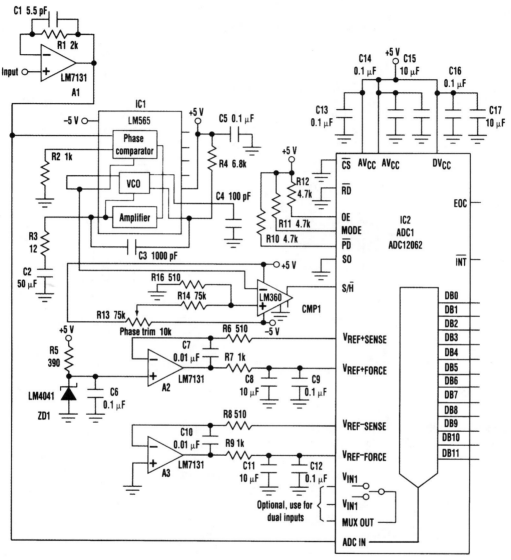

ELECTRONIC DESIGN

Fig. 64-2

Digital radio techniques promise to dramatically enhance receiver capabilities. Directly converting the IF (intermediate frequency) to a digital signal facilitates digital signal processing of demodulated signals. The example described here features an AM radio detector circuit for the typical IF of 455 kHz. For RF demodulation, the input is sampled at a rate exactly equal to the input-signal frequency. The digital output signal, which is actually an alias of the input, is the recovered modulation.

65

Moisture- and Fluid-Detector Circuits

The sources of the following circuits are contained in the Sources section, which begins on page 1043. The figure number in the box of each circuit correlates to the entry in the Sources section.

WATER DETECTOR

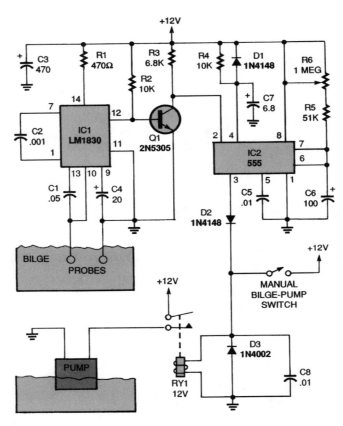

Fig. 65-1

The heart of the circuit is the LM1830 fluid detector. It contains all the circuitry needed to sense fluid levels and activate an external device (relay, etc.). The IC generates an ac signal that is passed through two probes in the fluid. The IC's detector circuit senses the presence or absence of fluid by comparing the resistance between the probes with its internal reference resistance. When the probes detect the presence of water, the LM1830 will trigger Q1, and that in turn will trigger IC2 (an LM555), which starts a timing period. The output from pin 3 of IC2 closes the relay, and the bilge pump is activated for the duration of the timing period. The timing period is adjustable using R6, a 1-MΩ potentiometer. For the values shown, and depending on the setting of R6, the timing period is about 5 to 120 s. Components R4, D1, and C7 are tied to pin 4 of IC2 to hold the timer at RESET when power is first applied.

MOISTURE DETECTOR

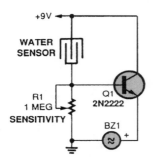

This simple circuit can save your house from water inundation. Also, it can save your electrical equipment from water damage. When raindrops fall on the water sensor, a small current flows, turning on Q1 and the buzzer. Alternatively, you could replace the buzzer with a relay to drive a pump. The water sensor can be made from an etched circuit board.

POPULAR ELECTRONICS

Fig. 65-2

SOIL MOISTURE MONITOR

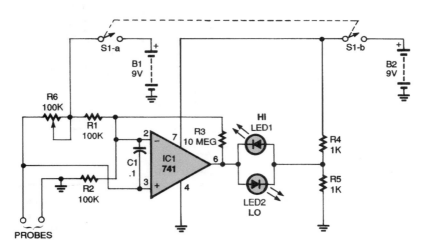

POPULAR ELECTRONICS

Fig. 65-3

In the circuit, when R_6 equals the value of the resistance of the soil between the two probes, the bridge circuit is in balance. To set up the circuit, bring the soil that you wish to monitor to the proper moisture level and insert the two probes into that soil. Then adjust R_6 until the two LEDs are both off, or at least dim to the same level. The two LEDs serve as a null indicator. That is, LED2 is on when the ground resistance is higher than the preset resistance of R_6, and LED1 is on when the ground resistance is lower than the preset value. When both LEDs go dark or dim to the same brilliance, the bridge circuit is in balance. At balance, R6 equals the ground resistance. The probes are two 6-inch by $\frac{1}{16}$-inch round stainless-steel rods mounted about 1 inch apart in an insulated handle.

RAIN SENSOR

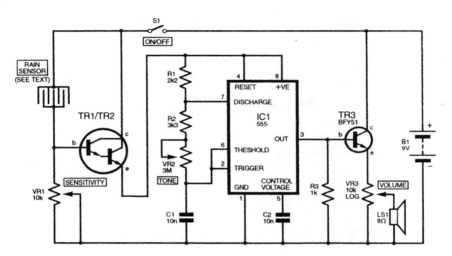

Fig. 65-4

The figure shows a simple rain sensor design that incorporates pitch and volume controls in the alarm signal. Whenever the sensor is bridged by droplets of water, the Darlington transistor TR1/TR2 will conduct. This enables IC1, a 555 astable tone generator, to function, powering a small loud-speaker through a driver transistor (TR3). VR2 determines the pitch of the audio tone, which can be anything from about 25 Hz to 18 kHz, while VR3 adjusts the volume. The sensitivity of the circuit is set by VR1. The Darlington pair could be constructed with two separate ZTX300s or a single TIP122. For the sensor, use a small piece of stripboard, linking alternate strips into an interlocking design. The circuit will operate from a standard 9-V PP3-type battery.

AC-OPERATED WATERING-CONTROL SYSTEM

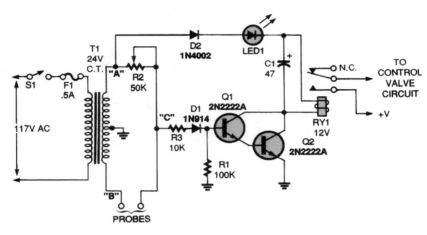

POPULAR ELECTRONICS

Fig. 65-5

Ac operation of the circuit reduces possible electrolysis effects over time. Two probes are inserted into the soil. When the soil is wet, it is more conductive and Q1 conducts, cutting Q2 off. As the soil dries, Q1 loses bias, allowing Q2 to turn on, actuating the relay (RY1). RY1 closes, operating the watering-system control valve. When the soil is sufficiently moist, Q1 turns on, cutting off Q2, RY1, and the water valve. R2 sets the trip level.

WATERING CONTROL WITH RF MOISTURE SENSING

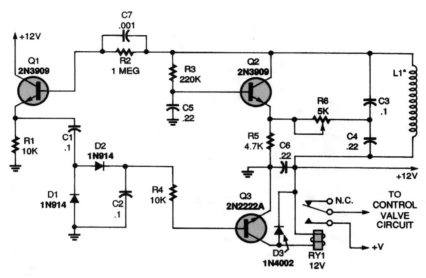

POPULAR ELECTRONICS

Fig. 65-6

The moisture sensor does not depend on the ohmic resistance of the soil to activate the water valve. Instead, a VLF oscillator circuit with its tuned inductor buried in the soil senses the moisture content by absorption. Transistor Q2, L1, C3, and C4, make up a simple Hartley oscillator. The circuit oscillates at about 16 kHz. Transistor Q1, in an emitter-follower configuration, isolates the output circuit from loading or influencing the oscillator circuit. Diodes D1 and D2 convert the RF signal to dc to supply bias current for Q3, which operates the water-valve relay as long as the oscillator's output signal is sufficient to turn it on. The oscillator's sensitivity to varying soil conditions is set by R6, a 10-turn trimmer potentiometer. The inductor, L1, consists of a 100-ft length of #26 plastic-covered wire wound in a 4-inch loop and kept together with electrical tape. The loop is one area where much can be gained by experimenting. A pancake loop might be more sensitive or a different-shaped loop might work better. Also try different oscillator frequencies.

WATER-LEVEL MONITOR

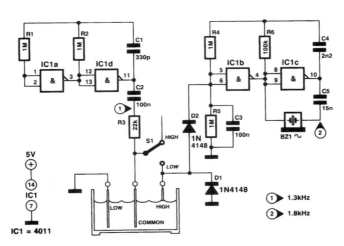

ELEKTOR ELECTRONICS

Fig. 65-7

This circuit was devised to produce an audible alarm when a water tank was approaching empty. The circuit consists of two conventional CMOS oscillator gate pairs, as found in the textbooks. With the switch in the LOW position, gates IC1a and IC1d provide an alternating current flow through C2, R3, the common electrode, the water, and the low electrode, to earth. When the water level drops, the circuit to earth is interrupted, and the ac then fed via S1 to the junction of D1 and D2. This charges C3 so that pin 5 of IC1b goes positive, causing gates IC1b and IC1c to start oscillating. The audible transducer is an ac buzzer. The switch is then moved to the HIGH position. Pin 6 of IC1 is then at ground potential via R5, waiting for the tank to fill and connect the common and high electrodes via the water. The ac then flows through D2 to once again sound the alarm. The pump is turned off and S1 is restored to sense empty once more. The rate of charge of capacitor C3 can be changed by altering the value of R_3. The power supply for the circuit can be from 5 to 12 V.

SPRINKLER GUARDIAN

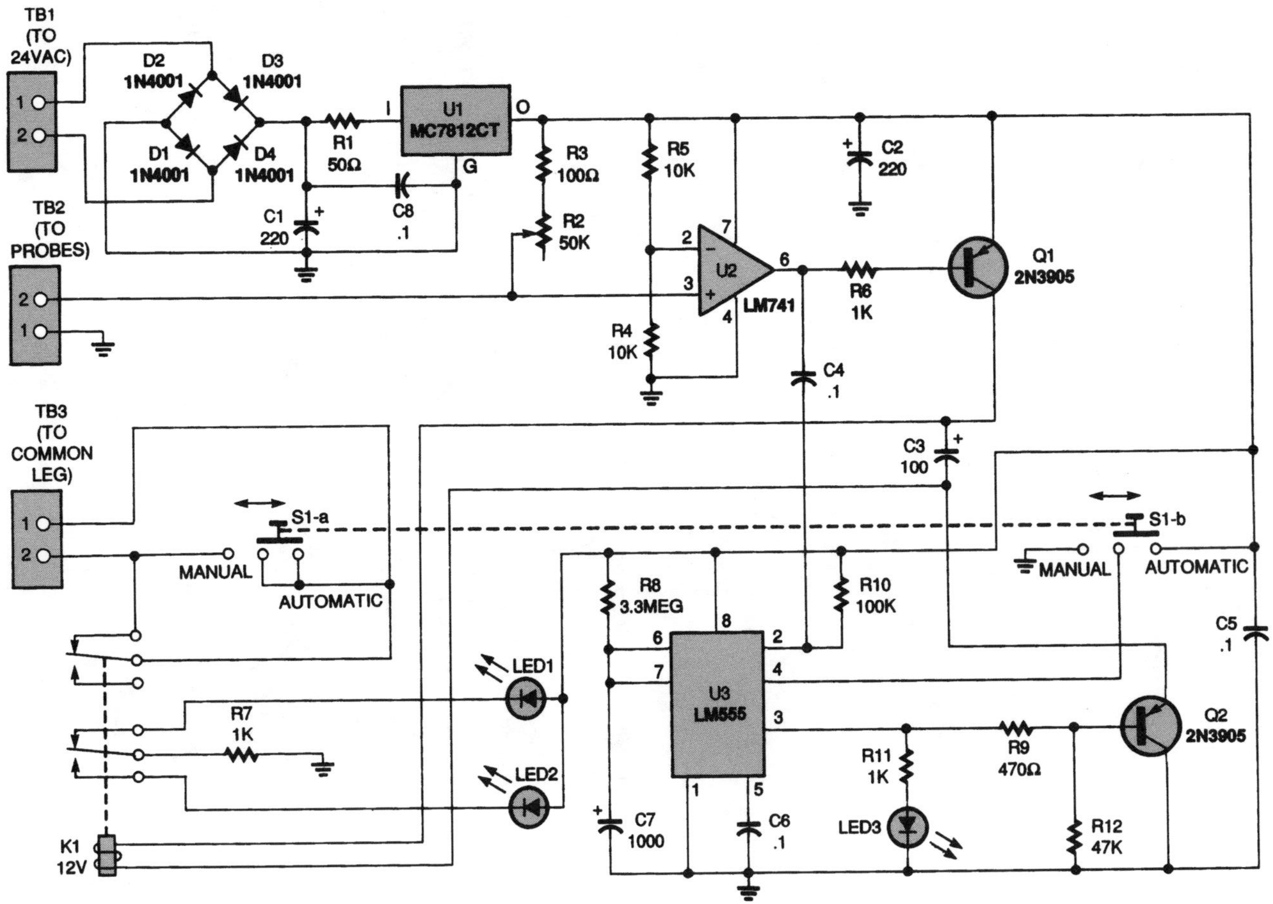

Fig. 65-8

ELECTRONICS HOBBYISTS HANDBOOK

SPRINKLER GUARDIAN *(Cont.)*

The Sprinkler Guardian can be used in conjunction with any sprinkler controller that uses standard 24-Vac valves. The add-on unit senses the moisture content at the ground surface; if the ground is dry, the unit allows the controller to execute all the ON times preset by the controller clock. The unit obtains its power supply from the controller that it is used with. The output of U1 supplies a regulated 12 V to the rest of the circuit. The ground probes are connected to terminals 1 and 2 of TB2. Resistor R2 is used to balance the resistance of the earth between the two probes so that op amp U2, which is configured as a comparator, outputs a high when the ground is dry. PNP transistor Q1 is turned off, K1 is not energized, and terminals 1 and 2 of TB3 are shorted. Under those conditions, if the controller calls for watering the lawn, watering will commence. Now, assume that the ground is wet. The resistance between the probes drops, causing the output of U2 to go low, which turns on Q1. That energizes K1. Under those conditions, even when the controller is programmed to water, the valves will not open. Now assume that the soil is dry. Relay K1 is off. When the programmed watering time arrives, the water valve is opened, starting the timed watering cycle. The probes are wetted immediately, and U2 goes low. As the voltage output of U2 drops, a negative-going pulse is delivered through C4 to pin 2 of U3, a 555 timer, which triggers the timer into operation. The time output from pin 3 turns Q2 off, which keeps K1 from closing during the time cycle determined by R8 and C7. That allows time for the sprinkler system to complete the watering cycle. A DPDT switch, S1, is used to place the circuit in either the "automatic" or "manual" mode. In the automatic mode, the circuit operates as previously described. In the manual mode, the terminals of TB3 are shorted, effectively removing the Guardian from the system.

AUTOMATIC WATERING SYSTEM CONTROLLER

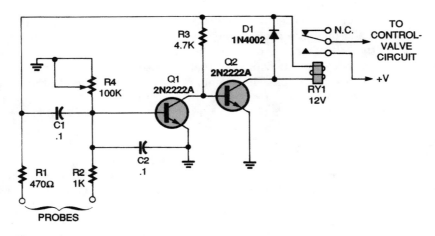

POPULAR ELECTRONICS

Fig. 65-9

Two probes are inserted into the soil. When the soil is wet, it is more conductive and Q1 is forward-biased, cutting Q2 off. As the soil dries, Q1 loses bias, allowing Q2 to turn on, actuating the relay RY1. RY1 closes, operating the watering-system control valve. When the soil is sufficiently moist, Q1 turns on, cutting off Q2, RY1, and the water valve.

66

Motor-Control Circuits

The sources of the following circuits are contained in the Sources section, which begins on page 1043. The figure number in the box of each circuit correlates to the entry in the Sources section.

Robotic Motor-Control Circuit
Electronic-Latch Dc Motor-Controller Circuit
H-Bridge Ac Motor-Drive Circuit
Fan-Speed Control
H-Bridge Motor-Drive Circuit
Incubator Motor Controller
Fan-Speed Controller
Stepper Motor Driver
Solid-State Drill-Speed Control
Ac Motor Control
PM Dc Motor-Speed Control
Light-Activated Motor-Drive Circuit

ROBOTIC MOTOR-CONTROL CIRCUIT

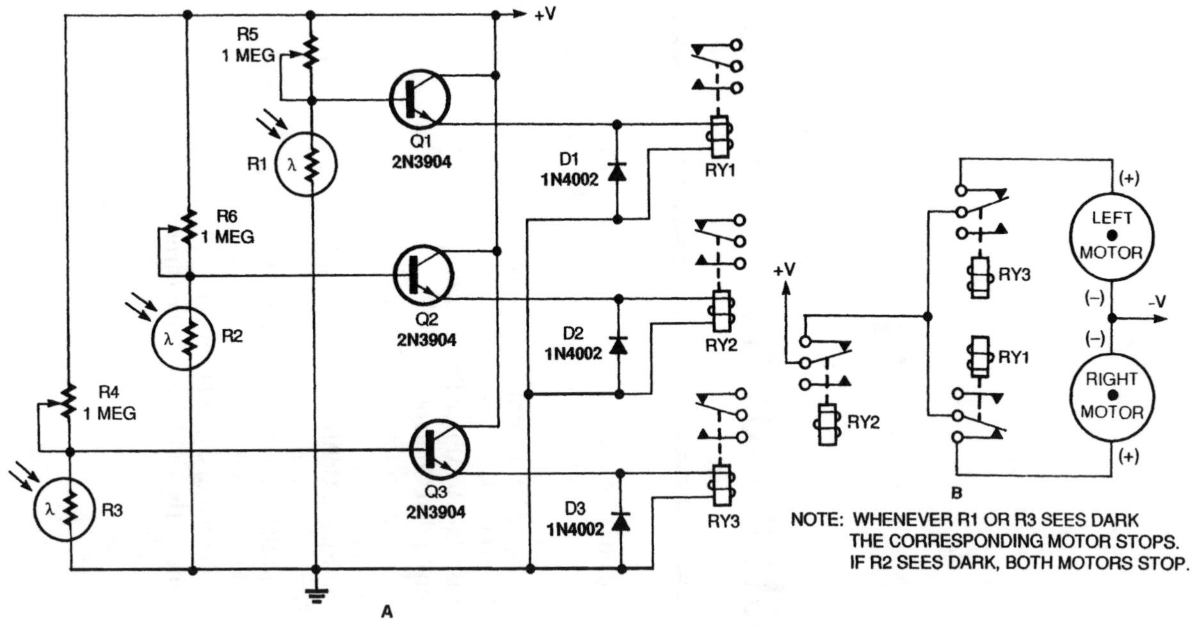

NOTE: WHENEVER R1 OR R3 SEES DARK THE CORRESPONDING MOTOR STOPS. IF R2 SEES DARK, BOTH MOTORS STOP.

Fig. 66-1

This circuit, with three light sensors and three relays, was used to allow a robot to follow a light beam. Loss of the beam causes relay actuation, which controls a motor that produces corrective action.

ELECTRONIC-LATCH DC MOTOR-CONTROLLER CIRCUIT

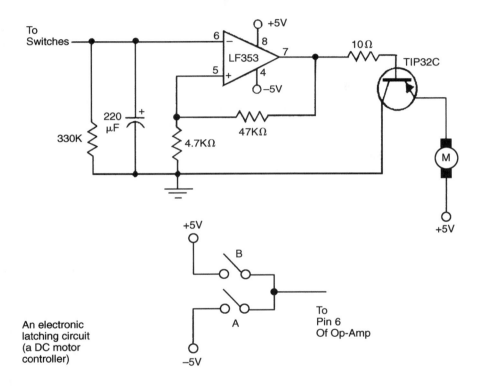

An electronic
latching circuit
(a DC motor
controller)

Fig. 66-2

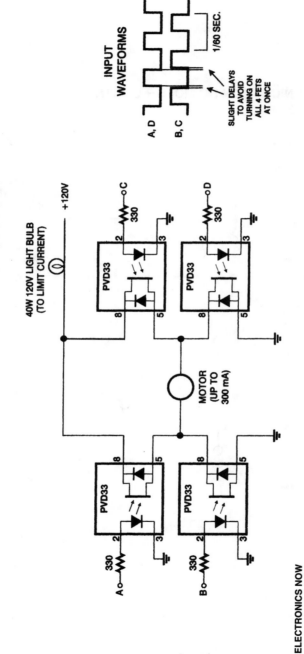

H-BRIDGE AC MOTOR-DRIVE CIRCUIT

INPUT WAVEFORMS

1/60 SEC.

A, D

B, C

SLIGHT DELAYS
TO AVOID
TURNING ON
ALL 4 FETS
AT ONCE

+120V

40W 120V LIGHT BULB
(TO LIMIT CURRENT)

330   °C

PVD33   330

PVD33   °D

MOTOR
(UP TO
300 mA)

8   5   8   5

PVD33   PVD33

2   3   2   3

330   330

A°   B°

ELECTRONICS NOW

Fig. 66-3

This circuit can be used to drive ac motors from a dc source. The PVD33 devices are by International Rectifier and are high-powered optoisolators.

FAN-SPEED CONTROL

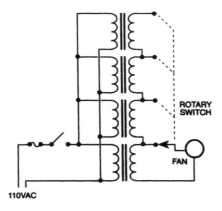

ROTARY
SWITCH

FAN

110VAC

NUTS AND VOLTS

Fig. 66-4

Several 24-V 1.5-A transformers can be used as a speed control, where a triac would be usable, such as on an induction motor. Connect the primaries in parallel and the secondaries in a series, as shown.

H-BRIDGE MOTOR-DRIVE CIRCUIT

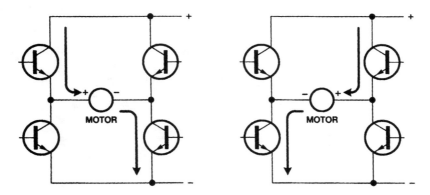

MOTOR

MOTOR

ELECTRONICS NOW

Fig. 66-5

An H bridge can run a dc motor in either direction by turning on two transistors at a time. An H bridge is a circuit that supplies power to a motor and can reverse its polarity. It is usually used to run a dc motor forward and backward.

INCUBATOR MOTOR CONTROLLER

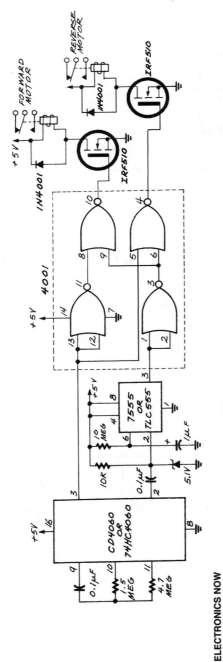

Fig. 66-6

The circuit has a 4060 CMOS long-period timer, which consists of an oscillator plus a series of binary counters that divide the frequency by 2, 4, 8, etc. up to 16,384. The oscillator runs at a comfortable 2 Hz (2 cycles per second), controlled by a capacitor and two resistors, and the binary counter divides this to produce a square wave that has only one cycle every 2 hours, as well as another one that oscillates twice as fast. The faster of the two square waves determines when the motor runs, and the slower one determines its direction. Here's how it's done: First, the faster square wave goes through a capacitor, resistor, and diode, which convert it into a narrow negative-going pulse that occurs once per hour. The 555, connected as a monostable, stretches the short pulse so that it is 10 s long (the length of time the motor should run). Logic gates then steer that pulse to one of two relays so that the motor runs either forward or backward, depending on which state the slow square wave is in at the time. Because the forward and reverse pulses are generated by the same 555 circuit, they are the same length—this is important so that the reverse cycle will exactly undo the motion of the forward cycle.

632

FAN-SPEED CONTROLLER

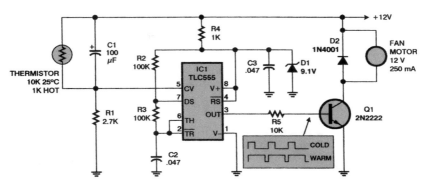

ELECTRONICS NOW

Fig. 66-7

The speed of the fan motor increases as the thermistor gets warmer. The motor is fed with pulses whose duty cycle increases from 34 to 100 percent.

STEPPER MOTOR DRIVER

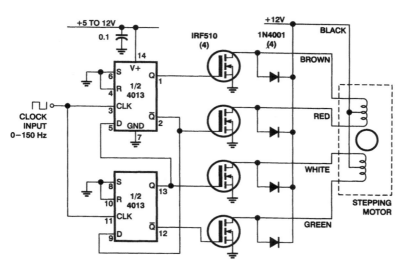

ELECTRONICS NOW

Fig. 66-8

In this circuit, four IRF510 power FETs are driven by a CMOS counter to generate the necessary two-phase drive quadrature. Although ICs are available to do this, this approach is handy for the experimenter because it uses commonly available parts. The stepper motor was taken from a discarded floppy-disk drive.

633

SOLID-STATE DRILL-SPEED CONTROL

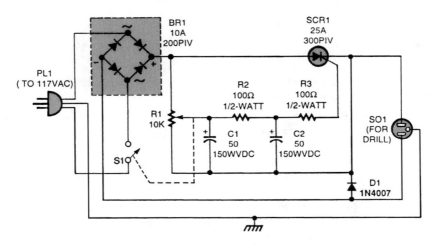

POPULAR ELECTRONICS

Fig. 66-9

Add this circuit to a single-speed drill to make the speed variable. The bridge rectifier (BR1) provides the full-wave pulsing direct current for the SCR switch (SCR1); BR1 should be rated at 200 PIV and have a current rating of 10 A, while SCR1 should have a PIV of 300 V and a current rating of 25 A. Diode D1 is used to counter the back voltage developed by the drill motor; D1 can be rated at 2 A. The speed of the drill is varied by R1.

AC MOTOR CONTROL

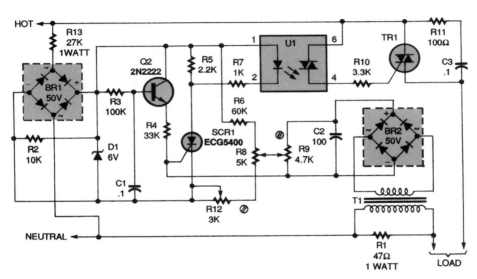

POPULAR ELECTRONICS

Fig. 66-10

This circuit controls fans and other ac motors that have less than $\frac{1}{4}$ horsepower and lack a centrifugal starting switch. It works by controlling the effective voltage in the ac circuit and governs both starting and varying load conditions. A ramp voltage is developed across capacitor C1. The voltage across C1 will vary the delay in turning on SCR1. The amplitude of that voltage is controlled manually by R8 for adjusting the motor speed, and by a preadjusted potentiometer (R9) to provide governing action (R9 should be adjusted to provide maximum regulation). The level of the ramp, relative to the firing voltage of Q1, is set by R12. The value of R_1 must be chosen to accommodate the load; a lower value might be more satisfactory for a larger motor. Transformer T1 can be wound with a primary of 25 turns of 26-gauge wire and a secondary of 200 turns. Alternatively, any small transformer with a turns ratio of approximately 1:10 will do.

PM DC MOTOR-SPEED CONTROL

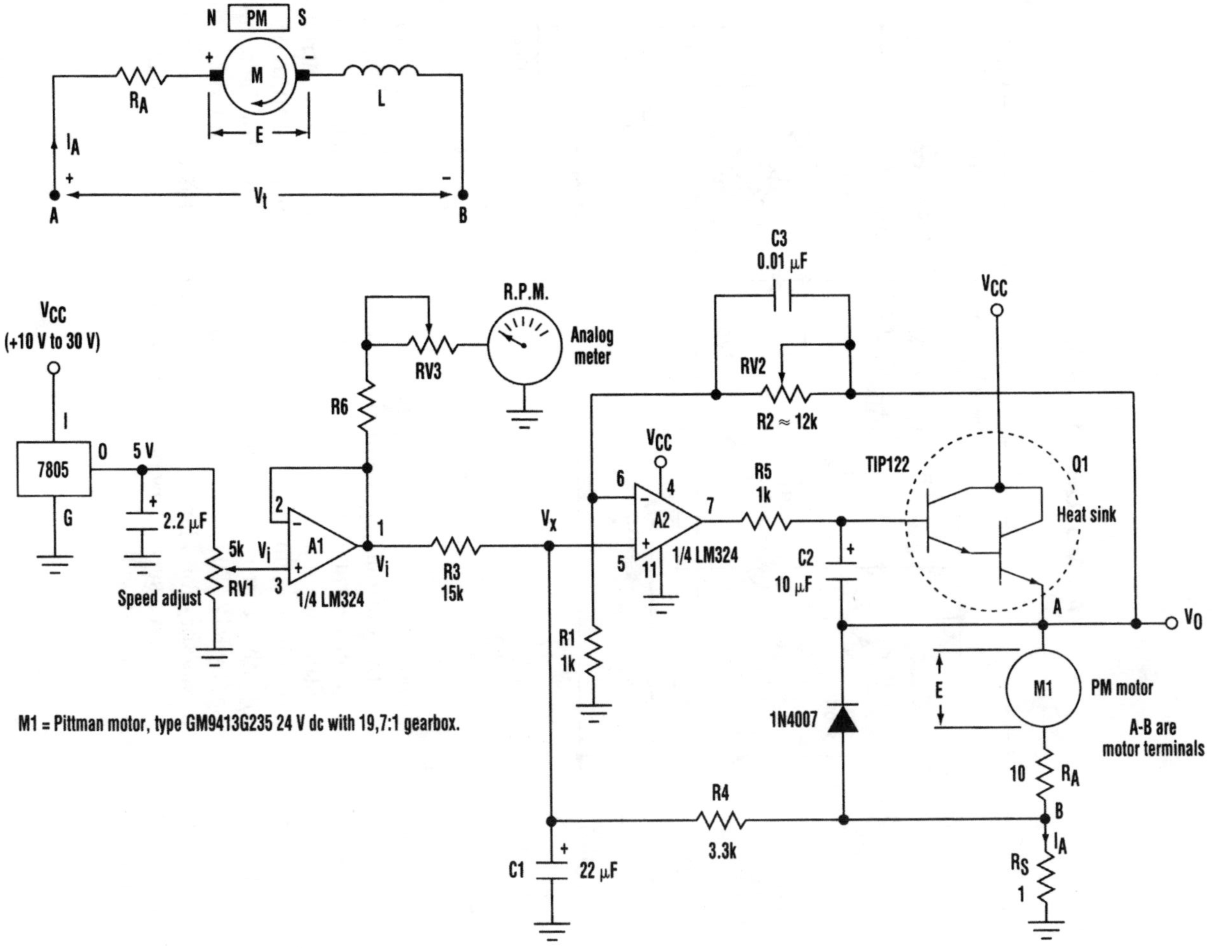

M1 = Pittman motor, type GM9413G235 24 V dc with 19.7:1 gearbox.

ELECTRONIC DESIGN

Fig. 66-11

PM DC MOTOR-SPEED CONTROL *(Cont.)*

Speed control of permanent-magnet (PM) dc motors with the aid of optical or dc tachometers is generally inconvenient and difficult, particularly on motors with integral gearboxes. The high-speed shaft of the motor that drives the gearbox isn't always accessible and the speed of the geared-down shaft often is too low for tachometers. Described here is a single-supply regulating speed-control circuit that doesn't require a tachometer. It keeps the motor torque high under load by using positive feedback to compensate for the drop caused by armature resistance. In unregulated variable-speed PM dc motor systems, the drop in speed under load is particularly pronounced at low motor-supply voltages. The positive feedback generates a negative resistance that compensates for the nonlinear effects caused by armature resistance. It thereby ensures that the speed-control input voltage (V_i) linearly controls the speed of the motor. Armature resistance compensation is achieved if:

$$R_S = R_A / [\text{gain}(R_3) / (R_3 + R_4)] - 1$$

The divider action of R3 and R4 together with the gain reduces the value required for R_S to minimize the power dissipation. C1 and R4 dampen the positive-feedback signal's response time, but they also form a low-pass filter and attenuate the motor current noise fed to the A2 input. The maximum output voltage swing from A2 is approximately $V_{CC} - 2$ V, and there is a 1.2-V V_{be} loss by T1. This implies that the supply voltage (V_{CC}) should be about 5 V above the maximum desired motor voltage in order to allow for extra output drive to the motor under heavy load conditions. A reasonable choice for R_S is approximately $R_A/10$, and the gain of A2 should be trimmed with RV2 to ensure that the motor's speed does not drop when loaded.

LIGHT-ACTIVATED MOTOR-DRIVE CIRCUIT

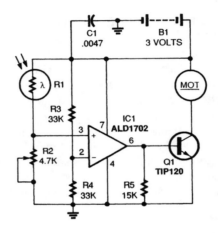

Fig. 66-12

The circuit is a basic light-activated on/off switch for the motor. A CMOS op amp (IC1) is used as a voltage comparator. The comparator monitors the levels of two input voltages and switches its output on or off, depending on which input voltage is greater. The input on pin 2 is set to a reference voltage of about half the supply voltage by R3 and R4. The input on pin 3 is connected to a voltage divider made up of a cadmium sulfide photocell (R1) and potentiometer R2. The resistance of a photocell changes depending upon the amount of light shining on it, so the intensity of light is indicated by the voltage on pin 3 of IC1. The light level at which the circuit turns on is adjusted by R2. When the voltage on pin 3 of IC1 is greater than that on pin 2, the output (pin 6) turns on. The output of IC1 drives Q1 directly. That transistor acts like a current amplifier for the op amp. The transistor switches the motor on and off.

67

Motorcycle Circuits

The sources of the following circuits are contained in the Sources section, which begins on page 1043. The figure number in the box of each circuit correlates to the entry in the Sources section.

Motorcycle Burglar Alarm
Motorcycle Turn-Signal System
Motorcycle Intercom

MOTORCYCLE BURGLAR ALARM

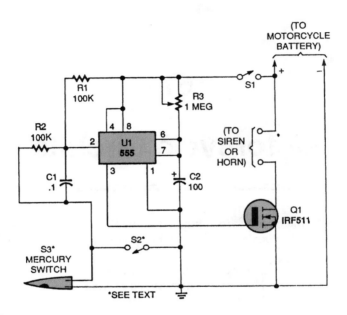

Fig. 67-1

A 555 IC is connected in a one-shot timer circuit that turns on an FET and either a siren or the bike's horn for a preset time period. Switch S1 is used as an on/off switch. Closing either of two switches, S2 or S3, will trigger the IC. When either switch closes, pin 2 of U1 goes low. That triggers the IC to produce a positive output at pin 3 and sounds the alarm for the time period set by R3. The mercury switch, S3, is the switch that activates the alarm if someone moves your bike. Switch S2 can be used as a panic switch if you ever feel threatened. The IRF511 N-channel FET (Q1) will handle currents up to 4 A. If you need a higher-current device, an IRF530, which is rated at 14 A, can be substituted.

MOTORCYCLE TURN-SIGNAL SYSTEM

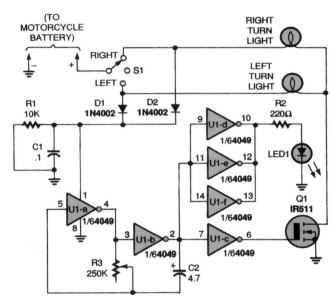

POPULAR ELECTRONICS

Fig. 67-2

A low-frequency oscillator made up from a CMOS hex inverter drives an LED indicator and a power FET to act as a flasher. R3 and C2 control the flash rate. S1 controls L or R side lights for turn-signaling purposes.

641

MOTORCYCLE INTERCOM

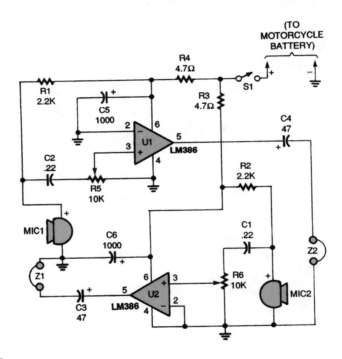

POPULAR ELECTRONICS

Fig. 67-3

Why yell at your passenger when you can talk? Use this two-way intercom to make communicating on the open road a lot easier.

68

Music Circuits

The sources of the following circuits are contained in the Sources section, which begins on page 1043. The figure number in the box of each circuit correlates to the entry in the Sources section.

Analog Delay/Flanging Unit
Music Vision Circuit

ANALOG DELAY/FLANGING UNIT

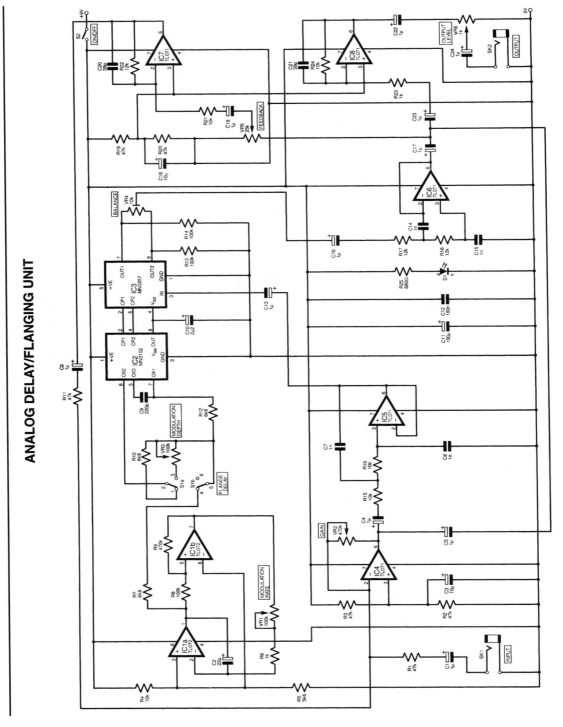

Fig. 68-1

EVERYDAY PRACTICAL ELECTRONICS

ANALOG DELAY/FLANGING UNIT *(Cont.)*

Audio from the guitar is fed first to a preamp and LP filter, then through an analog delay line IC. The delay is a function of clock frequency. The clock frequency, and, therefore, the delay time of the line, can be modulated by a low-frequency oscillator. This gives the characteristic sound to the guitar audio.

MUSIC VISION CIRCUIT

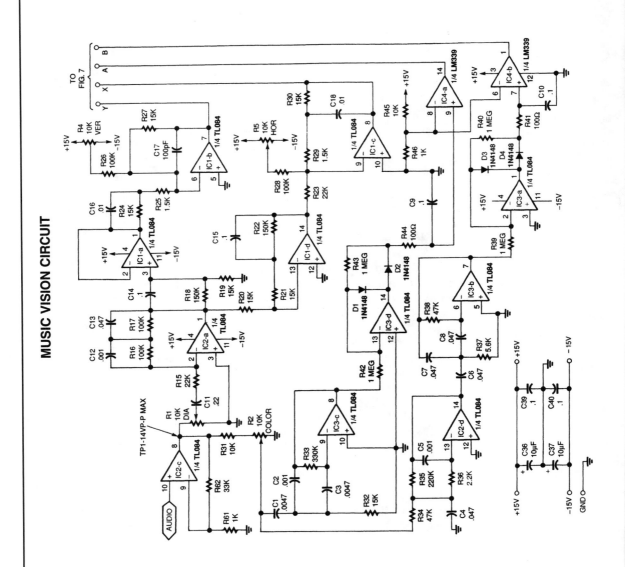

MUSIC VISION CIRCUIT (Cont.)

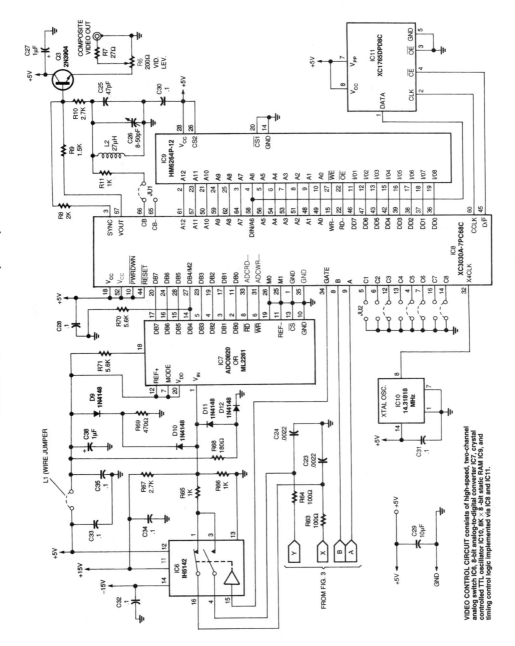

VIDEO CONTROL CIRCUIT consists of high-speed, two-channel analog switch IC6, 8-bit analog-to-digital converter IC7, crystal controlled TTL oscillator IC10, 8K × 8 -bit static RAM IC9, and timing control logic implemented via IC8 and IC11.

MUSIC VISION CIRCUIT (Cont.)

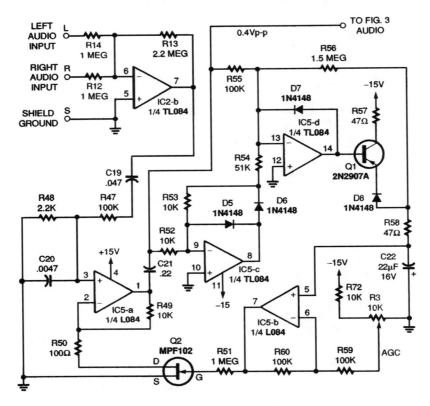

Fig. 68-2

The basic concept involves processing an audio signal input into two components (x and y) that are phase-shifted approximately 90° relative to each other. The signals are then digitally sampled and stored in a video RAM buffer as an array of data bits. Bit locations in RAM correspond with the horizontal (x value) and vertical (y value) positions of points to be illuminated on the TV screen. The video RAM is periodically read and erased in synchronism with video frame scanning. The resulting serial video data is combined with color burst and sync signals to generate an NTSC-compatible composite video output signal. The music vision circuit essentially converts the standard raster-scanned TV screen into an oscilloscope-type display in which horizontal and vertical positions are directly controlled by x and y signals. The 90° phase shift between the two signals generates complex and dynamic two-dimensional patterns on the TV screen when music, voice, or other audio signals are applied to the input. The audio-frequency content of the input signal determines the color of the video pattern being displayed.

69

Noise Circuits

The sources of the following circuits are contained in the Sources section, which begins on page 1043. The figure number in the box of each circuit correlates to the entry in the Sources section.

Periodic 60-Hz Noise Eliminator
V_{CC} Audio Noise Reducer

PERIODIC 60-Hz NOISE ELIMINATOR

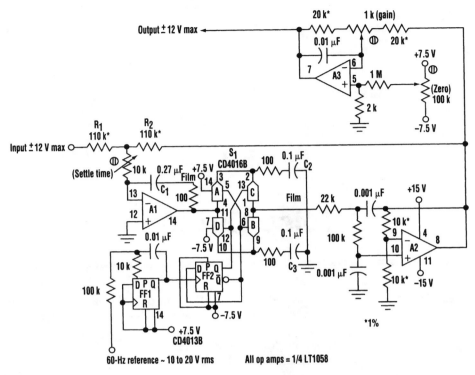

ELECTRONIC DESIGN

Fig. 69-1

This analog/digital synchronous-averager circuit implements a robust "comb" filter that (theoretically) infinitely attenuates all 60-Hz harmonics. It does this independently of precision component tolerances. Signal components with frequencies below 20 Hz are passed virtually undisturbed. In the circuit, A1 continuously integrates and inverts the sum of the input signal and the output of buffer amplifier A2. Depending on the state of FF2, either switches S1A and S1B or S1C and S1D will conduct. Because flip-flop FF2 toggles once each 60-Hz cycle, A1 always integrates the difference between the instantaneous input voltage and the integral of the input taken over the preceding cycle. The transfer function of such a piecewise integration is characterized by a series of impulses that occurs at f (the fundamental frequency of the integration cycle) and at all integer multiples of f. This extreme attenuation of harmonic noise is not limited by component tolerance. The filter's setting time for an input step, however, does depend upon the trimming of potentiometer P1. Optional unity-gain inverter A3 undoes the signal inversion performed by A1 and incorporates trimmer pots P2 and P3 for precise adjustment of overall filter gain and zero offset.

V_{CC} AUDIO NOISE REDUCER

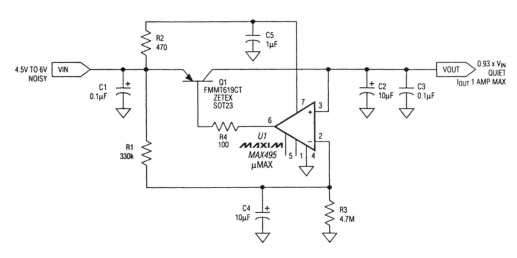

MAXIM

Fig. 69-2

The circuit reduces noise and ripple voltage by 40 dB over the 100-Hz to 20-kHz audio range. It provides a clean source of 5-V power for driving audio circuits in portable applications such as cellular phones and multimedia notebook computers. Most linear regulators reject noise only up to 1000 Hz or so, and the bulk of a low-frequency passive filter is unwelcome in portable applications. The circuit shown accepts noisy V_{CC} in the 4.5- to 6-V range and produces quiet V_{CC} at a dc level 7 percent lower. For example, it produces 4.65 V at 1 A from a nominal 5-V source, with only 200 μA of quiescent current. The largest capacitor is 10 μF and the resistors can be 0.1 W or surface-mount 0805 size. When operating, the circuit acts as a wide-bandwidth buffered voltage follower (not a regulator) whose dc output level is 7 percent below V_{in}. R1 and R3 form a voltage divider that provides the 7 percent attenuation, and C4 helps to form a 93-percent filter replica of V_{in} at the op amp's inverting input. The op amp's small input bias current (25 nA typical) allows large resistor values for R1 and R3, yet limits the maximum dc error to only 20 mV. The result is a low-pass filter with a 2-Hz corner frequency that provides 20 dB of attenuation at 20 Hz.

70

Operational-Amplifier Circuits

The sources of the following circuits are contained in the Sources section, which begins on page 1043. The figure number in the box of each circuit correlates to the entry in the Sources section.

OP-AMP DC FEEDBACK STABILIZATION

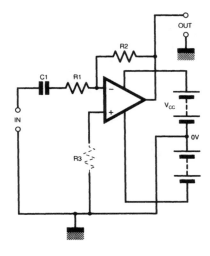

EVERYDAY PRACTICAL ELECTRONICS

Fig. 70-1

Blocking capacitor C1 ensures that dc negative feedback is maximized.

OP-AMP DC COUPLING AND OFFSET

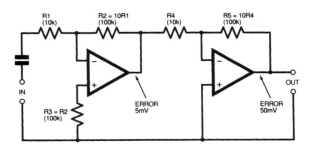

EVERYDAY PRACTICAL ELECTRONICS

Fig. 70-2

Dc interstage coupling might be tolerable in terms of offset errors.

OP-AMP DC OVERALL FEEDBACK

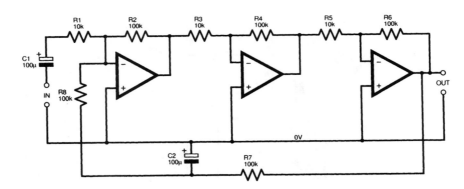

EVERYDAY PRACTICAL ELECTRONICS

Fig. 70-3

Overall dc feedback stabilizes the whole amplifier.

OP-AMP TESTER

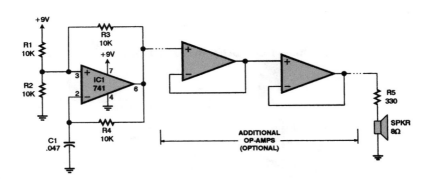

ELECTRONICS NOW

Fig. 70-4

The circuit is a simple "dead or alive" test for an op amp. It is an oscillator that produces a 1500-Hz square wave, audible as a tone in the speaker. By looking at the output with an oscilloscope, you can measure the rise time. Several op amps can be tested at once by cascading them, as shown in the diagram.

71

Optical Circuits

The sources of the following circuits are contained in the Sources section, which begins on page 1043. The figure number in the box of each circuit correlates to the entry in the Sources section.

Optically Isolated Analog Multiplexer
Optically Isolated Precision Rectifier

OPTICALLY ISOLATED ANALOG MULTIPLEXER

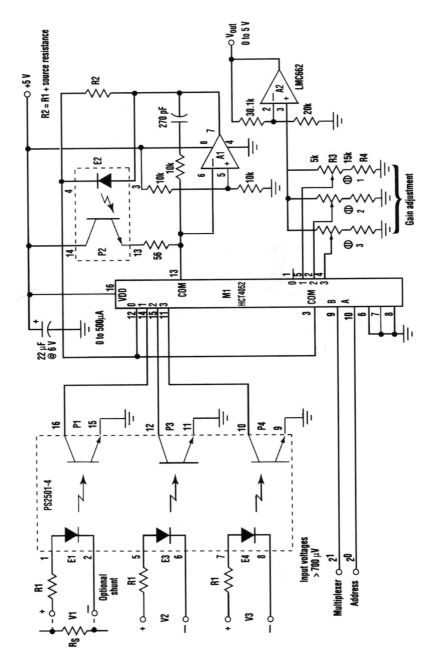

Fig. 71-1

ELECTRONIC DESIGN

OPTICALLY ISOLATED ANALOG MULTIPLEXER *(Cont.)*

The circuit described here provides three channels of optically isolated input that will work for many precision signal-acquisition applications. The only power required is a single 15-V rail with ground common to the analog-to-digital converter. Multiplexer operation is based on an ordinary quad-channel optoisolator (PS2501-4). Each LED (such as E1) in combination with input-scaling resistor R1 will, in response to V_{in}, pass a current of $I_1 = (V_{in} - V_{LED})/(R_1 + R_s)$, where V_{LED}=the forward voltage drop of the LED and R_s=the signal source's internal resistance. R1 is chosen to establish a full-scale LED current near 500 μA. Assuming (for example) that channel 1 is selected, the resulting photocurrent in P1 will tend to pull the summing point of A1 low. In response, A1 will forward-bias E2 to generate a balancing photocurrent in P2. The E2 current required to maintain this balance will closely track I_1, and the V_{LED} of E2 will be close to the V_{LED} of E1. The current sourced through M1 to the gain-adjustment network is directly proportional to the floating input voltages. Therefore, it is independent of the LED voltage drops.

OPTICALLY ISOLATED PRECISION RECTIFIER

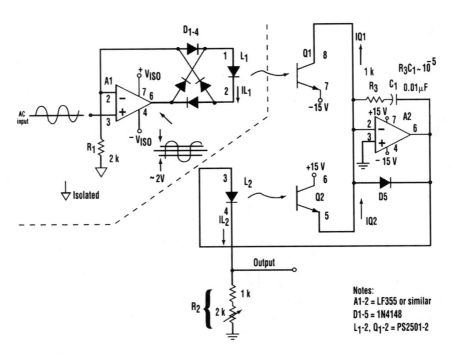

Fig. 71-2

An isolation amplifier and precision rectifier can be combined in one topology, as demonstrated here. It achieves excellent rectification symmetry, zero stability, and good linearity (better than 1 percent) and frequency response (>>10 kHz), with a minimum of precision components. A1 acts as a voltage-to-current converter by serving the current through the D1 to D4 bridge and L1. Therefore, the voltage developed across R1 equals the instantaneous input voltage. The diode bridge's full-wave rectification causes L1 to be forward-biased, regardless of the polarity of the input voltage. The magnitude of the bias controls the intensity of optical coupling between L1 and Q1, and, thereby, the magnitude of Q1's collector current. A2 servos the current through L2 and R2 so that the current passed by Q2 balances that passed by Q1. Because of the good tracking of elements of the PS2501-2 dual optoisolator, a constant ratio exists between the L1 and L2 currents. Consequently, R2 can be adjusted so that the output voltage across R2 is equal to the rectifier's isolated input voltage. R3 and C1 provide frequency compensation for the L2-Q2 feedback loop. D5 prevents potentially destructive reverse bias of L2.

72

Oscilloscope Circuits

The sources of the following circuits are contained in the Sources section, which begins on page 1043. The figure number in the box of each circuit correlates to the entry in the Sources section.

Delayed Sweep Adapter
Portable Scope Calibrator
Scope Calibrator
Voltage Cursor Adapter
Dual-Scope Adapter
Scope Mixer
Dual-Trace Converter

DELAYED SWEEP ADAPTER

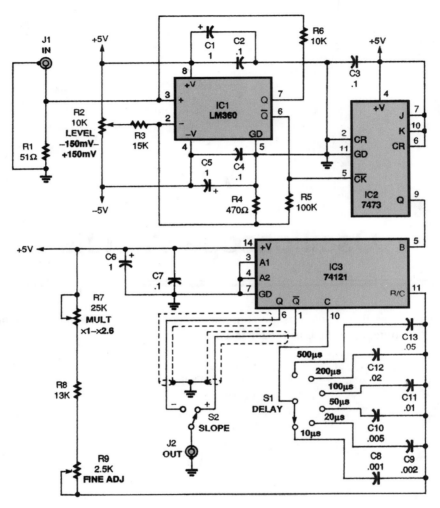

Fig. 72-1

The schematic for the delayed sweep adapter is shown in the figure. Power for the circuit is provided by a dual-polarity, 5-V supply. An LM360 high-speed comparator (IC1) with complementary TTL outputs is at the heart of the circuit. A trigger-level control, potentiometer R2, allows the adapter to trigger on any part of the waveform being displayed. With the values shown, the circuit functions well with a Hitachi V-212 oscilloscope, whose channel 1 output is about 25 mV per vertical division of the signal display. Assuming a normal display of about six divisions, the level control provides a range of ±150 mV. You can change that range to suit your scope output by adjusting the ratio

DELAYED SWEEP ADAPTER *(Cont.)*

of R3 (15,000 Ω) to R4 (470 Ω); however, maintain the ratio of R5 (100,000 Ω) to R4 (470 Ω) for proper hysteresis. Resistor R1 (51 Ω) "terminates" the 50-Ω cable from the channel 1 output jack. IC2, a 7473 TTL dual J/K flip-flop, is configured in its toggle mode to divide the input frequency by 2. That ensures that IC3, a 74121 TTL monostable multivibrator, will function accurately over the input waveform's full time period. The pulse output from IC3 is coupled to the trigger-input jack of your oscilloscope, and the "slope" is selected via S2 to match the slope selected on your oscilloscope. Potentiometer R9 allows fine positioning of the trace and potentiometer R7 allows the output pulse width selected to be multiplied by a factor from "×1" to over "×2.5." Switch S1 lets you select the desired pulse width by switching in C8 through C13, in a 1–2–5 sequence. The adapter circuit was designed to operate on an external ±5-V, 100-mA power supply to avoid 60-Hz pickup in the unit.

PORTABLE SCOPE CALIBRATOR

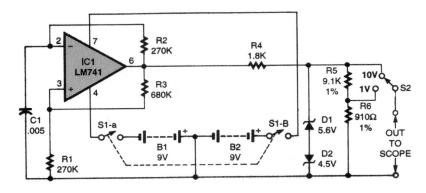

POPULAR ELECTRONICS

Fig. 72-2

This simple circuit can be used to calibrate scopes or other equipment.

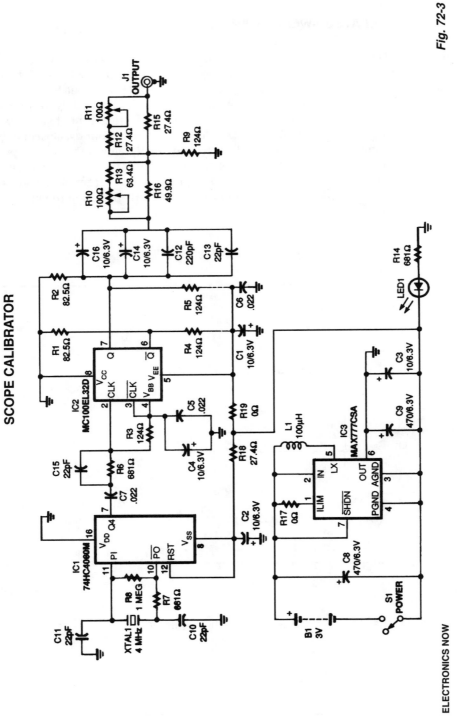

SCOPE CALIBRATOR

ELECTRONICS NOW

Fig. 72-3

A 74HC4060 HCMOS oscillator/divider integrated circuit (IC1) was chosen for generating the master clock. Choosing that part lets us use an inexpensive 4-MHz crystal. The divide-by-16 output of IC1, at 250 kHz, is applied to a Motorola MC100EL32 divide-by-2 ECLips flip-flop (IC2) through level-shift and decoupling components C7, C15, and R6. Resistor R3 sets the input impedance seen by IC2 and divides the amplitude of IC1's output down to the ECLips input levels. The MC100EL32 has transition times of both outputs specified at less than 350 ps (0.35 ns) and temperature-compensated output levels. The divide-by-2 action

SCOPE CALIBRATOR *(Cont.)*

of the device provides 125 kHz at approximately 800 mV p-p to the amplitude adjustment circuit and output stage. The output of IC2 is terminated with loads consisting of resistor pairs R1-R4 and R2-R5. The values shown provide an equivalent load voltage of about −2 V and a load impedance of 50 Ω. The output of IC2 is ac-coupled by capacitors C12, C13, C14, and C16 to an adjustable T-attenuator network. The various values of the coupling capacitors combine to transmit the fat rise-time edges (C12 and C13) while preventing droop at the 125-kHz base frequency (C14 and C16).

VOLTAGE CURSOR ADAPTER

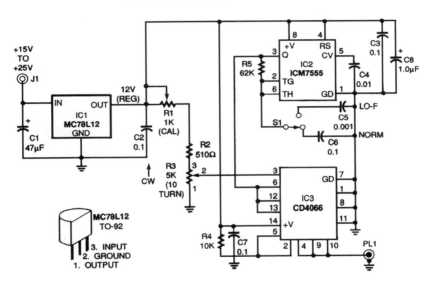

Fig. 72-4

This is the schematic diagram for the voltage cursor adapter. The required 15- to 25-V power to the circuit can be supplied either by batteries or by a wall-mounted ac to 15-Vdc adapter. The MC78L12 voltage regulator (IC1) supplies regulated 12 Vdc to the rest of the circuit. The ICM7555 timer (IC2) drives the CD4066B, a CMOS bilateral switch (IC3). This drive frequency can either be a normal frequency (NORM) of 100 kHz or a low frequency (LO-F) of 10 kHz, depending on the setting of switch S1. Set S1 to LO-F for inputs below 500 Hz. The dc reference voltage supplied to pin 3 of IC3 is set by R3, a 10-turn, 5000-Ω precision potentiometer. The voltage can be read directly from a turns counter dial coupled directly to the potentiometer's wiper. The accuracy of this reading can be 1 percent or better. Trimmer potentiometer R1 permits the voltage to R3 to be calibrated to precisely 10 V. The circuit is calibrated by setting the digital reading on the turns counter of R3 to the full clockwise position and adjusting R1 for a reading of 10 V at the wiper of R3 with a digital voltmeter. Bilateral switch IC3 converts the dc reference to a square wave with exactly the same wiper amplitude. The square-wave output appears on the common pins 4, 9, and 10 of IC3 and coaxial plug PL1.

DUAL-SCOPE ADAPTER

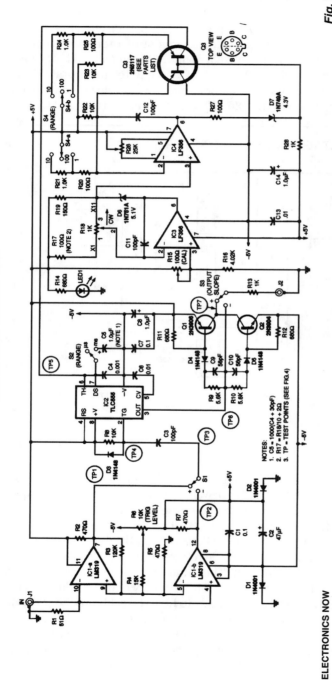

Fig. 72-5

ELECTRONICS NOW

The input stage is the trigger amplifier based on the two channels of an LM319 dual comparator (IC1-a and IC1-b). The input signal is taken from the channel 1 output jack of your oscilloscope, and it is applied to J1 on the adapter. The input slope is selected by switch S1, and the signal is coupled to C3 and IC2, the multivibrator stage. Capacitor C3 provides the proper negative-going pulse to trigger IC2, a TLC555 CMOS timing IC. Switch S2 selects the desired range—either milliseconds or microseconds—and switch S3 selects the polarity of the output slope. Transistors Q1 and Q2 form a fast (less than 10-ns delay) inverting output stage. The output from switch S3 is ±5 V, which is sufficient for excellent contrast in intensity modulation. It is also satisfactory for stable triggering in the adapter's delayed-sweep mode. Precise input current to capacitors C4 and C5 is supplied from the collector of dual transistor Q3, a 2N5117. It is a dual-matched PNP pair in one package. Operational amplifier IC3, an LF356, is configured as a noninverting follower with gain. Transistor Q3 is configured as an unusual "current mirror" that unloads the relatively slow operational amplifier, allowing the second half of Q3 to demonstrate its fast dynamic

DUAL-SCOPE ADAPTER *(Cont.)*

response. A reference voltage is developed across both sets of emitter resistors (R20 to R25) to produce the constant charging current for capacitors C4 and C5. The two sections of switch S4 (S4-a and S4-b) select the range—either 1.0, 10, or 100—in conjunction with switch S2 (milliseconds or microseconds). Zener diodes D6 and D7 permit the operational-amplifier outputs to operate in their linear regions. Variable resistor R18 (1 kΩ, 10 turns) is the range multiplier.

SCOPE MIXER

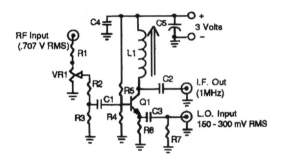

ScopeMixer Parts List	
R1	75 Ω
R2	910 Ω
R3, R7	51 Ω
R4, R6	1 k
R5	4.7 k
VR1	200 Ω Linear Composition
C1, C2, C3	.001 µF
C4	0.1 µF
C5	10 µF
L1	Broadcast Ferrite Loopstick*
*Antique Electronic Supply Cat. #PC-70-A (or equivalent)	

Fig. 72-6

This circuit permits use of a low-frequency scope to view RF signals. The signal to be viewed is mixed with a LO signal about 1 MHz away and the IF output is viewed on the scope.

DUAL-TRACE CONVERTER

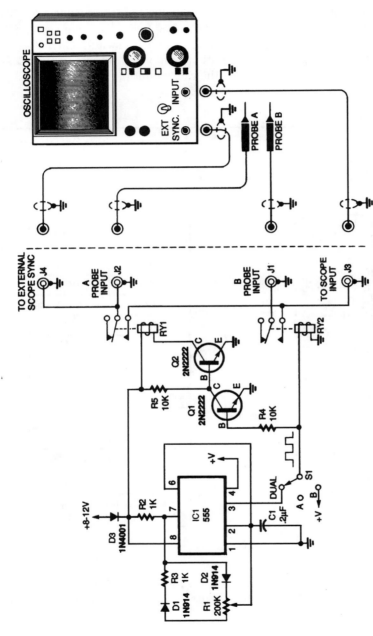

Fig. 72-7

ELECTRONICS NOW

The probe switching circuit is shown in the figure. Two SPST reed relays, RY1 and RY2, switch the signal from two probes to a single scope input. Reed relays provide smooth and noise-free switching. The input side of each relay is connected to a scope probe, and the outputs of the relays are tied together and connected to the scope's single-channel input. A 555 timer, IC1, is configured as a variable-duty-cycle square-wave oscillator. Potentiometer R1 varies the oscillator's duty cycle to compensate for response-time variations that different relays might have. The potentiometer is adjusted to give equal brightness to both of the signals being displayed on the oscilloscope. If the relays you choose work well with a 50-percent duty cycle, the potentiometer

666

can be replaced with two fixed resistors wired in series, with the junction between the two resistors connected to pin 2 of IC1. With the parts values shown, the timer has an output frequency of about 30 Hz. Switch S1 allows independent viewing of each oscilloscope trace. When S1 connects pin 3 of IC1 to the coil of RY2 and R4, both input traces will be displayed. When S1 connects +V to the coil of RY2 and R4, only channel B will be displayed. With the switch in the center OFF position, only the channel-A signal is displayed.

73

Oscillator Circuits—Audio

The sources of the following circuits are contained in the Sources section, which begins on page 1043. The figure number in the box of each circuit correlates to the entry in the Sources section.

Audio Generator Circuit
Three-Tone Oscillator
Three-Frequency Audio Oscillator
Wien-Bridge Oscillator Circuit
Center-Tapped Transformer Audio Oscillator
Crystal-Controlled Audio Generator
Twin-Tee Tone Oscillator
UJT Oscillator

AUDIO GENERATOR CIRCUIT

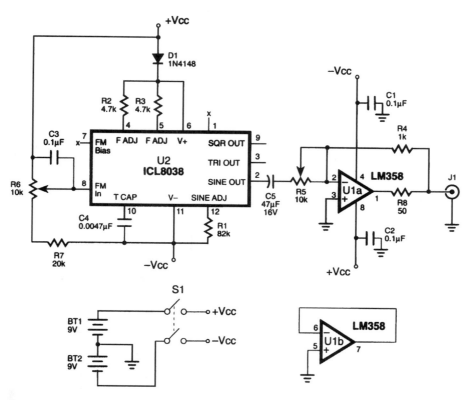

Fig. 73-1

A very compact audio signal generator can be constructed using a precision waveform generator IC (ICL8038) made by Harris Semiconductor. This IC is unique in that it will generate three waveforms: sine, triangle, and square waves. The frequency output is adjustable between 0.01 Hz and 300 kHz, and the sine-wave output distortion is less than 1 percent. The schematic diagram illustrates the application of the ICL8038 IC to construct a battery-powered mini-audio-signal generator that will provide 500 Hz to 1.5 kHz with adjustable-amplitude output. The output frequency of the generator is set by the combination of R2, R3, and C4. The output of the generator IC is ac-coupled to op amp U1 through C5. U1 buffers the output of U2 and provides amplitude output adjustment in conjunction with R5. Frequency control is accomplished through the adjustment of R6. Bipolar bias for the circuitry is supplied by two 9-V batteries.

THREE-TONE OSCILLATOR

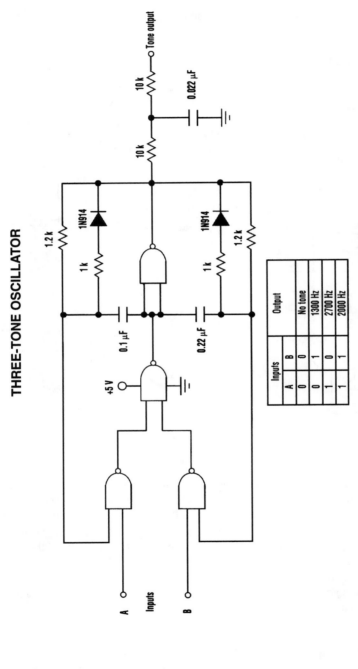

Inputs		Output
A	B	
0	0	No tone
0	1	1300 Hz
1	0	2700 Hz
1	1	2000 Hz

Fig. 73-2

ELECTRONIC DESIGN

This TTL tone oscillator can generate one of three tones. The tone generation depends on the status of the two input control lines. The tone generation depends on the status of the two input control lines. The design consists of two standard TTL audio tone oscillators merged together using one 7400 TTL chip. Two input lines control the output status. With both inputs low, no output is generated. When input A is high, the high-tone oscillator is gated on. When input B is high, the low-tone oscillator is gated on. With both A and B high, both oscillators are gated on and the output tone generated is midway between the high and low tones. The 1N914 diode in series with a 1-kΩ resistor imposes a 50-percent duty cycle on the output tone. A low-pass filter on the output removes harmonics and "mellows" the output tone.

THREE-FREQUENCY AUDIO OSCILLATOR

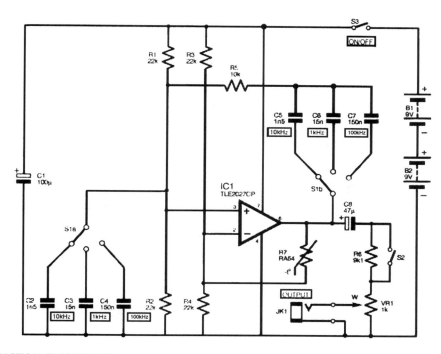

Fig. 73-3

The full circuit diagram for the audio sine-wave generator is shown. Resistors R1 and R2 bias the noninverting (pin 3) input of IC1, and their parallel resistance forms one element of the Wien network. They are the equivalent of R1 in Fig. 1. The other resistor in the Wien network is R5. Three switched pairs of capacitors (C2 to C7) provide the unit with its three different output frequencies. Resistors R3 and R4 bias the inverting (pin 2) input of IC1, and their parallel resistance also acts as one element of the negative-feedback network. Thermistor R7 is the other section of the negative-feedback circuit. An RA53 thermistor is the normal choice for this application, but because of its lower cost, an RA54 is used in this circuit. Potentiometer VR1 is the variable-output attenuator. Opening switch S2 introduces losses through resistor R6 that reduce the output by about 20 dB. Use a value of 100 kΩ for R6 if a reduction by about 40 dB is preferred.

WIEN-BRIDGE OSCILLATOR CIRCUIT

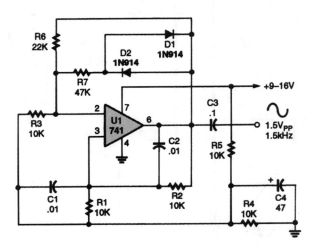

POPULAR ELECTRONICS

Fig. 73-4

A 741 op amp is connected to a Wien-bridge, audio-sine-wave oscillator configuration, with C1, C2, R1, and R2 determining the circuit's operating frequency. With the use of NPO capacitors and metal-film resistors, the oscillator's frequency is stable enough for use in tone-control applications. The fixed-frequency oscillator shown can easily be converted into a tunable oscillator by substituting a dual-gang linear potentiometer for R1 and R2. Different frequency ranges can be covered by using other matched values for C_1 and C_2. Larger values produce lower frequencies, and vice versa for small values.

CENTER-TAPPED TRANSFORMER AUDIO OSCILLATOR

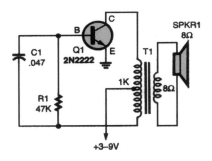

POPULAR ELECTRONICS

Fig. 73-5

The oscillator circuit uses a center-tapped transistor output transformer to serve as a load for the collector of Q1, supply a feedback signal for the base, and serve as the output winding for driving the speaker. R1 supplies dc bias and C1 completes the ac path from the transformer to Q1's base. You can play around with the values of C1 and R1 to change the oscillator's output level and tone, but don't reduce R1's value too much or the transistor will draw excessive collector current.

CRYSTAL-CONTROLLED AUDIO GENERATOR

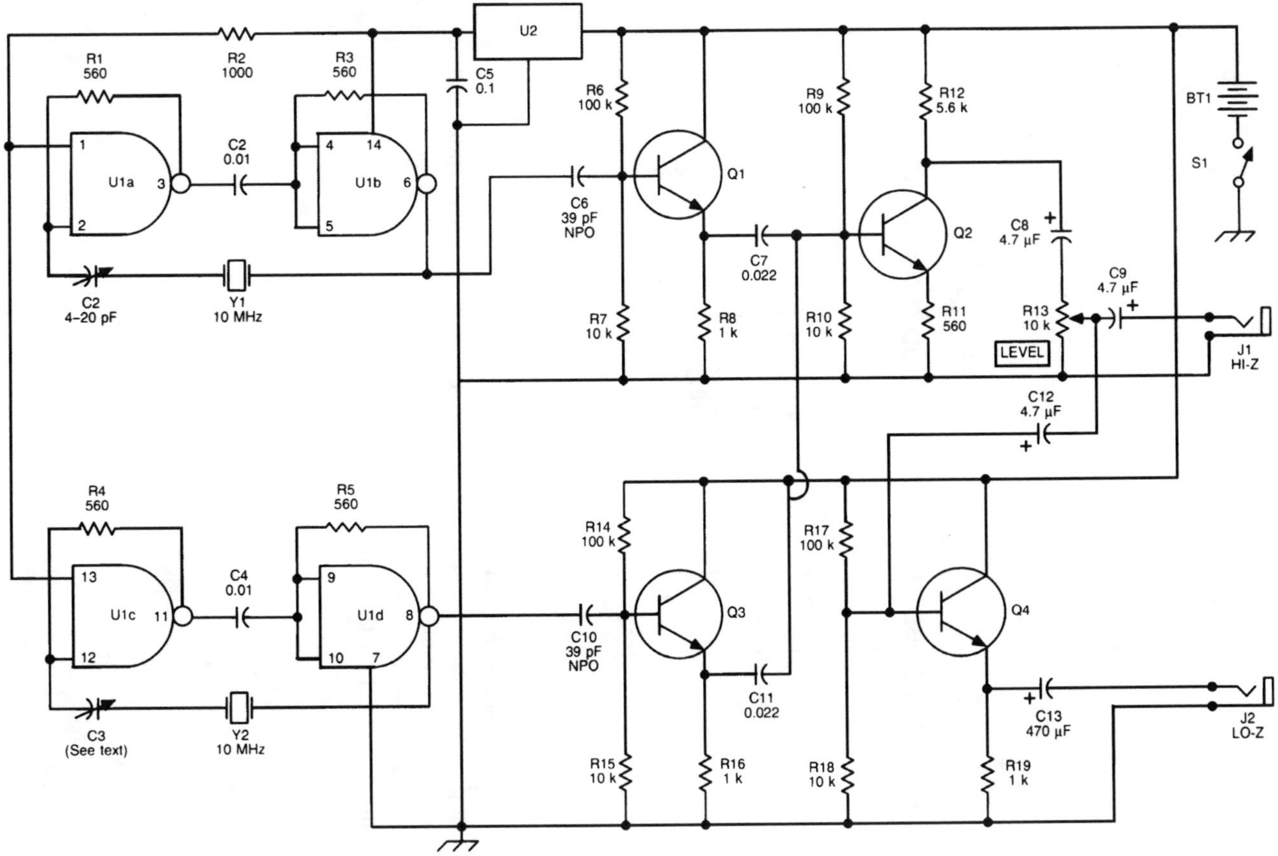

73 AMATEUR RADIO TODAY

Fig. 73-6

In this circuit, two signals from crystal oscillators operating at nearly the same frequency are mixed in Q2 and the audio difference frequency is used. C3 is a small air padder capacitor of 3 to 40 pF and can be fitted with a dial that is calibrated in the audio frequency produced. C2 is used for an initial calibration of the fixed oscillator to exactly 10.000 MHz.

TWIN-TEE TONE OSCILLATOR

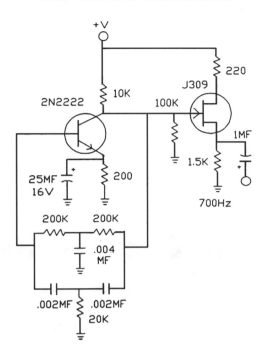

73 AMATEUR RADIO TODAY

Fig. 73-7

A 2N2222 transistor is used in a twin-tee oscillator, followed by a FET buffer circuit.

UJT OSCILLATOR

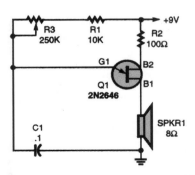

When power is connected to the circuit, C1 charges through R1 and R3. When the voltage reaches the UJT gate's (G) threshold voltage (usually around 0.5 to 0.8 of the supply voltage), the dynamic resistance between B2 and B1 drops to a very low value. Capacitor C1 then discharges through the emitter-base junction and the 8-Ω speaker. The frequency of the UJT oscillator is determined by the values of C1, R1, and R3.

POPULAR ELECTRONICS

Fig. 73-8

74

Oscillator Circuits—Hartley

The sources of the following circuits are contained in the Sources section, which begins on page 1043. The figure number in the box of each circuit correlates to the entry in the Sources section.

NE602 HARTLEY VFO CIRCUIT

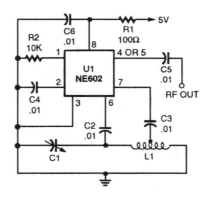

POPULAR ELECTRONICS

Fig. 74-1

In this circuit, L1 is tapped at about one-third to one-fourth from the ground end. L1 is about $10/f$ MHz, while C1 is chosen to resonate L1 to the desired frequency.

HARTLEY OSCILLATOR

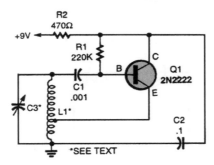

POPULAR ELECTRONICS

Fig. 74-2

The oscillator's operating frequency is determined by the values of L1 and C3. The feedback level is determined by the location of the tap on L1, which normally will be about one-fifth to one-fourth of the total turns. As an example, the Hartley oscillator will operate in the 5-MHz range with the following LC values: L1 is 20 turns of 18 enameled copper wire close wound on a 1-inch plastic form with a tap up five turns from the bottom. Capacitor C3 can be any small variable capacitor with a maximum capacitance value of 100 pF. With the proper LC values, the Hartley oscillator can operate from audio to UHF.

VOLTAGE-TUNED HARTLEY JFET VFO

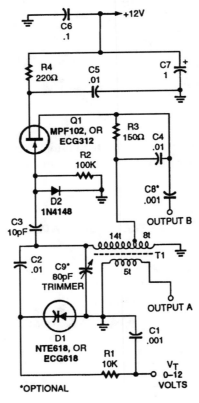

Fig. 74-3

This voltage-tuned JFET Hartley oscillator can be made to sweep its entire frequency range by applying a sawtooth waveform that rises from 0 to 12 V to the V_t input. In this circuit, a varactor is used for the tuning element instead of an air-core inductor. The frequency range is around 5 MHz.

MODIFIED HARTLEY OSCILLATOR

(a)

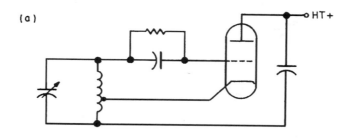

(b)

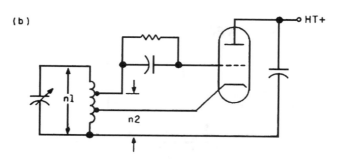

Fig. 74-4

A. F. Lampkin, in 1939, showed that a high-C Hartley oscillator (as typically used in the once popular ECO VFO) could be improved by a factor of about 10 times simply by tapping the grid (gate or base) connection down the coil.

NE602 HARTLEY OSCILLATOR

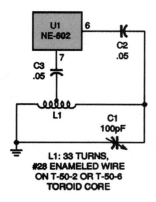

In this Hartley VFO, feedback is provided via a tap on the inductor.

POPULAR ELECTRONICS

Fig. 74-5

HARTLEY VFO WITH BUFFER AMPLIFIER

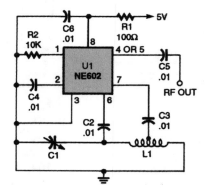

POPULAR ELECTRONICS

Fig. 74-6

The figure shows a Hartley oscillator with buffer amplifier. This circuit can be tuned via C1, or if C1 is removed and the varactor network is connected to point A, a tuning voltage can be used instead. For best stability, L1 can be replaced with an air-core unit. Frequency range is 2 to 5 MHz, but the circuit can operate 1 to 10 MHz with suitable values of L1, C2, C3, C4, and C5.

HARTLEY JFET OSCILLATOR

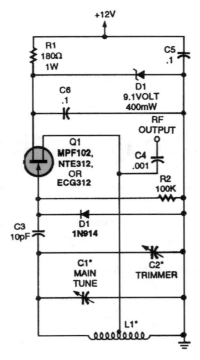

POPULAR ELECTRONICS

Fig. 74-7

This simple Hartley JFET oscillator has a built-in zener-diode regulator to condition the power-supply voltage down to 9.1 V. L1 depends on the desired frequency. Typical values are C1=5 to 50 pF variable, C2=100 pF plus a 3- to 30-pF trimmer, and L1=14 μH (for 3.5 MHz). L1 should be mechanically rigid, should have an air core (do not use ferrite cores because of drift effects), and should have $Q \geq 200$ for best stability. The tap will be typically 10 to 25 percent of the total turns.

75

Oscillator Circuits—Miscellaneous

The sources of the following circuits are contained in the Sources section, which begins on page 1043. The figure number in the box of each circuit correlates to the entry in the Sources section.

MMIC AMPLIFIER/OSCILLATOR

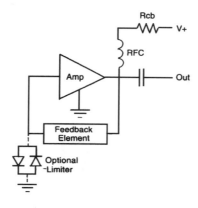

RF DESIGN

Fig. 75-1

The figure shows a basic oscillator circuit using an MMIC amplifier.

SPARK-GAP OSCILLATOR

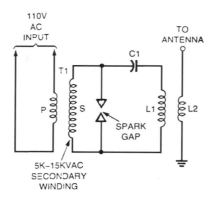

POPULAR ELECTRONICS

Fig. 75-2

A high-voltage current-limiting transformer (T1) supplies power to the basic LC tuned circuit. As C1 charges to near the transformer's maximum output voltage, the spark gap's air space breaks down, completing the circuit between the inductor and capacitor, L1 and C1. The tremendous inductive kick in the circuit is caused by the inductive field collapse when the spark gap shorts out the LC series circuit. The LC tuned circuit oscillates in a very broadband manner.

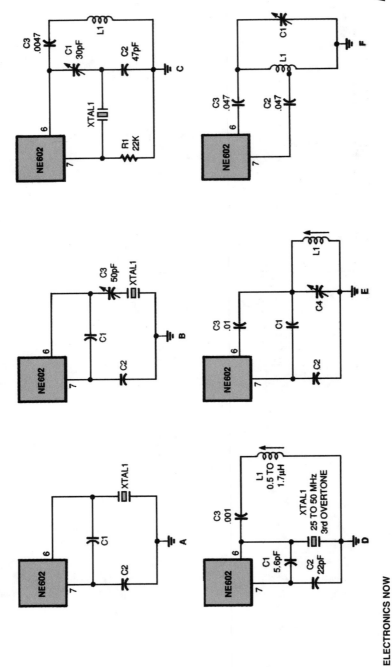

NE602 LOCAL-OSCILLATOR CIRCUITS

Fig. 75-3

The local oscillator for the NE602 can take the form of either a crystal-controlled (*a* through *d*) or a resonant-tank circuit (*e* and *f*).

NE602 OSCILLATOR CIRCUIT

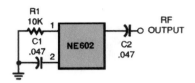

If the LO signal of the NE602 is sent directly to the output pins, the device can be used as a low-cost, high-frequency oscillator.

ELECTRONICS NOW

Fig. 75-4

OPERATIONAL TRANSCONDUCTANCE AMPLIFIER/OSCILLATOR

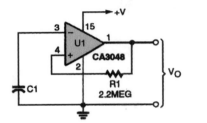

The only components in the simple circuit shown are a CA3048 OTA, a feedback resistor (R1), and a timing capacitor (C1). The output frequency is approximately given by:

$$f_o \approx 1 / (2\pi R_1 C_1)$$

Timing resistor R1 should be from 1 to 3.9 MΩ. When R1 is of a greater value, the circuit sometimes stops oscillating, depending on the specific CA3048 used.

ELECTRONICS HOBBYISTS HANDBOOK

Fig. 75-5

32-kHz DMOS OSCILLATOR

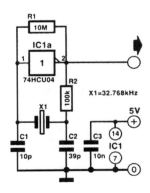

Capacitors C1 and C2 and the crystal form a π section that causes a phase shift of 180°. Circuit IC1 is an inverter, which also causes a phase shift of 180°. Thus the total phase shift is 360°, which is necessary if the oscillations are to be sustained.

ELEKTOR ELECTRONICS

Fig. 75-6

NE602 AM-MODULATED OSCILLATOR CIRCUIT

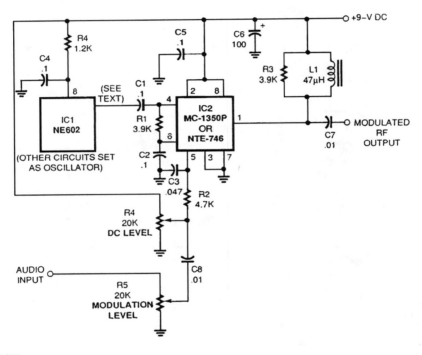

Fig. 75-7

By using an MC-1350P modulator IC, the output of the NE602 can easily be amplitude-modulated.

GATE DIP OSCILLATOR

TABLE 1—COIL WINDING DATA

Band (MHz)	Turns	Wire Size/Type
3.5 6.5	45	32-enameled
6.5 11	32	26-enameled
11 19	14	20-enameled
15 24	10	20-enameled
21 36	7	insulated connection wire
32 56	4	insulated connection wire
60 110	U-shaped*	16 enameled

*1.8-inches long

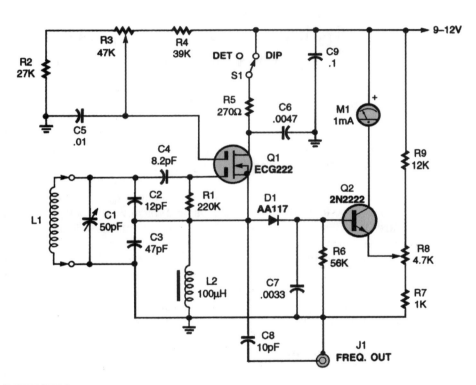

POPULAR ELECTRONICS

Fig. 75-8

This dip oscillator, based on the classic vacuum-tube circuit, is very useful for checking tuned circuits and antennas, and is a valuable tool for the RF experimenter. Coil forms are $\frac{1}{2}$-in diameter plastic tubing.

ACCURATE RING OSCILLATOR

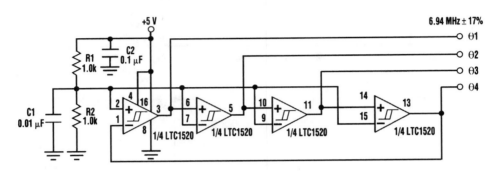

Fig. 75-9

Four square-wave outputs can be produced with a single IC using this simple circuit setup. It takes advantage of the tightly controlled propagation delay of the LTC1520 quad line receiver to produce a stable set of output waveforms. The circuit as shown uses all four comparator stages to deliver quadrature output waveforms (90° phase increments). Because t_{PLH}/t_{PHL} skew (the difference between low-to-high and high-to-low propagation delays) is typically only 500 ps, the waveforms have duty factors that are very close to 50 percent. Channel-to-channel skew is usually only 400 ps, holding phase error between outputs to around 5°. An eight-phase oscillator (with 45° phase increments) can be constructed by inverting ϕ 1 to ϕ (4 through another four LTC1520 comparators. It is also possible to build two-phase and three-phase ring oscillators by using two- and three-comparator sections, respectively. Just be sure that an odd number of inversions are around the loop, otherwise, the oscillator becomes a latch! Each LTC1520 input has a Thevenin input resistance of ±18 kΩ to $\frac{2}{3}$ V_{CC}. For a more accurate 50-percent duty factor, the dc threshold should be biased closer to $\frac{1}{2}$ V_{CC}. This is easily done by connecting an external resistor divider to pull the switching threshold near $\frac{1}{2}$ V_{CC}, as is done with R1 and R2 in this figure. The measured temperature stability of the depicted oscillator is very good, normally varying less than 5 percent from 0° to 70°C.

HIGH-STABILITY OSCILLATORS

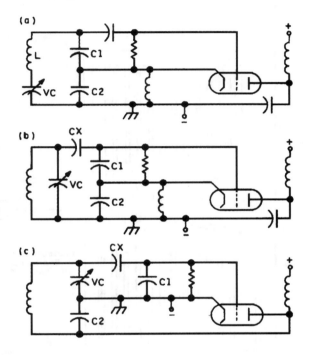

Fig. 75-10

These three high-stability oscillator circuits show the minor, but significant, differences between them: (*a*) Gouriet-Clapp oscillator with series resonance; (*b*) Seiler low-C Colpitts oscillator with parallel resonance; (*c*) the Vackar oscillator, in which the ratios $C_2:V_C$ and $C_1:C_X$ should both be about 1:6.

STRIP-LINE OSCILLATOR

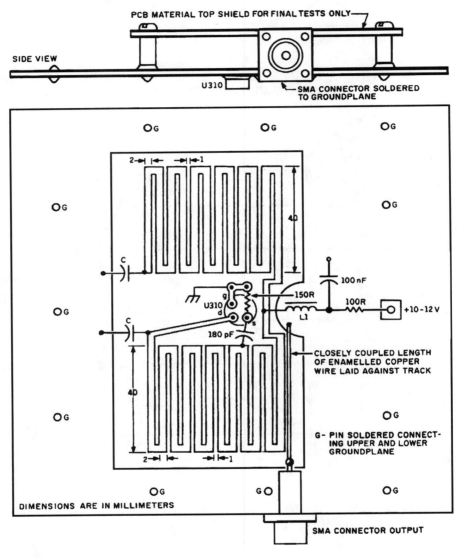

Fig. 75-11

The figure shows a prototype strip-line oscillator that is capable of providing an operating range of 3.5:1 before oscillation ceases. Length A: 40 mm, width B: 2 mm, width C: 1 mm. To change the frequency range, alter the length A and the length of the output coupling wire. Typical frequencies with $C=0$ pF are about 145 MHz; with 27 pF, about 80 MHz; with 56 pF, about 60 MHz; and with 150 pF, about 36 MHz (plate-ceramic capacitors). With $\frac{1}{16}$-inch double-sided PC board, note that the bottom ground plane covers the whole of the board. The output is about 0 dBm. The U310 FET is pushed through from the bottom side of the PC board until it touches the ground plane and is soldered directly to the ground plane.

STABLE INDUCTIVE-TUNED OSCILLATOR

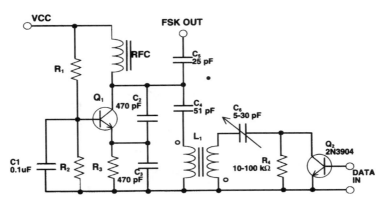

RF DESIGN

Fig. 75-12

In frequency-shift keying (FSK) applications, when modulating high data rates, coupling data to the varactor can pose a problem. A duplex filter is necessary to prevent RF loading or leaking of the oscillator tank while transferring the fast data edges to the varactor. These two constraints contradict one another—especially when the edge's rise time approaches that of the carrier frequency. If an RF choke or a large-value resistor is used to inject data to the varactor, then the edges of the data will be severely distorted by RC or inductive-capacitive (LC) time constants. The loss of these rise times can result in duty-cycle distortion as seen by the threshold level detector in the receiver. To improve upon the traditional VCO techniques, varying the inductor provides increased tuning sensitivity and better resonator Q. Inductive tuning uses the coupling or leakage inductance characteristics of a transformer. Thus, tuning capacitor C4 can be changed to a fixed, high-quality capacitor. Use of a negative-temperature-coefficient capacitor here can balance out the active device's capacitance variation with temperature. The tuning inductor (L1) is now a transformer. At about 48 MHz, it is a nine-turn, center-tapped coil of AWG28 wound on a T25-10 Micrometals core. Other transformers also can be implemented, depending on operating frequencies and performance requirements. The Q for this inductor is about 100, which is more than sufficient to isolate the active device to prevent its pulling the oscillator tank over the extended temperature range of -40 to $85°C$. The inductance variation or stability with temperature is dependent on the core material of L1.

LOW-NOISE OSCILLATOR

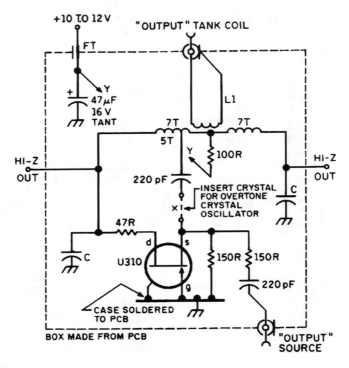

Fig. 75-13

This oscillator is suitable for low-noise applications. L is wound on a 6-mm form, and C is 15 to 220 pF. The frequency range is 20 to 80 MHz. The output is 0 dBm into 50 Ω. For use of the circuit as a crystal-controlled oscillator, a crystal can be inserted as shown.

TTL RING OSCILLATOR

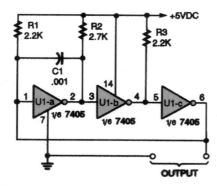

ELECTRONICS HOBBYISTS HANDBOOK

Fig. 75-14

A TTL ring-oscillator circuit uses an odd number of inverters in an astable configuration.

FRANKLIN OSCILLATOR

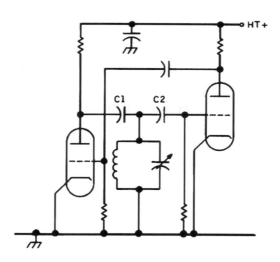

COMMUNICATIONS QUARTERLY

Fig. 75-15

Shown in tube form, this circuit shows the principle and the configuration of the Franklin oscillator.
The basic Franklin master oscillator, first described in 1930, uses very low-value capacitors, C1 and C2 (about 1 to 3 pF), imposing a very light load on the high-Q tuned circuit. Further advantages include the two-terminal inductance and the earthing of one end of the resonant circuit. This form of oscillator can be readily adapted for use with MOSFET devices.

LOW-NOISE DC REGULATOR FOR OSCILLATOR CIRCUITS

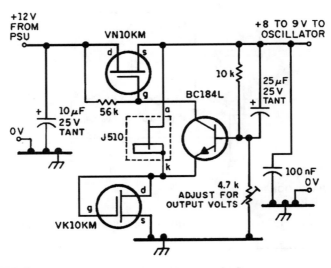

COMMUNICATIONS QUARTERLY

Fig. 75-16

This regulator uses FET devices for low-noise dc output needed for use with variable-frequency free-running oscillators.

NEON LAMP OSCILLATOR

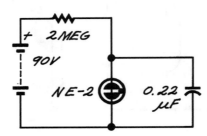

ELECTRONICS NOW

Fig. 75-17

The neon lamp has negative dynamic resistance—the voltage across it falls while conduction is increasing. As a result, it flashes on and off.

The frequency depends on the time constant RC and the supply voltage, as well as lamp characteristics.

76

Oscillator Circuits—Square Waves

The sources of the following circuits are contained in the Sources section, which begins on page 1043. The figure number in the box of each circuit correlates to the entry in the Sources section.

ADJUSTABLE-DUTY-CYCLE SQUARE-WAVE OSCILLATOR

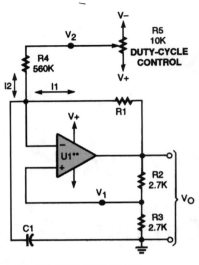

ELECTRONICS HOBBYISTS HANDBOOK

Fig. 76-1

Use R5 to adjust this variable-duty-cycle square-wave generator. That potentiometer controls I2 to vary the timing.

SQUARE-WAVE GENERATOR

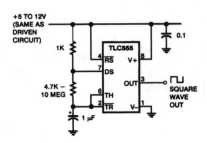

ELECTRONICS NOW

Fig. 76-2

A square wave for driving a stepper motor driver can be obtained from this circuit, which uses a CMOS timer in an oscillator circuit. A potentiometer can be used to obtain variable-frequency output.

ARBITRARY-DUTY-CYCLE SQUARE-WAVE OSCILLATOR

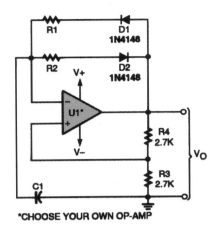

ELECTRONICS HOBBYISTS HANDBOOK

Fig. 76-3

This op-amp-based square-wave generator uses diode switching to produce a fixed duty cycle other than 50 percent.

CMOS INVERTING SQUARE-WAVE OSCILLATOR

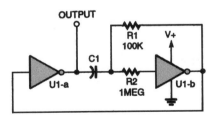

ELECTRONICS HOBBYISTS HANDBOOK

Fig. 76-4

This CMOS inverting square-wave oscillator can be varied over a wide range. Note that R2 should be 10 times the value of R1, as shown.

IC can be a CMOS inverter, or else a NOR or NAND gate wired as an inverter. The frequency is approximately $1/(2.2R_1C_1)$.

SQUARE-WAVE OSCILLATOR

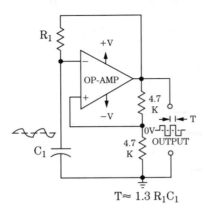

$$T \approx 1.3\, R_1 C_1$$

WILLIAM SHEETS

Fig. 76-5

An op amp can be used as a square-wave generator, as shown.

555 SQUARE-WAVE GENERATOR

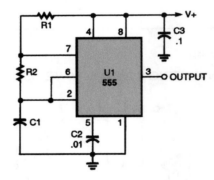

ELECTRONICS HOBBYISTS HANDBOOK

Fig. 76-6

The 555 can be connected in either monostable or astable configurations, but for continuous square waves, the astable circuit is used. The square-wave signal appears at pin 3 of the 555. The output frequency of the 555 astable multivibrator is found from

$$f_o = 1.44 \,/\, [(R_1 + 2R_2)C_1]$$

The duty cycle of the 555's square-wave output is determined by the relationship between R1 and R2, and is given by:

$$Duty\ cycle = (R_1 + R_2) \,/\, R_2$$

RC OSCILLATOR

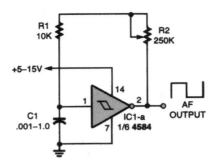

POPULAR ELECTRONICS

Fig. 76-7

A single gate of a 4584 chip is connected in a simple RC square-wave audio-oscillator circuit. C1, R1, and R2 set the oscillator's frequency. The frequency range can be modified easily by changing the value of C1. The larger you make the capacitance, the lower the frequency range, and vice versa. Mylar, polystyrene, or a similar low-leakage capacitor works best in these high-impedance RC circuits.

CD4047 CMOS OSCILLATOR

This CMOS square-wave oscillator is based on the 4047 multivibrator circuit, which can be used for either monostable ("one-shot") or astable applications. In the configuration shown, the 4047 is an astable multivibrator. There are three outputs from the 4047. The first is the oscillator (OSC) output, which is connected directly to the internal oscillator circuit. The other two outputs, Q and $\overline{Q}$, are complementary to each other and operate at one half the frequency of the internal oscillator. The output frequency is set by timing components R1 and C1:

$$f_o = 1/(4.4R_1C_1)$$

R_1 should be between 10,000 Ω and 1 MΩ, while C_1 should be 100 pF or more (the maximum capacitance is not limited theoretically, but a practical limit exists when the leakage resistance of C1 is of the same order of magnitude as R1).

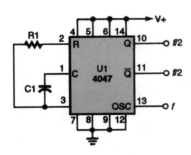

ELECTRONICS HOBBYISTS HANDBOOK

Fig. 76-8

CMOS SCHMITT-TRIGGER SQUARE-WAVE OSCILLATOR

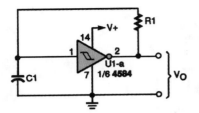

A CMOS, inverting, Schmitt-trigger square-wave oscillator requires very few components to function. Frequency=1.4/(R1C1).

ELECTRONICS HOBBYISTS HANDBOOK   *Fig. 76-9*

SCHMITT-TRIGGER RC OSCILLATOR

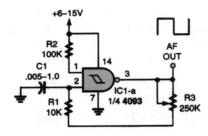

POPULAR ELECTRONICS   *Fig. 76-10*

Here's another square-wave oscillator that uses a CMOS IC with input hysteresis. In this *RC* oscillator circuit, a single gate of a quad two-input NAND Schmitt-Trigger IC is the active element.

SQUARE-WAVE OSCILLATOR

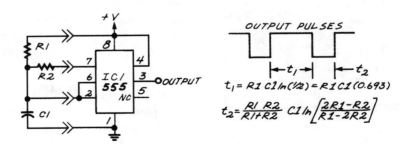

$$t_1 = R1\ C1\ ln(\tfrac{1}{2}) = R1C1\ (0.693)$$

$$t_2 = \frac{R1\ R2}{R1+R2}\ C1\ ln\left[\frac{2R1-R2}{R1-2R2}\right]$$

ELECTRONICS NOW   *Fig. 76-11*

This circuit will produce a 50-percent duty cycle from a 555.

77

Oscillator Circuits—VCO

The sources of the following circuits are contained in the Sources section, which begins on page 1043. The figure number in the box of each circuit correlates to the entry in the Sources section.

TRANSMISSION-LINE RESONATOR VCO CIRCUIT

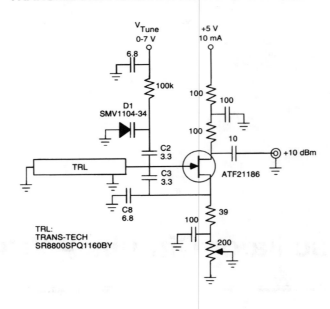

Fig. 77-1

The figure shows a typical VCO circuit using a transmission-line resonator and varactor tuning. This circuit is for 855 to 890 MHz.

1700-MHz VCO CIRCUIT

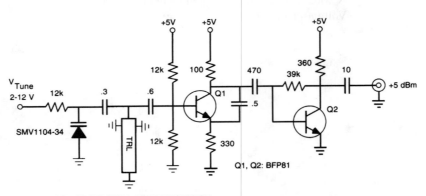

TRL: TRANS-TECH SR8800SPQ1995BY

Fig. 77-2

This VCO circuit is for 1700 MHz and uses a ceramic coaxial resonator.

NE602 VOLTAGE-TUNED LOCAL-OSCILLATOR CIRCUITS

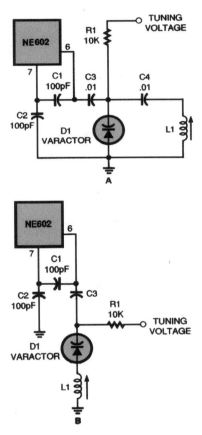

ELECTRONICS NOW

Fig. 77-3

The LO circuit for the NE602 can also be voltage controlled. Here are two different voltage-tuned local-oscillator circuits.

VOLTAGE-TUNED CLAPP OSCILLATOR

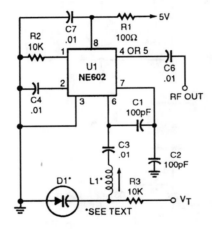

POPULAR ELECTRONICS

Fig. 77-4

This voltage-tuned Clapp oscillator uses a varactor diode to set its operating frequency. D1 is an NTE614. L1 is about $7/f$ (MHz) microhenries. With the circuit shown, the frequency range is about 6 to 15 MHz.

78

Oscillator Circuits—VFO

The sources of the following circuits are contained in the Sources section, which begins on page 1043. The figure number in the box of each circuit correlates to the entry in the Sources section.

JFET VFO

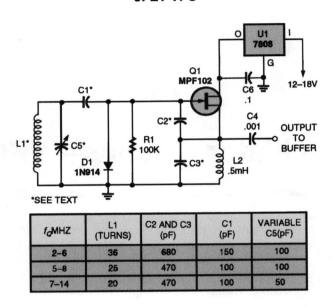

f_o MHZ	L1 (TURNS)	C2 AND C3 (pF)	C1 (pF)	VARIABLE C5(pF)
2–6	36	680	150	100
5–8	25	470	100	100
7–14	20	470	100	50

POPULAR ELECTRONICS

Fig. 78-1

This circuit is a simple JFET-based, variable-frequency oscillator that can be used in receiver or transmitter circuits. The circuit is very stable; hence, if good-quality components and appropriate construction techniques are used, there will be very little drift in frequency. The component values listed in the table will give you a starting point to set up the oscillator to cover a desired frequency range. Inductor L1 is wound on a T50-6 toroidal core using #26 or #28 enamel-coated copper wire. The oscillator's frequency can be increased a small amount by removing a turn from the coil's winding; by the same token, the frequency can be decreased by adding a turn or two. After you have achieved the desired frequency range, cover the coil with a clear plastic coating or coil dope to keep the coil wires from moving around and causing the oscillator's frequency to shift.

NE602 VOLTAGE-TUNED CLAPP VFO

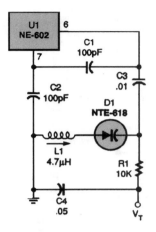

Here is a series-resonant, voltage-tuned Clapp oscillator circuit.

Fig. 78-2

VFO TUNING CIRCUITS

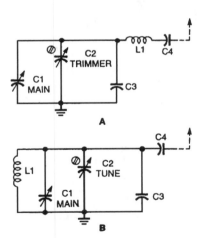

Fig. 78-3

These are typical tuning circuits. The one in B is parallel resonant and the one in A is series resonant.

VACKAR VFO

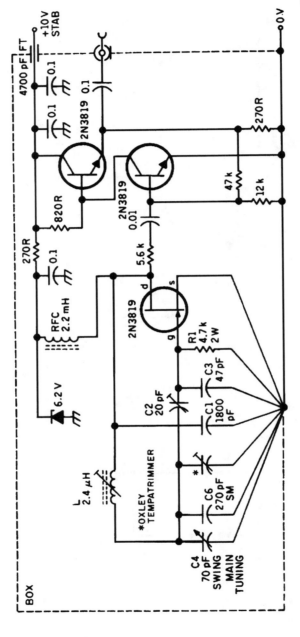

Fig. 78-4

COMMUNICATIONS QUARTERLY

This high-stability FET Vackar oscillator, covering 5.88 to 6.93 MHz, was developed by G3PDM in the late 1960s, but is still a valid design for HF VFOs for receiver, transmitter, and transceiver applications.

NE602 VARACTOR-TUNED COLPITTS VFO

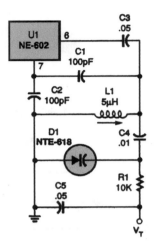

POPULAR ELECTRONICS

Fig. 78-5

You can use a varactor diode (D1) to tune a VFO circuit electronically. This figure shows a parallel-resonant, voltage-tuned Colpitts oscillator.

HIGH-STABILITY VFO

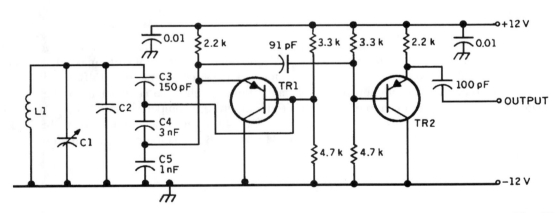

COMMUNICATIONS QUARTERLY

Fig. 78-6

This circuit was published in the 1960s, but with modern devices, it is still a useful design approach. The transistors are both 2N384.

4.0- TO 4.3-MHz VFO CIRCUIT

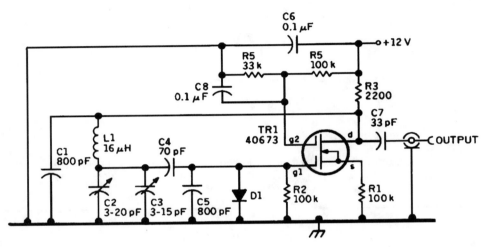

Fig. 78-7

This is a recent 4.0- to 4.3-MHz oscillator design by K6BSU. It seems likely that performance could be improved by increasing the value of $C_2 = C_3$ and decreasing that of L1 to make the device a true Vacker oscillator. The following component information is recommended by K6BSU: C1 and C5, 800-pF polystyrene; $C2$, air trimmer for setting calibration, $C3$, air variable for main tuning; C4, 70-pF, NPO ceramic and silver mica to provide required temperature compensation; C7, 33-pF NPO ceramic; $C6$ and C8, 0.1-μF 25-V monolithic capacitor; TR1, 40673 or SK3050 dual-gate MOSFET; D1, 1N4153 or 1N914 silicon signal diode; L1, 16-μF, 34 turns #26 enameled on ¾-in diameter ceramic form (no slug core), winding length 0.6 inch; all resistors 0.25 W, 5 percent.

STABLE HF VFO

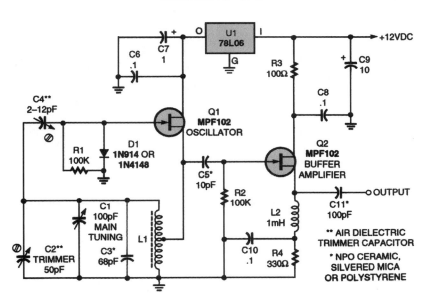

POPULAR ELECTRONICS

Fig. 78-8

This circuit is a Hartley oscillator, as identified by the fact that the feedback to the JFET transistor (Q1) is supplied by a tap on the tuning inductance (L1). The position of the tap is usually between 10 and 35 percent of the total coil length. A buffer amplifier is included, and it provides two basic functions: It boosts the low output power from the oscillator to a higher level and it isolates the oscillator from changing load impedances. The main tuning capacitor (C1) is a 100-pF air-dielectric variable unit of heavy-duty double-bearing construction. The trimmer (C2) is used to set the exact frequency, especially when using a dial that must be calibrated. Note that several of the fixed capacitors are indicated as being NPO ceramic-disk, polystyrene, or silvered mica (in order of preference). Also notice that a voltage regulator (U1) serves the oscillator. This generates very little heat; nonetheless, it is still a good idea to mount the regulator away from the actual oscillator circuit. The small trimmer capacitors (C2 and C4) are air-dielectric types, rather than mica or ceramic. The purpose of C4 is to provide dc blocking to the transistor-gate circuit. It has such a small value because we want to lightly load the LC-tuned circuit. That trimmer is adjusted from a position of minimum capacitance (i.e., with the rotor plates completely unmeshed from the stator plates), and is then advanced to a higher capacitance as the oscillator is turned on and off. The correct position is the lowest value that allows the oscillator to start immediately every time that power is applied.

NE602 COLPITTS VFO

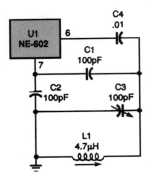

In this Colpitts variable-frequency oscillator (VFO), an inductor/capacitor combination is used to set the frequency.

POPULAR ELECTRONICS

Fig. 78-9

COLPITTS VFO

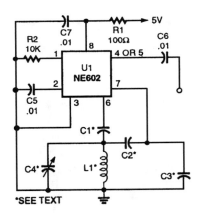

POPULAR ELECTRONICS

Fig. 78-10

Three capacitors (C1, C2, and C3) are used in this NE602AN IC Colpitts oscillator circuit, rather than two, because of the need for dc blocking. The values of the components are as follows:

$$C_1, C_2, C_3 \approx 2400 / f \text{ (MHz)}$$
$$C_4 = 1 / [(2\pi f)^2 L_1]$$
$$L_1 \approx 7 / f \text{ (MHz) microhenries}$$

VFO AGC CIRCUIT

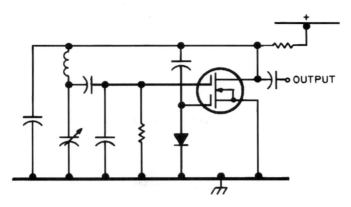

+

OUTPUT

Fig. 78-11

This is a possible simple AGC system for use with a Vackar oscillator, which should be an improvement over a diode connected directly between gate and source. D1 can be any good-quality RF silicon diode.

TEMPERATURE-COMPENSATING VFO CIRCUITS

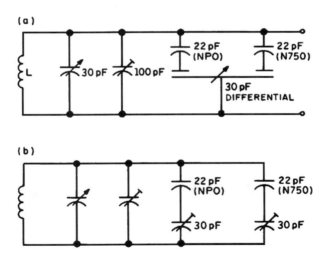

(a)

L

30 pF   100 pF

22 pF (NPO)   22 pF (N750)

30 pF DIFFERENTIAL

(b)

22 pF (NPO)   22 pF (N750)

30 pF   30 pF

Fig. 78-12

Here are two ways in which adjustable temperature compensation can be achieved: (*a*) with a differential capacitor, and (*b*) with two conventional trimmers.

CLAPP JFET VFO

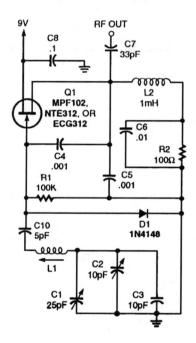

Fig. 78-13

The JFET VFO circuit shown here can be identified as a Clapp oscillator by its series-tuned LC network. This circuit as shown can be used from 0.5 to 7 MHz.

79

Photography-Related Circuits

The sources of the following circuits are contained in the Sources section, which begins on page 1043. The figure number in the box of each circuit correlates to the entry in the Sources section.

DARKROOM EXPOSURE TIMER

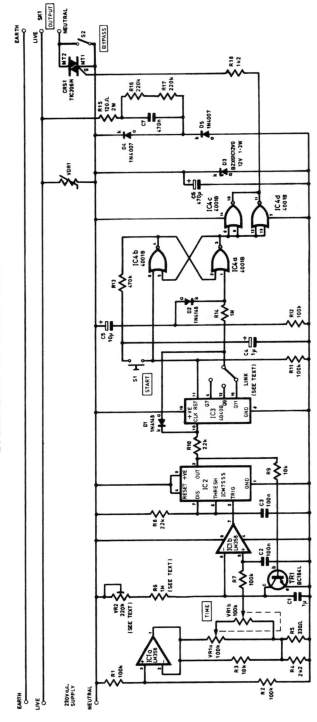

Fig. 79-1

EVERYDAY PRACTICAL ELECTRONICS

In this circuit, op-amp IC1b and timer IC2 form a voltage-controlled oscillator circuit with its frequency varied by the dual-ganged potentiometer (VR1). Resistor R3 effectively swamps any variations in VR1 value, preventing them from altering the lowest control voltage, as set by R4 and R5. The values of these resistors are chosen to give a ratio between the highest and lowest voltages of 32.5:1, which sets the timing range from 0.4 to 13 minutes, ensuring cover of the intended range of 0.5 to 12 minutes. Op-amp IC1b is used as a comparator. Capacitor C1 charges in exponential fashion from VR2 and R6 until the voltage across it reaches the selected control voltage from VR1b. When it does so, the output of IC1b goes low, triggering timer IC2, which then delivers a 2.5-ms pulse from output pin 3. This is used to discharge C1 via transistor TR1. VR2 provides the overall trimming adjustment for calibration. From IC2, the output pulses are applied through resistor R10 to the CLOCK input of binary counter IC3. This counts the pulses until the selected output goes high, blocking the CLOCK signal via diode D1. Pin 15 of IC3 is the 11th counter output, so 1024 clock pulses are counted. The final output of the circuit is controlled by quad NOR gate IC4. When the cir-

716

DARKROOM EXPOSURE TIMER *(Cont.)*

cuit is first energized, a positive pulse from C5 via D2 to IC4a pin 1 causes IC4a output pin 3 to go low and IC4b output pin 4 to go high, the OFF state. The high output of IC4b pin 4 charges C4 through R13. When START switch S1 is pressed, output pin 3 goes high, the ON state. At the same time, the pulse resets counter IC3 via its input, pin 11. Capacitor C4 will not be recharged until the output returns to the OFF state, ensuring that if switch S1 is held down or accidentally pressed again before timing is complete, the output time will not be affected. Timing ceases when the link-selected output of IC3 goes high and resets IC4a through R14.

SIMPLE SYNCHRONIZED SLAVE PHOTO FLASH

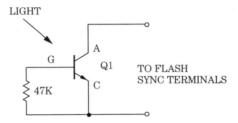

WILLIAM SHEETS

Fig. 79-2

A light-activated silicon-controlled rectifier (Q1) is triggered by a flash of light from the master flash unit.

ENLARGER TIMER STEP CIRCUIT (0 TO 59 s)

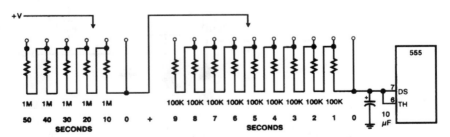

ELECTRONICS NOW

Fig. 79-3

This circuit shows how to set up a NE555 timer to produce timing steps from 0 to 59 s in 1-s intervals.

ENLARGER TIMER

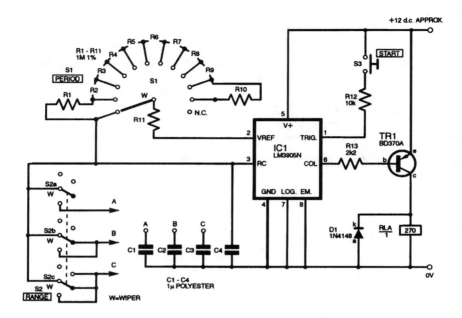

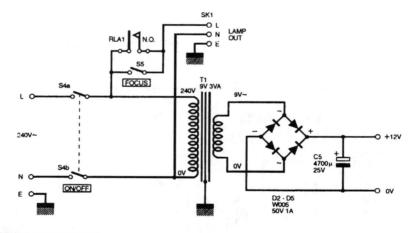

Fig. 79-4

This fairly accurate design is also suitable for color work. Unlike many simple timer designs, this unit is based around the LM3905N 8-pin D.I.L. precision timer IC, which repeats the timing period very consistently with little drift, although the accuracy of the timer depends on the tolerance of the components used. In the circuit diagram, IC1 is an LM3905N timer whose period is determined by an

ENLARGER TIMER *(Cont.)*

external RC network selected by switches S1 and S2. The formula for the time period is $t=RC,$ where t is in seconds, R is in ohms, and C is in farads. Switch S1 is a single-pole, 12-way rotary type which selects among timing resistors R1 to R11. S2 is a three-pole, four-way rotary switch that progressively adds C1 to C3 *in parallel* with C4, thereby adjusting the timing capacitance range. In conjunction with the 1-percent resistor network (R1 to R11), each 1μ/1M pair represents a 1-s period; hence, the maximum possible period is 44 s. Timing is initiated by closing switch S3. A floating output transistor *within* IC1 drives a PNP buffer, which, in turn, operates relay RLA. The circuit requires roughly a 12-V rail, and the relay coil is chosen accordingly. The IC will operate from 4.5 to 40 Vdc, and a standard full-wave power supply can be used. The relay contacts RLA1 switch on the enlarger bulb, and a separate FOCUS switch (S5) can be included across the relay contacts (or perhaps drive the relay coil manually), if desired.

CONSTANT-EXPOSURE ENLARGER TIMER CIRCUIT

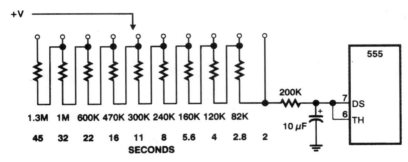

ELECTRONICS NOW

Fig. 79-5

This circuit shows how to set up a NE555 timer to produce time steps proportional to f/stops. Shown are values for switch-selected times of 2, 2.8, 4, 5.6, etc., through 45 s for constant exposure with various enlarger lens settings.

SLIDE-PROJECTOR STEPPER CIRCUIT

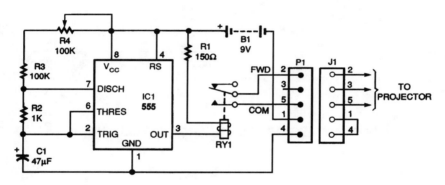

ELECTRONICS NOW

Fig. 79-6

This circuit uses a 555 timer to drive a relay connected to a slide projector to provide automatic advance. R4 sets the time between slide changes.

80

Power-Supply Circuits—Ac to Dc

The sources of the following circuits are contained in the Sources section, which begins on page 1043. The figure number in the box of each circuit correlates to the entry in the Sources section.

135-VDC SUPPLY

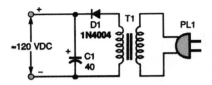

POPULAR ELECTRONICS

Fig. 80-1

This supply can be used for vacuum-tube experiments. It uses a 120:120 Vac isolation transformer to isolate the ac line from the dc output. T1 can be 30- to 100-mA capacity and C1 is a 40-μF 250-V electrolytic.

VACUUM TUBE EXPERIMENTER'S SUPPLY

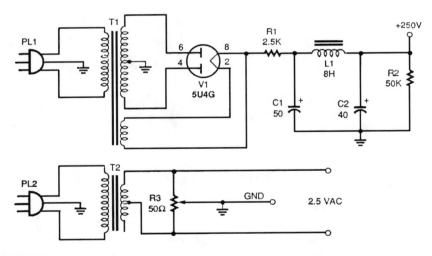

POPULAR ELECTRONICS

Fig. 80-2

This supply furnishes +250 V at about 60 mA, 2.5 Vac at 4.5 A, and it has a balance pot for the filament (B-return) center tap. T1 is a power transformer with 350-0-350 Vac, center tapped, 5 Vac at 3 A. T2 is 2.5 Vac at 4.5 A.

10-W CW TRANSCEIVER POWER SUPPLY

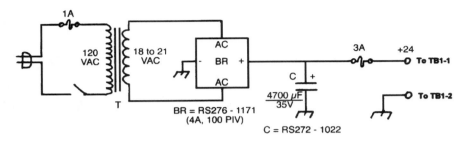

73 AMATEUR RADIO TODAY

Fig. 80-3

This power supply delivers about 24 Vdc at 1 A and was originally used for supplying power to a 10-W CW ham rig.

SERIES-REGULATED LINEAR POWER SUPPLY

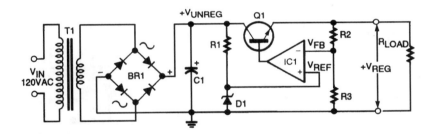

ELECTRONICS NOW

Fig. 80-4

The ac line is isolated from the power supply by transformer T1. Full-wave diode bridge BR1 delivers unregulated dc with ripple on it to the filter capacitor (C1.) The filtered dc is delivered to pass transistor Q1, shown here in series with the load. The series-pass regulator with a transistor pass element regulates the voltage to ensure that a constant output level is maintained, despite variations in the power line voltage or circuit loading. The basic linear series regulator consists of transistor Q1, reference resistor R1, sensing resistors R2 and R3, voltage-reference zener diode D1, and operational amplifier IC1, organized as an error amplifier. The zener diode D1 provides a fixed reference voltage at the positive input to amplifier IC1. The output voltage of the supply establishes the emitter voltage and provides a feedback voltage for the negative terminal of the amplifier IC1. The equation for the regulated output voltage is $V_{reg} = V_{ref}(1 + R_1/R_2)$.

OFF-LINE 24-V 100-W DC SUPPLY

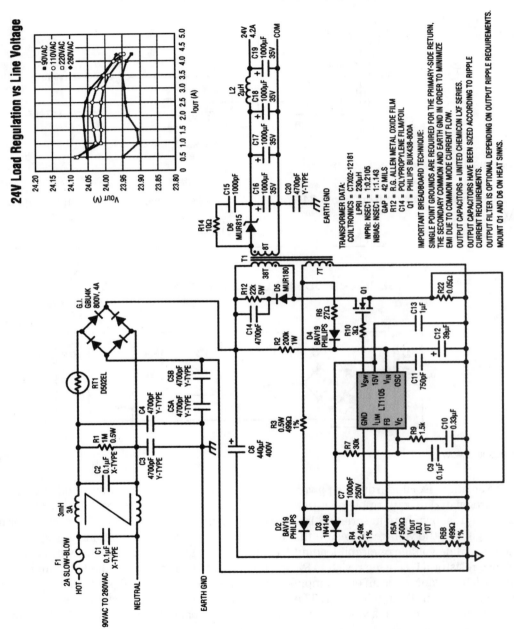

Fig. 80-5

OFF-LINE 24-V 100-W DC SUPPLY *(Cont.)*

Today's electronic systems often rely on distributed power to support a myriad of control functions. This contrasts with historical approaches, which provided a regulated dc voltage to each of the subsystems with an off-line converter, often an expensive off-the-shelf box with less-than-optimal reliability. Shown here is a discrete approach that simplifies the task of distributed power design, saving time and money during system design. The LT1105 current-mode PWM control IC makes a simple, low-cost, yet highly reliable distributed power supply with the added advantage of a customizable footprint. The 24-V output is regulated to better than ±1 percent over a line range of 90 to 260 Vac and an 8:1 dynamic load range. A maximum output current of 4.2 A is available for a total of 100 W. All components, including the transformer, are specified in this design. The transformer meets international safety standards UL 1950 and IEC950. This transformer and others that provide 36 or 48 V at 100 W are available off the shelf. The LT1105 uses unique design techniques, eliminating the optocoupler feedback normally found in off-line supplies, yet enabling the regulator to provide tight line and load regulation. A totem-pole output drives the gate of an external high-voltage FET (Q1), and switch current is monitored by sense resistor R22. Short-circuit protection is provided by "burp" mode operation, whereby the LT1105 will continuously shut down and restart during the fault condition.

SMART AC POWER STRIP

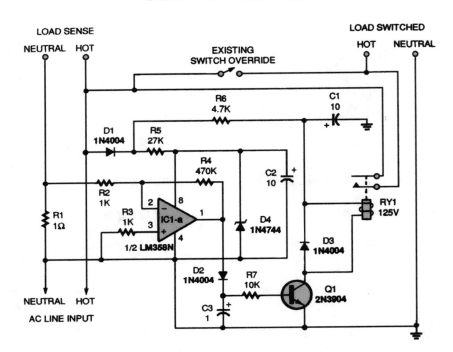

POPULAR ELECTRONICS

Fig. 80-6

The schematic of the Smart Strip circuit is shown. The ac line input is connected directly to the 117-Vac line of a power strip. The voltage is rectified by diode D1 and filtered by capacitor C2. The load-sense lines are connected to the ac socket in the power strip that will contain the device that will be used to turn the others on. When the load sense device is turned on, current flows through R1, a 1-Ω, 10-W resistor. To limit the power in R1 to 5 W, therefore, a maximum load of no more than 5 A should be connected to the load sense outlet. The resulting voltage drop across R1 is fed to one section of an LM358N op amp, IC1-a, through resistors R2 and R3. Zener diode D4 limits the supply for the op amp to 15 Vdc. The voltage drop across R1 could be very small if the device plugged into the load sense socket does not draw much current. To ensure that the circuit is sensitive enough to detect such small-load devices, the gain of IC1-a is set at 470 by resistors R2 and R4. Because the output IC1-a is halfway rectified, diode D2 and capacitor C3 are used to form a peak-hold circuit. As long as C3 is charged to 0.7 V or more (when a powered-up load sense device is detected), transistor Q1 will be on, and relay RY1 will close. When those normally open contacts close, the hot line is connected to the "load-switched" sockets, effectively turning on any devices that are connected to those outlets. Diode D1, resistor R6, and capacitor C1 provide a dc supply for the 12-V coil of the relay; diode D3 acts as a clamping diode.

ANTIQUE RADIO POWER SUPPLY

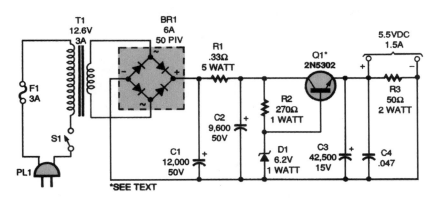

POPULAR ELECTRONICS

Fig. 80-7

This supply was intended for powering an antique radio using 01A triodes needing dc filament voltage. It is a straightforward zener regulator and pass transistor. The large values of capacitance used are for reducing 120-Hz hum to a minimum.

IR TRANSMITTER POWER SUPPLY

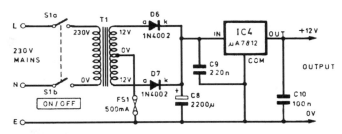

EVERYDAY PRACTICAL ELECTRONICS

Fig. 80-8

This supply provides 12 Vdc for an IR transmitter or other circuit. The current is about 260 mA. A 120-V transformer can be substituted for the 230-V unit shown, if desired.

5-V OFF-LINE REGULATOR

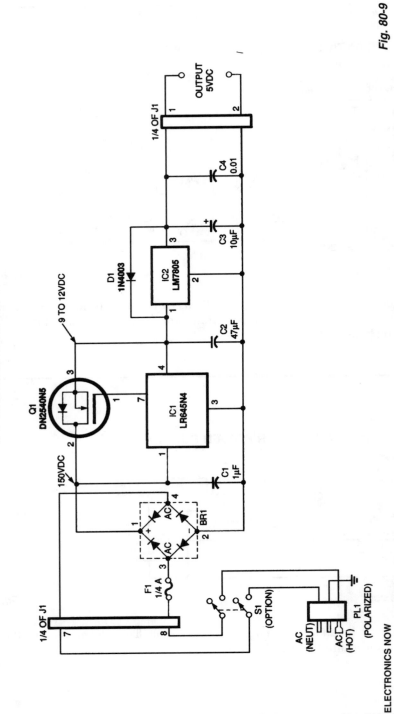

Fig. 80-9

ELECTRONICS NOW

This circuit produces 5 Vdc using an off-line regulator. An input from 12 to 120 Vac is applied to pins 7 and 8 of the eight-pin circuit-board-mounted header (J1). Although the LR645N4 (IC1) accepts voltages up to 450 V, the input of this circuit is limited by the 200 PIV (peak inverse voltage) rating of the bridge rectifier (BR1). The full-wave-rectified input is applied to the 1-µF filter capacitor (C1). The filtered 150 Vdc is applied directly to pin 1 of IC1. Filter capacitance is unusually low because IC1 has a ripple rejection ratio (V_{in}/V_{out}) of 60 dB at 120 Hz with no load. The filtered input dc is also applied to pin 2 of the N-channel, depletion-mode MOSFET Q1, made by Supertex. It was designed to be compatible with LR6 linear regulators. MOSFET Q1 conducts

728

5-V OFF-LINE REGULATOR *(Cont.)*

up to 100 mA from the high-voltage source through the gate control line of the IC1, thus bypassing it. The combined regulated voltage and current source from Q1 and IC1 is then filtered through capacitor C2 to stabilize the regulated output. The regulated output of IC1 is applied to the LM340 (7805) voltage regulator (IC2) that regulates the output voltage of the combination of IC1 and Q1 to a logic-compatible 5 Vdc. The total available current at the 5-Vdc output is 100 mA, more than adequate current for most logic circuitry. The 5-Vdc is also filtered by 10-μF tantalum capacitor C3 for additional circuit stability under full current load.

WARNING: There is no isolation between line and load, and a potentially dangerous shock hazard exists. Do not use this circuit if the potential for accidental contact exists with the circuit and any other person or device.

OFF-LINE REGULATOR

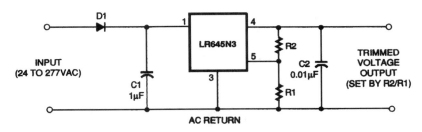

ELECTRONICS NOW

Fig. 80-10

This circuit produces 8 to 12 Vdc using an off-line regulator. The ratio of the resistors determines the output voltage. The best value for R_1 plus R_2 is 250 kΩ, although the data sheet allows values between 200 and 300 kΩ. The 250-kΩ value minimizes loading, permitting the circuit to provide more accurate output voltage.

WARNING: There is no isolation between line and load, and a potentially dangerous shock hazard exists. Do not use this circuit if the potential for accidental contact exists with the circuit and any other person or device.

DUAL 5-V SUPPLY

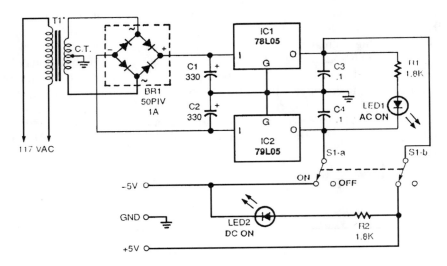

Fig. 80-11

This is a circuit for a simple dual 5-V supply for experimental use. T1 is a 12-V center-tapped 0.3-A transformer.

VACUUM-TUBE AUDIO-AMPLIFIER POWER SUPPLY

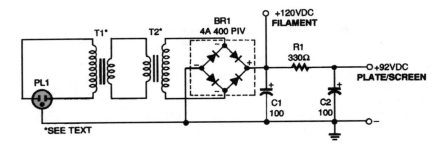

Fig. 80-12

A suitable power supply that provides isolation from the 120-Vac line is shown. The circuit uses two filament transformers connected back to back. The secondary voltages can be almost any value, as long as they are equal. A 1:1 isolation transformer can be substituted, if available. The transformers should be rated at least 25 VA or more.

SIMPLE DIGITAL-PANEL METER POWER SOURCE

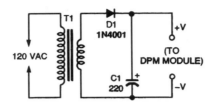

POPULAR ELECTRONICS

Fig. 80-13

This source will supply 9 to 12 Vdc with floating ground, needed by many digital panel meters.

5-V POWER SUPPLY

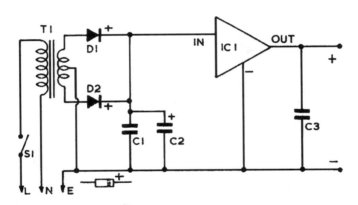

ELECTRONIC EXPERIMENTERS HANDBOOK

Fig. 80-14

The 5-V power supply shown here will provide voltage-regulated power for digital ICs needing up to 500 mA. With the 5-V power supply constructed as a separate unit, any circuit to be operated can be plugged in, and this allows it to provide power for various projects. It can also be used where a 5-Vdc, 500-mA supply is suitable, so it can, in some cases, be pressed into service where a 4.5- or 6-V battery would otherwise be fitted.

DUAL 51-V SWITCHING POWER SUPPLY

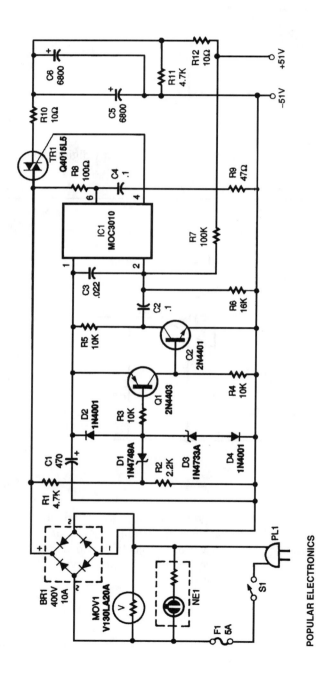

Fig. 80-15

The ac voltage taken from PL1 feeds into bridge rectifier BR1, which delivers a full-wave output of approximately 165 Vdc. The network composed of R1, R2, D1, D3, and D4 generates a series of 5-V pulses that perform two important functions: First, the pulses are used as a 5-V power source for the pulse shaping and monostable network through D2 and C1. Second, the pulses trigger optoisolator IC1 and power triac TR1 via the pulse-shaping network, composed of Q1, Q2, and R3 to R5, and the monostable circuit, made of C2 and R6. Resistor R2 sets the maximum pulse width and, therefore, the maximum output voltage. Without feedback, the unfiltered peak voltage is approximately 90 V. To obtain the required 51-V output, the feedback network (composed of R6, R7, and C3) reverse-biases the optoisolator whenever the output voltage exceeds 51 V. That, then forces TR1 to turn off as the unfiltered voltage goes to 0 V. The RC feedback network, therefore, regulates the output voltage by actively mod-

732

ifying IC1's conducting state. Resistor R8 limits the current through the optoisolator, and C4 and R9 ensure that the optoisolator's operation is stable and safe. Also, R10 limits surge current through TR1 when the supply is first turned on. Capacitors C5 and C6, along with R10, form a low-pass filter stage that minimizes ripple current. Resistor R11 discharges C5 and C6 when the power supply is turned off.

HIGH-CURRENT POWER SUPPLY

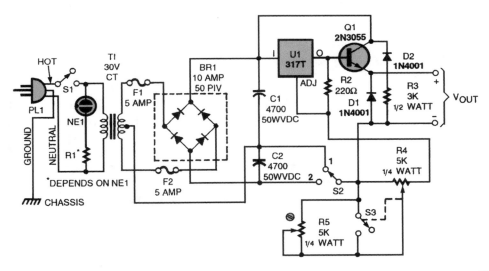

Fig. 80-16

This 10-A power supply uses a 24- to 30-V center-tapped power transformer that is capable of delivering 5 to 10 A. The voltage output is controlled by the circuit consisting of R4, R5, and S3; note that S3 is part of R4. For a fixed-voltage output, R4 should be set for zero ohms (fully counterclockwise). In that position, switch S3, will open. Trimmer potentiometer R5 should then be adjusted so that the circuit produces a 12-V output (or whatever output your application demands). For an adjustable output, R4 is turned clockwise, closing S3, and removing R5 from the circuit. The output voltage is then controlled by the resistance offered by R4 alone. When SPDT switch S2 is in position 1, the maximum output current is achieved, with both halves of T1 providing current to the filter section to double the overall current output. However, the maximum output voltage is halved in that position. That is a more efficient setting because the power transistor need not drop as large a voltage. In position 2, the maximum voltage almost equals the rating of T1. D1 and D2 are included in case power is turned off with an inductive load attached.

15-V RECEIVER POWER SUPPLY/CHARGER

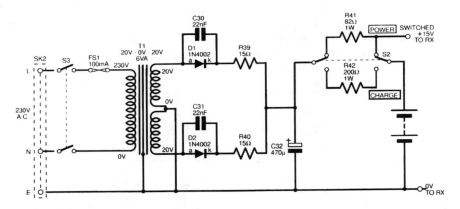

EVERYDAY PRACTICAL ELECTRONICS

Fig. 80-17

This power supply is for powering a receiver or similar load at 15 V, 100 mA. The NiCd battery pack is a 600-mAh AA cell assembly and is charged at 50 to 60 mA. A 120-V primary transformer with the same secondary voltage can be substituted for the 230-V unit shown.

SIMPLE DUAL-VOLTAGE POWER SUPPLY

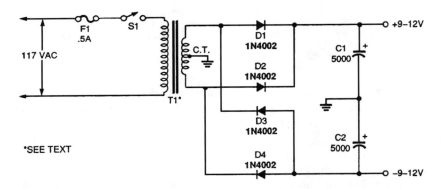

POPULAR ELECTRONICS

Fig. 80-18

This circuit can be used to power op-amp circuits or other applications that require dual plus and minus supplies. This is a most basic circuit and is not regulated. T1 is a 120:12-V transformer of 100 mA (or more) capacity.

81

Power-Supply Circuits—
Buck Converter

The sources of the following circuits are contained in the Sources section, which begins on page 1043. The figure number in the box of each circuit correlates to the entry in the Sources section.

Negative Buck Regulator
Basic Buck Converter Circuit

NEGATIVE BUCK REGULATOR

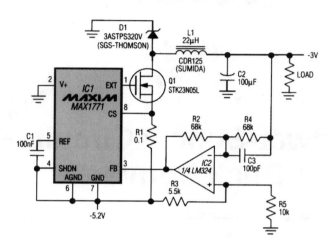

MAXIM

Fig. 81-1

This circuit adopts a step-up (boost) dc-to-dc controller for use in a negative buck-regulator application. The supply voltage must be negative, and it must deliver 160 to 750 mA. Although the boost-regulator IC operates in a buck-regulator circuit, its standard connections enable proper control of Q1. The output voltage, however, must be inverted by an op amp for proper voltage feedback: The load is referred to the most positive supply rail instead of to IC1's ground terminal, so the controller must increase its duty cycle as V_{out} (referred to that terminal) increases. The op amp therefore inverts the feedback signal and shifts it to match the 1.5-V threshold internal to IC1. IC1 is configured in its nonbootstrapped mode, which provides an adequate gate-drive signal (ground to −5.2 V) for the external MOSFET (Q1). With V_{out} set to −3 V and the output current ranging from 160 mA to above 700 mA, the circuit's conversion efficiency ranges from 84 percent to as high as 87.5 percent.

BASIC BUCK CONVERTER CIRCUIT

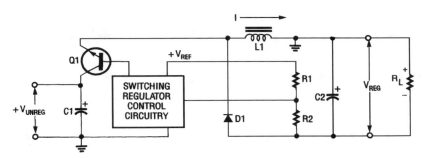

Fig. 81-2

The buck regulator is an alternative single-transistor converter. Series transistor Q1 chops the input voltage and applies the pulse to an averaging inductive-capacitive filter, consisting of L1 and C1. The output voltage of this simple filter is lower than its input voltage.

82

Power-Supply Circuits—Dc-to-Dc

The sources of the following circuits are contained in the Sources section, which begins on page 1043. The figure number in the box of each circuit correlates to the entry in the Sources section.

STEP-UP REGULATOR

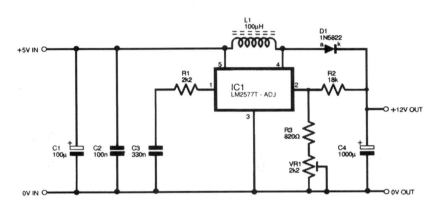

EVERYDAY PRACTICAL ELECTRONICS

Fig. 82-1

Refer to the figure for the circuit diagram of the step-up regulator. This circuit is based on an LM2577T-ADJ, which is designed to give a combination of high performance and simplicity. It can provide output currents of up to 3 A, and it has both overcurrent and thermal-protection circuits built in. It requires a minimum input potential of 3.5 V, and it shuts down automatically if the input voltage is inadequate. The maximum input voltage is 40 V. Capacitors C1 and C2 provide supply decoupling at the input, and C3 and resistor R1 provide frequency compensation. Inductor L1, diode D1, and an internal switching transistor of IC1 form a standard step-up output stage, with output smoothing provided by capacitor C4. Resistors R2 and R3 and preset VR1 form a potential divider that controls the output voltage. IC1 has an internal 1.2-V reference generator, and the voltage at pin 2 is, therefore, stabilized at this figure. With the specified values in the potential divider circuit, the output voltage range is around 9.5 to 25 V.

NE602 DC SUPPLY INTERFACE

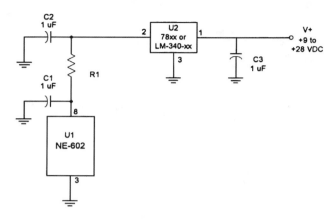

73 AMATEUR RADIO TODAY

Fig. 82-2

This is one method of deriving the 4.5- to 8.0-Vdc supply needed by the NE602 chip.

STEP-DOWN REGULATOR

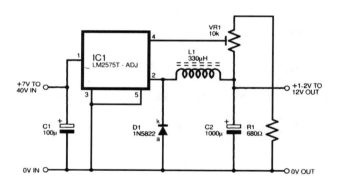

EVERYDAY PRACTICAL ELECTRONICS

Fig. 82-3

There are several versions of the LM2575T. The device used here has an "ADJ" suffix, which indicates that it has an adjustable output voltage. Pin 4 connects to the inverting input of the internal error amplifier. This amplifier's noninverting input is fed from a 1.2-V reference source. A discrete potentiometer connected across the output and feeding into pin 4 enabling the output voltage to be set at any figure above 1.2 V. In this case, the potentiometer is formed by VR1 in series with resistor R1. The adjustment range provided by VR1 is from 1.2 V to a little over 16 V. Higher output voltages can be accommodated by making R1 lower in value, but a circuit of this type is mainly used where there is a large voltage difference between the input and the output. Consequently, it is unlikely to be used with output voltages of more than about 16 V. The absolute maximum input voltage is 40 V.

NEGATIVE SUPPLY GENERATOR

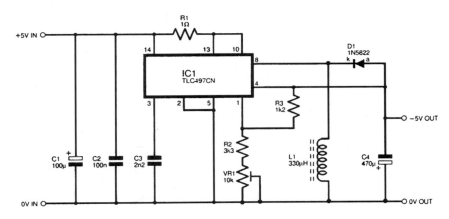

EVERYDAY PRACTICAL ELECTRONICS

Fig. 82-4

The TLC497CN is very useful for applications that do not involve high output currents. It provides a cost-effective means of providing a negative supply of up to about 150 mA. A circuit diagram for an inverter based on this device is shown. The efficiency of the circuit is typically a little over 50 percent with an input potential of 10 V, but is significantly under 50 percent with an input supply of 5 V. Capacitors C1 and C2 are supply-decoupling components on the input supply. Resistor R1 is the series resistance in the current-limiting circuit at the input to IC1. This is an essential safety feature because the circuit will often be fed from a high-current supply that could almost instantly "fry" IC1 in the event of an overload on the device's output. Capacitor C3 is the timing component in the oscillator section of the PWM. The TLC497CN uses a fixed pulse width and a variable clock frequency to control the average output voltage.

MINI POWER SOURCE

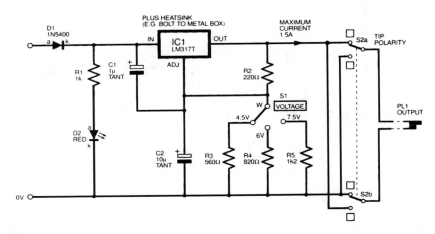

EVERYDAY PRACTICAL ELECTRONICS

Fig. 82-5

The figure shows a unit capable of offering different voltages to operate low-voltage appliances from a 12-V battery. Regulators of the LM317 type are current-limiting, short-circuit-proof, and overheat-proof, although they are still not indestructible. This circuit will provide up to 1.5 A. IC1 should be heatsinked, perhaps by bolting it with an insulating kit to a metal box.

1.3-V SUPPLY

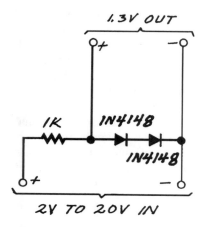

This is a simple 1.3-V regulator to power the ZN414 IC.

ELECTRONICS NOW　　　　　　　　*Fig. 82-6*

3-A LOW-DROPOUT REGULATOR WITH 50-μA QUIESCENT CURRENT

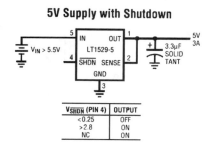

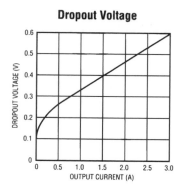

LINEAR TECHNOLOGY　　　　　　　　*Fig. 82-7*

The LT1529 is a 3-A, 0.6-V dropout regulator with 50-μA no-load quiescent current and only 16 μA I_Q in shutdown. This makes the LT1529 useful for battery-powered applications where both long battery life and high-load current surges are expected. At 1.5-A load current, dropout voltage is just 0.43 V. The LT1529 needs no external diodes to protect against reverse battery or reverse output current faults. This allows the LT1529 to be used for backup power situations in which the output is held high while the input is at ground or reversed.

2-V REFERENCE

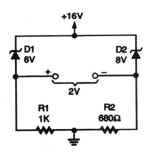

+16V

D1
6V

D2
8V

2V

R1
1K

R2
680Ω

POPULAR ELECTRONICS

Fig. 82-8

Zener diodes that operate below the 3- or 4-V level do not perform as well as higher-voltage zeners, and are not normally used for low-voltage references. The 2-V reference uses two higher-voltage zeners, D1 and D2, to obtain a stable operating reference voltage. Any possible voltage changes from temperature variations are almost completely canceled out with the two-zener circuit, making it a more accurate reference source than a circuit with a single zener. Other low-reference-voltage sources can be created by substituting different-valued zeners for D1 and D2.

20-μA QUIESCENT CURRENT REGULATOR

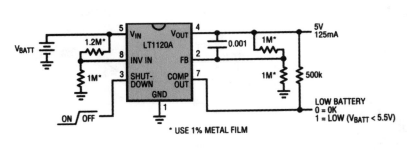

Dropout Voltage

LINEAR TECHNOLOGY

Fig. 82-9

Knowing when the battery voltage is getting low allows equipment to arrange time to save valuable information. The LT1120A not only provides micropower, low-dropout regulation, but also includes a low-battery detection comparator, all in an eight-lead DIP or SO package. The LT1120A supplies up to 125 mA at regulated voltages from 2.5 to 20 V. The low dropout voltage of 0.6 V (max.) lowers the battery input voltage requirements, and the small 40-μA (max.) quiescent current extends battery life. The LT1120A has yet more functions. The logic-compatible SHUTDOWN pin removes power to the load. The precision 2.5-V reference has the ability to both sink and source current to supply low-power backup circuits. Also, the LT1120A's design prevents excessive current draw when in dropout.

INVERTER OUTPUT STAGE OF A DC-TO-DC CONVERTER

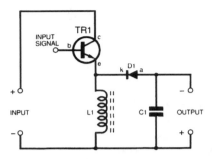

EVERYDAY PRACTICAL ELECTRONICS

Fig. 82-10

This figure shows the circuit diagram for an inverter output stage.

INDUCTORLESS BIPOLAR SUPPLY GENERATOR

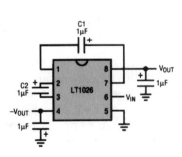

Output Voltage

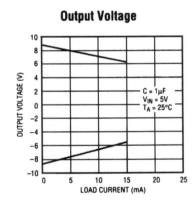

LINEAR TECHNOLOGY

Fig. 82-11

Localized generation of bipolar supply rails is useful in many op-amp, data-conversion, and interface applications. To avoid the design time and effort of putting together a switching regulator circuit, switched-capacitor conversion should be considered as a simple, effective alternative. The LT1026 is shown here in a useful application circuit that can generate rail voltages of both polarities from a 5-V logic supply. Output voltage will vary with load (as shown on the graph), from ±9 V at minimal load currents to ±6 V at a load current of 15 mA. These boosted rails can be used to extend common-mode input and output ranges of circuits without doing level shifting, ac coupling, or signal clipping. Because the LT1026 is supplied in an eight-lead SO package, the entire circuit can be surface-mounted using tantalum or ceramic capacitors.

STEP-UP SWITCHING IC/INVERTER

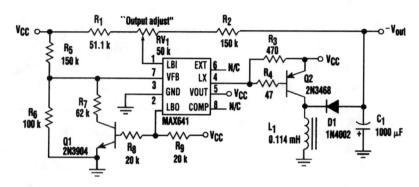

ELECTRONIC DESIGN

Fig. 82-12

A popular step-up switching IC can be made to supply a negative voltage. The MAX641BCPA is a boost converter intended for use with 3-V batteries. To get the 641 to function as an inverter, some of its pin functions must be altered. Pin 1 is used as the voltage-feedback input, and pin 2 is used as the overvoltage-detector output. When the voltage on pin 1 falls below $+1.31$ V, pin 2 becomes a low-impedance path to ground. This will force Q1 to cut off and allow the voltage on pin 7 to rise. When the voltage on pin 7 exceeds $+1.31$ V, the internal oscillator shuts off. At this point, the output voltage is equal to or less than the set level. When C1 discharges to a point where the voltage on pin 1 rises above $+1.31$ V, pin 2 will change to a high impedance, allowing Q1 to saturate and pull pin 7 below $+1.31$ V. This causes the internal oscillator to turn on. Pin 4, the output of the internal oscillator, will pump C1 negative while the oscillator is running. For the values shown in the figure, the adjustable output voltage range is approximately -4 to -13 V. The magnitude of the drivable load depends on how much ripple can be tolerated. The values shown will easily drive a 50- to 75-mA load and provide good regulation.

BASIC STEP-UP OUTPUT STAGE OF A DC-TO-DC CONVERTER

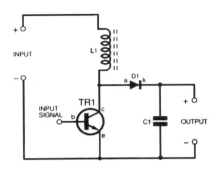

EVERYDAY PRACTICAL ELECTRONICS

Fig. 82-13

This figure shows the basic circuit configuration for a step-up output stage.

CHARGE PUMP BOOST CONVERTER

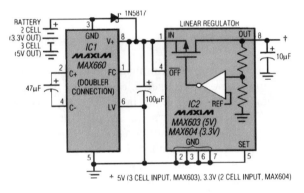

MAXIM

Fig. 82-14

The circuit overcomes the charge pump's lack of regulation by adding a regulator externally. Another option—if load currents are modest—is to add regulation on the chip. Regulation in a monolithic chip is generally accomplished either as linear regulation or as charge-pump modulation. Linear regulation offers lower output noise; therefore, provides better performance in (for example) a GaAsFET-bias circuit for RF amplifiers. Charge-pump modulation (which controls the switch resistance) offers more output current for a given die size (or cost) because the IC need not include a series pass transistor.

GaAsFET BIAS SUPPLY

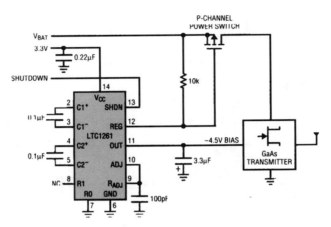

LINEAR TECHNOLOGY

Fig. 82-15

Cellular telephones and other mobile communications gear use depletion-mode GaAsFET RF output transistors, which require a negative bias voltage for proper operation. This LTC1261 circuit generates a regulated negative voltage using no inductors, and will give plenty of drive for GaAsFET bias circuits. The LTC1261CS8 inverts the supply voltage (5 V to −4.5 V), whereas the LTC1261CS can invert or double the supply voltage (3 V to −4 or −5 V). Both parts regulate the generated output voltage. The quiescent current is just 600 μA, and the shutdown current is only 5 μA. The combination of high oscillator frequency, which reduces the switched capacitor size, and the variety of fixed-output regulated negative voltages allows for a minimum-space power supply. The LTC1261 comparator output controls a P-channel MOSFET to ensure that the drain current for the GaAsFET is switched off until the regulated negative output voltage is valid. This ensures that the gate voltage is sufficient to keep the GaAsFET off during power-up, preventing unsaturated operation and excessive operation.

3.5-V 7-A LINEAR REGULATOR SUPPLY

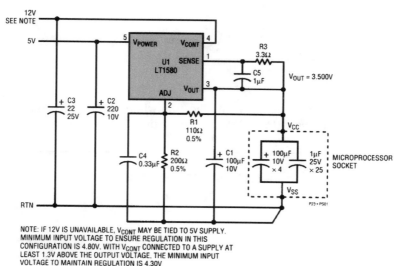

NOTE: IF 12V IS UNAVAILABLE, V$_{CONT}$ MAY BE TIED TO 5V SUPPLY.
MINIMUM INPUT VOLTAGE TO ENSURE REGULATION IN THIS
CONFIGURATION IS 4.80V. WITH V$_{CONT}$ CONNECTED TO A SUPPLY AT
LEAST 1.3V ABOVE THE OUTPUT VOLTAGE, THE MINIMUM INPUT
VOLTAGE TO MAINTAIN REGULATION IS 4.30V

LINEAR TECHNOLOGY

Fig. 82-16

The LT1580 linear regulator circuit shown achieves ±2 percent dc output-voltage accuracy at up to 7 A load current. This circuit restricts the output-voltage transients because of 200-mA to 4-A load-current steps to 65 mV p-p. The LT1580 has a separate remote sense input that maintains the dc voltage at the load accurately, independent of the load current. Output-voltage variation over the full load-current range, known as *load regulation,* with remote sense connected ($R_3 = 0$ Ω) is very close to ±0 percent. The LT1580 transient response to 3.8-A-load current steps is 88.7 mV p-p ($R_3 = 0$ Ω). Resistor R3 is added to intentionally introduce some dc load regulation. The feedback resistors are chosen to set the no-load output voltage slightly higher than 3.5 V. At 4-A load current, the output voltage is regulated slightly below 3.5 V.

DC-TO-DC CONVERTER FOR DIGITAL PANEL METERS

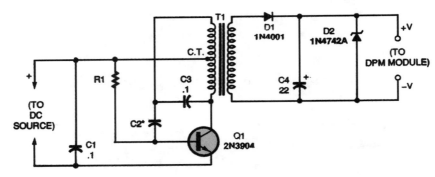

POPULAR ELECTRONICS

Fig. 82-17

Often a digital panel meter is used in an application where the meter is on the hot side of a circuit. This supply will provide the 9- to 12-Vdc floating supply for the meter.

NEGATIVE REGULATOR

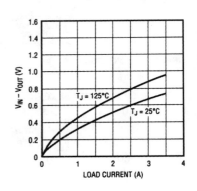

Dropout Voltage

*CURRENT LIMIT = 15k/R_LIM = 3.5A

LINEAR TECHNOLOGY

Fig. 82-18

Regulating negative voltages with minimal dropout is now possible with the LT1185 universal voltage regulator. It supplies up to 3 A of output current with a dropout voltage guaranteed to be less than 1.2 V. The five-lead TO-220 package includes a pin that allows accurate current-limit adjustment for lower-current applications. Although aimed primarily at negative regulation applications, the LT1185 works equally well as a floating positive regulator. Output voltage is programmable from 2.3 to 30 V.

LM317 CIRCUIT

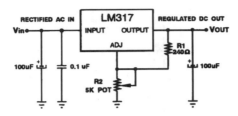

Fig. 82-19

The figure shows a typical simplified wiring diagram. Note that the output voltage is determined by resistors R1 and R2 (the value of R2 is recommended to be around 240 Ω). The regulated output voltage is given by the formula $V_{out} = 1.25 \ (1 + R_2/R_1)$.

FLASH MEMORY V_{PP} GENERATOR

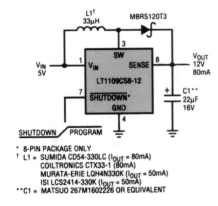

* 8-PIN PACKAGE ONLY
† L1 = SUMIDA CD54-330LC (I_{OUT} = 80mA)
    COILTRONICS CTX33-1 (80mA)
    MURATA-ERIE LQH4N330K (I_{OUT} = 50mA)
    ISI LCS2414-330K (I_{OUT} = 50mA)
** C1 = MATSUO 267M1602226 OR EQUIVALENT

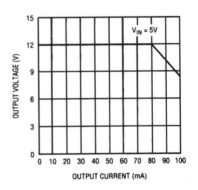

Fig. 82-20

Generating 12 V for flash memory programming is a common requirement in portable systems and PCMIA cards. The LT1109-12 dc-to-dc converter simplifies this task and uses only 0.75 in^2 of PC board space. The LT1109-12 is offered in an eight-lead SO package and requires only three other surface-mount components to construct a complete 12-V V_{PP} generator. At 12 V, 60 mA of programming current is produced, enough for simultaneous programming of two flash memories. The circuit draws 320 μA (max.) of standby current while shut down and provides a clean transition from 5 to 12 V at its output with no overshoot.

3.3-V 7-A SUPPLY

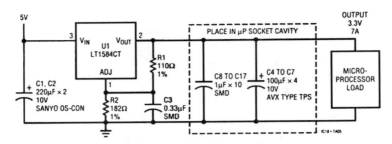

Fig. 82-21

Many microprocessor systems use 3.3-V microprocessors, cache RAM, and chip sets. This system configuration increases the current requirements of the 3.3-V supply. In addition, many of these microprocessors have a stop-clock feature for power savings, which introduces a load-current step to the power supply. Adjustable regulators are recommended for microprocessors that have power-saving (stop-clock) modes. The circuit shown has good transient response to load steps for most 3.3-V microprocessors. An external capacitor at the ADJUST pin can reduce the total filter capacitance required by one half to take care of large load transients.

GaAsFET BIAS SUPPLY

−4.1V Generator with 1mV$_{P-P}$ Noise

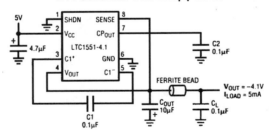

Fig. 82-22

The LTC1550 and LTC1551CS8-4.1 are switched-capacitor voltage inverters that generate a regulated −4.1-V output at up to 20-mA load current. An internal linear postregulator reduces the output-voltage ripple to less than 1 mV, making the LTC1550 and LTC1551CS8-4.1 excellent for use as bias-voltage generators for transmitter GaAsFETs in portable RF and cellular telephone applications. The single supply voltage can range from 4.5 to 6.5 V (7 V absolute maximum). The charge pump uses four small external capacitors and operates at 900 kHz, eliminating interference with the 400- to 600-kHz IF signals commonly used in RF systems.

1-V 600-kHz SWITCHING SUPPLY

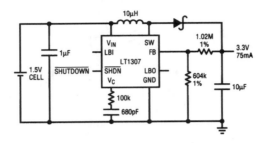

Fig. 82-23

The LT1307 chip will produce 5 V at 40 mA or 3.3 V at 75 mA from a single AA-cell power source. It operates at 600 kHz and uses 60-μA standby current, and it has a low-battery detector.

7-A 2.5-V SUPPLY

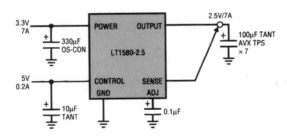

Fig. 82-24

The latest microprocessors use low-voltage processes that allow the clock frequencies to increase dramatically. Increased clock frequencies result in higher core-supply currents. Several next-generation microprocessors will use a 2.5-V supply voltage and require greater than 5-A supply current. The LT1580-2.5 in the circuit shown has the lowest dropout of any 7-A linear regulator, only 0.6 V typical. This allows conversion from a standard 3.3-V main supply down to 2.5 V. In order to achieve this low dropout performance, a second low-current control supply 1.3 V greater than the 2.5 V output is needed. A system 5-V supply conveniently provides this voltage, and only 200 mA is required from the control supply. The LT1580-2.5 also has fast transient response to load-current steps, minimizing required bulk output capacitance. An internal reference voltage with ±0.5 percent initial tolerance and Kelvin sense input make the output regulation very tight.

VOLTAGE REGULATOR SINKS AND SOURCES

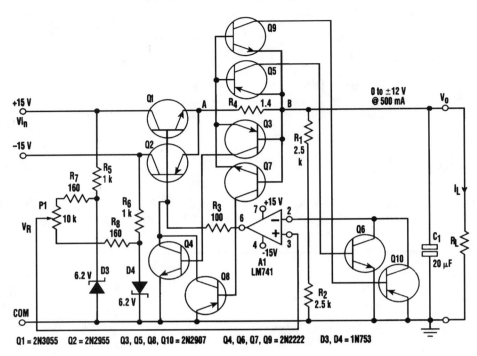

Q1 = 2N3055   Q2 = 2N2955   Q3, Q5, Q8, Q10 = 2N2907   Q4, Q6, Q7, Q9 = 2N2222   D3, D4 = 1N753

ELECTRONIC DESIGN

Fig. 82-25

It is possible to build a circuit that provides a variable output from -12 to $+12$ V (passing smoothly through 0 V) that can source or sink at any voltage. The basic regulator consists of an op amp (A1), series pass transistors Q1 and Q2, a reference voltage from P1, and a voltage divider (R1 and R2). The rest of the elements provide short-circuit protection for the regulator. The reference voltage is generated by zener diodes D3 and D4. With a 10-turn potentiometer (P1), the reference voltage (V_R) for the op amp can be varied from -6 to $+6$ V. The output voltage is given by $V_O = V_R(1 + R_1/R_2)$. Because $R_1 = R_2$, the output can be varied from 0 to ±12 V. When V_O is positive and the regulator is sourcing current (I_L positive), the base of Q1 is at $V_O + 0.7$m and Q1 is conducting. When V_O is positive and R_L is terminated in a supply voltage higher than V_O, the regulator is forced to sink current (I_L negative). At this time, Q2 conducts and sinks the current, and A1 maintains the base of Q2 at $V_O - 0.7$. Similar arguments apply when the output voltage is negative. C1, a nonpolarized electrolytic capacitor, prevents oscillations. R4 is a current-sensing resistor for short-circuit protection and limits the output current to 55 mA. For a positive V_O, if I_L is positive (sourcing) and reaches 500 mA, the voltage drop V_{AB} across R4 approaches 0.7 V and forward-biases the E-B junction of Q3. Q3 conducts and drives the base of Q4. Because Q4 goes into saturation as a result of the drop across R3, it clamps the base voltage of Q1 to ground and the output voltage drops, limiting the current in

Q1 to 500 mA. Similarly, under a positive V_o, if I_L is negative (sinking) and reaches 500 mA, V_{AB} across R4 approaches -0.7 V and forward-biases the E-B junction of Q5. Q5 drives Q6 into saturation, and the inverting input of the op amp is clamped to ground. Because the noninverting input is held at V_R (which is still positive), the output starts climbing toward $+15$ V. This prevents Q2 from sinking more current than 500 mA. Under a negative V_o, with I_L negative, Q7 and Q8 provide short-circuit protection. With V_o negative and I_L positive, Q9 and Q10 provide short-circuit protection.

EFFICIENT REGULATED STEP-UP CONVERTER

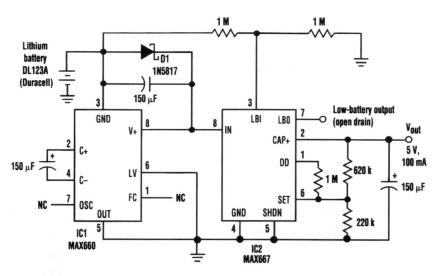

ELECTRONIC DESIGN

Fig. 82-26

Two eight-pin ICs form a regulator circuit that can convert a lithium battery's 3-V output to 5 V and deliver load currents as high as 100 mA. It operates without inductors or transformers, and draws only 200 μA of quiescent current. At $V_{in} = 3$ V, it offers 81-percent efficiency with a 100-mA load and 84 percent with a 20-mA load. Efficiency will increase as V_{in} falls. For example, at $V_{in} = 2.7$ V (the cell's loaded output for most of its operating life), efficiency for a 40-mA load current is 90 percent. Voltage from the lithium battery (a 2/3-A size Duracell DS123A) is doubled by the high-current charge pump (IC1). The Schottky diode (D1) is included to assure startup in this configuration. D1 won't affect efficiency because it doesn't conduct load current during normal operation. IC2 is a linear regulator with a dropout voltage of only 40 mV at $I_{load} = 40$ mA. This load, allowed to drain the battery until $V_{out} = 4.5$ V, yields a battery life of 16 hours. Reducing the load to 20 mA extends the battery life to 36 hours.

LT1580 CIRCUIT

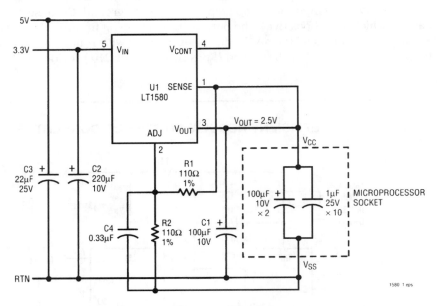

Figure 1. LT1580 Delivers 2.5V from 3.3V at Up to 6A

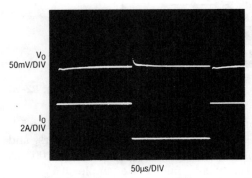

50µs/DIV

Figure 2. Transient Response of Figure 1's Circuit with Adjust-Pin Bypass Capacitor. Load Step Is from 200mA to 4 Amps

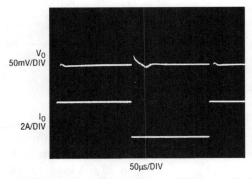

50µs/DIV

Figure 3. Transient Response Without Adjust-Pin Bypass Capacitor. Otherwise, Conditions Are the Same as in Figure 2

LINEAR TECHNOLOGY

Fig. 82-27

756

LT1580 CIRCUIT *(Cont.)*

Figure 1 shows a circuit designed to deliver 2.5 V from a 3.3-V source with 5 V available for the control voltage. Figure 2 shows the response to a load step of 200 mA to 4.0 A. The circuit is configured with a 0.33-μF ADJUST-pin bypass capacitor. The performance without this capacitor is shown in Fig. 3. The difference in performance is the reason for providing the ADJUST pin on the fixed-voltage devices. A substantial savings in expensive output decoupling capacitance can be realized by adding a small ceramic capacitor at this pin.

ONE-CELL CONVERTER

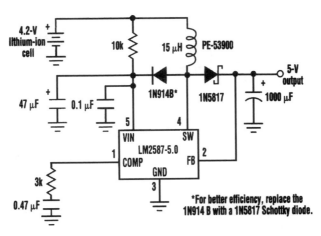

Fig. 82-28

The regulator generates a 5-V output from a single lithium-ion cell. Initially, the lithium-ion cell is at 4.2 V, which is greater than the 3.75-V undervoltage-lockout (UVL) limit of the LM2587 Simple Switcher. Once it starts up, the boost circuit will continue to regulate—even when the battery voltage drops lower than the UVL limit. The level that the input voltage can drop to depends on the maximum load current desired. Of course, the source can be any type of battery—three alkaline, four NiCd, or even two lithium batteries. When power is applied to the circuit, the 10-kΩ resistor charges up the 47-μF input capacitor and supplies the startup current to the LM2587. After startup, the LM2587 input and the input capacitor draw current from the switch node through the 1N914B bootstrap diode. When that happens, the IC input pin gets charged up to the output voltage minus a diode drop. In the circuit, with a 5-V output, the IC input voltage is 4.5 V after startup. It stays at 4.5 V—even when the input voltage drops to below 3.75 V. The input voltage can drop to 1.25 V when the regulator is supplying 250 mA. At 2.4 V, the regulator is 84 percent efficient if the 1N914B bootstrap diode is replaced with a 1N5817 Schottky diode.

5-V MICROPOWER LINEAR REGULATOR

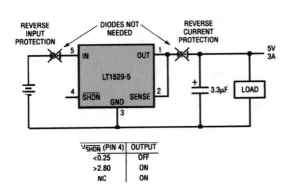

VSHDN (PIN 4)	OUTPUT
<0.25	OFF
>2.80	ON
NC	ON

Dropout Voltage

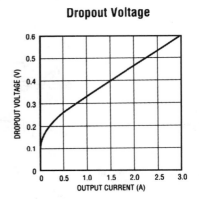

LINEAR TECHNOLOGY

Fig. 82-29

In battery systems where high load current is required intermittently and the system must remain in standby mode between uses, a linear regulator with micropower quiescent current and shutdown capability is needed. These situations arise in powering laptop disk drives, portable radio transmission modems, and peripheral motors, which are intermittently used. The LT1529 has 3 A current capability, 50 μA quiescent current, and just 16 μA shutdown current. As shown in the circuit and chart, the LT1529-5 is a low-component-count solution with a very low dropout voltage of 0.6 V at 3-A output current. This low dropout is ideal for battery-powered systems in which extracting the maximum energy from the battery is important. In dropout, the output voltage will decrease smoothly, following the input. The quiescent current of the LT1529 increases only slightly in dropout, unlike the situation with many other low-dropout PNP regulators.

83

Power-Supply Circuits—High Voltage

The sources of the following circuits are contained in the Sources section, which begins on page 1043. The figure number in the box of each circuit correlates to the entry in the Sources section.

LASER RECEIVER PHOTOMULTIPLIER TUBE SUPPLY

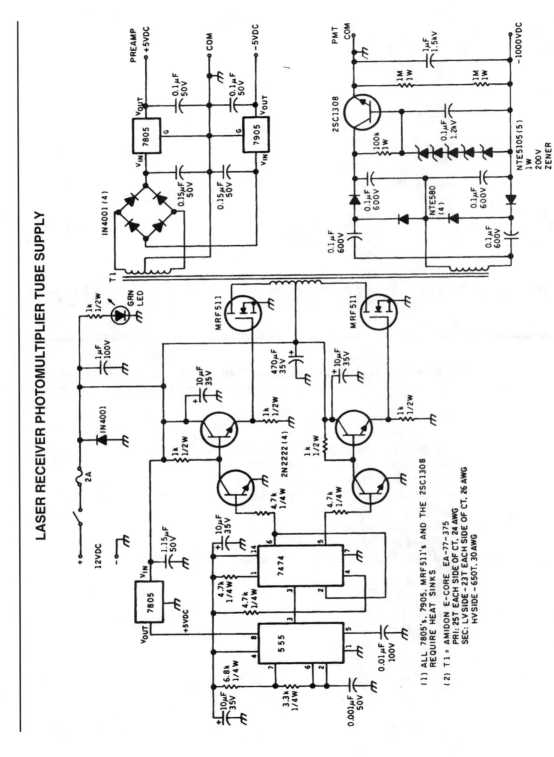

(1) ALL 7805's, 7905, MRF511's AND THE 2SC1308 REQUIRE HEAT SINKS

(2) T1 = AMIDON E-CORE EA-77-375
PRI: 25T EACH SIDE OF CT, 24 AWG
SEC: LV SIDE – 23T EACH SIDE OF CT, 26 AWG
HV SIDE – 650T, 30 AWG

Fig. 83-1

LASER RECEIVER PHOTOMULTIPLIER TUBE SUPPLY *(Cont.)*

The figure shows the circuit diagram of a regulated 1000-Vdc power supply driven from a 12-Vdc source. A low-voltage secondary regulated power source supplying power to the PMT video pre-amplifier is included. Again, follow the caution regarding PMT power supplies. All precautions must be observed when working with the PMT high-voltage supply.

HIGH-VOLTAGE POWER SUPPLY

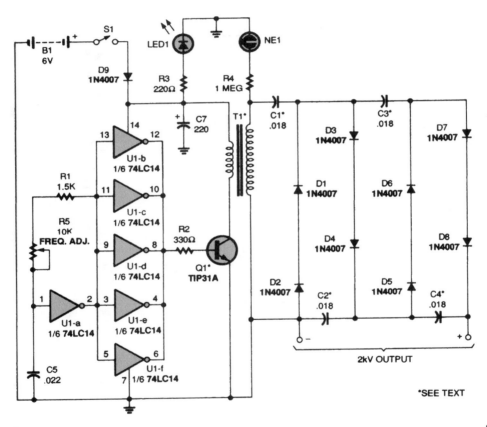

Fig. 83-2

This circuit uses a transformer to generate a high-voltage output (up to 20,000 V) from just four C cells. T1 can be a 50 to 100:1 turns ratio unit or can be made from a flyback transformer from a junked B/W TV.

HELIUM-NEON LASER SUPPLY WITH HV MULTIPLIER IGNITION

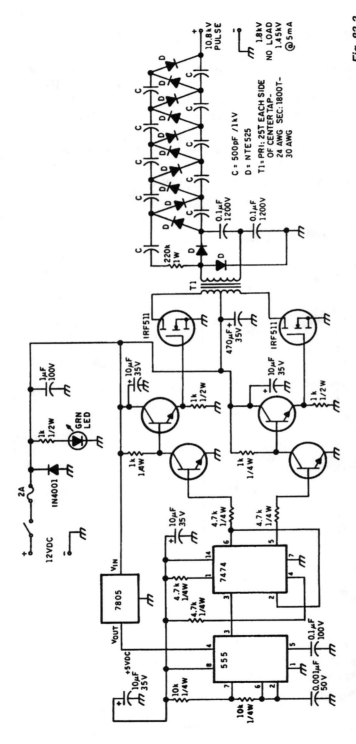

C = 500pF /1kV

D = NTE525

T1= PRI: 25T EACH SIDE
OF CENTER TAP–
24 AWG   SEC:1800T–
30 AWG

10.8kV
PULSE

1.8kV
NO LOAD
1.45kV
@5mA

Fig. 83-3

COMMUNICATIONS QUARTERLY

This figure illustrates a way in which a momentary 10-kV voltage pulse can be generated to initiate the plasma discharge across the laser tube. A Diode voltage-multiplier circuit is connected in series with the main supply and obtains its input power across one of the voltage-multiplier diodes in the main supply. The voltage across this diode is typically 1.8 kV p-p. With a ten-section multiplier, the output voltage is approximately 9 kV in series with a 1.8-kV sustaining supply. The capacitance value of the capacitors in the multiplier chain is much smaller than that of the capacitors for the main supply. When the power supply is first turned on, the capacitors in the voltage multiplier charge up to the ignition voltage. As the plasma forms, the laser tube draws more current and the multiplier capacitors cannot maintain their charge. As a result, the voltage immediately drops to that of the main sustaining supply, with all the diodes in the multiplier chain forward-biased. In this manner, an HV pulse is generated to ionize the gases.

GEIGER COUNTER 700-V LOW-CURRENT SUPPLY

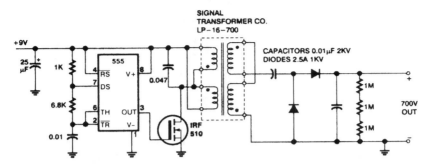

Fig. 83-4

This 700-V, 0.25-mA power supply converts a +9-Vdc input to 10-kHz ac, then uses the setup transformer and the voltage-doubler circuit to produce the required output.

REGULATOR FOR 700-V LOW-CURRENT SUPPLY

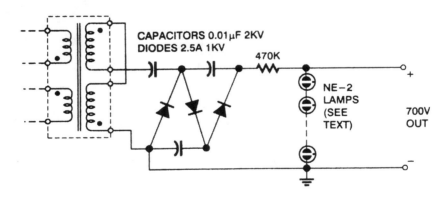

Fig. 83-5

If you need a regulated output, change the voltage doubler to a tripler, and use neon lamps as a regulator. The number of lamps you will need will depend on the characteristics of the lamps used, and will have to be found by experimentation. Each lamp has approximately a 55- to 70-V the breakdown voltage. The lamps should be shielded from light (or painted black) because light can influence the breakdown voltage.

HELIUM-NEON LASER POWER SUPPLY WITH HV PULSE IGNITION

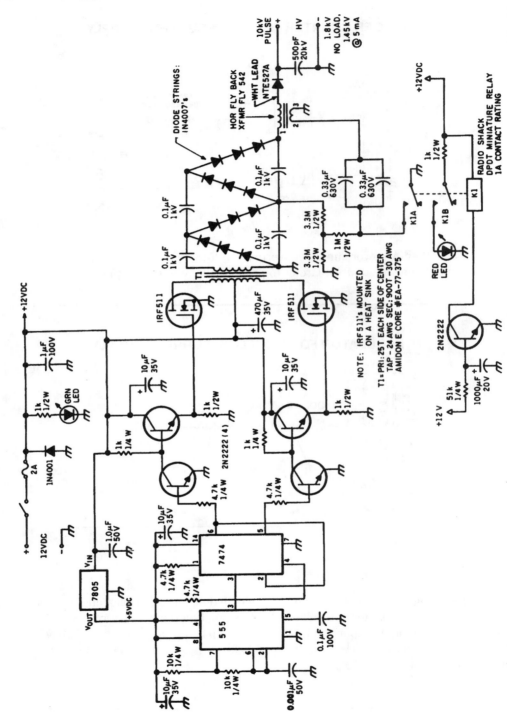

Fig. 83-6

HELIUM-NEON LASER POWER SUPPLY WITH HV PULSE IGNITION *(Cont.)*

The helium-neon laser requires two voltages: a voltage in the range of 10 kV that starts the laser and then turns off once the discharge begins, and a lower-voltage power supply to sustain the discharge. The circuit diagram illustrates one method of generating the laser ignition voltage. A fraction of the main supply voltage is used to charge up a capacitor. When triggered, it is discharged through a high-voltage ignition transformer in series with the main power supply output. When the supply is first turned on, a delay circuit allows the main supply to stabilize at full voltage and charge up the capacitor. A nonlatching relay operates to discharge the capacitor into the ignition coil on a one-shot basis after the timing delay. The HV diode in series with the ignition coil and the capacitor across the supply rectifies the damped oscillatory waveform out of the ignition coil, producing a single positive pulse across the laser tube to ionize the gas in the laser. Once the laser is ignited by the HV pulse, the main power supply maintains the discharge.

PHOTOMULTIPLIER TUBE SUPPLY

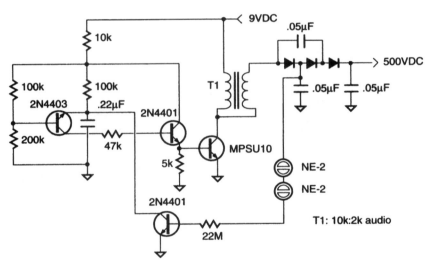

NUTS AND VOLTS

Fig. 83-7

This circuit is quite efficient, drawing virtually no current when the load is a photomultiplier (PM) used as a scintillation detector. Be sure that the voltage-divider resistors on the PM tube are very high to conserve power—the higher the better. If the high-voltage circuit uses a diode voltage-multiplier chain, then connect the chain to the transformer, just as the three-diode doubler is connected. Lower the regulated voltage by removing one of the neons or by selecting an appropriate zener. If a higher voltage is needed, simply extend the diode multiplier.

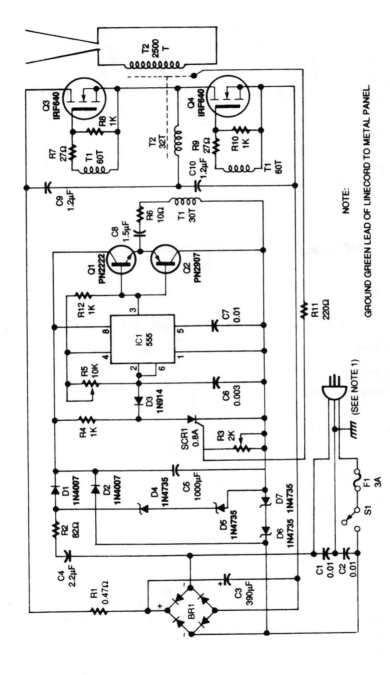

JACOBS LADDER HV SUPPLY

ELECTRONICS NOW

Fig. 83-8

NOTE:
GROUND GREEN LEAD OF LINECORD TO METAL PANEL

The operation of the Jacobs's Ladder depends on the current-limited power supply that delivers 12,000 V at 40 mA from the 120-Vac line. The full-wave bridge BR1 rectifies the 120-Vac line input, and resistor R1 limits the dc charging current of capacitor C3 to a safe value. The drive circuit power is obtained by dropping the 120-Vac line through capacitor C4 and current-limiting resistor R2. Capacitor C4 and resistor R2 present a complex impedance, so most of the ac line voltage is dropped across the reactance. The 555 timer IC is configured as a square-wave oscillator. The output frequency of its square waves is determined by the setting of trimmer potentiometer R5 and capacitor C6. The frequency is about 25 kHz for the values of R5 and C6 shown.

766

JACOBS LADDER HV SUPPLY *(Cont.)*

Potentiometer R5 can be used to adjust the circuit's power output. Increasing the frequency reduces the circuit's output by increasing the inductive reactance of the transformer leakage inductance. The output of IC1 on output pin 3 appears at the bases of the current "source," NPN transistor Q1, and "sink," PNP transistor Q2. The emitters of this transistor pair are ac-coupled through capacitor C8 and resistor R6 to drive the primary of driver-isolation transformer T1. This drive prevents dc from flowing through the primary. Transformer T1 is wound on a high-permeability core with as few turns as possible to eliminate leakage inductance. The gate circuits of MOSFETs Q3 and Q4 contain 27-Ω resistors (R7 and R9) to slow their switching times. This eliminates possible parasitic oscillations that could occur if the MOSFETs were switched at their speed limit. The primary of output transformer T2 contains 32 turns, but its secondary contains 2500 turns. The ratio of these turns is approximately 1 to 78. When this is multiplied by the rectified line voltage of 160 Vdc, an output of about 12,000 peak volts is obtained across the secondary. This 12,000-V output is the peak open-circuit voltage of the system, and it produces a short-circuit current of approximately 40 mA. This current is limited by the leakage inductance caused by the loose magnetic coupling between the primary and secondary circuits of transformer T2.

HV POWER SUPPLY

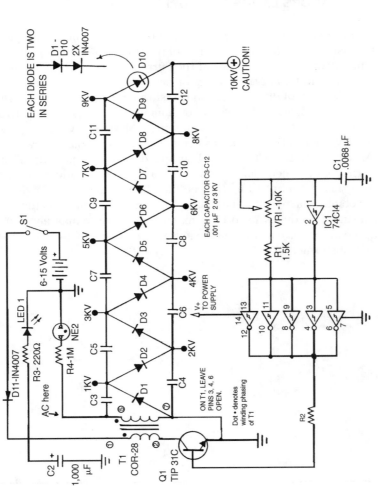

PARTS LIST

T1 — 12V to 1,000V step-up transformer, Allegro #COR-2B or equivalent

IC1 — Hex Schmitt Trigger integrated circuit, Allegro #74C14 or equivalent

D1-D10 — 1N4007 1-A, 1,000 PIV rectifier (two rectifiers in series for each listed, 20 in total)

D11 — 1N4007 1-A, 1,000 PIV rectifier

l1 — Ultrabright red LED

l2 — NE2 neon indicator lamp

C1 — .0068 uF, polyester or Mylar capacitor

C2 — 1,000 uF, 16 WVDC aluminum electrolytic capacitor

C3-C12 — .001 uF, 2 or 3 KVDC ceramic capacitor

R1 — 1,500 ohm, 5% carbon film resistor

R2 — See text — will be 220, 390, or 470 ohm depending upon supply voltage

R3 — 220 ohm, 5% carbon film resistor

R4 — 1 megohm, 5% carbon film resistor

VR1 — 10K potentiometer

F1 — 2A fast-blow fuse

S1 — SPST switch

1 — Special 14-pin integrated circuit socket w/integral bypass capacitor, Allegro #ED2124 or similar

1 — TO-3 heatsink and #6-32 mounting hardware

Fig. 83-9

NUTS AND VOLTS

This power supply uses a Cockcroft-Walton HV multiplier circuit to produce up to 10 kV from a 12-V source. Up to 5 W is available from this circuit.

NEON TUBE HIGH-VOLTAGE POWER SUPPLY

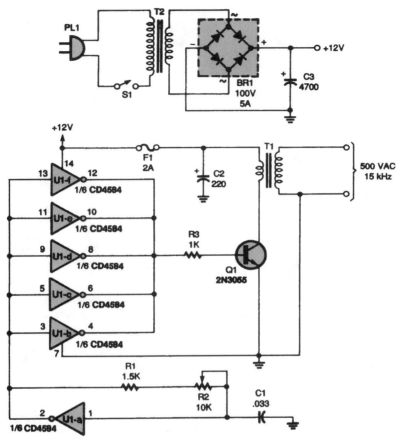

Fig. 83-10

Power for the circuit is supplied by a step-down transformer (T2) and a full-wave bridge recti-fier (BR1), which convert 117-Vac to 12-Vdc. Capacitor C3 acts as a filter. The heart of the circuit is a Schmitt-trigger hex inverter, U1. Capacitor C1, resistor R1, and potentiometer R2 are connected to one section of the inverter, U1-a, to form an oscillator that runs at approximately 15 kHz (because U1 is a Schmitt trigger, the output square wave is very clean). Adjusting R2 varies the frequency. The output of the oscillator is fed into the remaining five inverters (U1-b through U1-f), which are con-nected in parallel to create a buffer that is capable of increasing the drive current of the oscillator. The square-wave output is used to drive the switching transistor (Q1), which, in turn, switches the primary windings of the ferrite-core transformer (T1). Approximately 500-Vac at 15 kHz are output from the secondary of the transformer. T1 is made from a ferrite pot core. Wind 500 turns of #30-gauge wire for the secondary. Coat it with insulating varnish and add a layer of insulating tape, over which a 20-turn winding of #22 wire is wound to form the primary winding.

84

Power-Supply Circuits—
Multiple Output

The sources of the following circuits are contained in the Sources section, which begins on page 1043. The figure number in the box of each circuit correlates to the entry in the Sources section.

DUAL SUPPLY

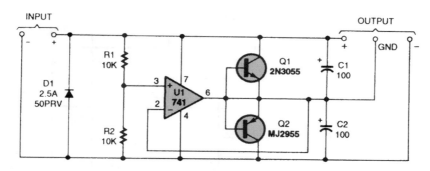

POPULAR ELECTRONICS

Fig. 84-1

The key idea is that in a dual-supply powered circuit, "ground" is midway between the positive and negative supply voltages. Unfortunately, ordinary voltage regulators can't do the job. The reason is that they can only source current, not sink it, and the ground terminal in a dual supply might have to either source or sink current, depending on which half of the load is drawing more current at the time. In the circuit, R1 and R2 divide the input voltage in half. Op amp U1 reproduces that voltage at the "ground" output terminal by making either Q1 or Q2 conduct as much as necessary. Capacitors C1 and C2 hold the output voltage steady when the load changes suddenly. Diode D1 is present to protect the circuit's input from reversed polarity by blowing the fuse of the main power supply. If your power supply has no fuse, place D1 in series with the circuit. That will afford the same protection with only slightly reduced regulation and voltage output.

SPLIT 12-V POWER SUPPLY

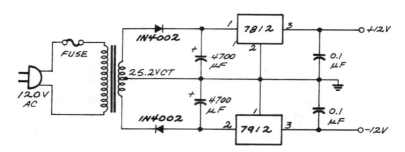

ELECTRONICS NOW

Fig. 84-2

Two IC regulators and a bridge circuit supply ±12 V.

DUAL-OUTPUT VOLTAGE REGULATOR

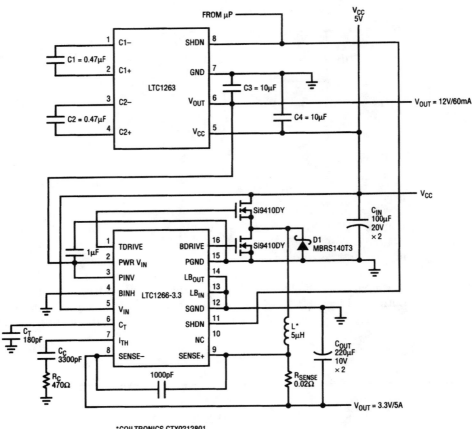

*COILTRONICS CTX0212801

LINEAR TECHNOLOGY

Fig. 84-3

The LTC1266-3.3 and the LTC1263 are perfect complements for one another. The combination of the two parts provides two regulated outputs of 3.3 V/5 A and 12 V/60 mA from an input range of 4.75 to 5.5 V. The LTC1263, using only four external components (two 0.47-μF charge capacitors, one 10-μF bypass capacitor, and one 10-μF output capacitor), generates the regulated 12-V/60-mA output from a 5-V input using a charge-pump tripler. During every period of the 300-kHz oscillator, the two charge capacitors are first charged to V_{cc} and then stacked in series, with the bottom plate of the bottom capacitor shorted to V_{cc} and the top plate of the top capacitor connected to the output capacitor. As a result, the output capacitor is slowly charged up from 5 to 12 V. The 12-V output is regulated by a gated oscillator scheme that turns the charge pump on when V_{out} is below 12 V and turns it off when it exceeds 12 V. The LTC1266-3.3 then uses the 5-V input, along with the 12-V output from the LTC1263 and various external components, including bypass capacitors, sense resis-

DUAL-OUTPUT VOLTAGE REGULATOR *(Cont.)*

tors, and Schottky diodes, to switch two external N-channel MOSFETs and a 5-μH inductor to charge and regulate the 3.3-V/5-A output. The charging scheme for this part, however, is very different from that of the LTC1263. The LTC1266-3.3 first charges the output capacitor by turning on the top N-channel MOSFET, allowing current to flow from the 5-V input supply through the inductor. The amount of current flow in the inductor is monitored with a sense resistor, and the 3.3-V output is regulated by turning on and off the top and bottom N-channel MOSFETs to charge and discharge the output capacitor.

±15-V 100-mA DUAL POWER SUPPLY

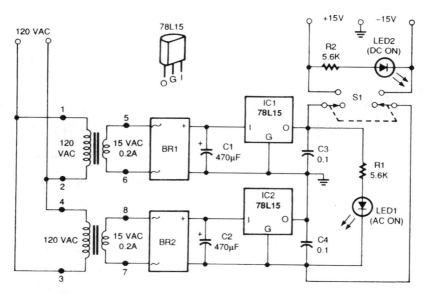

ELECTRONICS NOW

Fig. 84-4

A simple supply such as this is useful for many small projects, op amp circuits, etc.

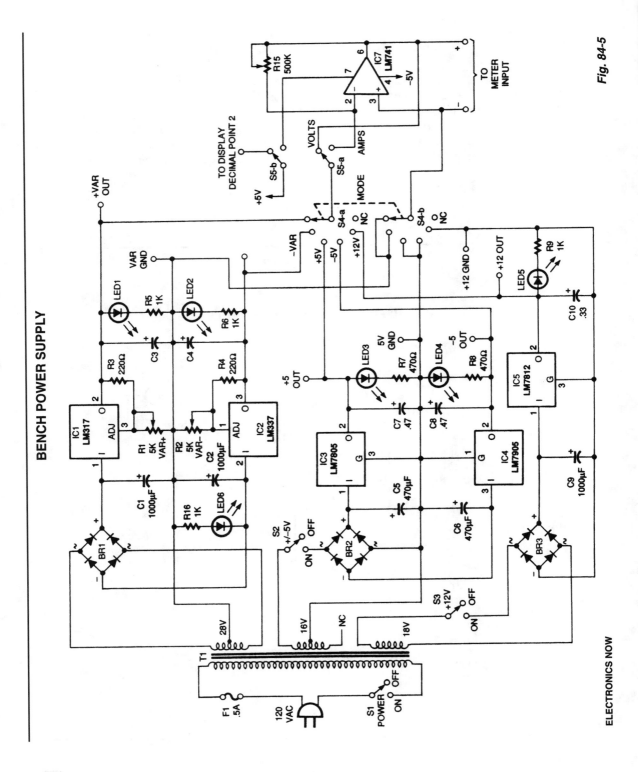

BENCH POWER SUPPLY

Fig. 84-5

ELECTRONICS NOW

BENCH POWER SUPPLY *(Cont.)*

The ac voltage from transformer T1 is rectified by bridges BR1, BR2, and BR3 and filtered by capacitors C1, C2, C5, C6, and C9. Voltage regulators IC1 to IC5 reduce the voltage to the desired fixed or variable levels. The LM317 regulator, IC1, provides a positive variable output from 1.2 to 28 Vdc, and an LM337, IC2, provides a negative variable output with the same range. The LM7805 regulator, IC3, supplies a fixed +5 V, and the LM7905, IC4, supplies a fixed −5 V. The LM7812 regulator, IC5, supplies a fixed +12 V. Capacitors C3, C4, C7, C8, and C10 improve transient response and prevent oscillation. Resistor networks R1 through R3 and R2 through R4 for IC1 and IC2, respectively, provide the necessary feedback to obtain the variable output voltages. An LED and current-limiting resistor are wired across each output to indicate when each output voltage is present. The main power indicator for the entire unit consists of LED5 and R18. Switch S1 controls ac power to the transformer primary, and switches S2 and S3 connect the secondary voltages to the 5- and 12-V regulator circuits. The ±5-V supply powers the voltage meter and display circuitry, so this section must be turned on to power the output meter. Switch S4 is a two-pole, six-throw (2P6T) rotary unit. Pole A switches the positive input to the voltmeter and pole B switches the ground input. (Three separate grounds are in the circuit.) Pole A of S4 is connected to one pole of DPDT switch S5 so that the signal can be routed directly to the meter for voltage readings or to the LM741 current-to-voltage converter IC7 for current readings. The second pole of S5 is connected to the +5-V supply. This pole is switched to the second decimal point of the display in the Voltage mode or to pin 7 of the LM741 in the Current mode.

DUAL-POLARITY LOW-CURRENT POWER SUPPLY

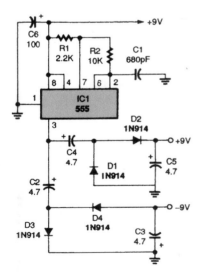

The 555 is operating in a free-running oscillator at about 120 kHz, producing a near-9-V square-wave output at pin 3. Diodes D1 and D2, along with capacitors C4 and C5, make up the rectifier circuit for the positive supply, while D3, D4, C2, and C3 make up the rectifier for the negative supply.

POPULAR ELECTRONICS   *Fig. 84-6*

POWER SUPPLY FOR ±12 V AND ±5 V

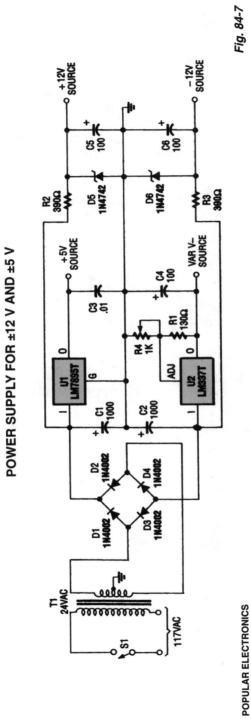

POPULAR ELECTRONICS

Fig. 84-7

This supply is suitable for small projects that need up to 30 mA at ±12 V and 0.5 A at ±5 V.

HANDY HOBBY POWER SUPPLY

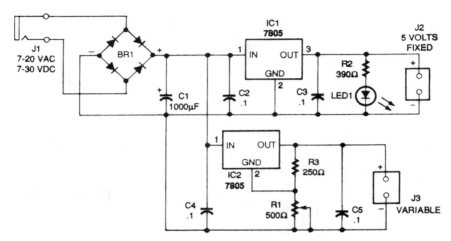

Fig. 84-8

The input voltage to the power supply at mini phono jack J1 must be from 7 to 20 Vac or from 7 to 30 Vdc. You can use any transformer or ac-to-dc wall adapter that meets those input requirements. An ac input at J1 is rectified by bridge rectifier BR1; a dc input passes through half of the rectifier unmodified, except that its value drops by the sum of two diode voltage drops. Two MC7805 5-V regulators are in the circuit: IC1 provides a fixed 5-Vdc output at J2, while IC2 has a variable dc output at J3. The variable output ranges from +5 Vdc to 2 V less than the input voltage to the power supply. The output of the fixed regulator is made variable by varying the voltage at pin 2 with potentiometer R1. (Pin 2 is normally grounded to produce the fixed-voltage output.) Each voltage regulator can safely handle up to 1 A of current, provided that the transformer or power adapter can handle the demand and that the regulator is properly heatsinked. The voltage regulators must be heatsinked if more than a few milliamperes is to be drawn from the supply. Power indicator LED1 is connected across the fixed 5-V output; it lights whenever the supply is powered.

17-W, 5-V AND 3.3-V POWER SUPPLY

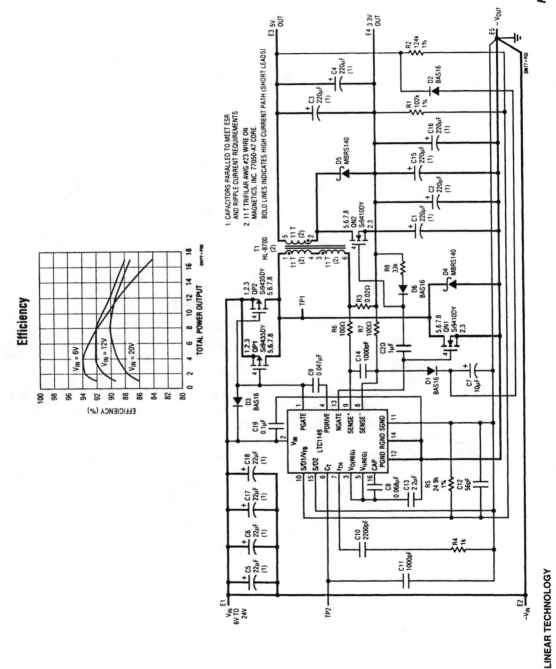

Efficiency

Fig. 84-9

LINEAR TECHNOLOGY

17-W, 5-V AND 3.3-V POWER SUPPLY *(Cont.)*

The low concerns of today's notebook computer designs do not have to restrict designers to low-performance solutions. The logic power supply is usually required to furnish excellent regulation at extremely high efficiency, but also carry a small price tag. This LTC1149 circuit addresses these requirements by using a single synchronous rectifier controller IC to generate both 3.3- and 5-V logic supplies. The 3.3-V output is generated in normal step-down regulator fashion using synchronous switches and an inductor. The key to circuit operation is the extra winding on the inductor and its corresponding synchronous rectifier, QN2. These provide the 5-V output by using transformer action. Cross regulation with this circuit is excellent because of the use of a trifilar-wound inductor and a "split feedback," which combines both outputs into the same feedback network. Efficiency exceeds 90 percent over most of the operating range. In addition, the 3.3- and 5-V outputs are inherently synchronous in switching frequency, and they will reach their rated voltage at the same time after power-up. Another bonus is that is if one logic supply is short-circuited, the other output will be disabled.

SWITCHABLE LINEAR VOLTAGE REGULATOR

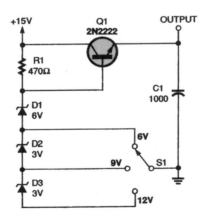

POPULAR ELECTRONICS

Fig. 84-10

This three-voltage regulated supply can be added to for more voltage ranges.

5- AND 15-V DUAL-POLARITY SUPPLY

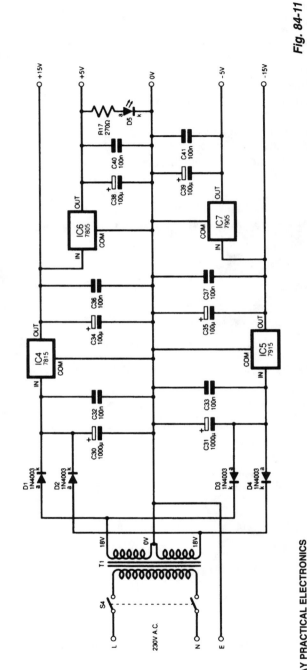

EVERYDAY PRACTICAL ELECTRONICS

Fig. 84-11

This supply will provide 5- and 15-V positive and negative voltages. It was used to power a MAX038 function generator, but it has other applications as well.

FIVE-OUTPUT CONVERTER

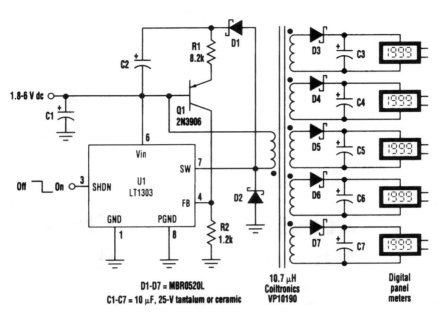

D1-D7 = MBR0520L
C1-C7 = 10 µF, 25-V tantalum or ceramic

10.7 µH
Coiltronics
VP10190

Digital
panel
meters

ELECTRONIC DESIGN

Fig. 84-12

Digital panel meters (DPMs) make excellent displays for instruments and test equipment, but they suffer from one major flaw: They require a floating power supply, usually in the form of a 9-V battery. This circuit powers up to five meters from a single 1.8- to 6-V source. Each of the five outputs is fully floating, isolated, and independent in every respect. The converter is based on a flyback design, and it uses a micropower, high-efficiency regulator (LT1303) and an off-the-shelf, surface-mount inductor. The coil has six identical windings and is high-voltage-tested to 500 V rms—more than adequate isolation for the application. Operation is as follows: Feedback is extracted from the primary by Q1, which samples the flyback pedestal during the switch-off time. Typical DPMs draw approximately 1-mA supply current. The primary also is loaded with 1 mA for optimum regulation and ripple. Snubbing components, a necessity in most flyback circuits, are obviated by the action of C1 and C2. The converter also can be used with a battery. In this case, a sixth panel meter can be powered by the primary across C2. All that is required to balance the load current is to increase R_1 and R_2 by a factor of 10. Although this circuit is set up for 9-V output (actually 9.3 V), some DPMs need 5 or 7 V. As a result, a 4.3- or 6.2-kΩ resistor should be used in place of R1 for these voltages. The output voltage is set by $R_1 = (V_{out} - 0.7)/1$ mA. If any outputs aren't needed, omit the associated components and parallel the unused winding with the primary, observing the phasing. With each output loaded at 1 mA, the input current is 16.5 mA on a 5-V supply. This figure rises to about 45 mA on a 1.8-V (two-cell) input.

85

Power-Supply Circuits— Transformer-Coupled

The sources of the following circuits are contained in the Sources section, which begins on page 1043. The figure number in the box of each circuit correlates to the entry in the Sources section.

BASIC FORWARD-CONVERTER CIRCUIT

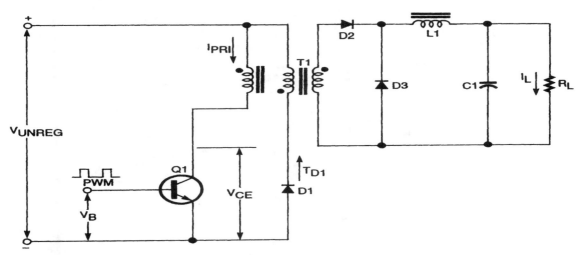

Fig. 85-1

The figure is a simplified schematic for a transformer-coupled forward converter. Based on the buck converter, it is similar to the flyback converter. Inductor L1, rather than transformer T1, stores the energy. When switching transistor Q1 turns on, current builds in the primary winding of transformer T1, storing energy. Because the secondary winding has the same polarity as the primary winding, energy is transferred forward to the output. Energy is also stored in inductor L2 through forward-biased diode D2. At this time, flywheel diode D3 is back-biased. When Q1 turns off, the transformer winding voltage reverses, back-biasing diode D2. Diode D3 becomes forward-biased, causing current to flow through R_L and delivering energy to the load through inductor L1. The third winding and diode D1 return transformer T1's magnetic energy to the dc input when Q1 is off.

FEEDBACK CONTROL FOR MAX253

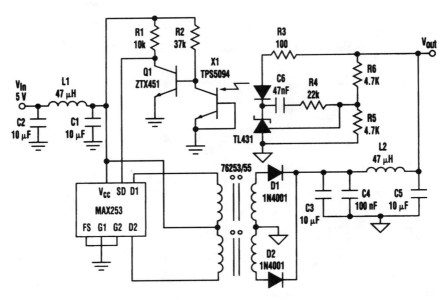

ELECTRONIC DESIGN

Fig. 85-2

The MAX253 and associated 76253 transformers were designed to allow easy implementation of simple dc-to-dc converter circuits. The main problem with the basic circuit is that there is no feedback, resulting in no control over the output voltage at light loads. When transformer isolators designed for 1-W power levels are used, output voltages can be observed rising above 5.5 V at less than 20-percent load. The circuit provides feedback control from the output via an optoisolator and voltage-reference device in a single pack (TPS5094). This results in an output voltage that is static at light loads. The circuit uses the internal TL431 reference devices of the TPS5904 to set a value at which the optoisolator is switched (V_{ref}=2.51 V). Using two 4.7-kΩ resistors (R5 and R6) gives an ideal output voltage of 5 V; with only 500 μA of drain current, the measured value was 4.95 V. The output of the optoisolator then controls the shutdown pin (SD) on the MAX253, via the ZTX451 buffer transistor, and switches the device off when the output voltage exceeds the predefined value. Therefore, the circuit works in a burst mode at light loads, with the switching being turned on and off as necessary. As the load increases, the device eventually reaches the stage where switching is occurring all the time.

BASIC PUSH-PULL CONVERTER CIRCUIT

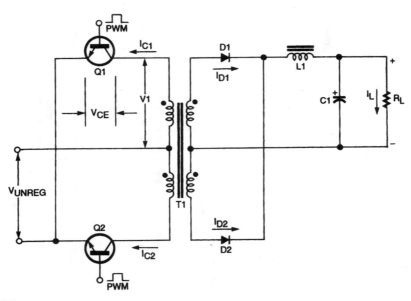

ELECTRONICS NOW

Fig. 85-3

The push-pull converter consists of two forward converters working 180° out of phase. Both halves of the push-pull converter deliver current to the load at each half cycle, and the inductor stores energy. Diodes D1 and D2 conduct simultaneously between transistor conduction, effectively short-circuiting the secondary isolation transformers. Acting as flywheel diodes, these diodes deliver useful power to the output.

5-V SUPPLY

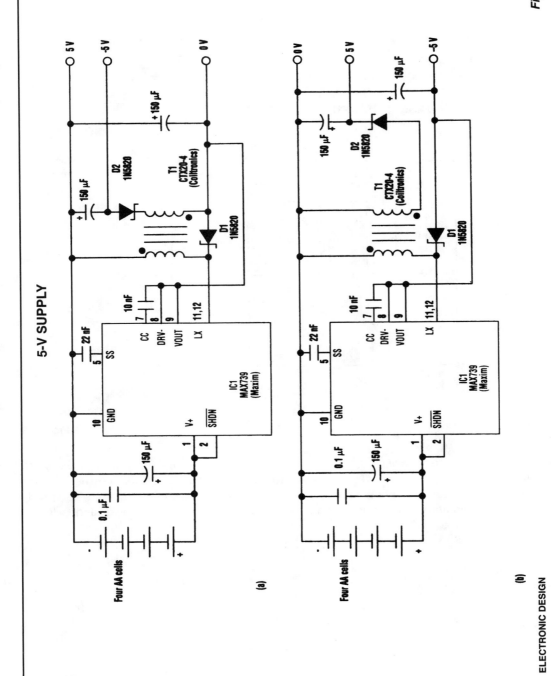

(a)

(b)

Fig. 85-4

ELECTRONIC DESIGN

5-V SUPPLY *(Cont.)*

The inverter circuit substitutes a transformer with two matched windings for the usual inductor (see figure a). When IC1's internal switch turns off, the circuit impresses V_{out} plus a diode drop across each winding. With the reference connection properly chosen, as shown, the second (right-hand) winding can generate an additional supply voltage (-5 V, in this case). V_{out} (pin 8) is the feedback connection. For stability, the regulated output (5 V, in this case) should have the heavier load. It usually does because the negative rail in most systems is only a bias supply. But if the system demands more load current from the -5-V output, the second winding should be reconnected to produce the 5-V output (see figure b). The transformer should have side-by-side bifilar windings for best coupling. The $V-$ value (nominally -5 V) depends on load currents and the transformer turns ratio (which can deviate from 1:1). Loads of 5 to 50 mA at 5 V, for example cause a $V-$ change of less than 300 mV—less than that expected from a charge pump. When unloaded, $V-$ increases because of rectification of the ringing that occurs when D2 turns on.

BASIC HALF-BRIDGE CONVERTER CIRCUIT

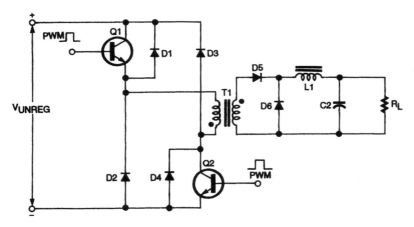

Fig. 85-5

A popular and established switching power-supply topology is the half-bridge converter. The same converter will work from either a 120- or 240-Vac input; it is simply necessary to change terminals. It is cost-effective over the 150- to 500-W range, and offers very good output noise characteristics and excellent transient response.

V_{OUT} BOOSTER

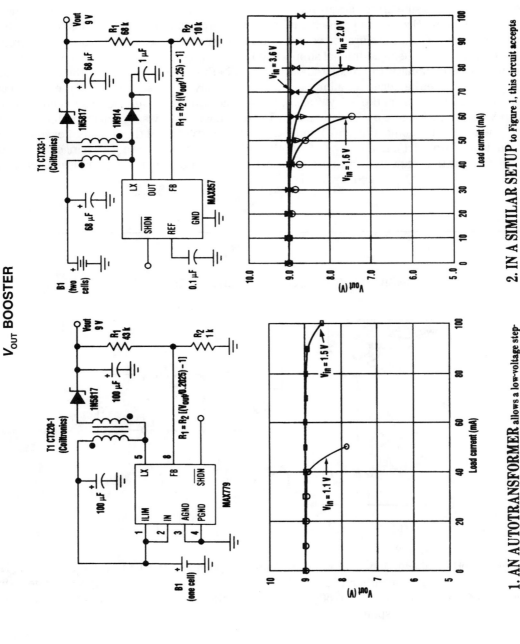

1. AN AUTOTRANSFORMER allows a low-voltage step-up converter to boost single-cell inputs as high as 10 V.

2. IN A SIMILAR SETUP to Figure 1, this circuit accepts two-cell inputs and generates regulated outputs as high as 10 V.

ELECTRONIC DESIGN

Fig. 85-6

788

V_{OUT} BOOSTER *(Cont.)*

Step-up dc-to-dc converters that operate from small input voltages often possess correspond-ingly low maximum breakdown voltages of 5 to 6 V. This limits the maximum output voltage available from such devices. However, by adding an autotransformer, the output voltage (V_{out}) can be doubled without exceeding the IC's breakdown voltage. A properly wound center-tapped inductor acts like a transformer with a 1:1 turns ratio. Combined with an IC that typically boosts single-cell inputs as high as 6 V, it produces a regulated 9-V output with no more than 4.5 V across the IC (Fig. 1). The circuit can be applied in smoke alarms as well as in other battery-operated equipment. It delivers an output of 30 mA at 9 V from a 1.1-V input, and as much as 90 mA at 9 V from a 1.5-V input. A similar circuit setup for two-cell inputs delivers 30 mA at 9 V from 1.6 V, and a current of 80 mA at 9 V from 3.6 V (Fig. 2).

BASIC FULL-BRIDGE CONVERTER CIRCUIT

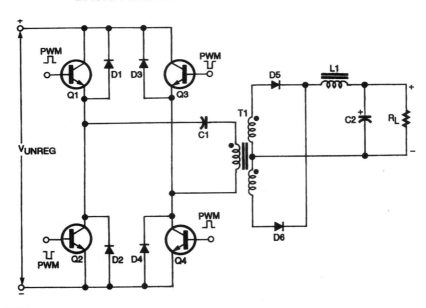

ELECTRONICS NOW

Fig. 85-7

Today, most switching power supplies rated for more than 500 W are variations on the full-bridge converter topology shown. This design has four transistors; because diagonally opposite transistors are on at the same time, each transistor must have an isolated base drive. Full-bridge converters are usually manufactured as enclosed modules for such applications as powering mainframe computers and supercomputers.

STEP-UP/STEP-DOWN REGULATOR

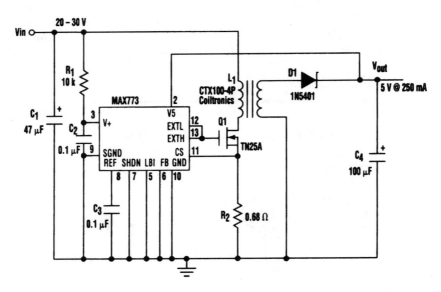

Fig. 85-8

Adding a transformer to a step-up dc-to-dc regulator allows inputs of 20 V and higher to be accepted while operating in a step-down mode. The circuit handles inputs to 30 V, but is easily modifiable for high specific voltages. The transformer's 1:1 turns ratio simplifiers procurement by allowing the use of a standard product (here, the Coiltronics CTX100-4P). Its 1:1 ratio also enhances stability by producing a duty cycle well below 50 percent. An ideal 1:1 transformer would generate $V_{in} + V_{out}$ at the bottom of the primary, but real transformers produce somewhat higher voltages. That voltage appears across Q1, so Q1's minimum breakdown voltage should be approximately $2V_{in} + V_{out}$. R2 limits the peak current (through Q1 and L1) to 0.33 A. The internal shunt regulator is a zener diode that is biased by R1 at approximately 2 mA. To cope with a wide range of input voltages, R1 can be replaced with a constant-current source.

2- TO 6-V CCFL POWER SUPPLY

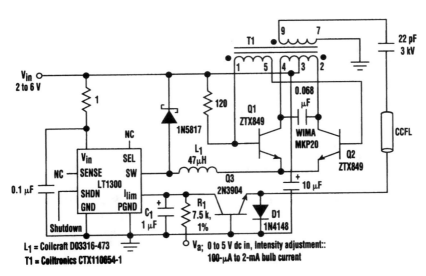

Fig. 85-9

The power supply shown operates from 2 to 6 V. It can drive a small (75-mm) CCFL over a 100-μA to 2-mA range. An LT1301 micropower dc-to-dc converter is used in conjunction with a current-driven Royer-class converter, consisting of T1, Q1, and Q2. When power is applied along with intensity-adjust voltage V_a, the LT1301's I_{lim} pin is driven slightly positive, causing maximum switching current through the IC's internal switch pin (SW). L1 conducts current that flows from transformer T1's center tap, through the transistors, into L1. L1's current is taken in switched fashion to ground by the regulator's action. The Royer converter oscillates at a frequency primarily set by T1's characteristics (including its load) and the 0.068-μF capacitor. LT1301 drives L1, which sets the magnitude of the Q1-Q2 tail current, and thus creates T1's drive level. The 1N5817 diode maintains L1's current flow when the LT1301's switch is off. The 0.068-μF capacitor combines with L1's characteristics to produce sine-wave voltage drive at the Q1 and Q2 collectors. T1 provides voltage step-up and about 1400 V p-p appears at the transformer's secondary.

LCD CONTRAST SUPPLY

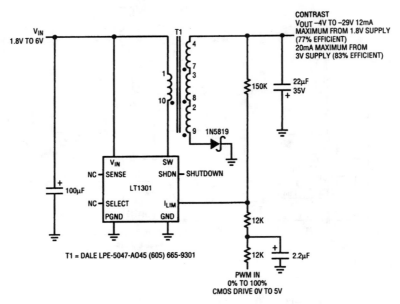

LINEAR TECHNOLOGY

Fig. 85-10

The LT1301 circuit shown generates a negative contrast voltage that may be varied from −4 to −29 V based on a PWM control signal generated by the system microprocessor. The LT1301 is a micropower switching regulator that needs only 120 μA quiescent current and can be shut down to draw just 10 μA. The input supply can range from 1.8 to 6 V, making it ideal for operation from a portable battery supply. Transformer T1 is a standard four-winding transformer with independent terminal connections for each winding. Three windings are connected in series to create a 1:3 turns ratio flyback transformer. Voltage feedback is applied to the current limit input I_{LIM} to control operation of the oscillator. A logic-level input, such as PWM generated at a microprocessor output pin or PWM circuit, sets the output voltage. As indicated, the efficiency is very good, even at low input voltages.

LCD AUXILIARY BIAS CIRCUIT

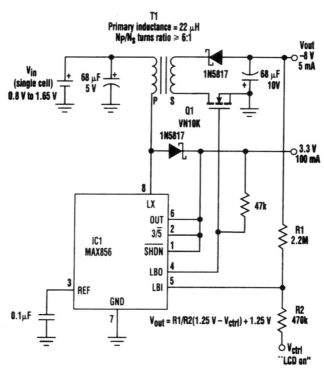

ELECTRONIC DESIGN

Fig. 85-11

This single-cell boost converter can generate the low supply voltages commonly needed in pagers and other portable instruments with small, graphic LCDs. The first is a regulated 3.3 V at 100 mA, and the other is a regulated negative output, suitable for use as an LCD bias voltage. The overall efficiency is about 80 percent. The main 3.3-V supply is provided by a boost converter (IC1). The auxiliary bias voltage is provided by an extra flyback winding (the T1 secondary), and is regulated via Q1 and the low-battery detector internal to IC1. As the battery discharges, its declining terminal voltage causes a decline in the voltage in the flyback winding. At minimum battery voltage (0.8 V), the T1 primary "sees" 3.3 V−0.8 V=2.5 V, so the 6:1 turns ratio produces 6(2.5)=15 V in the secondary. At maximum battery voltage (1.65 V), the primary sees only 1.5 V, producing 9.9 V in the secondary. MOSFET Q1 stabilizes this output by interrupting the secondary current, introducing the regulation necessary to generate a constant negative output. The regulator uses IC1's low-battery detector (a comparator/reference combination) as an on/off controller for Q1. In this circuit, the R1/R2 divider holds LBI between V_{ctrl} (normally 3.3 V) and the LCD bias output (normally −8 V). R_1 and R_2 are chosen so that LBO turns off when the LCD bias becomes too negative (and pulls the LBI voltage below 1.25 V). Load current then causes the LCD bias to drift upward (toward 0 V) until LBI exceeds 1.25 V, which causes Q1 to turn on again. A logic signal at the "LCD ON" terminal provides a means to enable and disable the negative output.

5- AND 3.3-V SUPPLY

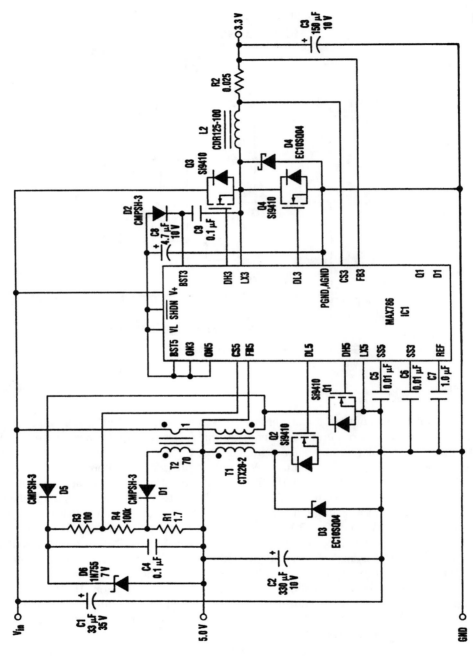

Fig. 85-12

ELECTRONIC DESIGN

5- AND 3.3-V SUPPLY *(Cont.)*

The IC shown is popular for generating 5 and 3.3 V because it includes two controllers that are highly efficient (typically >90 percent). However, the IC has a step-down (buck) topology that usually can't generate voltages equal to or higher than V_{in}. A four-cell NiCd or NiMH battery, for example, presents a problem because its terminal voltage can be above or below 5 V, depending on the state of its charge. This problem can be solved by designing in a flyback transformer, which allows V_{in} to range from 4 to 7 V. To ensure a proper gate drive to the external switching MOSFET (Q1), LX5 should be connected to ground and BST5 to the internal 5-V supply (V_L) as shown. When Q1 turns on, the T1 primary current increases and stores energy in the T1 core. When Q1 turns off, the synchronous-rectifier MOSFET Q2 turns on and enables current flow to the 5-V output. For flyback circuits, I_{out} flows only when the rectifier conducts. Yet, IC1 is a current-mode buck regulator for which I_{out} must be sensed while Q1 is on. The current-sense transformer (T2), therefore, measures the T1 primary current when Q1 is on, steps down the result with a 70:1 turns ratio, and develops a voltage across resistor R1. To ensure that synchronous rectifier Q2 remains on while Q1 is off, a simple charge pump (C4 and D5) and voltage divider (R3 and R4) provide a slight offset to the current-sense signal. Thus, Q2 remains on because the IC does not detect zero output current. V_{out} is regulated to 5 V, ±5 percent, and the maximum I_{out} is 1 A over the entire V_{in} range.

SWITCHING REGULATOR WITH TRANSFORMER-ISOLATED FEEDBACK

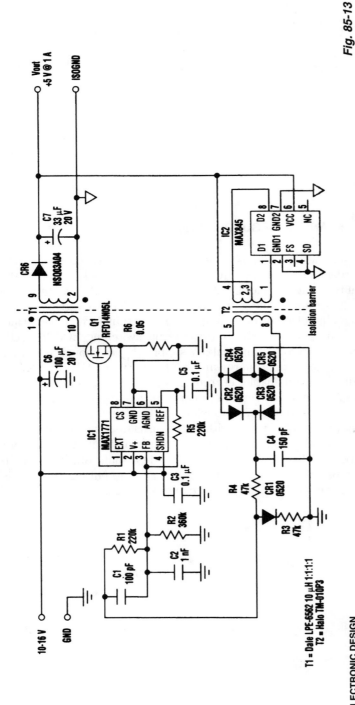

Fig. 85-13

ELECTRONIC DESIGN

The zero (nonexistent) line regulation of a push-pull, surface-mount transformer and driver (T2 and IC2) produces an isolated feedback signal (to pin 3 of IC1) proportional to the regulator's nominal 5-V output. What results is a fully isolated dc-to-dc converter without the bandwidth constraints and aging characteristics associated with an optoisolator. By alternately grounding each end of T2's center-tapped primary, the transformer driver (IC2) generates an ac signal proportional to the desired 5-V feedback voltage. A diode bridge (CR2 to CR5) and a capacitor (C4) convert this transformer's output to dc, and a diode-resistor network (CR1, R3, and R4) compensates for the temperature coefficient of the diode bridge. The result is a zero-T_C voltage slightly less than $\frac{1}{2}V_{out}$. In response to a 5.00-V output, the feedback network produces an isolated 2.404 V (at IC1, pin 3) and introduces about 250 ns of delay at 100 kHz—the equivalent of 9° of phase shift. This bandwidth is sufficient for the control loop in most

796

SWITCHING REGULATOR WITH TRANSFORMER-ISOLATED FEEDBACK *(Cont.)*

switching converters. Supply current for IC2 and the temperature-compensation network together is about 6 mA. Starting with a 5-V, nonisolated, transformer flyback converter in which V_{out} connects directly to the top of C1 and R1, you can insert the isolated-feedback circuit (bottom of the figure) between V_{out} and C_1/R_1. The only modification needed to accommodate this extra isolated-feedback circuit is to reduce the value of R_1, which ensures that the R1/R2 divider voltage is comparable to IC1's internal feedback reference (1.5 V). Performance of the isolated converter is virtually identical to that of the nonisolated converter, except for the power consumed by the isolated-feedback circuit. T2 provides an isolation of 500 V rms (transformers with 1500 V rms also can be obtained).

BASIC FLYBACK CONVERTER CIRCUIT

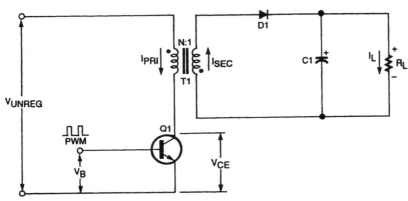

Fig. 85-14

    The figure shows a flyback converter, a variation on the boost regulator with a single switching transistor Q1 that eliminates the input inductor L1. This kind of off-line flyback switcher typically includes a bridge rectifier that converts the 120-Vac line voltage to 150 Vdc for the switching section. Current increases at a linear rate through the primary winding of transformer T1, which behaves like an inductor by storing energy in its core. As soon as Q1 cuts off, the magnetic field begins to collapse, and the winding polarities reverse. During the second half (flyback period), when Q1 is off, the energy is transferred to the secondary of transformer T1, charging capacitor C1 and feeding the output load. A PWM loop controls Q1's conduction by comparing output voltage to a set reference. If the load demands more current, the ON time is increased; if it demands less current, the ON time is decreased.

SNUBBER NETWORK ENERGY SAVER

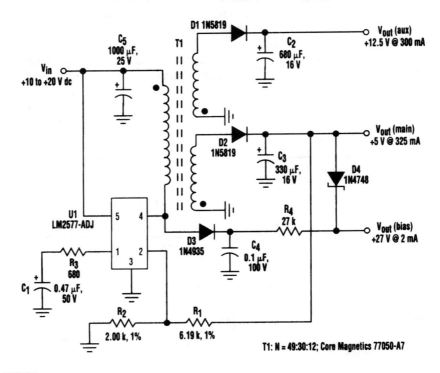

T1: N = 49:30:12; Core Magnetics 77050-A7

ELECTRONIC DESIGN

Fig. 85-15

A flyback regulator offers the advantage of providing multiple output voltages with a single magnetic structure. Therefore, it is very compact and cost-effective. This particular circuit has a main +5-V output and a +12.5-V auxiliary output. The device being driven also requited a "bias" voltage of +27 V with a few milliamperes of current. The heart of the regulator is a National Semiconductor LM2577-ADJ "simple switcher" controller IC, with resistors R1 and R2 providing the feedback for the main +5-V output. The auxiliary +12.5-V output is regulated by the intrinsic tight coupling of a discontinuous-mode flyback topology. R3 and C1 are compensation devices. Although another winding could have been used in the transformer to provide the +27-V bias output, a "free" output can be realized from the transfer of the voltage spikes in the primary winding to the reservoir capacitor (C4) via diode D3. The charge in the capacitor is drawn by the current of both the bias load and the shunt zener regulator (D4). Enough charge is depleted from the capacitor to allow the next voltage spike to almost fully dump its energy in the next cycle. In a sense, this is a modified snubber network in which the energy is being put to good use, instead of being wasted as heat on a resistor.

86

Power-Supply Circuits—
Variable Output

The sources of the following circuits are contained in the Sources section, which begins on page 1043. The figure number in the box of each circuit correlates to the entry in the Sources section.

Variable-Voltage Source
Experimenter's Power Supply
Dc Power Supply
Simple Adjustable Dc Supply
7.5-A 12- to 16-Vdc Regulated Power Supply

VARIABLE-VOLTAGE SOURCE

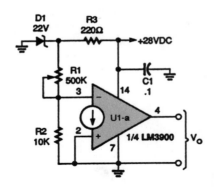

POPULAR ELECTRONICS

Fig. 86-1

This variable-voltage source can provide between 0.6 and 20 V, depending on the setting of R1.

EXPERIMENTER'S POWER SUPPLY

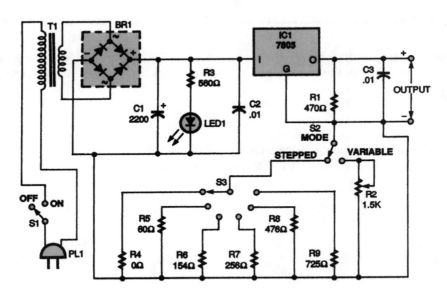

POPULAR ELECTRONICS

Fig. 86-2

The power-supply circuit has an output that can be varied linearly by potentiometer R2 or switched to specific levels by rotary switch S3.

DC POWER SUPPLY

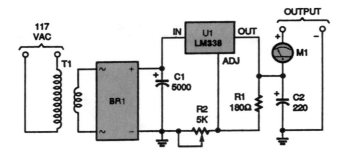

POPULAR ELECTRONICS

Fig. 86-3

Here's the schematic for The Crusher dc power supply. The outputs are front-panel-mounted banana jacks. T1 is a 24- to 26-V transformer with 1- to 5-A capacity. C1 should be a minimum of 1000 μF per ampere of dc. C2 should be at least 100 μF (or larger). C2 helps maintain good transient response for transients that are too fast for the regulator, and 220 to 470 μF is recommended. BR1 is a 50-V bridge rectifier. U1 must be heatsinked. M1 is an ammeter of 0–5 A or 0–10 A full scale. The output voltage can be adjusted from about 1.2 to 30 V via R2. $V_{out} = 1.26 (R_2 / R_1 + 1)$ volts.

SIMPLE ADJUSTABLE DC SUPPLY

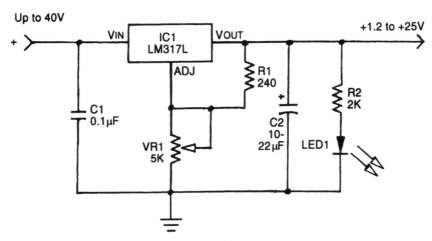

NUTS AND VOLTS

Fig. 86-4

Using an LM317L IC, this regulator will find broad application where a simple adjustable supply is needed. The IC is a surface-mount type, but a different package style can be used.

7.5-A 12- TO 16-VDC REGULATED POWER SUPPLY

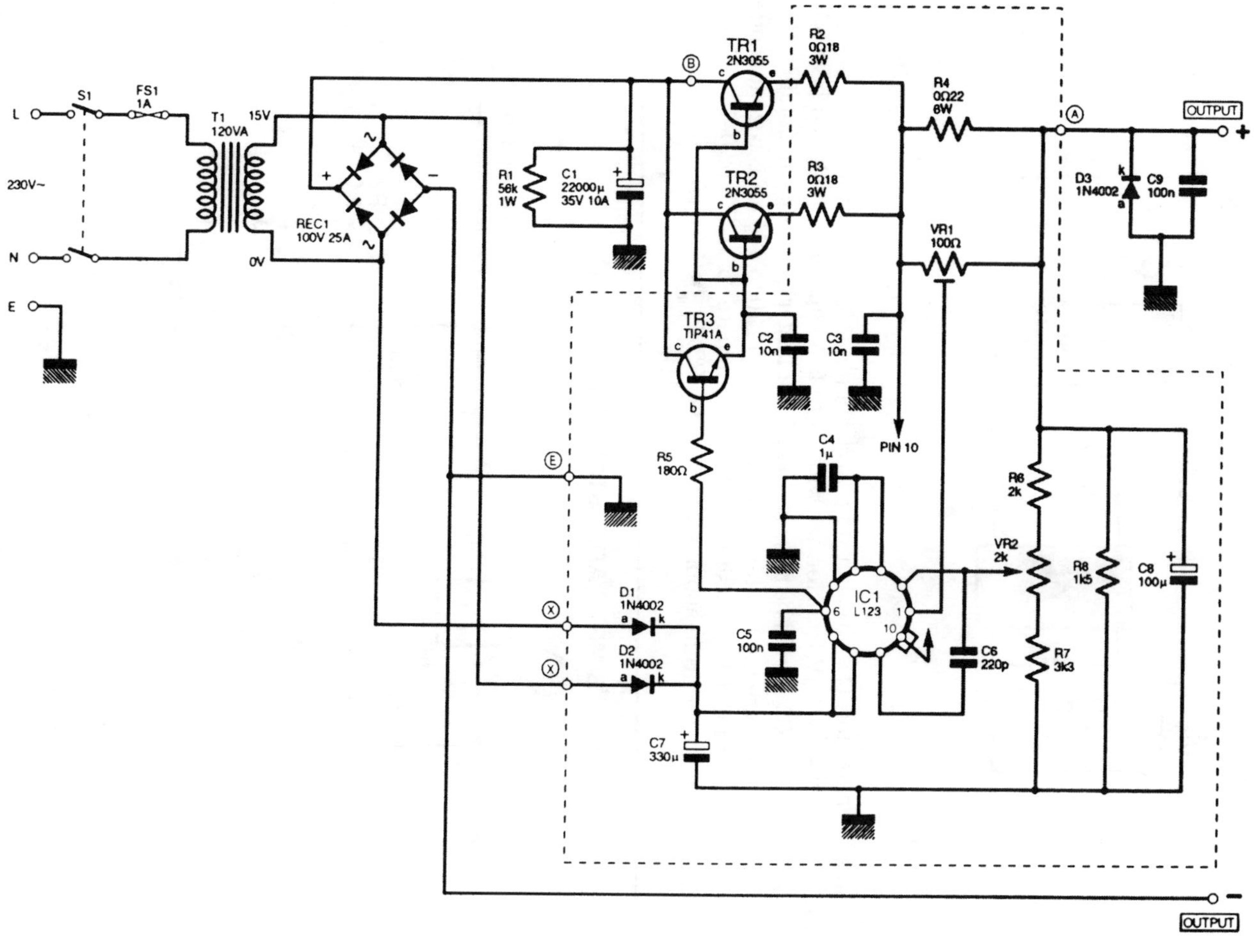

Fig. 86-5

EVERYDAY PRACTICAL ELECTRONICS

7.5-A 12- TO 16-VDC REGULATED POWER SUPPLY *(Cont.)*

This supply uses an L123 regulator IC and a pair of 2N3055 power transistors as series-pass regulators. VR1 is the current limit, and VR2 is the voltage control. The 2N3055 transistors and the TIP41A should be heatsinked because up to 50 W of dissipation can occur.

87

Probe Circuits

The sources of the following circuits are contained in the Sources section, which begins on page 1043. The figure number in the box of each circuit correlates to the entry in the Sources section.

ACTIVE HIGH-Z PROBE

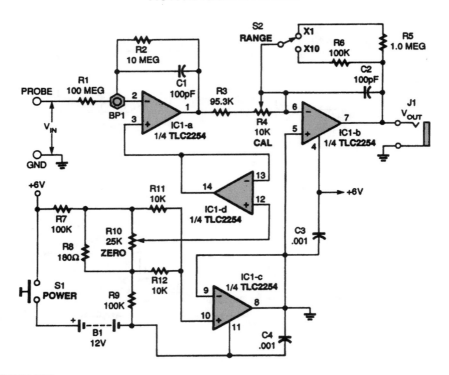

POPULAR ELECTRONICS

Fig. 87-1

This probe has a 100-mΩ input impedance and uses a TI TLC2254 quad op amp operating from a 12-V battery. It can be used with a DVM or an analog multimeter for measurements in the high-impedance circuits. BP1 is a Teflon standoff to ensure low leakage at the terminal.

INFRARED LOGIC PROBE

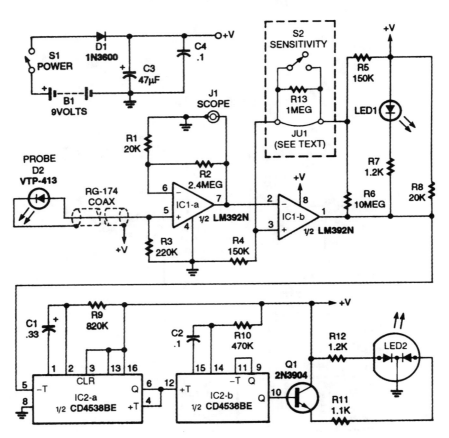

Fig. 87-2

Infrared light detected by photodiode D2 is amplified by IC1-a. The value of R2 can be changed to decrease the sensitivity of the circuit, if your application demands it. Connector J1 provides an output to an oscilloscope for the display of the amplified photodiode signal. This is handy when checking the pulsed emitters in most remote controls. Voltage comparator IC1-b squares up signals from IC1-a to digital logic levels for IC2-a. LED1 and current-limiting resistor R7 indicate the presence of steady-state infrared and also function with pulsed emitters if the duty cycle is appropriate. Monostable multivibrator IC2 conditions pulse trains with any period shorter than the time constant of R9 and C1 into a low-frequency waveform with a very high duty cycle. This provides pulses for LED2 that are constant in frequency and duty cycle, regardless of the high input frequency to IC2-a. Any frequency input to IC2-a with a period longer than the time constant of R9 and C1 creates IC2-b output pulses with the same width as before at the input frequency. Tricolor LED2 (a dual red/green device) functions as a pilot lamp and indicator for pulsed infrared sources. LED2 will always

INFRARED LOGIC PROBE *(Cont.)*

glow red and pulse amber (red+green) when infrared pulses are detected. The power source for the circuit is a 9-V battery. Alkaline batteries will provide many hours of operation because the circuit has low-power integrated circuits in place of S2 for most emitters. For certain devices, such as slotted optical switches, CD laser diodes, and reflective sensors, more sensitivity might be desirable. If you plan to use the probe for LEDs that operate below 0.5 mW, install S2 and R13—if not, you can install a wire jumper on the board instead of the switch.

SEVEN-SEGMENT LOGIC PROBE

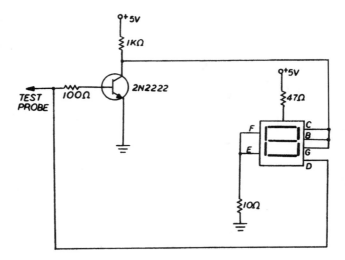

Fig. 87-3

This circuit is capable of indicating the "HI" or "LO" status of a digital circuit. The circuit in the figure will display the letter *H* for HI or +5 V and *L* for LO or 0 V. The 2N2222 NPN transistor serves as a driver for turning on the appropriate segments of the display, thereby producing the letter *H*. This condition will occur with a +5-V signal applied to the base of the transistor. If a 0-V signal is detected at the base of the transistor, the letter *L* will be displayed, indicating that the transistor driver is turned off.

LOGIC PROBE

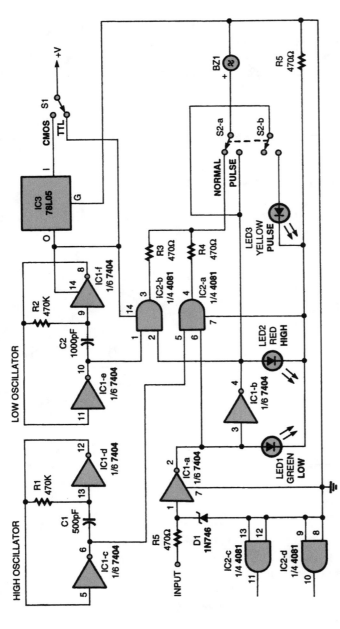

Fig. 87-4

POPULAR ELECTRONICS

The figure shows the schematic for a versatile logic probe. The zener diode clamps the input signal just above the TTL inverter's 2.2-V trigger voltage. Zener diode D1 can be left out if the probe is going to be used only on TTL circuits. The logic probe is based mainly on inverters IC1-a and IC1-b, which are sections of a 7404 integrated circuit. Depending on whether the input is high or low, the inverters enable AND gate IC2-a or IC2-b, each of which is a section of a 4081. Each AND gate is connected to an oscillator, one low-frequency, the other high. When an AND gate is made high, it passes the frequency of its oscillator to the piezo buzzer (BZ1). Whenever a high is at the input, the buzzer will produce the high tone; whenever a low is at the input, the buzzer will sound the low tone. With switch S2 in the PULSE position, if pulses are present at the input, a yellow LED will light and the

LOGIC PROBE *(Cont.)*

buzzer will sound at the frequency of the pulses. The CMOS/TTL switch selects the voltage from the circuit under test or the voltage from the 78L05 regulator. Note that when this switch is in the TTL position, it can work only with 5-Vdc, but when the switch is in the CMOS position, it can work with from 7.5 to 35 Vdc.

LOGIC PROBE CIRCUIT

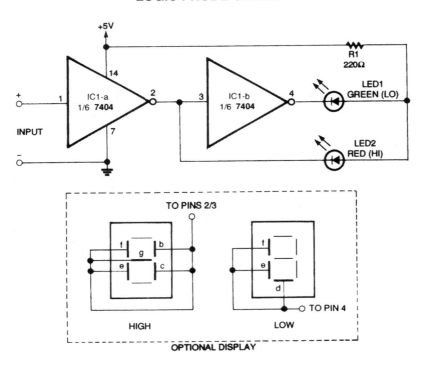

Fig. 87-5

In this logic probe circuit, the red LED indicates logic high and the green indicates logic low. The optional use of an LED seven-segment readout is also shown.

MINI HIGH-VOLTAGE PROBE

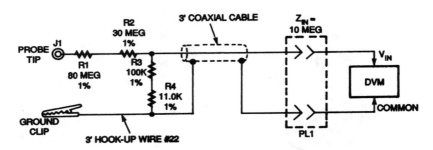

ELECTRONICS NOW

Fig. 87-6

The schematic diagram of the probe reveals its simple circuit. It is a standard voltage divider made up of resistors with a 1-percent tolerance. Resistors R1 and R2 are the key elements here. They are rated at 15,000 V and 10,000 V, respectively. The 7500-Vdc peak specification for the assembled probe must not be exceeded under any circumstances! The values for series resistors R3 and R4 are in parallel with the 10-MΩ resistance of the DVM, thus providing a 1000-to-1 voltage divider. Voltage measurements made by the probe should be multiplied by a factor of 1000. If your DVM uses an input impedance different from the standard 10 MΩ, you can adjust the R3/R4 series combination, as needed.

SCOPE PROBE CIRCUITS

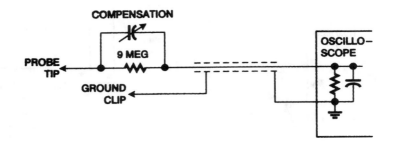

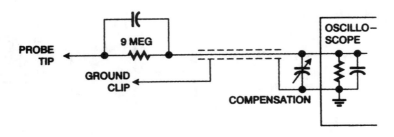

Fig. 87-7

Assuming a 1-MΩ scope input impedance and input capacitance C_{in}, $C_{compensation} = (R_{probe} / 1\ M\Omega)$ C_{probe} or $C_{probe} / C_{input} = R_{input} / R_{probe}$.

FREQUENCY PROBE

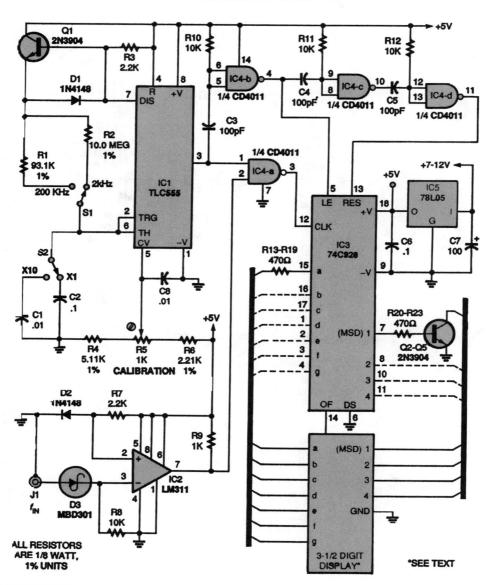

FREQUENCY PROBE *(Cont.)*

This design is a simplified $3\frac{1}{2}$-digit frequency counter with four ranges: 2000 Hz, 20.00 kHz, 200.0 kHz, and 2.000 MHz. An effort was made to miniaturize the circuit so that it would fit in a standard logic-probe-type case; the complete circuit was assembled on a perforated board cut down to $1\times4\frac{3}{4}$ inches. The normal crystal time base with all its dividers was eliminated in favor of one eight-pin DIP: IC1, a TLC555 CMOS oscillator. It provides gate timings of 1000 ms, 100 ms, 10 ms, and 1 ms using R1, R2, C1, and C2 as the crucial timing components. Select C1 to be exactly 1/10 the value of C_2. Calibration is done with trimmer potentiometer R5. Integrated circuit IC2 (an LM311) provides input conditioning for any waveform of ±0.9 to ±30 V, whether triangle, sine, or square wave. Integrated circuit IC4 (a CD4011 or 74C00) provides proper pulse delays to IC3, a 74C928 $3\frac{1}{2}$-digit counter chip that directly drives a miniature $3\frac{1}{2}$- or 4-digit, common-cathode LED display. The digit drivers shown, Q2 to Q5 and R20 to R23, can be replaced with a single DIP package (such as a 7549, 75492, etc.). A 78 regulator, IC5, provides the +5 V for the circuit, and is mounted on the board with the other components. It can be driven externally via a standard ac wall adapter of 7 to 12 V at 30 mA, or from a 9-V battery. Be sure to bypass the supply pins on each IC with a 0.1-μF monolithic capacitor for noise-free performance. When accuracy is important, a crystal-controlled time base should be used.

LOGIC PROBE

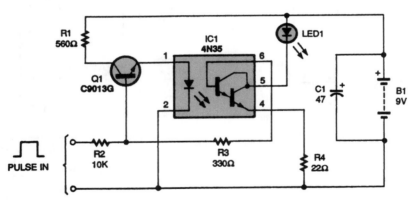

POPULAR ELECTRONICS

Fig. 87-9

The circuit uses a 4N35 optocoupler (IC1) and a C9013G transistor (Q1). A positive pulse at the base of Q1 turns on the transistor. The 10,000-Ω resistor (R2) limits the incoming pulse so as not to feed too much current to the base of Q1, preventing the transistor from being damaged. With the base of Q1 at a positive potential, electron current flows from the emitter to the collector junction. The incoming pulse also pulls pin 6 high, lighting up LED1 by allowing current to flow between pins 4 and 5. The 330-Ω resistor (R3) provides current limiting for pin 6, although R4 is the current limiter for pin 4. While testing a digital circuit, if the LED lights, that portion of the circuit is in a high state; if the LED remains off, a low is present. The circuit can be placed inside a small enclosure if its connections are made as short as possible.

WIRELESS DC PROBE

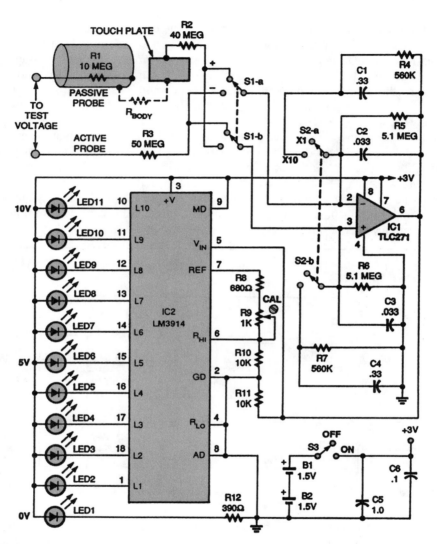

Fig. 87-10

This probe for low voltages only (<100 V) uses the body to complete the dc circuit in a novel way. The schematic for the dc voltmeter is shown. Power for the circuit is provided by two 1.5-V N-type cells, B1 and B2, which are wired in series. For the circuit to work, it must be completed or closed. That is accomplished when the passive probe is held in one hand and the active probe is held in the other and the points of the two probes are placed across two points in a circuit under test. The resistor labeled R_{body} is simply the resistance of the user's body. For R_{body} to be "swamped out," or kept from interfering with the reading, extremely high input-circuit impedances are used in the circuit. All of the people tested by the author for this project indicated a body resistance from hand to hand in the range of 200,000 to 500,000 Ω. Therefore, an input impedance of 50 MΩ, resistor R3, was added. That also provided an adequate safety isolation feature. The reduction of RFI and 60-Hz pickup is accomplished by the use of a symmetrical difference amplifier (IC1) at the input; the inherent common-mode rejection takes care of most the ac-related problems. The amplifier used for IC1 is a TLC271 CMOS op amp. Switch S1 is used to select polarity. Note that 50 MΩ of resistance are present at both of IC1's differential inputs (R3 at one input and the combined series resistance of R1 and R2 at the other). The user is placed in series with the 100-MΩ loop containing a 10-MΩ resistor (R1) as the protective device in the passive-probe section. Capacitors C1 to C4 provide additional filtering of residual ac noise and allow IC2, an LM3914 bar-graph display driver, to step up sequentially. Switch S2 selects between two selectable input ranges: X1, where each step equals 1 V, and X10, where each step equals 10 V. The display is made up of LED1 to LED11; LED2 through LED11 indicate the individual steps. LED1, the 0-V indicator, is also the power-on indicator.

HIGH-VOLTAGE PROBE

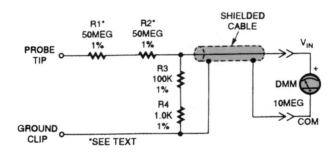

POPULAR ELECTRONICS

Fig. 87-11

This high-voltage probe allows your DMM to take high-voltage measurements to 10,000 V peak, and acts as a high-impedance (100-MΩ) probe for high-impedance circuit testing. The probe was designed to plug into a standard 10-MΩ input DMM. If your DMM has a different input, you can change R3 and R4 to suit it. The ratio of $R_1 + R_2 + R_3 + R_4$ should be exactly 1000 to 1; thus, a 10.00-V input will read 10.00 mV on your DMM. Keep in mind that the impedance of your DMM is in parallel with R_3/R_4 when you calculate new values.

88

Protection Circuits

The sources of the following circuits are contained in the Sources section, which begins on page 1043. The figure number in the box of each circuit correlates to the entry in the Sources section.

DIODE RELAY DRIVER PROTECTION

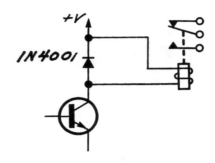

ELECTRONICS NOW

Fig. 88-1

The diode protects the switching transistor from the voltage spike caused by the relay coil.

DELAYED ACTION MOBILE RADIO-PROTECTION CIRCUIT

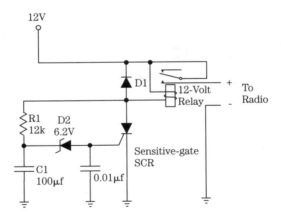

NUTS AND VOLTS

Fig. 88-2

This circuit uses an SCR and an RC circuit to delay application of power to a mobile radio or other equipment. This is to provide protection from starting motor- and alternator-induced transients.

OVER/UNDERVOLTAGE PROTECTOR

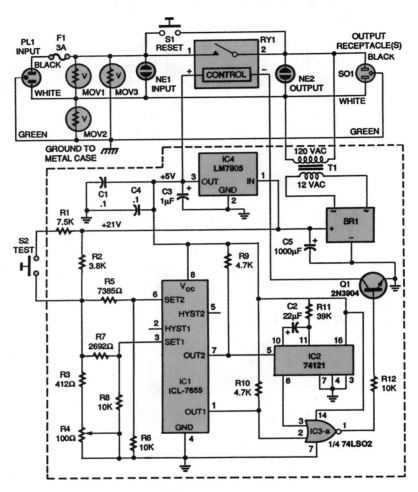

Fig. 88-3

The ICL7665 over/undervoltage detector manufactured by Maxim (IC1) monitors the line voltage and disconnects the load if the input voltage goes below 95 V or above 130 V. The over/undervoltage detector (IC1) monitors the filtered, unregulated dc voltage developed across C5, which changes with the input line voltage. The input voltage levels to IC1 are set by resistors R2 through R8. As long as the voltage at pin 3 is above 1.3 V and the voltage at pin 6 is lower than 1.3 V, both outputs (pins 1 and 7) are low and RY1 is turned on and its contacts are closed. The average line voltage varies at different locations. Trimmer potentiometer R4 provides a small adjustment range so

OVER/UNDERVOLTAGE PROTECTOR *(Cont.)*

that you can set the window's center for your particular line voltage. When IC1 detects an overvoltage at pin 3, it immediately opens the relay by sending a high output to pin 2 of NOR gate IC3-a, which, in turn, turns off Q1. When IC1 detects an overvoltage at pin 6, the output at pin 7 goes high. Monostable multivibrator IC2 provides a ½-s delay to keep RY1 turned off long enough for the voltage across C5 to return to normal. When pin 7 of IC1 goes high, it triggers IC2, and the high output at pin 6 of IC2 causes the output of NOR gate IC3-a to go low, which turns Q1 off and deenergizes RY1. Switch S2 lets you quickly check the circuit's operation at any time. Pushing S2 raises the voltage at the junction of R2 and R5 and turns RY1 off. RESET switch S1 restores normal operation. Switch S1 turns the circuit on initially. An LM7805 voltage regulator (IC4) provides a constant 5-V supply for the integrated circuits and the relay control current.

ELECTROSTATIC PROTECTOR

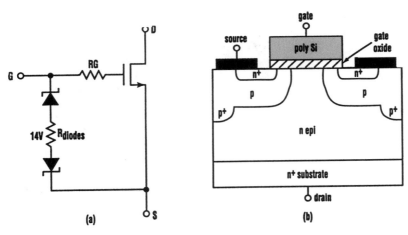

ELECTRONIC DESIGN

Fig. 88-4

A simple way to protect the gate of a power MOS device is to place back-to-back zener diodes between the device's gate and source (*a*). The breakdown voltage of the zeners is chosen to be less than the oxide-rupture voltage so that the ESD transient cannot harm the gate. The resistance of the clamp diodes must be kept to a minimum because ESD transients can have high peak currents (in amps), and the voltage appearing at the gate will be the sum of the zener breakdown voltage and the IR voltage drop across the diode resistance. The machine model (MM) is particularly stressful because there is no resistance to limit the current. Along with the diodes, the resistor (R_G) is added to improve MM performance. R_G prevents the gate of the MOSFET from charging to a dangerous voltage level during the time the ESD transient is dissipated by the diodes. A machine model here is defined as a 200-pF capacitor with no series resistance.

SIMPLE CROWBAR CIRCUIT

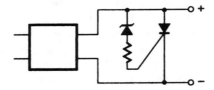

NUTS AND VOLTS

Fig. 88-5

This circuit uses a zener diode and an SCR. When the supply voltage exceeds the zener breakdown voltage plus the gate turn-on voltage of the SCR, current flows in the gate circuit, turning on the SCR and effectively shorting the supply. A fuse is generally placed in the supply line; it is designed to blow when the SCR fires, protecting and shutting down the supply.

SHORT-CIRCUIT SHUTDOWN CIRCUIT

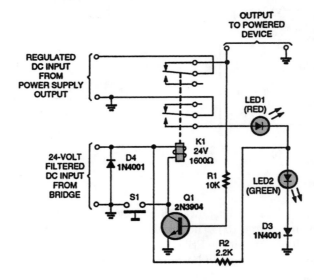

POPULAR ELECTRONICS

Fig. 88-6

When a short or severe undervoltage develops, Q1 is cut off, deenergizing Q1, and disconnecting the load from the power supply.

MOV SURGE-PROTECTION CIRCUIT

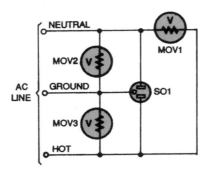

POPULAR ELECTRONICS

Fig. 88-7

Three MOV devices (connected as shown) provide superior protection against line surges because all three legs are protected against excessive potential differences.

SIMPLE REVERSE-POLARITY PROTECTION CIRCUITS

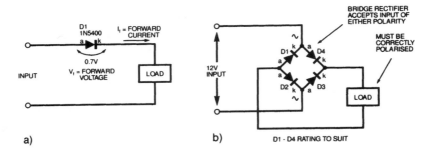

a)

b)

EVERYDAY PRACTICAL ELECTRONICS

Fig. 88-8

Two methods to avoid damage to circuitry from reverse voltages using diodes are shown.

REVERSED-POWER-SUPPLY PROTECTOR CIRCUIT

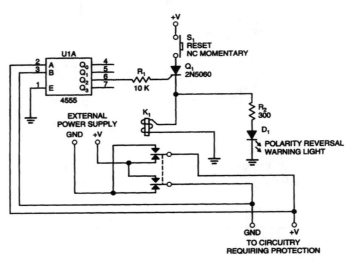

Fig. 88-9

This circuit is designed to sense a power-supply output (to other circuitry) and detect a re-versed-polarity condition. The circuit consists of a complementary metal-oxide semiconductor (CMOS) decoder (CD4555B), which monitors the output of an external power supply for proper po-larity. If the power-supply polarity is incorrect, the CD4555B detects that condition and triggers a 2N5060 SCR (Q1). After triggering, the SCR supplies current to a relay (K1), which, upon activation, changes the external power-supply connections (to the external circuitry) to the correct polarity and illuminates a light-emitting diode (LED) to provide a visual warning of the polarity-reversal condi-tion. After the reversed-polarity condition is remedied, pressing the contact switch (S1) momentar-ily will reset the circuit to its original state. In its present configuration, the circuit can accommodate standard positive power supplies with dc voltages of +5, +10, +12, and +15 V. The decoder was de-signed to operate at these different voltages. With some modifications, the circuit can monitor and correct anomalous power-supply outputs from +3 V, +28 V, or negative-polarity power supplies, or voltages that are either too high or too low.

SIMPLE CIRCUIT TO DISCONNECT LOAD FROM BATTERY

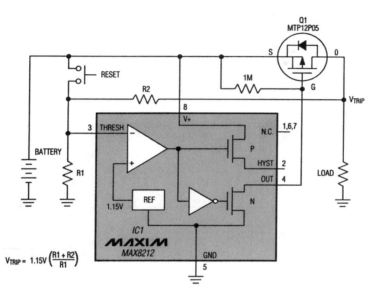

$$V_{TRIP} = 1.15V \left(\frac{R1 + R2}{R1}\right)$$

MAXIM

Fig. 88-10

To prevent battery damage, the circuit disconnects the load at a predetermined level of load voltage. This level (V_{trip}, closely proportional to the battery voltage) is determined by R1 and R2 so that the voltage at pin 3 of IC1 equals 1.15 V: $V_{trip} = 1.15$ V$(R_1 + R_2) / R_1$. The allowed range for V_{trip} is 2 V to 16.5 V. The load-battery connection remains open until the system receives a manual reset command. Pressing Reset (or pulling pin 3 above 1.15 V with a transistor) reconnects the load after the battery is recharged or replaced. Battery drain with the load disconnected is only 5 μA, so the circuit can remain in that state for an extended period without causing a deep discharge of the battery. Choose Q1 for a minimal voltage drop (source to drain) at the required load current.

POWER-SUPPLY PASS-TRANSISTOR PROTECTION CIRCUIT

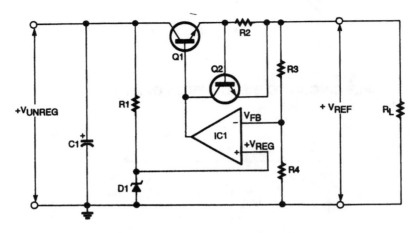

ELECTRONICS NOW

Fig. 88-11

A current-limiting transistor and resistor protect the pass transistor and rectifier bridge in this linear supply.

MODEM-PROTECTION CIRCUIT

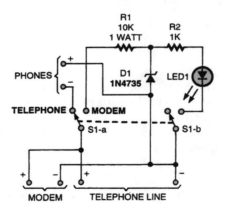

POPULAR ELECTRONICS

Fig. 88-12

A common problem in modem communications is that family members often pick up extensions during a modem call. Forget about having your downloads messed up by someone lifting a phone. This circuit completely disconnects all phones on a line when you are using a modem.

UNDERVOLTAGE DISCONNECT SWITCH

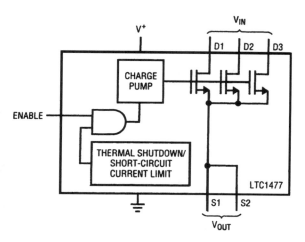

Fig. 88-13

In an effort to conserve energy, simple shutdown schemes are incorporated into many battery-operated circuits. Not all circuits lend themselves to direct control, however. Instead, the supply must be turned off by a switch. The LTC1477 high-side switch is designed for this purpose and includes short-circuit current limiting and thermal shutdown to guard against faulty loads. The figure shows the LTC1477 and LTC699 conjoined in an undervoltage disconnect application. The LTC699 microprocessor supervisor disables the LTC1477; hence, the load is disabled whenever the input voltage falls below 4.65 V. An external logic signal applied to the gate of Q1 can also disable the LTC1477. When enabled, the LTC1477 output ramps over a period of approximately 1 ms, thereby limiting the peak current in the load capacitor to 500 mA. This prevents glitches on the 5-V source line that might otherwise affect adjacent loads.

ELECTRONIC CIRCUIT BREAKER

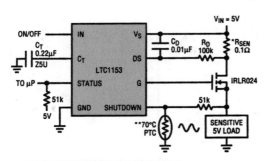

Trip Delay

ALL COMPONENTS SHOWN ARE SURFACE MOUNT.
* IMS026 INTERNATIONAL MANUFACTURING SERVICE, INC. (401) 683-9700
** RL2006-100-70-30-PT1 KEYSTONE CARBON COMPANY (814) 781-1591

LINEAR TECHNOLOGY

Fig. 88-14

Fault conditions can quickly put stress on power sources and also on the loads themselves. Without a protective device to interrupt power during fault conditions, the supply can be overloaded or a load can overheat. The LTC1153 is a high-side MOSFET switch controller device with built-in circuit breaker, which turns off the MOSFET switch in the event of a fault. There is a built-in automatic reset, along with an input for thermal shutdown. The trip delay and reset times are fully adjustable. Because of its low (20 μA max) standby current, the LTC1153 can also be used in battery-powered equipment to protect the battery without reducing battery life. The switch can also be controlled with logic input. A status indicator reports fault conditions to a processor.

AMPLIFIER PROTECTOR

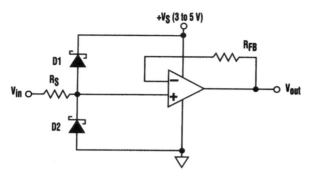

Fig. 88-15

Unprotected amplifiers in noninverting, single-supply, voltage-follower stages might need external protection if input signals can exceed the rail voltages. Low-threshold Schottky diode clamps D1 and D2 provide safe clamping while R_S limits the fault current.

SHORT PROTECTION AND SHUTDOWN CIRCUIT

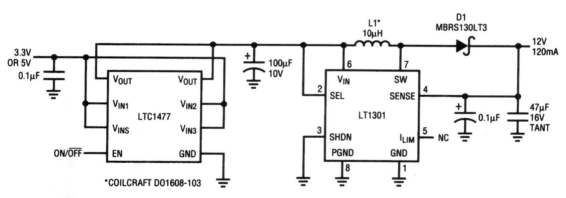

Fig. 88-16

An LT1301 is used in the figure to boost a 3.3-V or 5-V input to 12 V, as for V_{PP} for flash memory. Although the LT1301 features a shutdown control, the input supply can still feed through to the output through L1 and D1. Similarly, a short circuit on the output could drag down the input supply. With the addition of the LTC1477, the circuit furnishes 100-percent load shutdown and output short-circuit protection.

FAILURE MONITOR

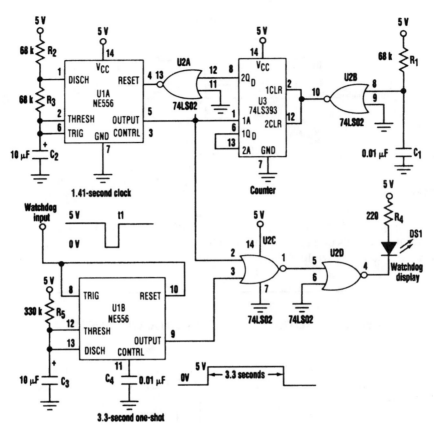

ELECTRONIC DESIGN

Fig. 88-17

Some time ago, a situation arose in which a watchdog timer was needed to monitor an embedded computer. The computer in question required approximately 2½ minutes to boot up and run the application software, so a monitor circuit was created that could wait 3 minutes before indicating failure. U1A, U2A, U2B, and U3 form a 3-minute timer. U1A operates as a free-running oscillator with a period of 1.41 s. The square-wave pulses are counted by U3, which generates an oscillator reset via U2A that effectively stops the clock. During the 3-minute interval, U2C and U2D send the oscillator signal to DS1, causing the indicator to flash. This flashing shows that the equipment is in its power-up cycle. When the reset occurs, the output of U1A is held low. U2B is a power-up reset for the counter. After 3 minutes, the indicator will come on continuously unless the watchdog input signal is present. U1B is configured as a 3.3-s-duration one-shot. When the watchdog input is pulsed low, the output switches high and keeps the LED from turning on. If the next pulse is not received in at least 3.3 s, the output will remain low. As a result, the LED will turn on, indicating failure. Should the

FAILURE MONITOR *(Cont.)*

watchdog input begin pulsing prior to the 3-minute timer, the watchdog will take precedence and turn off the flashing LED.

What happens if the watchdog circuit fails? One of the characteristics of the NE556 timer is the inability to fully discharge the timing capacitor if the timer is retriggered prior to finishing a timing cycle. Applying a 10-Hz, 100-ms-wide pulsed input to the watchdog causes the capacitor charge to accumulate for approximately 8 s. This results in a 100-ms "heartbeat" flash of the LED every 8 s, which indicates that the circuit is still alive and that the LED hasn't burned out. The watchdog input is generated by an I/O line of the computer. Depending on the I/O board's design, the signal might need to be inverted to allow the watchdog input to remain normally high.

SIMPLE MOBILE RADIO PROTECTOR

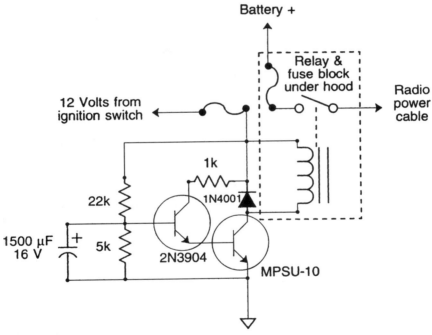

73 AMATEUR RADIO TODAY

Fig. 88-18

This circuit supplies power to a mobile radio only when the ignition is on and has a delay feature to prevent energizing the radio until the key has been on for several seconds, to avoid spikes and possible transients from the starter motor.

89

Receiving Circuits

The sources of the following circuits are contained in the Sources section, which begins on page 1043. The figure number in the box of each circuit correlates to the entry in the Sources section.

VARIOMETER RADIO

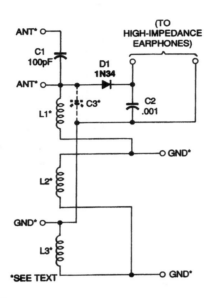

Fig. 89-1

The schematic of the complete variometer radio is shown in the figure. An antenna can be connected to the radio through either of the two points labeled *ANT*: either directly to the circuit or through a 100-pF capacitor. The ground connection can be made at any of the points marked *GND*. There is a reason for the preceding options: By varying the antenna capacitance, the ground connection, and the position of the sliding coil, the entire AM broadcast band can be tuned. Depending on the antenna and ground connections, it might be necessary to add a small capacitor, C3, at the point indicated in the schematic. If so, experiment with values between 25 and 200 pF (separately or in parallel) to find which gives the best result. If you build the variometer using Fahnestock clips (as explained later), adding the capacitor(s) after the radio is built if the need arises should be easy. When a signal is selected by adjusting the antenna, ground connection, and position of L2, the signal is passed on to the diode-detector part of the circuit, composed of D1, which demodulates the signal. The signal then goes through bypass capacitor C2 to the earphones. Only high-impedance earphones should be used with the variometer. L1 and L3 are 86 turns of #22 wire on a 1.25-inch form, and L2 is 74 turns of #22 wire on a 1.75-inch form.

VLF/HF LIGHTNING DETECTOR RECEIVER

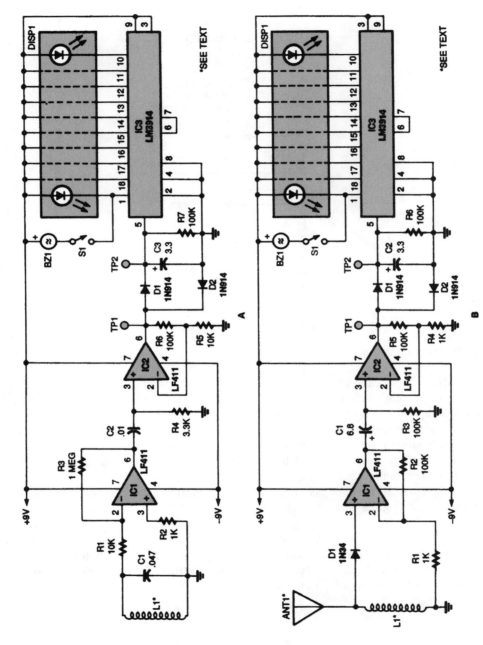

Fig. 89-2

POPULAR ELECTRONICS

832

VLF/HF LIGHTNING DETECTOR RECEIVER *(Cont.)*

It is known that cloud-to-ground lightning strikes produce far more VLF emissions than cloud-to-cloud strikes. By monitoring both VLF and HF, lightning can be analyzed for such things as strikes per flash, leader steps, amplitude or relative range, and the ratio of cloud-to-ground to cloud-to-cloud discharge. This pair of simple receivers separately monitor both VLF and HF. Note that they are both extremely similar, except for their respective antenna coils. The VLF-antenna coil, L1 in Fig. A, is made of 94 turns of #33 magnet wire wound on an 11-inch-diameter cardboard disk. The HF-antenna coil, L1 in Fig. B, is a common RF choke, or about 100 turns of very fine magnet wire on a ½-inch-long ferrite core (any similar junkbox choke should do). Antenna ANT1 is a 6-inch wire antenna. The first LF411 FET op amps (IC1 in both circuits) are preamps. The second LF411s (IC2 in both) add more gain. Test point 1 in either circuit can be used to connect high-impedance headphones or an oscilloscope. Test point 2 in either circuit is for a connection to chart recorders or event counters. The outputs of both circuits go to LM3914 bar-graph display drivers (IC3 in both circuits), which power both LED bar graphs (DISP1 in both) and low-voltage piezo buzzers (BZ1, again in both receivers). The negative leads of the buzzers, while shown connected to pin 1, can be connected to any of the bar-graph outputs. Both circuits are powered by split supplies.

REGENERATIVE SHORTWAVE RECEIVER

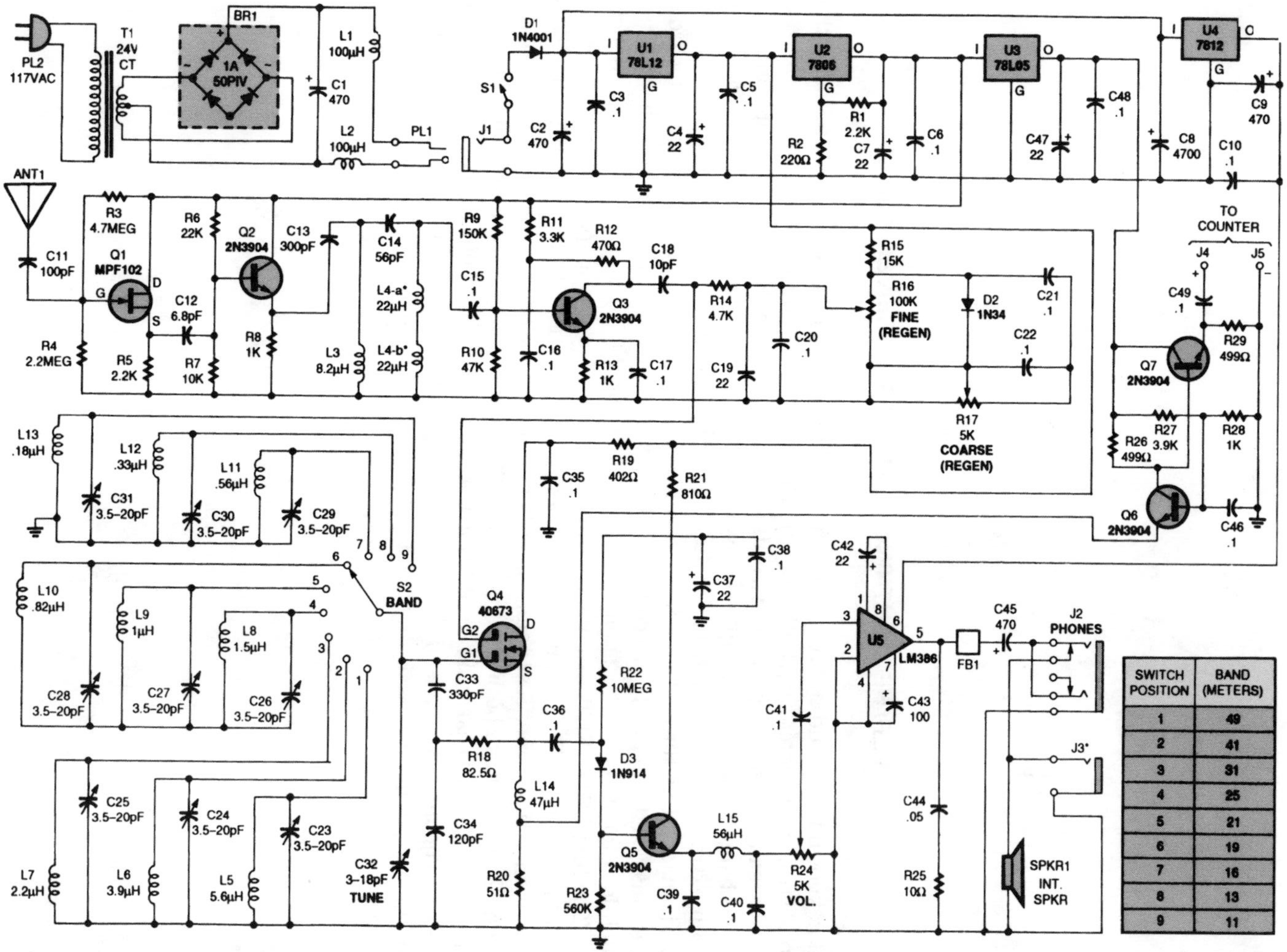

SWITCH POSITION	BAND (METERS)
1	49
2	41
3	31
4	25
5	21
6	19
7	16
8	13
9	11

Fig. 89-3

POPULAR ELECTRONICS

REGENERATIVE SHORTWAVE RECEIVER *(Cont.)*

The receiver is composed of several subassemblies: an active antenna, an amplifier (with a regeneration control and band-switching circuitry), an AM detector, a power amplifier, and some form of output device (internal speaker, external speaker, or phones), plus a multivoltage power supply. The schematic diagram of the receiver is shown in the figure, a dual-gate MOSFET, Q4 (which is analogous to a pentode vacuum tube), is used as the regenerative amplifier. The dual-gate MOSFET is configured as a Colpitts oscillator, rather than an Armstrong type. The MOSFET's feedback from source to gate is provided via R18 and C33. The circuit contains several standard fixed-value inductors, each of which is connected in parallel with a small variable capacitor. Those LC pairs along with a SP12T rotary switch (S2) are used for band selection. Regenerative amplification and AM detection are performed by two separate transistors, Q4 and Q5. The AM detector (which is called an *infinite-impedance detector*) has the advantage of not loading the RF stage appreciably. Because D3 is operated with a slight forward bias, either a germanium or a silicon diode can be used in that position. The dc voltage across R24 (the volume control) should be about 0.1 V. Regeneration is controlled by varying the voltage applied to gate 2 of Q4. The circuit also contains fine and coarse regeneration controls that allow delicate adjustments. The receiver uses an active-antenna circuit, consisting of transistors Q1, Q2, and Q3. Under most conditions, a short whip antenna is adequate.

RECEIVER INCREMENTAL TUNING CIRCUIT

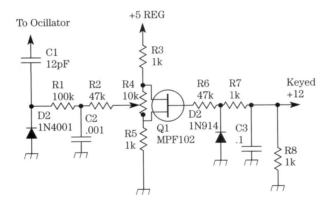

Fig. 89-4

D1 acts as a varactor diode coupled to the transceiver or receiver LO tuning circuit. On receive, Q1, the MPF102, is cut off and the varactor voltage is controlled by the 10-kΩ pot. On transmit, Q1 conducts; this effectively shorts the 10-kΩ pot, placing a fixed voltage on the varactor diode.

VACUUM-TUBE SW RECEIVER

VACUUM-TUBE SW RECEIVER *(Cont.)*

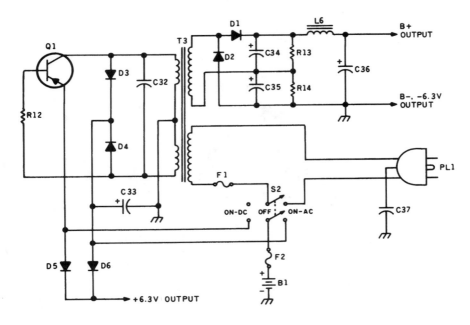

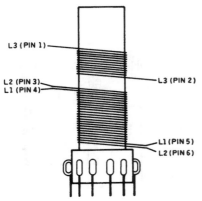

L1- 14 TURNS, CLOSEWOUND BETWEEN L2 TURNS.

L2- 15 TURNS, SPACED BY L1.

L3- 12 TURNS, CLOSEWOUND, WINDING SPACED 1/8" FROM TOP OF L1/2.

* USE NO.26 ENAMELLED WIRE FOR ALL 3 INDUCTORS.

** CORE IS 1/4" DIA. BY 1 1/8" LONG PHENOLIC, WITH 2 FERRITE SLUGS. CORE BASE SECTION IS 1/2" DIA. BY 3/8" LONG, WITH 6 SOLDER LUGS IMBEDDED.

This vacuum-tube receiver covers from 9.4 to 22 MHz and is of interest to those hobbyists who wish to experiment with vacuum-tube circuitry.

Fig. 89-5

IMPROVED REGENERATIVE SW RECEIVER

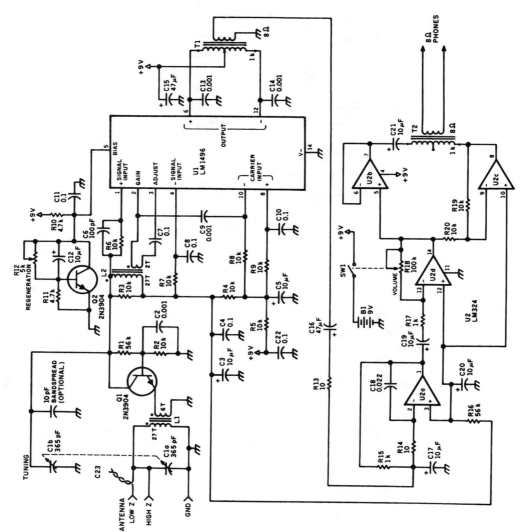

Fig. 89-6

IMPROVED REGENERATIVE SW RECEIVER *(Cont.)*

The figure shows the schematic of the circuit. A number of unique features give this design its trouble-free performance. The tendency for regenerative receivers to radiate oscillator-frequency signals is eliminated here by placing a transistor buffer Q1 between the antenna preselector circuit L1-C1a and the regenerative oscillator-tuned circuit L2-C1b. This buffer also nearly eliminates hand-capacitance effects. This design is also unique in that it uses an IC for the regenerative detector. U1, an LM1496 double-balanced mixer, is used here in a somewhat unorthodox manner. The differential SIGNAL INPUT amplifier transistors internal to the IC are used as a Hartley oscillator in conjunction with L2 and C1b. The regenerative feedback for this oscillator is supplied by the output of the GAIN and ADJUST pins of the LM1496. Some of the oscillator output is coupled to one of the CARRIER IN-PUT pins via C9, which allows the mixer section of U1 to act as an asynchronous detector, greatly improving the RF detection sensitivity over that of other regenerative circuits. The regeneration level is controlled by the voltage level applied to the BIAS pin of U1. The circuit containing R12 and transistor Q2 is used as a variable-voltage source, providing the regeneration level immunity from supply-voltage ripple. This bias level controls the quiescent current level through the SIGNAL IN-PUT amplifier transistors, which, in turn, determines the emitter output impedance of these transistors, controlling the amount of power delivered to the feedback winding of L2. This results in very smooth and predictable regeneration control. The outputs of U1 are coupled through audio transformer T1 into the first section of U2, an LM324 op amp. Volume control is achieved though U2d and variable resistor R18. Using a push-pull audio output stage, as is done here, also reduces susceptibility to audio oscillation. Although the use of switched or plug-in coils could have allowed multiband reception, in the interest of simplicity, this was not done. However, the values of L1 and L2 allow coverage of 5 to 15 MHz, where much of the shortwave action occurs.

ONE-CHIP AM RADIO

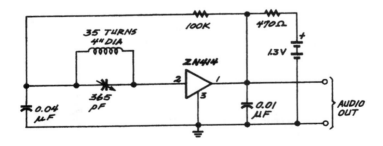

Fig. 89-7

The ZN414 IC contains a complete AM radio, with AGC, except for an audio section. Like an MMIC, it is powered through its output terminal.

WWV RECEIVER

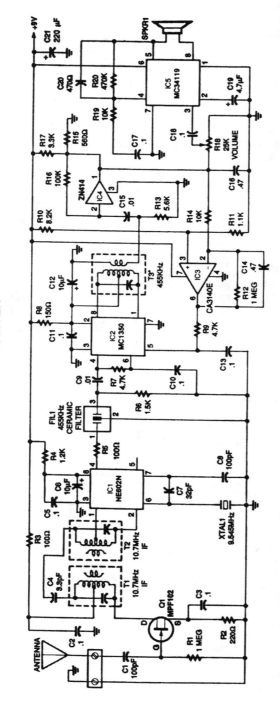

ELECTRONIC EXPERIMENTERS HANDBOOK

Fig. 89-8

This receiver is a crystal-controlled superheterodyne receiver with an MPF102 RF amplifier, NE602 mixer, ceramic IF filter, MC1350P IF amplifier, ZN414 AM detector, and MC34119 audio output. The AGC system uses a CA3140E op amp. A 9545-kHz crystal is used in the L.O.

SIMPLE 225- TO 400-MHz AIR-BAND RECEIVER

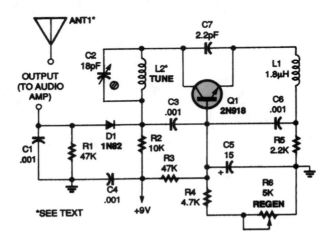

POPULAR ELECTRONICS

Fig. 89-9

Tuning coil L2 can be wound with 2 turns of #22 wire on a 5/32" drill bit. This circuit is well behaved as long as a good component layout is used. The component leads must be kept short and neat, especially the leads of the transistor. The lengths should not exceed ¾₆". Audio could be tapped off the tuning coil with a 5-μF (or so) electrolytic capacitor, but the 1N82 diode circuit seems to produce less signal loss. The RF signal from ANT1, an approximately 18" antenna, can be introduced through a small capacitor of 1 pF (or less) to the emitter of the 2N918. Other high-frequency transistors can be used, but might require different resistance in the regeneration circuit. The output of this circuit must go to an audio amplifier.

DIRECT-CONVERSION WWV RECEIVER

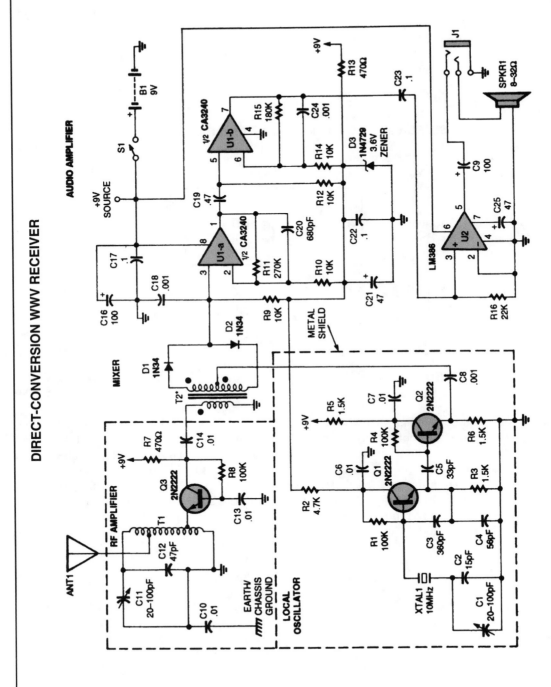

Fig. 89-10

ELECTRONICS HOBBYISTS HANDBOOK

DIRECT-CONVERSION WWV RECEIVER *(Cont.)*

Incoming RF is picked by the antenna (ANT1) and is coupled via an autotransformer to a grounded-base amplifier (Q3) before being applied to a diode-mixer network that comprises T2, D1, and D2. The best mixer performance is obtained when both secondary windings of T2 are identical, and D1 and D2 are matched. The output of the local oscillator (LO)—a grounded-collector Colpitts oscillator (built around Q1)—is applied to emitter-follower/buffer Q2, which provides a low-impedance drive signal for the mixer. The demodulated signal is coupled to a pair of high-gain op-amp stages (U1-a and U1-b). The op amps provide a 50-dB gain. Amplifier U2 provides a 20-dB gain, thereby producing sufficient output drive for an 8-Ω speaker or 32-Ω headphones. The volume is controlled merely by adjusting the length of the whip antenna. When driving 32-Ω headphones, the circuit consumes less than 25 mA; however, the current drain increases to 40 mA when driving an 8-Ω speaker.

TWO-IC MEDIUM-WAVE RECEIVER

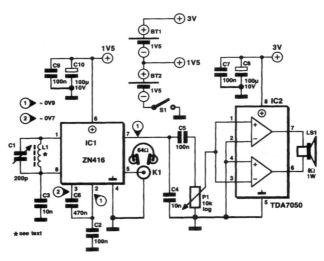

ELEKTOR ELECTRONICS

Fig. 89-11

The antenna is an inductor, L1, consisting of about 60 turns of 0.2-mm (SWG36) enameled copper wire on a ferrite rod with a diameter of 12 mm and a length of about 12 cm. The inductor is tuned with a 500-pF foil-dielectric variable capacitor, C1. The audio power amplifier, a TDA7050, is required only if you want to use a small loudspeaker instead of, or in addition to, the headphones. The AF power amplifier also adds the luxury of a volume control to the receiver. The receiver IC operates at 1.54 V from only one of two series-connected AA (penlight) batteries, which supply 3 V to the TDA7050. Current consumption is of the order of 8 mA.

SCA RECEIVER

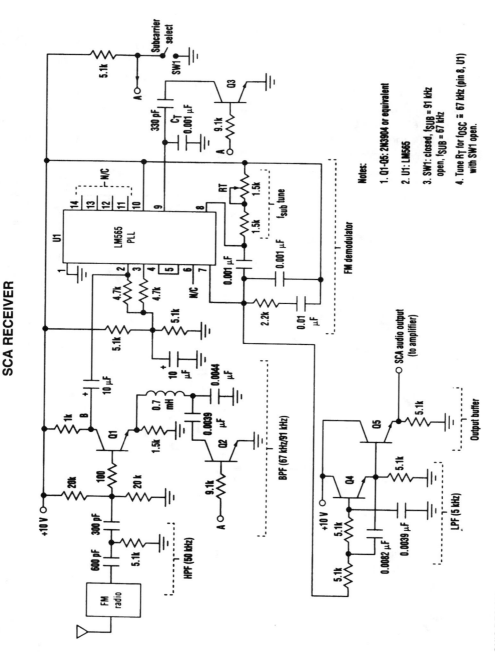

Notes:

1. Q1–Q5: 2N3904 or equivalent

2. U1: LM565

3. SW1: closed, $f_{SUB} = 91$ kHz open, $f_{SUB} = 67$ kHz

4. Tune R_T for $f_{OSC} \approx 67$ kHz (pin 8, U1) with SW1 open.

Fig. 89-12

ELECTRONIC DESIGN

SCA RECEIVER *(Cont.)*

A simple PLL-based FM demodulator can be made to demodulate two subcarriers and provide a source of background music. The input to this receiver is fed from the wideband audio output from an FM radio's first FM detector. This is the point that feeds the FM stereo demodulator in an FM receiver. Care should be taken to ensure that the signal is tapped off at a point which hasn't yet been filtered [the subcarrier(s) are still present]. This FM input signal is then passed through a second-order high-pass filter and peaker stage (Q1), which serves to bandpass and provide additional gain within the input spectrum prior to the FM demodulator input. The FM detection is accomplished by a simple LM565 PLL IC (U1) operating as an FM demodulator. The PLL's VCO is tuned to 91 kHz via R_T/C_T. The demodulated output signal is available at pin 7, which is followed by a second-order LPF/buffer combination (Q4). The characteristics of this filter can be modified to suit the user. The design shown has an audio corner frequency of about 5 kHz. The filtered output is the recovered audio output and is the input to an audio amplifier. To choose the second subcarrier (67 kHz), the peaker and VCO are gang-tuned by the Q2 and Q3 saturating switch transistors. These devices switch in appropriate-valued parallel capacitors to retune the peaker and VCO to the proper frequency for reception of the second subcarrier signal. Circuit values shown are for an FM level of about 50 to 300 mV rms at the input to the peaker stage. In addition, the PLL dynamic characteristics can be altered as desired by modifying the loop filter. The typical recovered audio level at pin 7 of U1 is 200 mV rms. To receive SCA signals, the FM receiver can simply be tuned to normal FM stations, and then the presence of either or both subcarriers can be checked.

VARIABLE-RESISTOR-CONTROLLED REGENERATIVE RECEIVER

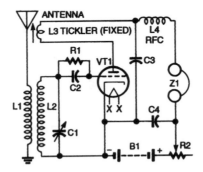

Fig. 89-13

R/C RECEIVER PREAMPLIFIER

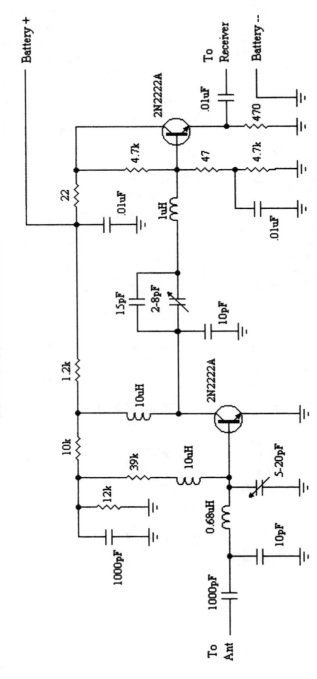

Fig. 89-14

NUTS AND VOLTS

The circuit shown is small, draws little current, and is easily fabricated on a perfboard—just keep the wires short and the by-pass capacitors close to their position on the schematic. To install, make a break in the wire from the antenna and insert the pre-amplifier. With proper trimming, this circuit yields about 20 dB of gain.

AM BROADCAST-BAND PRESELECTOR

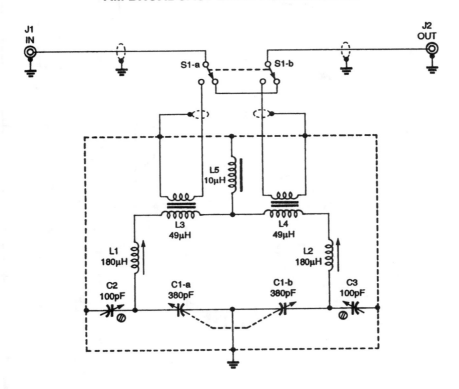

POPULAR ELECTRONICS

Fig. 89-15

The circuit for the AM-BCB preselector is completely bilateral; for that reason, J2 could just as easily be the input jack. That makes connecting the unit to your receiver a breeze. Note that switch S1 can be set so that the unit is bypassed.

80- AND 160-m DIRECT-CONVERSION RECEIVER

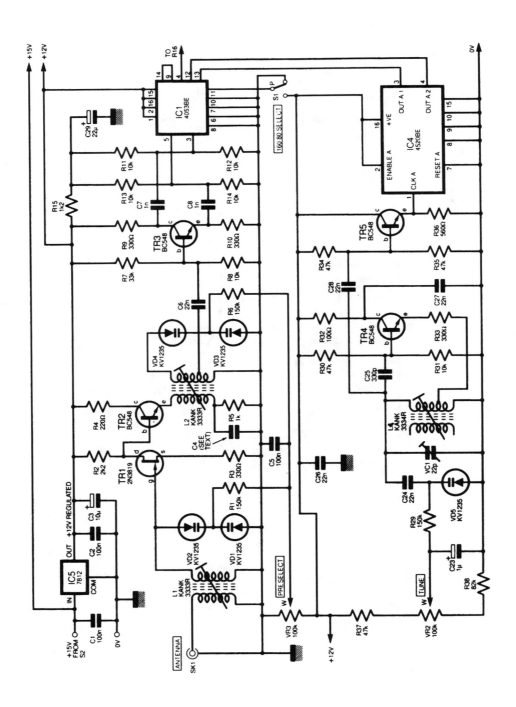

80- AND 160-m DIRECT-CONVERSION RECEIVER *(Cont.)*

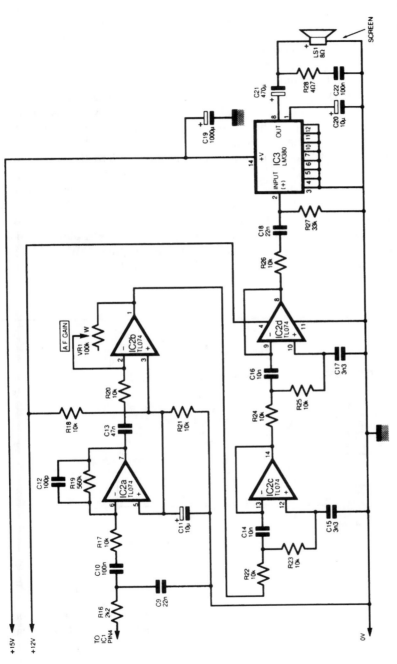

Fig. 89-16

EVERYDAY PRACTICAL ELECTRONICS

80- AND 160-m DIRECT-CONVERSION RECEIVER *(Cont.)*

The full circuit diagram, except for power supply, for the direct-conversion top-band and 80-m receiver is shown. Incoming RF from the antenna is fed to the primary side of coil L1. This transformer/coil is bought to resonance at the required frequency by a pair of back-to-back varicap diodes (VD1 and VD2). TR1, an RF, is coupled to the base (b) of transistor TR2, an emitter-follower. The output of TR2 feeds L2, the secondary being brought to resonance by varicaps VD3 and VD4. The variable-frequency oscillator (VFO) comprises a tuned circuit consisting of transformer L3, variable trimmer capacitor VC1, and varicap diode VD5. Variable bias to VD5 is provided by tuning control VR2. Transistor TR4 maintains oscillations at the desired frequency, and TR5 is an emitter-follower. The output of TR5 is sufficient to drive the CMOS divider chain formed by IC4. The VFO covers the frequency range 6.9 to 8.1 MHz, and this is divided by IC4 to produce 3450 to 4050 kHz for 80 m and 1725 to 2025 kHz for 160 m. The output of IC4 is fed to IC1, part of the product detector. The RF signal from the secondary tap on L2 is fed to a "phase splitter" formed around transistor TR3. Analog switch C1 operates at the selected VFO frequency and thus produces sum and difference frequencies of the VFO and the incoming RF. The output of IC1 at pin 4 is filtered to give resolved audio at this point. A high-gain inverting amplifier, with −3-dB points of approximately 300 Hz and 3 kHz, is formed by IC2a. This audio is now fed to IC3 with a maximum power output of 2 W. An LM380 is used in a standard configuration, and a fixed gain of around 30 dB is obtained.

RECEIVER FRONT-END RF ATTENUATOR

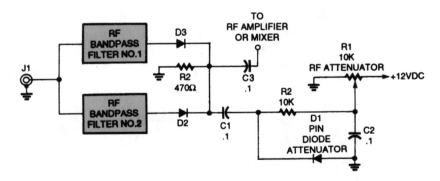

POPULAR ELECTRONICS

Fig. 89-17

This PIN-diode front-end RF attenuator circuit contains a simple shunt circuit. The dc voltage from the potentiometer sets the attenuation level. The PIN diodes can be MV3404 or similar types.

SOLID-STATE REGENERATIVE RECEIVER

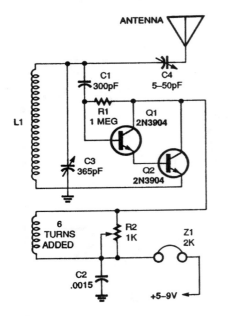
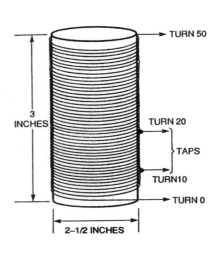

POPULAR ELECTRONICS

Fig. 89-18

You will need to modify the coil shown to use it with the regenerative receiver. Place a single layer of electrical tape around the bottom winding of the coil (L1); wind 6 turns of 19- or 20-gauge wire over the taped winding, leaving 4" pigtails. Tape the winding in place. Then connect the modified coil to the receiver circuit as shown in the figure. The two 2N3904 transistors are connected in a Darlington high-input-impedance circuit configuration to reduce loading of the tuned circuit. Potentiometer R2 controls the positive RF feedback. If the receiver seems dead or if only weak signals are heard, try reversing the leads on the six-turn feedback coil.

TWO-TRANSISTOR TRF RADIO RECEIVER WITH AUDIO AMPLIFIER

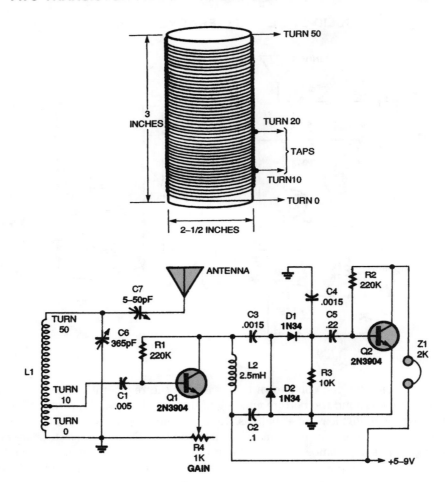

POPULAR ELECTRONICS

Fig. 89-19

This receiver allows you to select the gain of its RF amplifier by adjusting R4.

SIMPLE TRF RECEIVER

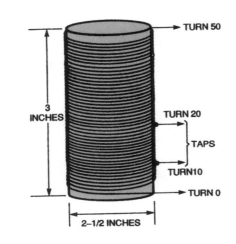

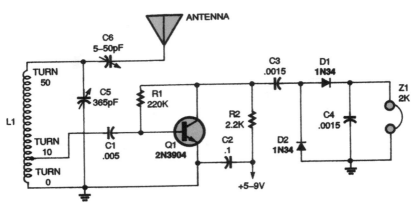

POPULAR ELECTRONICS

Fig. 89-20

Here is a receiver that contains an RF amplifier. The desired radio-frequency signal can be selected by tuning C5.

60- TO 72-MHz VIDEO AND SOUND RECEIVER AND IF SYSTEM

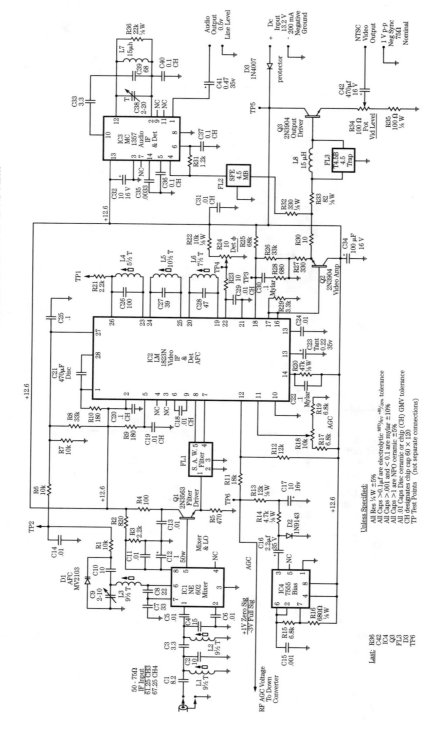

Fig. 89-21

WILLIAM SHEETS

60- TO 72-MHz VIDEO AND SOUND RECEIVER AND IF SYSTEM (Cont.)

This IF system uses a NE602 oscillator/mixer to convert VHF TV signals of 60 to 72 MHz (channel 3 or channel 4 under the U.S. NTSC) to 45 MHz. An SAW filter is used to provide bandpass filtering, followed by an LM1823 video IF/AGC/AFC/detector IC. Recovered video is fed to a sound trap and 4.5-MHz sound takeoff filter. Video is fed to the output emitter follower and is up to 1.5 V p-p into 75 Ω. An MC1357P FM IF/limiter/quadrature detector recovers audio information and provides 0.5 V p-p line-level audio. An LM555 oscillator and 1N914 diode act as a dc-to-dc converter system to derive a negative 3-V supply for the LM1823 AGC system. Both AFC and AGC voltages are available for interfacing with external downconverters used for 440, 915, and 1280 MHz amateur TV reception. Operation is from a 12-V supply at about 200 mA.

A complete kit of parts, including PC board, is available from North Country Radio, P.O. Box 53, Wykagyl Station, New Rochelle, NY 10804-0053A.

RECEIVER FRONT-END SELECTOR

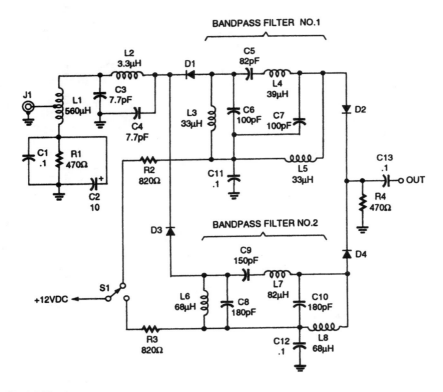

POPULAR ELECTRONICS

Fig. 89-22

Receiver front-end selection can be accomplished by using PIN-diode switches (as shown). The PIN diodes can be MV3404 or similar types.

TESLA COIL SECONDARY RECEIVER

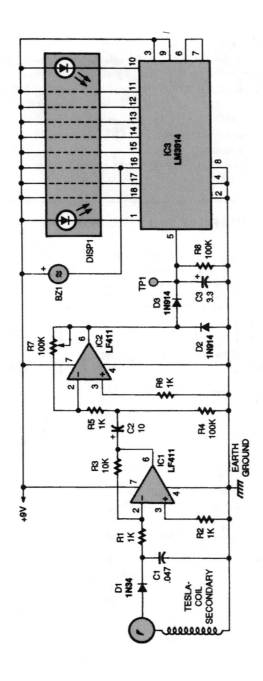

Fig. 89-23

POPULAR ELECTRONICS

Here's a Tesla-coil secondary to try: Wind 750 turns of 24-gauge enameled magnet wire onto an 18" long piece of 1.9" outer-diameter PVC pipe. The large coil has an inductance of about 2800 mH, with a self-capacitance of about 20 pF. One end of the coil should be earth grounded. Put a metal ball (a drawer pull knob or doorknob) at the other end. Now attach it to a crystal radio. The RF is detected by a 1N34 germanium diode, D1, and you will note that signal strength is high. What you should hear is one or possibly several AM radio stations, depending on the coil, the radio frequencies in your area, and distance to the transmitter(s).

OP-AMP VLF RECEIVER

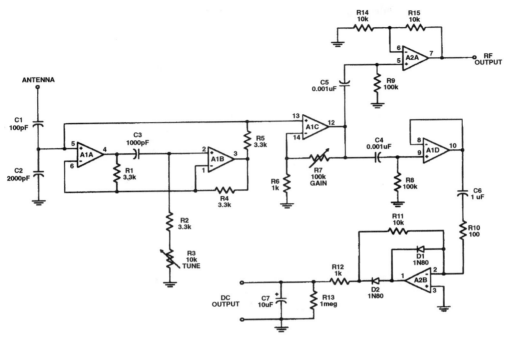

73 AMATEUR RADIO TODAY

Fig. 89-24

The figure shows the circuit diagram for an operational-amplifier VLF receiver. The circuit in the figure uses a virtual-inductor front end. A virtual inductor is a circuit that acts like an inductor, but isn't. Operational amplifiers A1A and A1B form a gyrator circuit. The inductance of this circuit is the product of the components shown between A1A and A1B [$C_3 \times 3300 \times (R_2 + R_3)$]. Capacitor C2 resonates with the virtual inductance produced by the gyrator circuit to tune the desired frequency. For the values of C2, C3, R2, and R3 shown, the circuit will tune from about 15 kHz to more than 30 kHz. Resistor R3 is the tuning control. It is a potentiometer, and should be either a multiturn knob via a vernier-reduction drive. The receiver front-end amplifier consists of amplifier A1C, which has a maximum gain of ×101 [i.e., $(R_7/R_6)+1$]. The output of A1C is an RF signal with a frequency equal to that tuned by the gyrator and C2. This signal is coupled to the RF output stage (A2A) through capacitor C5. The RF output stage shown here is a noninverting operational amplifier circuit with a gain of ×2. The dc output circuit consists of a precise rectifier (A2B). The precise rectifier works like a regular rectifier, but does not have the low-voltage "knee" between 0 and 0.6 V (for silicon) or 0 and 0.2 V (for germanium). The pulsating dc from the precise rectifier is filtered and smoothed to straight dc, at a value proportional to the signal strength, by an RC integrator consisting of R12 and C7. The buffer amplifier (A1D) is used to isolate the precise rectifier from the RF output amplifier.

ONE-CHIP FRONT-END AM RECEIVER

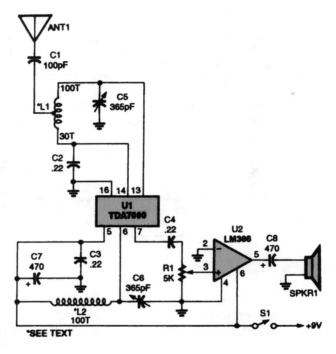

Fig. 89-25

The schematic diagram of the receiver circuit is shown. About all you have to do is wind two coils, connect a few components together, and tie the input to a simple wire antenna, and a receiver is born. The two coils, L1 and L2, each comprised 100 turns of #28 enamel-covered copper magnet wire wound on T80-2 toroid cores (with a tap at the 30th turn on L1).

90

Record and Playback Circuits

The sources of the following circuits are contained in the Sources section, which begins on page 1043. The figure number in the box of each circuit correlates to the entry in the Sources section.

Tapeless Record/Play Device
Tapeless Play-Only Device
1-min Record/Playback Circuit
Continuous-Loop Playback Device
Record/Playback Circuit with Automatic Power Down

TAPELESS RECORD/PLAY DEVICE

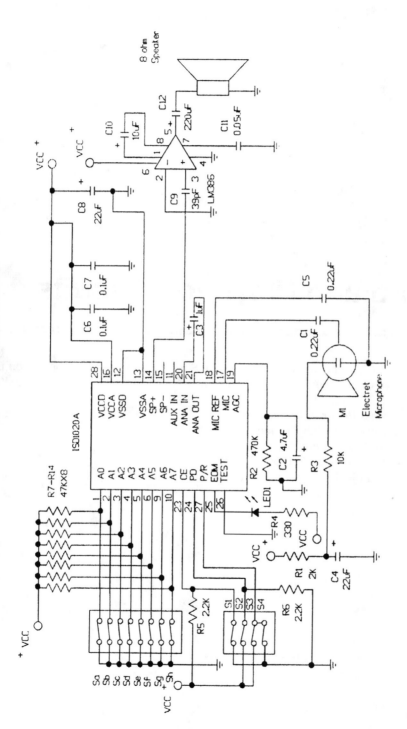

Fig. 90-1

NUTS AND VOLTS

The schematic for a record/play device with an audio power amplifier added is shown.

- A four-section DIP switch is used to activate the Chip enable, Power down, Playback, and Record functions.
- An LED detects an end-of-message condition. It remains off during Record or Play and turns on when a message ends.
- A three-terminal electret microphone is used for recording.
- The only off-board components are the microphone, speaker, and power supply. For optimum sound, the speaker should be baffled (placed in an enclosure). Four 1.5-V AA cells can be used as a power source.

TAPELESS PLAY-ONLY DEVICE

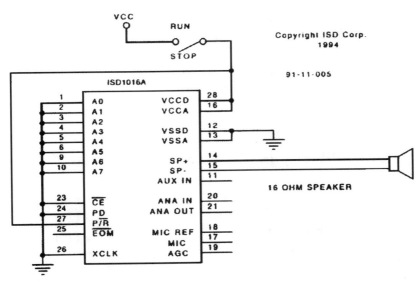

Fig. 90-2

This requires a prerecorded message to be preprogrammed in the ISD1020A. To play a message, set the switch to the +5-V position. The ISD1020A plays the entire contents one time, then stops. Because CE is tied low, the set end-of-message bits are ignored. Momentarily changing the switch to open, then back to +5 V makes the ISD1020A play again.

1-min RECORD/PLAYBACK CIRCUIT

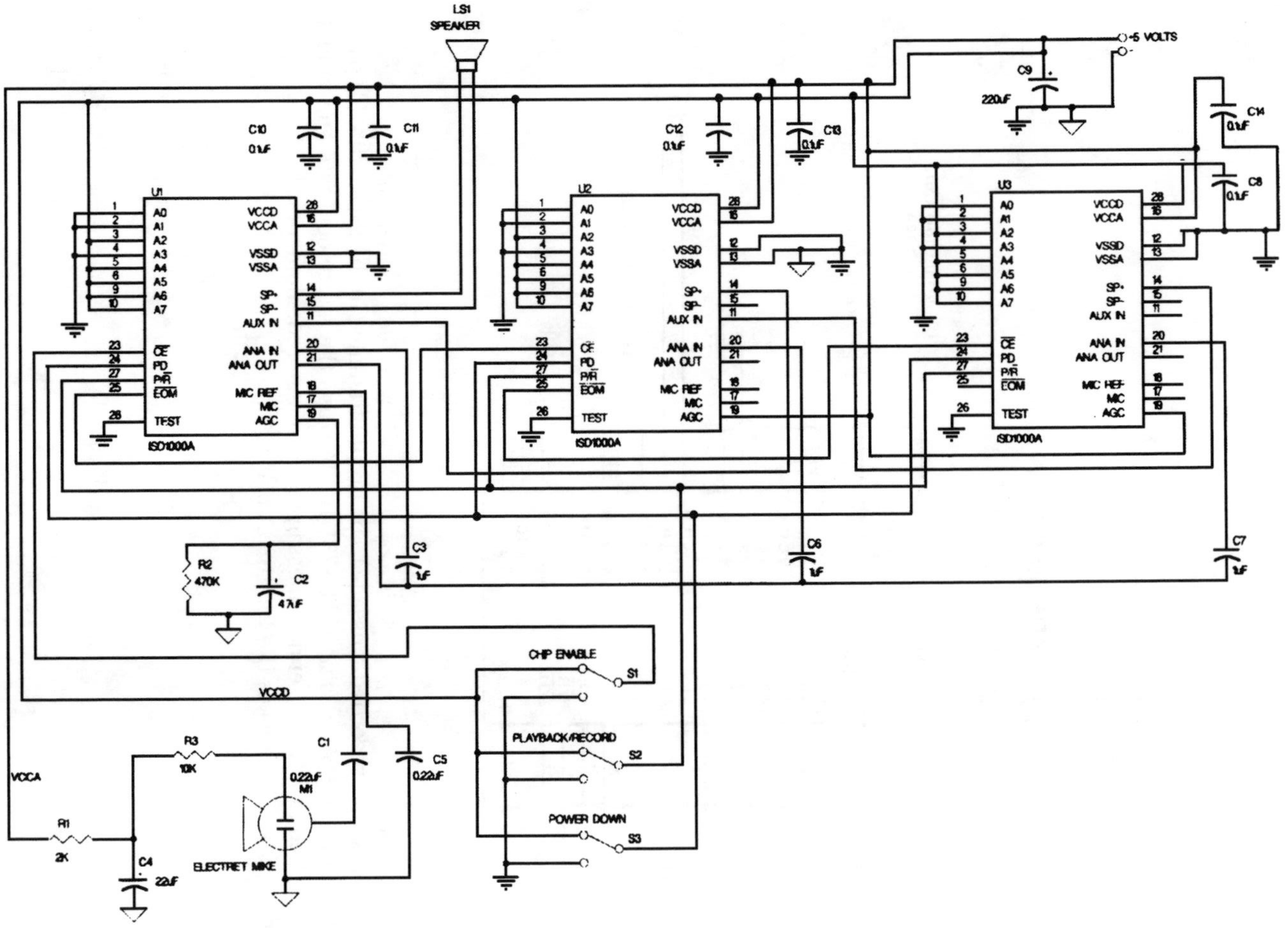

NUTS AND VOLTS

Fig. 90-3

1-min RECORD/PLAYBACK CIRCUIT *(Cont.)*

You can cascade several ISD1020As to increase a system's capacity. A circuit to create a 1-minute message is shown. With this configuration, a message is stored across chip boundaries in a manner transparent to the user. Changing from Record to Play resets the MSP to the beginning of the first message in the series. The next Play proceeds under control of CE. The circuit operates exactly like that for a single device.

CONTINUOUS-LOOP PLAYBACK DEVICE

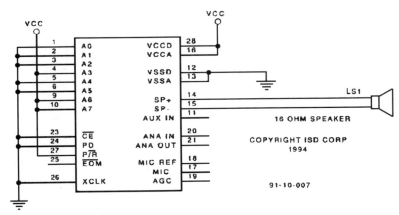

NOTE: CIRCUIT LOOPS BEGINNING AT 000 AND THE MESSAGE CAN'T REACH 160 (END)

NUTS AND VOLTS

Fig. 90-4

This requires a prerecorded message to be preprogrammed in the ISD1020A. If you want a message to repeat over and over again, automatically, the circuit in the figure will do the trick. It uses the Operation mode to accomplish playback looping as long as the message starts at address 0 (the beginning of the memory) and does not require the full 20 seconds of analog storage. A message is first recorded into an ISD1020A using the Yack/Yack project, with all address bits tied low. Next, connect the circuit in the figure. Note that address bits A6 and A7 are tied high to enable the Operation mode. Bit A3 is also tied high to enable continuous repeat. With PD low, P/R high, and CE held low, the beginning message in the ISD1020A will repeat.

RECORD/PLAYBACK CIRCUIT WITH AUTOMATIC POWER DOWN

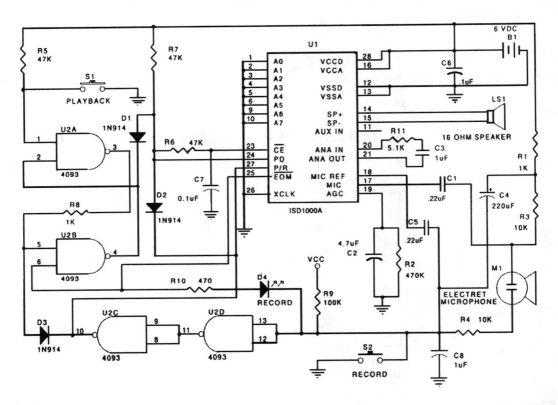

Fig. 90-5

Many applications for sound recording and reproduction, where battery power is necessary, require minimum power consumption. A circuit that achieves automatic power down in a record-and-play application is shown. The cross-coupled latch, consisting of a 4093 Schmitt trigger quad two-input NAND package (U2A and U2B), works to control the PD and CE pins for the Play mode. Circuitry for the Record mode consists of U2C and U2D plus several support components. To start the Record cycle, S2 supplies the ground to the microphone circuit and takes the input to U2D low. The Schmitt Trigger action of U2 debounces the logic input to the ISD1020A.

91

Relay Circuits

The sources of the following circuits are contained in the Sources section, which begins on page 1043. The figure number in the box of each circuit correlates to the entry in the Sources section.

50-μA Solid-State Relay
Polarity-Sensitive Relay Circuit
Solid-State "Latching-Relay" Circuit
Fault-Tolerant Relay Driver

50-µA SOLID-STATE RELAY

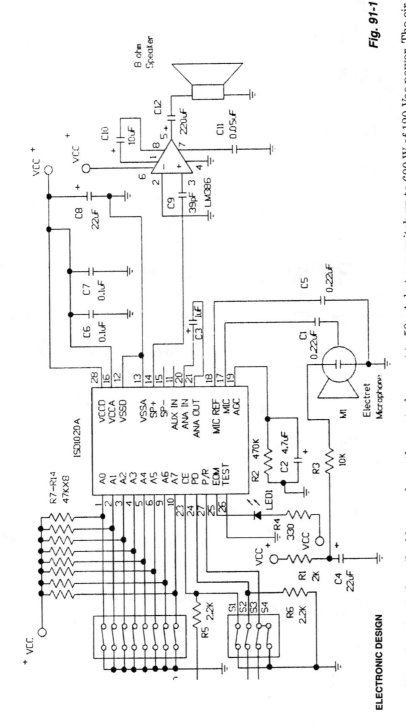

ELECTRONIC DESIGN

Fig. 91-1

The relay circuit described here reduces the control current to 50 µA, but can switch up to 600 W of 120-Vac power. The circuit uses a conventional 10-A, 400-V triac. The triac gate current is routed through a small bridge rectifier, a 15-V zener diode, a sensitive SCR (Q2), and a 180-Ω resistor (R2). The circuit allows a small SCR to control ac current through the triac's gate terminal. The voltage needed to control the SCR's gate terminal is developed by rectifying and filtering the ac voltage across the triac. Capacitors C2, C4, and C5, resistor R3, bridge rectifier BR2, and zener diode D2 form the rectifier circuit. C5 and D2 filter and limit the supply voltage to about 8 V, while the 470-Ω resistor R3 limits the charging current. To develop a slightly higher volt-

50-µA SOLID-STATE RELAY *(Cont.)*

age (up to 30 V p-p), a 15-V zener diode (D1) is inserted in series with the SCR. D1 delays the triac's conduction trigger point each half cycle and produces only a slight reduction in rms power to the load. The dc control voltage, produced by the rectifier circuit, is switched to the SCR's gate using a sensitive Darlington-type optoisolator (A1). Only about 50 µA of LED current in the isolator is needed to fully turn on the SCR.

POLARITY-SENSITIVE RELAY CIRCUIT

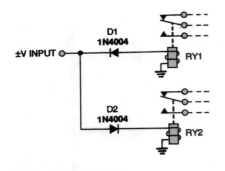

When a positive voltage is input to the circuit, relay RY1 is activated; when a negative voltage is input, RY2 is activated.

POPULAR ELECTRONICS

Fig. 91-2

SOLID-STATE "LATCHING-RELAY" CIRCUIT

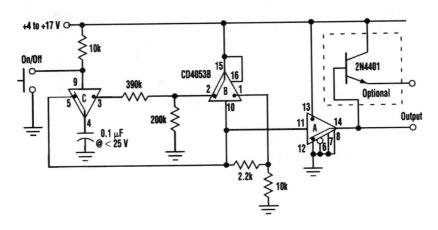

ELECTRONIC DESIGN

Fig. 91-3

A multitude of simple circuits perform switch debouncing because there are many ways to implement a toggling latch. The circuit illustrated here merges these functions into one topology, is extremely tolerant of power-supply variation (it works fine over more than a 3:1 range of supply voltage), and, most importantly, is unusual. The IC used in this case is the 4053B triple CMOS SPDT switch. Section B acts as a bistable latch and stores the current ON/OFF state of the relay. The 2.2-kΩ resistor between pins 1 (the ON terminal of the switch) and 10 (the control input) provides positive feedback that sustains the switch in whichever state it finds itself. Thus, if the latch is ON, pin 1 will be switched to pin 15 and, thereby, the positive rail. This will reinforce the positive state of pin 10 and the ON state of the switch. If, on the other hand, the latch is OFF, then the connection of pin 1 to pin 15 will be broken and the 10-kΩ pull-down resistor will drag pin 1 to ground. This will pull pin 10 low, in turn, holding the latch in the OFF state. Meanwhile, pin 2 will output the logic inverse of the signal appearing in pin 1. This signal communicates to pin 3 and, if the control push button is open, will connect to pin 4 and charge the capacitor to a state opposite that of the latch. There matters will rest until someone presses the button. This will connect the capacitor to pin 10 and toggle the Latch state. 4053B section A serves as an output buffer and has an output impedance of about 200 Ω when run with a 5-V supply and 100 Ω when run with a 12-V supply. If this is insufficient for the application, a transistor emitter-follower, like that illustrated, will boost output capability to beyond 0.1 A.

FAULT-TOLERANT RELAY DRIVER

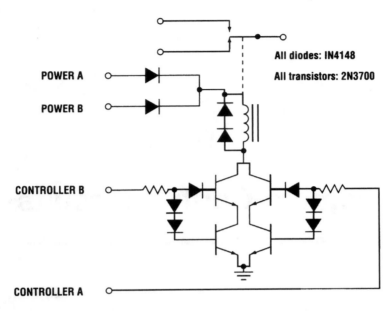

POWER A

POWER B

All diodes: IN4148

All transistors: 2N3700

CONTROLLER B

CONTROLLER A

ELECTRONIC DESIGN

Fig. 91-4

Mechanical relays are useful in remote-switching applications that require electrical isolation between control and switched circuits. The traditional approach for driving the relay coils uses a single-transistor common-emitter switch. The figure shows a circuit that can reduce the likelihood of having an unrecoverable failure in the relay driver circuit. Additional transistor switches, inserted in series with the original controlling transistors, maintain proper operation if a single transistor fails. The diodes connected to the upper transistors prevent reverse base-current flow if the collector-base junctions break down. The diodes connected to the lower transistor provide proper biasing for the circuit. The upper and lower transistors will have a similar V_{be} because they are the same type and have nearly identical currents. Therefore, the V_{ce} of the lower transistor will be about the same as one diode voltage drop, and the device will be operating in the active region. Power to the relays can be provided by one of two voltage sources connected together through diodes in a wired-OR configuration. The additional diode clamps the coil inductance voltage spike in the event that one of the suppression diodes fails shorted.

92

Remote-Control Circuits

The sources of the following circuits are contained in the Sources section, which begins on page 1043. The figure number in the box of each circuit correlates to the entry in the Sources section.

Simple Remote Control
Simple Remote-Control Interface Circuits

SIMPLE REMOTE CONTROL

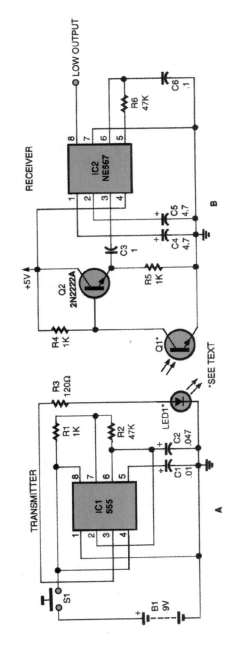

Fig. 92-1

POPULAR ELECTRONICS

Here is a remote-control transmitter and receiver. The theory of the circuit's operation is very simple. The transmitter (Fig. A) generates IR light pulses at a frequency of 320-Hz (set in part by R2). The pulses arrive at the collector of the phototransistor and are amplified by the 2N2222 transistor. An NE567, IC2, is tuned to 320-Hz (by R6), so if pulses of any other frequency arrive at the phototransistor, the NE567's output will remain high. When the 320 Hz signal enters the phototransistor, the NE567 recognizes this frequency and pulls its output low. If you wish to change the NE567's operating frequency, you will have to adjust R2 and R6 to the same value. Keep in mind that the NE567 will work between 100 Hz and 1 kHz. You can also add more resistors (for more frequencies) instead of R2, and more NE567s tuned at the desired frequencies in order to make a multichannel remote control system.

SIMPLE REMOTE-CONTROL INTERFACE CIRCUITS

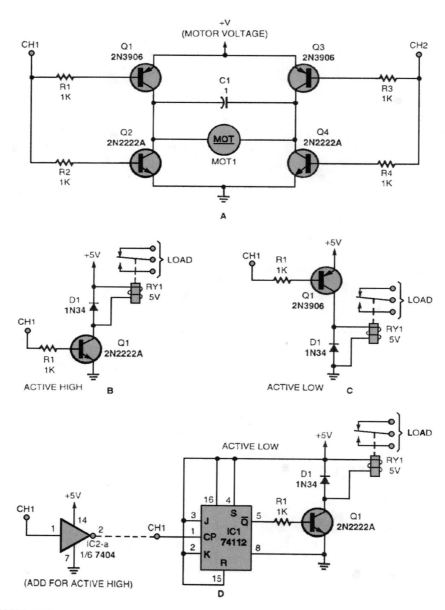

Fig. 92-2

SIMPLE REMOTE-CONTROL INTERFACE CIRCUITS *(Cont.)*

If you want to control a dc motor, a momentary load, or an on/off load, these are some useful interface circuits. Figure A shows a dc-motor controller that requires the outputs of two different channels (CH1 and CH2). Note that it doesn't matter if the channel outputs are active low or high; what matters is that CH1 must be different from CH2 for the motor to be energized. The circuits in Figs. B and C are momentary-load controllers for active high and active low channel outputs, respectively. The circuit in Fig. D is a toggled-load controller. If you have an active high channel output, you must add the inverter shown.

93

Robot-Control Circuit

The sources of the following circuits are contained in the Sources section, which begins on page 1043. The figure number in the box of each circuit correlates to the entry in the Sources section.

Universal Remote-Controlled Robot

UNIVERSAL REMOTE-CONTROLLED ROBOT

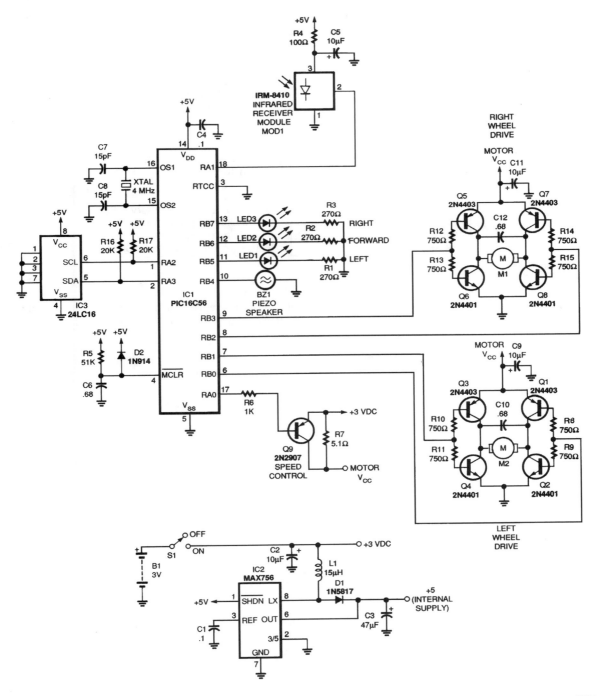

UNIVERSAL REMOTE-CONTROLLED ROBOT *(Cont.)*

RUNABOUT OPERATION

Universal Remote Key	Runabout Action
Up Channel	Move Forward
Down Channel	Move Reverse
Up Volume	Turn Right
Down Volume	Turn Left
Key 1	Single LED On (Left)
Key 2	Single LED On (Middle)
Key 3	Single LED On (Right)
Key 4	"Erratic Driver" Mode
Key 5	Beep (Horn)
Key 6	Dual Tones
Key 7	Rising Tones
Key 8	Change Speed
Key 9	Falling Tones
Key 0	Shift Key (Selects Memory 1-6)
Enter Key	Enter/Exit "Program" Mode
Power Key	Run Selected Program
Mute Key	Pause (In Program Mode Only)

Fig. 93-1

This project uses a microcontroller programmed with software that is posted on the Gernsback BBS at 516-293-2283 as part of RUNABOUT.ZIP. The robot can be controlled via a universal remote control.

94

RF Converter Circuits

The sources of the following circuits are contained in the Sources section, which begins on page 1043. The figure number in the box of each circuit correlates to the entry in the Sources section.

Simple Scanner Converter
VLF Converter
"Bat-Band" Converter
NE602 Direct-Conversion Output Circuit
NE602 Frequency Converter
31-m SW Converter for Auto Radios

SIMPLE SCANNER CONVERTER

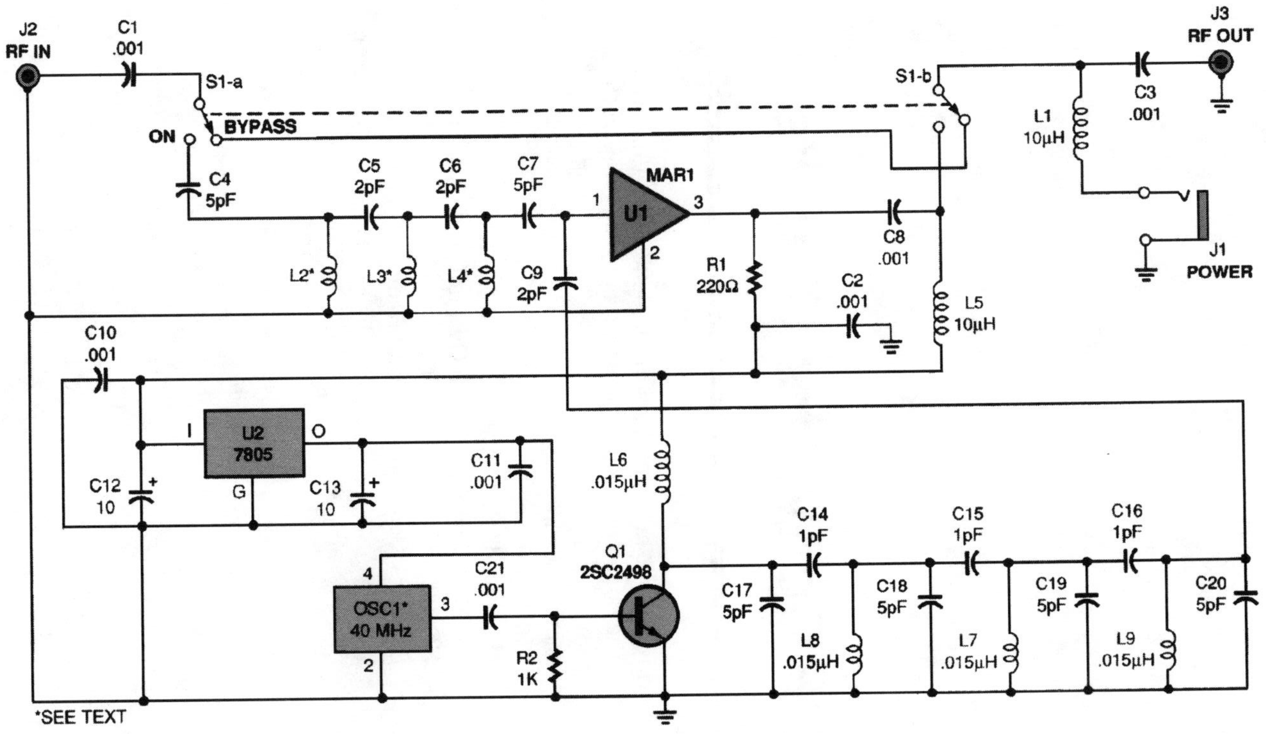

Fig. 94-1

POPULAR ELECTRONICS

878

SIMPLE SCANNER CONVERTER *(Cont.)*

This circuit can be powered from any 9- to 12-Vdc source, including a good alkaline 9-V battery. Switch S1 either puts the unit into its Bypass mode, where a scanner connected to J3 will receive its normal signals, or applies power to the circuit and down-converts all 800- to 950-MHz signals as follows: At the heart of the circuit is OSC1, a 40-MHz oscillator module. Transistor Q1 amplifies the oscillator's output, which is then bandpass filtered four times so that only the tenth harmonic at 400 MHz is presented to the input of U1, a Mini Circuits MAR1 wideband UHF/VHF amplifier. Signals from an antenna connected to J2 are high-pass filtered by capacitors C4 to C7 in conjunction with inductors L2 to L4. Those inductors are etched into the tracings on the PC board, making the exact PC board layout a necessity if the circuit is to function. Mixer U1 amplifies and mixes the two inputs—signals between 800 and 950 MHz and the 400-MHz local oscillator—and passes the 400- to 550-MHz output to J3. This converter is not suitable for areas where signal strength is low or areas where a large number of strong signals are present, as no RF stage is used and little input filtering is used before the mixer, making this circuit prone to spurious responses.

VLF CONVERTER

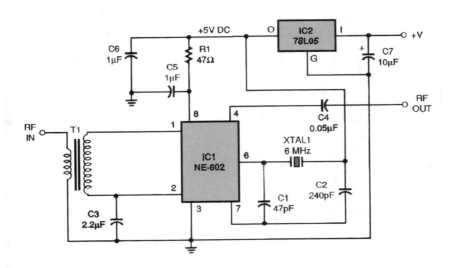

POPULAR ELECTRONICS

Fig. 94-2

This very simple converter permits reception of the VLF 3- to 30-kHz band on a shortwave receiver covering the 6-MHz (49-m) shortwave broadcast band. T1 is an audio transformer of about 8 to 1000 Ω. These can easily be purchased new or found in junked transistor radios.

"BAT-BAND" CONVERTER

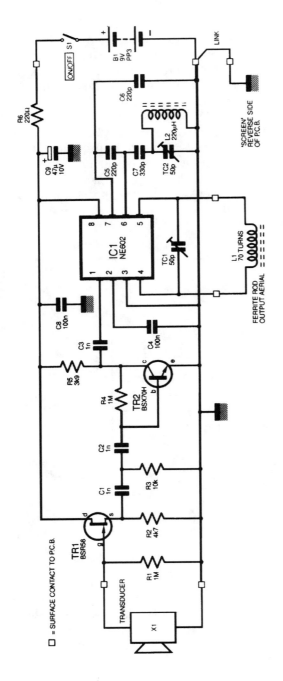

□ = SURFACE CONTACT TO P.C.B.

Fig. 94-3

EVERYDAY PRACTICAL ELECTRONICS

The full circuit diagram of the "bat-band" converter is shown. All components, with the exception of the transducer (X1), the ON/OFF switch, and the ferrite rod antenna, are surface-mount devices (SMDs). The signals from X1 are very small (microvolts, rather than millivolts). A low-noise FET input stage, TR1, has minimal loading effect on X1 and provides a low-impedance source for the passive high-pass filter. This simple filter consists of capacitors C1 and C2 and resistor R3. It attenuates signals below about 15 kHz, but the rolloff is very gradual. Mid-range and lower-frequency audio is strongly reduced. The falling sensitivity of transducer X1 further limits the audio frequencies reaching the mixer IC1. Despite this level of attenuation, a little audio still gets through the system, but it has no effect on the ultrasonic operation. When the receiver is tuned close to the 1-MHz local oscillator, a little audio feedback will be heard. This is a real help with tuning in because it readily identifies the low-frequency end of the "bat" band. The transistor TR2 is a low-noise amplifier, and the SMD-type BSX70H works well in this circuit. IC1 is a double-balanced mixer and Colpitts oscillator in one package. The signal inputs at pin 1 and pin 2 are used in unbalanced mode, with pin 2 decoupled by capacitor C4.

880

NE602 DIRECT-CONVERSION OUTPUT CIRCUIT

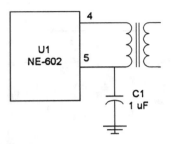

This figure shows a direct-conversion output circuit for an NE602.

Fig. 94-4

NE602 FREQUENCY CONVERTER

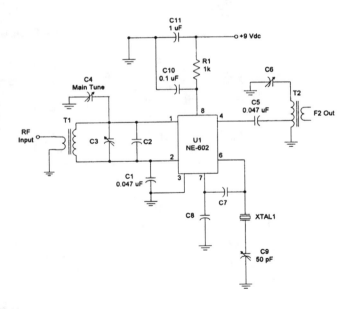

Fig. 94-5

The NE602 can be used as a frequency converter with this circuit.

31-m SW CONVERTER FOR AUTO RADIOS

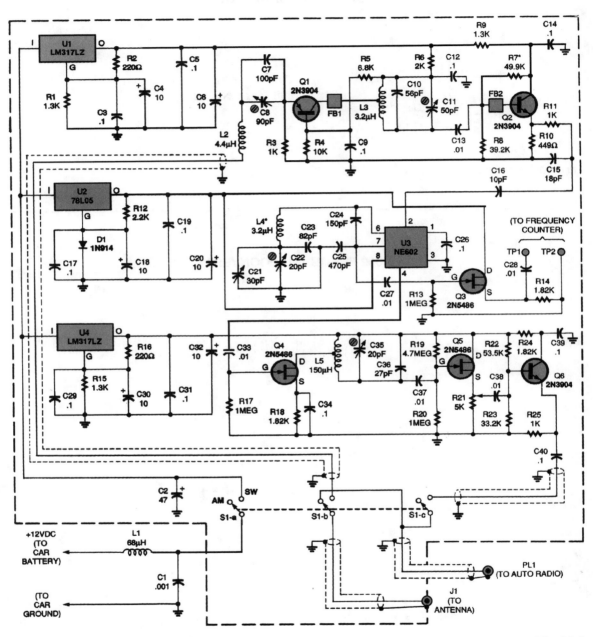

POPULAR ELECTRONICS

Fig. 94-6

31-m SW CONVERTER FOR AUTO RADIOS *(Cont.)*

Power for the circuit is taken from the car battery and is dropped to the proper voltages for three sections of the circuit by three separate regulator ICs: U1, U2, and U4. Inductor L2 and capacitors C7 and C8 act as the circuit's antenna tuner. The tuned signal is fed to an input bandpass filter composed of L3, C10, and C11. An NE602 oscillator IC, U3, is used as a combined mixer and oscillator. That configuration is known as a *series-tuned Colpitts* or *Clapp oscillator*, and is among the most temperature-stable variable oscillators. The 1710-kHz output filter consists of L5, C35, and C36. Each of the filters in the circuit was limited to a single LC section to simplify as much as possible the alignment of the converter. Transistor Q3 is a frequency-counter buffer that is used only during alignment. The gain of the converter is sufficient to overload the input of some receivers. Potentiometer R21 can be used to decrease the output level and prevent overload.

95

Rocket Circuits

The sources of the following circuits are contained in the Sources section, which begins on page 1043. The figure number in the box of each circuit correlates to the entry in the Sources section.

ROCKET LAUNCHER IGNITER CIRCUIT

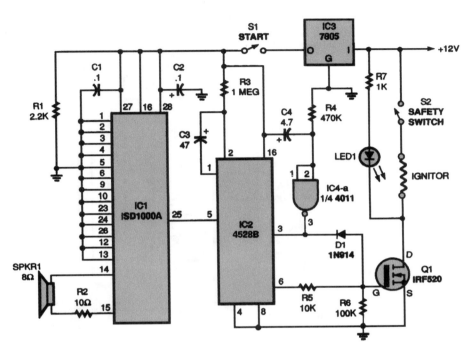

Fig. 95-1

An ISD1000A chip (IC1) containing a previously recorded message is connected in a simple playback-only circuit that operates when power is supplied by closing the START switch (S1). A 7805 voltage regulator, IC3, is used to drop the 12-V input voltage to 5 V. When the audio message is completed, IC1 produces a 16-ms low output pulse at pin 25. That low output pulse is fed to the input, pin 5, of a monostable multivibrator, IC2. That IC produces a long timed output pulse at pin 6, which turns Q1 on, supplying power to the igniter and firing the rocket engine. This circuit contains several safety features. Switch S2 removes power from the igniter fuse. One gate of a 4011 quad two-input NAND gate (IC4-a) keeps the 4528B monostable multivibrator from misfiring and igniting the rocket motor prematurely. When the Start switch is first activated, IC4-a's input goes high while C4 is charging up, and its output at pin 3 goes low, keeping IC2 from responding to any false input pulse at pin 5. IC4-a's pin 3 clamps the input of Q1 to ground, through D1, keeping it turned off. After a few seconds, C4 is fully charged and IC4-a's output goes positive, enabling IC2 and allowing Q1 to operate. Though not shown in the figure, the inputs of the quad IC's three unused gates (IC4-b to IC4-d, pins 5, 6, 8, 9, 12, and 13) must be tied to ground.

LED ROCKET COUNTDOWN LAUNCHER

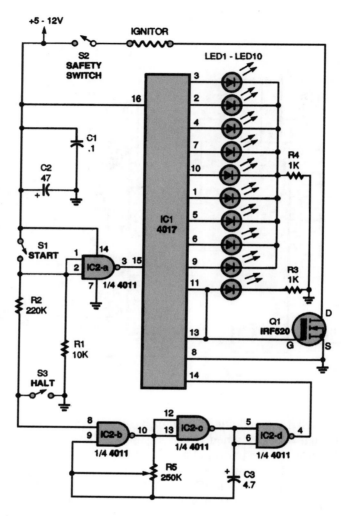

Fig. 95-2

A 4011 quad two-input NAND gate and a 4017 decade-counter/divider IC are the heart of the launcher. Transistor Q1, an IRF520 hexFET, sends the current through the igniter fuse to fire the rocket engine. Two gates of a 4011 NAND gate, IC2-b and IC2-c, are connected in a low-frequency oscillator circuit, with R5 and C3 setting the oscillator's frequency. Another gate of that IC, IC2-a, starts the countdown and sets pin 15 of the 4017 to the Run condition. The IC's fourth gate, IC2-d,

LED ROCKET COUNTDOWN LAUNCHER *(Cont.)*

inverts and buffers the oscillator's output and supplies the clock input to pin 14 of the 4017. The 10 LEDs indicate the count. When the last one (LED1) turns on, pin 11 goes high, turning Q1 on and firing the igniter fuse. The last LED, LED1, will remain on until S1 is switched off. The countdown can be halted by closing S3. Opening S3 continues the countdown.

ROCKET LAUNCHER VOICE COUNTDOWN RECORDER CIRCUIT

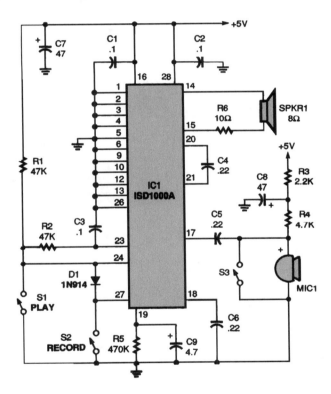

POPULAR ELECTRONICS

Fig. 95-3

Use this circuit to record your countdown sequence on the ISD1000A. Once you are satisfied with your recording, the circuit will no longer be needed. This circuit can be used as a stand-alone voice recorder circuit.

TIME-DELAYED MODEL ROCKET LAUNCH CONTROL

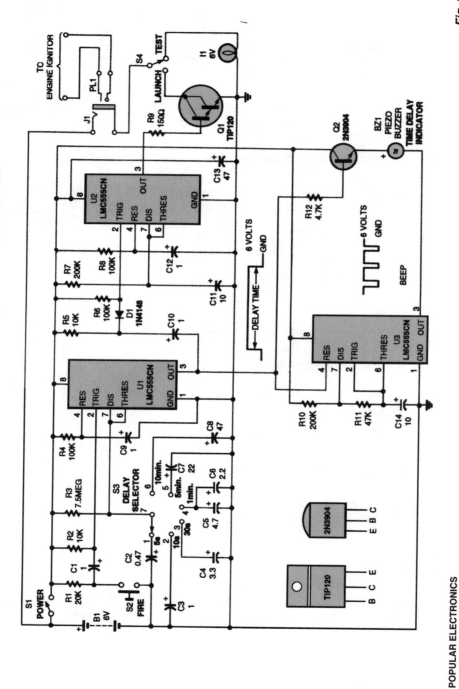

POPULAR ELECTRONICS

Fig. 95-4

The circuit, which gives a choice of six delay settings, consists of three 555 oscillator/timers (U1 to U3), a pair of transistors (Q1 and Q2), four switches (S1 to S4), a piezoelectric buzzer (BZ1), and a few support components. Power for both the control circuit and the rocket-engine igniter is provided by a 6-V power source that is composed of four AA-cell alkaline batteries. Closing switch S1 feeds power to the launch-control circuit, but does not initiate a launch sequence. A pair of series RC circuits

TIME-DELAYED MODEL ROCKET LAUNCH CONTROL *(Cont.)*

(R4-C9 and R8-C12, respectively) are used to debounce the RESET inputs (pins 4) of U1 and U2 (a pair of 555 oscillator/timer ICs). When S2 is closed, C1's negative terminal is connected to ground through the switch, momentarily pulling pin 2 of U1 (which is configured as a monostable multivibrator, or one-shot) low, activating it. Once triggered, U1's output goes high for an interval that is determined by R3 and one of six timing capacitors (C2 through C8). The timing capacitor is selected via DELAY SELECTOR switch S3. The high output of U1 at pin 3 causes Q2 to turn on, allowing the piezoelectric buzzer BZ1 to turn on. Monostable U1's output is also fed to the RESET input of U3 at pin 4, causing it to oscillate with a duty cycle of about 75 percent. As long as the output of U1 is high and the astable is oscillating, BZ1 beeps once every 4 s. The resistor discharges coupling capacitor C10 whenever U1's output goes high. At the end of the selected time delay, U1's output goes low. That low is coupled through C10 and D1 to the TRIGGER input (pin 2) of U2. Components R7 and C11 set U2's high-output interval to approximately 3 s. During that 3-s interval, U2's high output at pin 3 is fed to Darlington transistor Q1 through R9. With S4 in the LAUNCH position, Q1 grounds one end of the engine igniter, effectively connecting it to the battery's negative terminal.

MODEL ROCKET IGNITION CIRCUIT

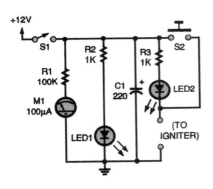

This circuit requires a 12-Vdc source. A battery pack of eight AA cells in series will work. Meter M1 gives you an indication of battery voltage; change the batteries when you see a noticeable drop in current. The circuit contains two LEDs: LED1 reminds you that the circuit is powered up, and LED2 lets you know that you have continuity through the igniter and are ready to fly. Capacitor C1 gives igniters that extra kick they occasionally need. Pushing the momentary push-button switch (S2) sends the rocket on its way. Use a red switch for that. The igniter leads are connected to the plugs marked *igniter*.

POPULAR ELECTRONICS   *Fig. 95-5*

96

Sawtooth Generator Circuits

The sources of the following circuits are contained in the Sources section, which begins on page 1043. The figure number in the box of each circuit correlates to the entry in the Sources section.

Simple Sawtooth Waveform Generator
Linear Sawtooth Generator
Linear Sawtooth Waveform Generator Circuit

SIMPLE SAWTOOTH WAVEFORM GENERATOR

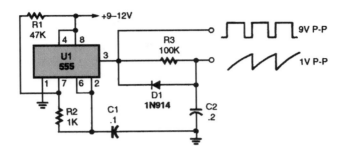

POPULAR ELECTRONICS

Fig. 96-1

This sawtooth generator is built around a 555 configured as an astable multivibrator.

LINEAR SAWTOOTH GENERATOR

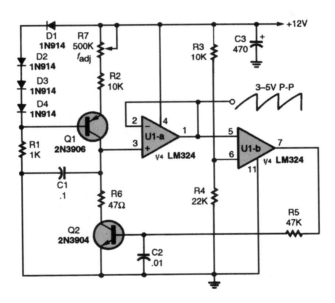

POPULAR ELECTRONICS

Fig. 96-2

Q1 is connected in a simple constant-current generator circuit. The value of Q1's emitter resistor sets the constant-current level flowing from the transistor's collector to the charging capacitor (C1). One op amp of an LM324 quad op amp IC, U1-a, is connected in a voltage-follower circuit. The input impedance of the voltage follower is very high and offers little or no load on the charging circuit. The

LINEAR SAWTOOTH GENERATOR *(Cont.)*

follower's output is connected to the input of U1-b, which is configured as a voltage comparator. The comparator's other input is tied to a voltage divider, setting the input level to about 8 V. The output of U1-b at pin 7 switches high when the voltage at its positive input, pin 5, goes above 8 V. That turns on Q2, discharging C1. The sawtooth cycle is repeated over and over as long as power is applied to the circuit. The sawtooth's frequency is determined by the value of C1 and the charging current supplied to that capacitor. As the charging current increases, the frequency also increases, and vice versa. To increase the generator's frequency range, decrease the value of C1, and to lower the frequency, increase the value of C_r. The output is about 3 to 5 V.

LINEAR SAWTOOTH WAVEFORM GENERATOR CIRCUIT

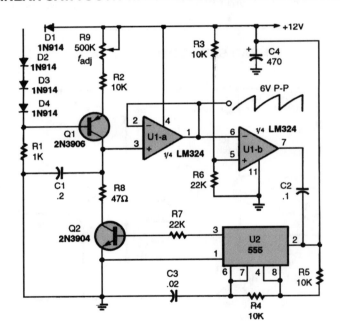

Fig. 96-3

In this circuit, Q1 is a constant-current source, whose output is controlled by R9. C1 charges linearly, and follower U1-a drives comparator U1-b. When the threshold set by R3 and R6 is reached, U1-b changes state and triggers pulse generator U2, gating on Q2 and discharging C1 through R8. The cycle then repeats. The period is approximately $8 \times C_1/I_1$, where I_1 is the collector current of Q1, as set by R9 and the base bias on Q1.

97

Seismic Radio Beacon Circuits

The sources of the following circuits are contained in the Sources section, which begins on page 1043. The figure number in the box of each circuit correlates to the entry in the Sources section.

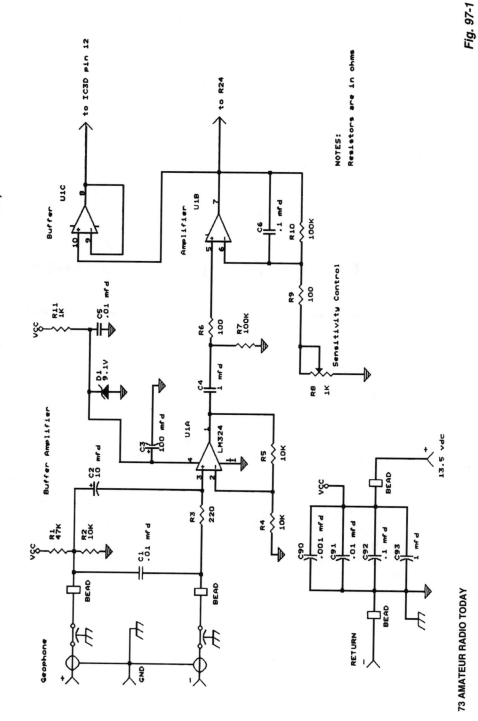

SEISMIC RADIO BEACON (GEOPHONE AMPLIFIER)

Fig. 97-1

This geophone amplifier and buffer was used as part of an amateur radio beacon that monitors seismic activity in an earth-quake-prone area of the United States.

894

SEISMIC RADIO BEACON (AUDIO VCO AND BUFFER)

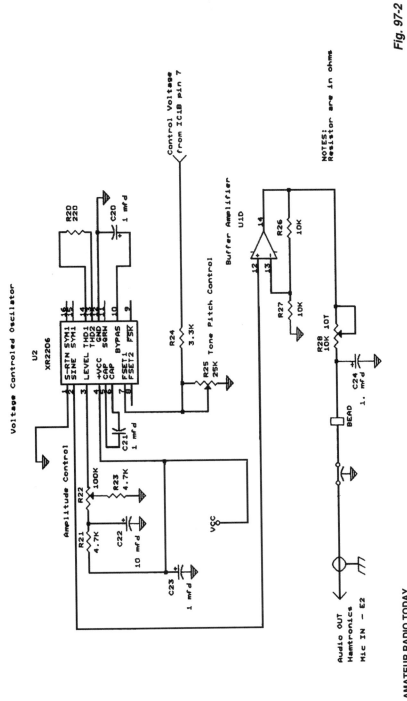

Fig. 97-2

This audio and VCO buffer is fed by a geophone amplifier and produces a tone modulated by seismic activity, as picked up by a geophone.

SEISMIC RADIO BEACON (THRESHOLD CIRCUIT)

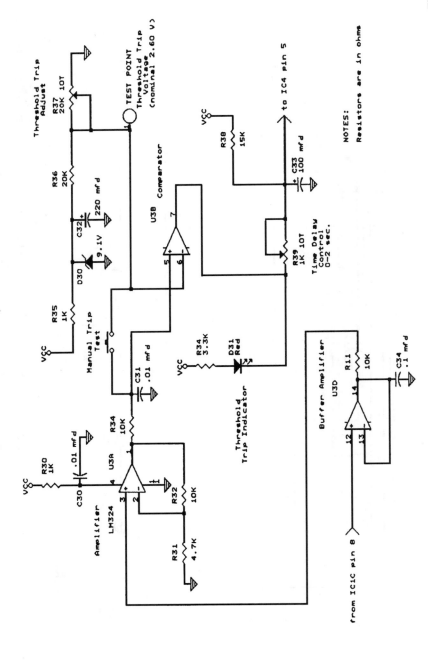

Fig. 97-3

73 AMATEUR RADIO TODAY

This threshold circuit is fed by a geophone amplifier and is actuated by seismic activity, as picked up by a geophone. The threshold is adjustable.

SEISMIC RADIO BEACON (TIMER AND SWITCHING)

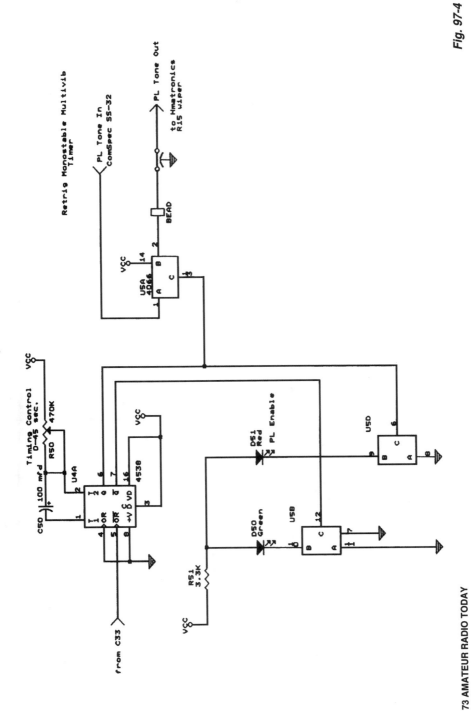

Fig. 97-4

This timing circuit is fed by a threshold circuit that is actuated by seismic activity, as picked up by a geophone. The timing is adjustable and controls a 123-Hz PL tone that feeds a transmitter.

SEISMIC RADIO BEACON (RF POWER AMPLIFIER)

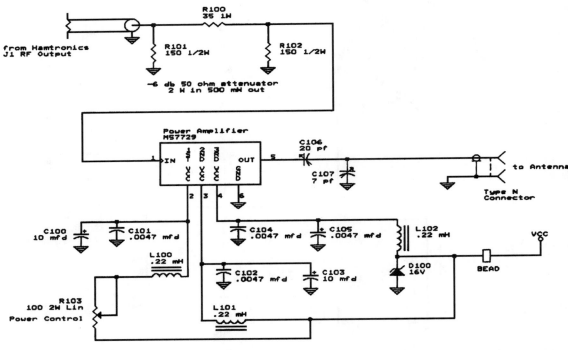

Fig. 97-5

This RF power amplifier operates at 432 MHz and was used as a power amplifier for a small transmitter used in an amateur radio beacon that also has seismic monitoring capability.

98

Shaft Encoder Circuits

The sources of the following circuits are contained in the Sources section, which begins on page 1043. The figure number in the box of each circuit correlates to the entry in the Sources section.

Shaft Encoder Pulse-Generating Circuit
Shaft Encoder

SHAFT ENCODER PULSE-GENERATING CIRCUIT

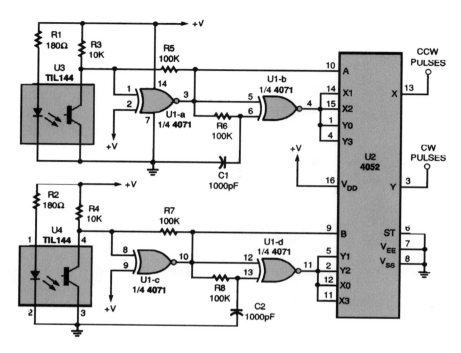

Fig. 98-1

This circuit produces one output pulse for every transition of the photosensor output or four pulses for every section or line pair. The exclusive-NOR gates U1-b and U1-c produce a pulse on every transition, and U2, a dual four-channel data selector, decodes them into "up" pulses or "down" pulses. The number of sections or lines that can be accommodated on a disk is determined by its diameter and the width of the sections. As an example, if the sections are made approximately 0.125" wide, 25 section pairs can put on a 2" diameter disk and 100 pulses will be generated with every revolution of the disk. The only limiting factor is that the sections must be wider than the photosensor's aperture at the point they pass in front of it. The alignment of the photosensor assembly is not critical. The only concern is that one sensor be over a light area while the other sensor is at a transition from light to dark or dark to light. Two sections of U1, U1-a and U1-c, act as buffers and waveshapers for the outputs of the sensor, while U1-b and U1-d generate a pulse at each transistor of the sensor's output. The 100-kΩ resistors between the output and the input of U1-a and U1-c provide positive feedback to speed up the rise and fall times and also generate hysteresis to eliminate noise during the transitions. The pulses produced by U1-b and U1-d at the transitions are the result of the delay produced by the RC circuits located at one input of each gate. For example, U1-b pin 5 follows the transition immediately, whereas pin 6 follows slowly. With the values shown, the two inputs are dif-

SHAFT ENCODER PULSE-GENERATING CIRCUIT *(Cont.)*

ferent for about 100 μs after every transition. While the inputs are different, the output is low. Therefore, U1-b generates a negative pulse for every transition of sensor U3 and U1-d generates a negative pulse for every transition of U4. The states of U3 and U4 determine whether the pulse is sent to the X output of the data selector (U2, pin 13) to indicate a CCW pulse or to the Y output (pin 3) to indicate a CW pulse.

SHAFT ENCODER

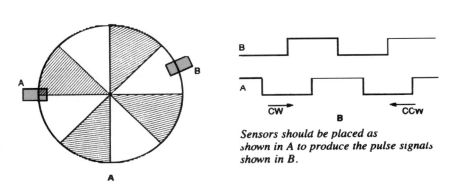

Sensors should be placed as shown in A to produce the pulse signals shown in B.

A

POPULAR ELECTRONICS

Fig. 98-2

    The electro-optical shaft encoder is a combination of encoding disk, photosensors, and counters. The encoding disk and photosensors generate a pulse train that can be counted to determine rate. The disk's direction of rotation can be sensed and used to determine count direction (for example, count up for clockwise and count down for counterclockwise). The disk need not be a precision part; it can be made from clear plastic with dark sections or lines painted on it. The number of lines on the disk determines how many pulses are generated per revolution. If two photosensors are used, two signals in quadrature are generated; from those, the direction can be sensed. A simple encoding disk is shown in A; note the alternating transparent and opaque sections. A pair of electro-optical photosensors are positioned (as shown) so that when one is centered over a section, the other is positioned over a transition. As the disk is rotated, each photosensor will be alternately illuminated and obscured and will produce outputs (as shown). A clockwise rotation is indicated when there is a positive transition (from dark section to light section) at sensor B while sensor A is low (obscured by a dark section), a positive transition at A while B is high (over a light section), a negative transition of B (going from light to dark) while A is high, and so forth. A counterclockwise rotation is indicated when there is a positive transition of B when A is high, a negative transition of A when B is high, etc.

99

Sine-Wave Generator Circuits

The sources of the following circuits are contained in the Sources section, which begins on page 1043. The figure number in the box of each circuit correlates to the entry in the Sources section.

SINE-WAVE DOUBLER

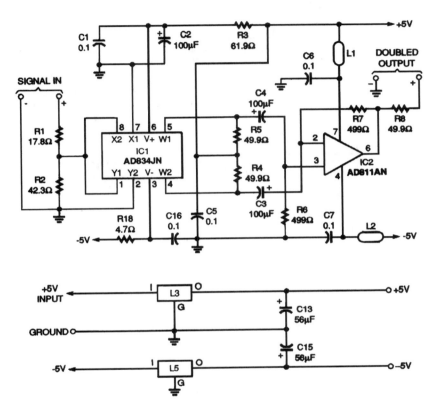

Fig. 99-1

This circuit uses a four-quadrant multiplier by Analog Devices to perform frequency doubling on a sine-wave input signal. By applying a sine wave to both x and y inputs, the multiplier multiplies the sine wave by itself, resulting in a sine-squared waveform. This waveform is equivalent to a dc level and a cosine component at twice the frequency and half the original amplitude, i.e., $\sin^2 x = (1 - \cos 2x)/2 - 3$ dB bandwidth is typically 30 MHz.

ONE-FILTER THREE-PHASE SINE-WAVE GENERATOR

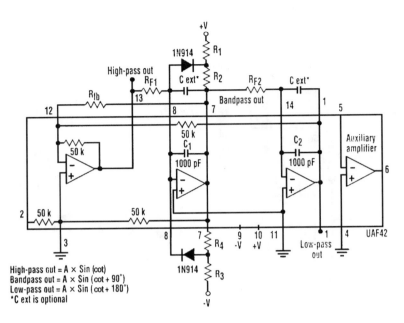

ELECTRONIC DESIGN

Fig. 99-2

It is possible to build a three-phase sine-wave oscillator using just one UAF42 state variable filter along with some resistors and diodes. Three output nodes are available: high-pass out, bandpass out, and low-pass out. The signals at the bandpass and low-pass out nodes are 90° and 180° out of phase, respectively, with those at the high-pass out node. An on-board auxiliary op amp is available for use as a buffer or gain stage. The frequency of oscillation is set with resistors RF1 and RF2 using the simple equation:

$$f_{osc} = 1/2\pi RC \quad (1)$$

where $R = R_{F1} = R_{F2}$, and $C = C_1 = (C_2) = 1000$ pF.

The maximum f_{osc} obtainable using the UAF42 state-variable filter is 100 kHz. However, distortion becomes a factor for frequencies above 10 kHz. Resistance R_1, R_2, R_3, and R_4 should be selected using the following equation to set the desired signal amplitude:

$$R_1/R_2 = R_3/R_4 = [(V_0 + V_{cc})/(V_0 - 0.15)] - 1 \quad (2)$$

Resistor R_{fb} provides a positive feedback path from the bandpass out node to the summing-amplifier input. This provides the necessary "startup" required to begin oscillation. Suggested values are as follows:

- If f_{osc} ±1 kHz, then $R_{fb} = 10$ MΩ.
- If 10 Hz $\leq f_{osc} <$ 1 kHz, then $R_{fb} = 5$ MΩ.
- If $f_{osc} <$ 10 Hz, then $R_{fb} = 750$ kΩ.

ONE-FILTER THREE-PHASE SINE-WAVE GENERATOR *(Cont.)*

To design a 1-kHz, 1.2-V peak oscillator, use Eq. 1 to calculate R_{F1} and R_{F2}:

$$R_{F1} = R_{F2} = 1/(2 \times \pi \times 1 \text{ kHz} \times 10^{-9}) = 159.2 \text{ k}\Omega$$

Use Eq. 2 to determine values for the signal-magnitude-setting resistances R_1/R_2 and R_3/R_4:

$$R_1/R_2 = R_3/R_4 = [(V_O + V_{CC})/(V_O - 0.15)] - 1$$

Assuming $V_{CC} = 15$ V, then

$$R_1/R_2 = R_3/R_4 = 15.4$$

Setting R_1 and R_3 equal to 15.4 kΩ and R_2 and R_4 equal to 1 kΩ would provide the proper resistor ratios. These resistors act as loads to the internal op amp

---

STABLE SINE-WAVE GENERATOR

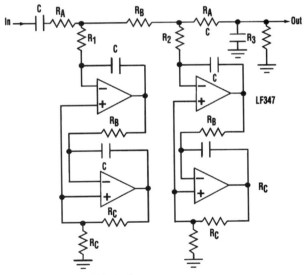

Notes: Values for 1-kHz signal
C = 0.01 μF, 1%*
R_A = 10 k, 1%*
R_B = 20 k, 1%
R_C = 20 k nominal; matched required
R_1 = 1.407 k calculated; 1.40 k, 1% used
R_2 = 506.6 Ω calculated; 511 Ω, 1% used
R_3 = 100 k nominal; for dc bias
*Looser tolerance with matching is okay if R_1 and R_2 are adjusted.

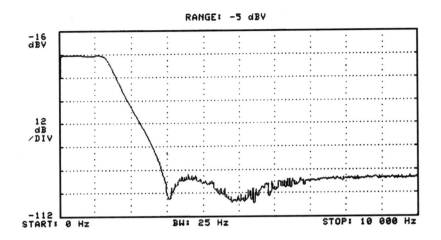

RANGE: -5 dBV

START: 0 Hz   BW: 25 Hz   STOP: 10 000 Hz

ELECTRONIC DESIGN

Fig. 99-3

Semidigital circuits (e.g., crystal oscillators and dividers) can create square waves of very stable amplitude and frequency. Removing the odd harmonics is a reasonable task for a filter. The obvious solution, a narrowband filter, isn't acceptable because analog types are notorious for poor stability. Digital and semidigital types (e.g., switched capacitor) are better in this respect, but they add their own noise and harmonics. The task can be accomplished using the filter shown. Without R1 and R2, it is an active version of a five-pole passive low-pass LC ladder. This type has excellent amplitude stability in the passband, 30 dB/octave slope outside the passband, low component sensitivity, and a capacitor to ground at the output, which ensures continuous high-frequency rolloff and minimizes stray noise pickup. The rejection would be inadequate at the third and fifth harmonics, but notches at these frequencies can be created with just two more resistors, R1 and R2. This turns the device into an elliptic-like filter.

100

Siren, Warbler, and Wailer Circuits

The sources of the following circuits are contained in the Sources section, which begins on page 1043. The figure number in the box of each circuit correlates to the entry in the Sources section.

Warble Tone Oscillator
Wailing Witch Noisemaker
Siren Generator

WARBLE TONE OSCILLATOR

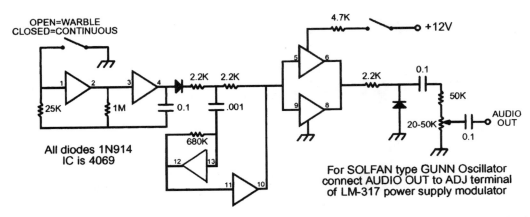

73 AMATEUR RADIO TODAY

Fig. 100-1

Originally used for microwave transmitter testing, this warble oscillator produces a tone that jumps between two frequencies.

WAILING WITCH NOISEMAKER

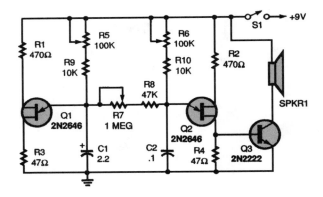

POPULAR ELECTRONICS

Fig. 100-2

Two unijunction transistors, Q1 and Q2, are connected in a dual-oscillator circuit. Transistor Q1 is connected in a very-low-frequency relaxation oscillator circuit, with C1, R1, R5, and R9 setting the operating frequency. Transistor Q2 is connected in a similar oscillator circuit that operates at a much higher audio frequency. That oscillator's frequency is set by R6, R10, and C2. Resistor R7, a 1-MΩ potentiometer, sets the mixing level of the two oscillators. By adjusting R5, R7, and R6, many strange sounds can be created.

SIREN GENERATOR

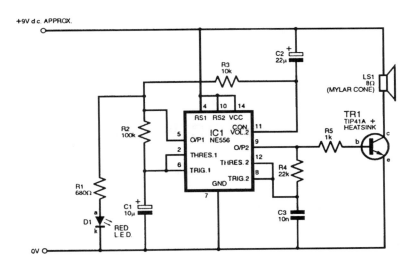

Fig. 100-3

An adaptable siren generator circuit with a multitude of uses is shown. It is based around a 556 twin-timer chip, IC1. An audio tone is created by one timer section and is directly coupled to the driver transistor, TR1. The other half of the timer is used to modulate the frequency of the audio tone using the CONTROL VOLTAGE pin terminal of the audio-oscillator section, pin 11. If capacitor C2 is omitted from the circuit, a twin-tone alarm generator will be created. However, with C2 in place as shown, a "wailing" tone is produced. Capacitor C1 governs the rate of tone change and LED D1 flashes for extra effect. A waterproof Mylar-coned 8-Ω loudspeaker was used for LS1, and the volume can be adjusted for different impedances by altering the value of resistor R5. The circuit operates from approximately a 9-V rail.

101

Sound-Effects Circuits

The sources of the following circuits are contained in the Sources section, which begins on page 1043. The figure number in the box of each circuit correlates to the entry in the Sources section.

Electronic Trombone
Electronic Noisemaker
Cricket Chirp Simulator
Electronic Parrot

ELECTRONIC TROMBONE

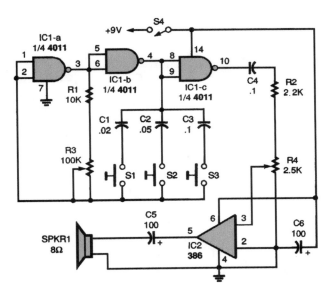

POPULAR ELECTRONICS

Fig. 101-1

The heart of the trombone is R3, a slide potentiometer. Two gates of a quad two-input NAND gate, IC1-a and IC1-b, make up a simple audio-oscillator circuit, with R1, R3, C1, C2, and C3 setting the oscillator's frequency. The oscillator's output is buffered by IC1-c, which also supplies the drive signal for the power amplifier, IC2. The trombone's output level is set by R4. A slide handle, made of plastic or wood, should be attached to the slider of R3. The complete circuit, enclosed in a small plastic cabinet, with the three push-button switches, S1 to S3, mounted in a convenient location for playing. Just press one or more of the tone-control switches, S1 to S3, and work the slide.

ELECTRONIC NOISEMAKER

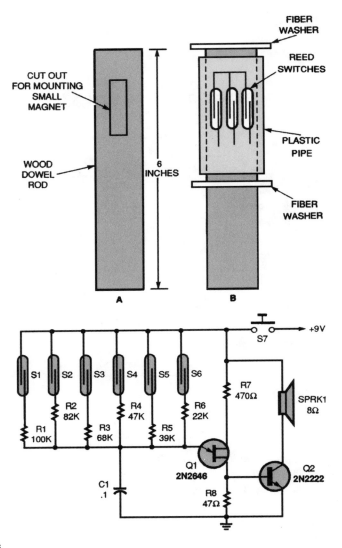

Fig. 101-2

A single unijunction transistor, Q1, generates the different tones, and a general-purpose NPN transistor, Q2, raises the level to drive a small speaker. The six reed switches, S1 to S6, are mounted around the outside of the plastic pipe (Fig. B). As the pipe turns over the magnet, the reed switches

ELECTRONIC NOISEMAKER *(Cont.)*

open and close, tying different-valued resistors to the oscillator's frequency-control circuitry. It is also possible for two reed switches to be closed at the same time. When that happens, the output tone will be much higher in frequency than when only one switch is activated. The oscillator's frequency range can be lowered by increasing the value of C_1, and raised by decreasing C_1. To operate, just grab the noisemaker by the dowel rod and give it a twist to start the plastic case turning.

CRICKET CHIRP SIMULATOR

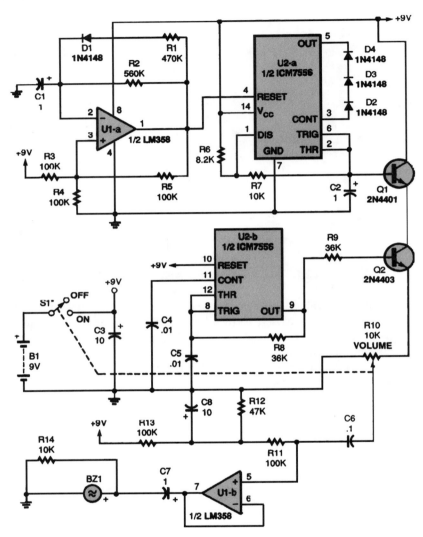

Fig. 101-3

Power is supplied from a 9-V battery, B1; current drain is a little under 2 mA, so an alkaline battery should last more than 250 hours. Switch S1, the power switch, is a part of potentiometer R10, which also acts as a volume control. A duty cycle is generated by op amp U1-a, which is configured as a pulse generator. Diode D1 establishes a fast charging rate that generates a 0.22-second period, which matches the duration of a cricket's seven-beat chirp. The discharge rate is established by R2 to generate a reset time of 0.35 second. One half of an ICM7556 CMOS timer, U2-a, is used to simu-

CRICKET CHIRP SIMULATOR *(Cont.)*

late the amplitude modulation of a chirp. Timer U2-b is configured as a conventional pulse generator with a symmetrical duty cycle and an output frequency of 2 kHz. The output of U2-b is used to switch Q2 on and off, which produces a voltage across volume control R10 that has the magnitude of the instantaneous voltage across C2 with a 2-kHz sampling rate. A portion of that voltage is applied to U1-b, which is configured as a unity-gain, noninverting buffer that drives the piezo element.

ELECTRONIC PARROT

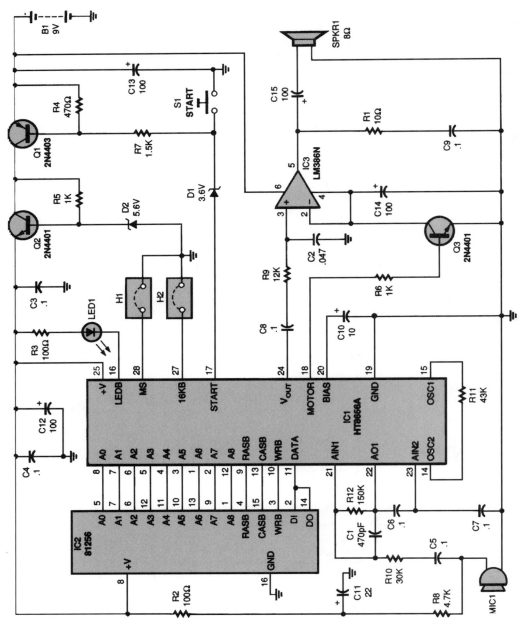

POPULAR ELECTRONICS

Fig. 101-4

ELECTRONIC PARROT *(Cont.)*

This circuit uses a Holtek HT8656 CMOS LSI chip and an 81256 256K × 1 dynamic RAM chip to form a voice record/playback circuit. The sampling rate is either 16 or 32 kbits per second, which allows 8- or 16-second recording time with the 256K DRAM. A 64K DRAM can be used if shorter times are suitable for your application. Power is via a 9-V battery.

102

Sound-/Voice-Activated Circuits

The sources of the following circuits are contained in the Sources section, which begins on page 1043. The figure number in the box of each circuit correlates to the entry in the Sources section.

SOUND-ACTIVATED SWITCH

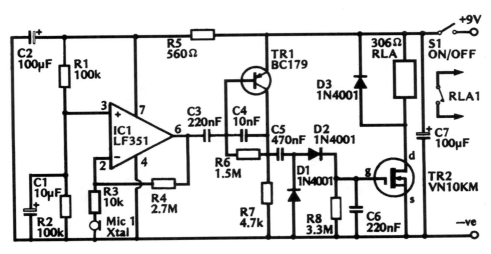

Fig. 102-1

The circuit uses IC1 as a low-noise preamplifier with a voltage gain of 270 times. This is followed by a second stage of amplification, which produces a similar voltage gain. The second stage uses TR1 in the common-emitter mode; this is a conventional arrangement, except for the inclusion of C4. This capacitor provides a considerable amount of high-frequency attenuation, which is necessary in order to prevent instability. Capacitor C5 couples the greatly amplified output of TR1 to a rectifier and smoothing circuit, which gives a positive dc output signal that is roughly proportional to the input-signal level. If the input signal is sufficiently strong, the bias voltage at the gate of VMOS device TR2 will be adequate to bias this transistor into conduction, and the relay that forms its drain load is then activated. A pair of normally open relay contacts are used to control whatever item of equipment is operated by the unit. Of course, the voltage at the gate of TR2 soon decays, as C6 discharges through R8 if the input signal ceases, and the relay is then switched off. The decay time is roughly 1 second, which is about the optimum time for most applications. The attack time of the circuit is only a fraction of a second, and the unit responds almost immediately when a sound is initially picked up by the microphone.

MELODY SOUND-TO-LIGHT CIRCUIT

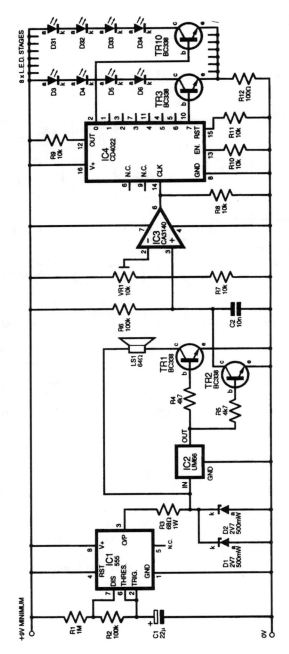

Fig. 102-2

EVERYDAY PRACTICAL ELECTRONICS

A method of synchronizing a simple light show to a musical chip is given by the circuit diagram shown. Here, a display of eight rows of four LEDs advances with each musical note. IC2 is a UM66 three-pin melody chip selected to generate the desired tune. When power is applied to IC2, the sequence plays once and then stops. Power must be reapplied to initiate a further rendition. IC1 is a 555 astable, which will constantly reenable the tune chip and repeat the melody. However, the maximum supply voltage of the UM66 range is 3.3 V; hence, a simple zener-diode supply based around two 500-mW 2V7 types (D1 and D2) is used. TR1 is a directly driven transistor amplifier and the melody is played through loudspeaker LS1. Each musical note, when present, also biases transistor TR2 on, which shunts capacitor C2 to ground. In the absence of a note, C2 charges, which causes op amp IC3 to change state at a point determined by preset potentiometer VR1. The output from IC3, pin 6, provides a clock pulse for IC4, a 4022 divide-by-8 counter. Its outputs drive a buffer transistor sequentially to illuminate a series of four LEDs. The light display then advances with each note.

SOUND-OPERATED SWITCH

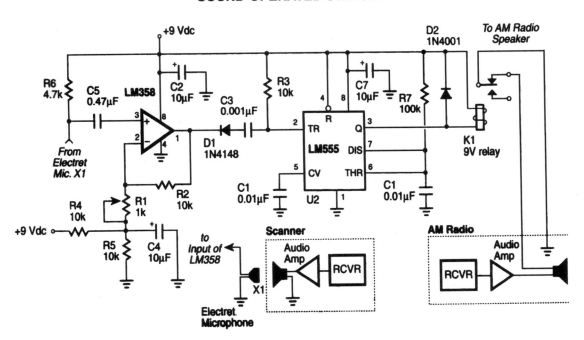

NUTS AND VOLTS

Fig. 102-3

An electret microphone is placed directly in front of the scanner speaker. When the microphone (X1) picks up sound (usually voice) from the speaker, the sound signal is amplified and detected, activates a retriggerable one-shot circuit, and controls the relay (K1). The relay controls the AM radio speaker, muting it when sound is detected from the scanner.

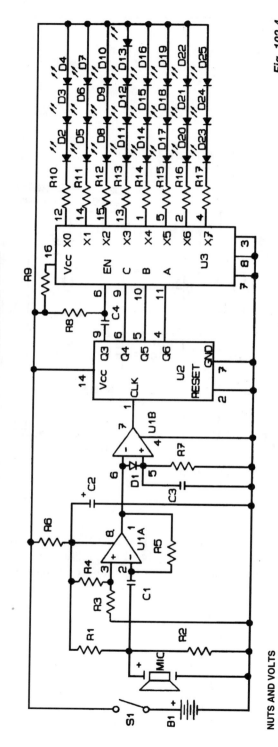

SOUND-ACTIVATED CHRISTMAS TREE

Fig. 102-4

Switch S1, an ON/OFF switch, directs power from the 9-V battery to the circuit. The first half of an LM358, U1A, is configured as a high-gain amplifier. Resistors R3 and R4 form a voltage divider that sets the bias voltage at pin 3, holding the amplifier in the Off state. Resistors R1 and R2, also forming a voltage divider, set the bias point for the microphone. When the audio input signal produced by the microphone exceeds the bias voltage, set by resistor R5, U1A turns on and operates as a high-gain amplifier (with the gain being set by resistor R5). Capacitor C1 couples the microphone's signal to the amplifier's inverting input, pin 2. Resistor R6 and capacitor C2 are used as a filter network for the op amp. The second half of U1, U1B, is configured as a peak signal detector. It is triggered by the negative half-cycle output of U1A. Each negative half-cycle of the signal produces a positive clock signal to the input of U2, the seven-stage binary ripple counter. Capacitor C3 stores the average peak level of the input signal, and resistor R7 determines C3's discharge time. Integrated circuit U2 produces a binary count on its Q3 through Q6 outputs that is proportional to the frequency of the audio input signal. The output of Q3 is connected, via capacitor C4, to the ENABLE pin of U3. When the input audio signal causes U2 to conduct, the relatively high-frequency output of Q3 enables U3. Resistor R3 maintains U3 in the Disabled mode in the absence of any audio input signal. U3's output lines, X0 through X7, are turned on and off in rapid sequence, causing the LEDs to display a random flashing effect that is proportional to the original audio signal.

SIGNAL-CONTROLLED SWITCH

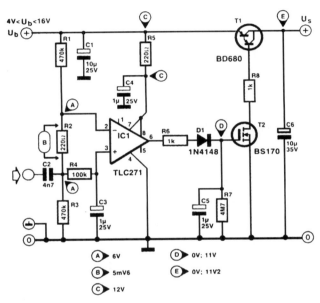

Fig. 102-5

This signal-controlled switch actuates or deactuates AF equipment, including preamplifiers, power amplifiers, and filters. It is particularly useful for battery-operated equipment because, owing to the low current drain of 12 to 14 µA, there is no need to switch the AF equipment off. When the switch has not detected any AF signal for 10 seconds, it switches off the supply to the equipment.

VOICE-ACTIVATED TAPE RECORDER SWITCH

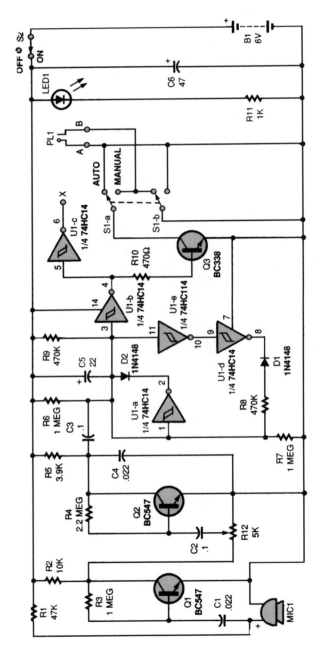

Fig. 102-6

POPULAR ELECTRONICS

A schematic of the tape recorder switch is shown. The circuit can be powered by B1, which is four AA batteries in series or any 6-Vdc supply. Audio signals from the electret microphone, MIC1, are coupled via capacitor C1 to two audio-amplification stages centered around transistors Q1 and Q2. A 5000-Ω potentiometer, R12, provides a sensitivity control that sets the audio level at which the switch will activate a tape recorder. The audio input to the digital switching section of the circuit is biased at half the supply voltage by two 1-percent, 1-MΩ, metal-film resistors (R6 and R7). Integrated circuit U1 is a 74HC14 hex Schmitt-trigger inverter. With a 6-V power supply, the Schmitt inverter gates will have a 1-V hysteresis gap between 2.2 and 3.2 V. That means that the input to a Schmitt inverter must rise above 3.2 V for the output to go low, but must fall below 2.2 V for the output to return to a high. The low on pin 3 of U1-b in the presence of audio is presented to pin 11 of U1-e and ends up at pin 8 of U1-d. That will pull pin 1 low in the absence of an audio signal from the microphone, making pin 2 high. Diode D2 blocks the dis-

VOICE-ACTIVATED TAPE RECORDER SWITCH *(Cont.)*

charging of C5 through its original charging path, so C5 starts to discharge through R9. After that delay, the voltage at pin 3 rises above 3.2 V, thus removing the connection from terminal A to ground and turning off the tape recorder. A continuous audio input to pin 1 will hold pin 4 high, but soon as it ceases, the time-out process begins. When power is applied to the circuit using switch S2, LED1 lights. However, if the ground connection to R11 is removed, and that end of the resistor is then connected to terminal X (pin 6 of U1-c), the LED will instead light only when the output of the circuit is active.

VOICE-EFFECTS SIMULATOR CIRCUIT

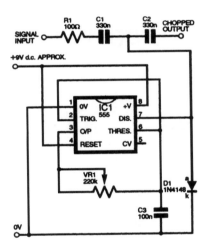

EVERYDAY PRACTICAL ELECTRONICS

Fig. 102-7

The circuit design shown produces a *Dr. Who* "Dalek voice" effect by chopping an audio signal at a low frequency. The best frequency seems to be around 50 to 90 Hz, and this is generated by IC1, a 555 timer. The arrangement shown is not an ordinary 555 astable multivibrator, but a "hysteresis" oscillator, which frees the 555 internal discharge transistor (pin 7) to act as a chopper, shunting the signal to 0 V internally. The "chopped" frequency is set by VR1, which is adjusted to give the most realistic sound. The input signal should be in the region of 50 to 150 mV rms from a low-impedance source—e.g., possibly a dynamic microphone—to avoid clipping the signal. The diode D1 is optional, and prevents the signal from losing symmetry, if overdriven. The output signal is fed to an external amplifier. The two dc-blocking capacitors C1 and C2 are optional and are needed only if any dc bias is present on the input signal side.

103

Stroboscope Circuits

The sources of the following circuits are contained in the Sources section, which begins on page 1043. The figure number in the box of each circuit correlates to the entry in the Sources section.

12-V STROBOSCOPE

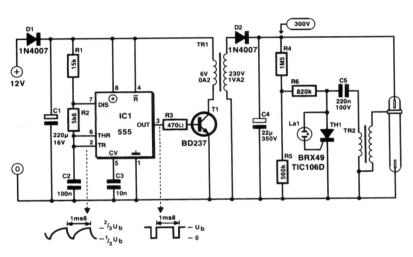

ELEKTOR ELECTRONICS

Fig. 103-1

Diode D1 is a polarity-protection device that can be omitted. A 555 is used as an astable with a frequency of about 0.7 kHz. When T1 is on, a current flows through the 6-V winding of TR1. When the collector voltage of T1 is nearly 0, the potential at the anode of D2 must be negative. When T1 becomes reverse-biased, its collector voltage rises to about 12 V and the potential at the anode of D2 must then be positive so that C4 can be charged. Never operate the converter without a load. It is, perhaps, advisable to shunt C4 with a 100-kΩ, 1-W resistor. The converter charges C4 to about 300 V. This causes a potential at junction R6-C5 of about 100 V. The neon lamp then comes on, so a gate current flows into the thyristor. This comes on and clears the way for C5 to discharge through TR2. This starting transformer produces a secondary voltage of a few thousand volts. This is sufficient for the xenon tube to strike and at the same time discharge C4. Then the operation can start again. The circuit draws a current of about 250 mA, but this depends on the flashing rate and the type of xenon tube.

STROBOSCOPE TIMING LIGHT

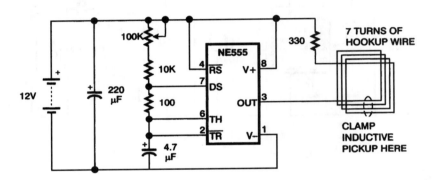

Fig. 103-2

The NE555 produces brief pulses of current in the wire, which is wound into a loop to increase the amount of induction. To the timing light, these pulses look like the pulses that flow through a spark-plug wire. Bear in mind that the maximum frequency of a timing light is about 20 Hz. Above that frequency, timing lights tend to skip pulses. You might be able to speed up a timing light by reducing the value of the high-voltage capacitor across the flashtube; the light will then be dimmer, but the capacitor will charge up more quickly for the next flash.

STROBOSCOPE FLASH UNIT

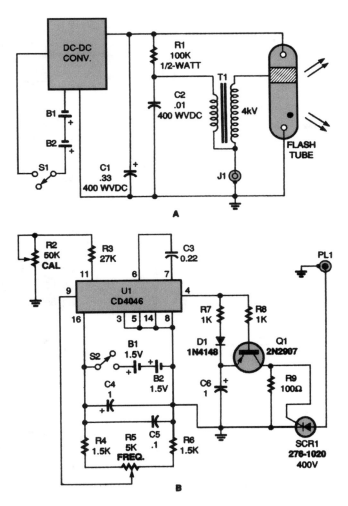

Fig. 103-3

If you retrofit a flash unit (shown in A), it can accept signals from the circuit in B to operate as a precision stroboscope.

FREEZE-FRAME STROBE

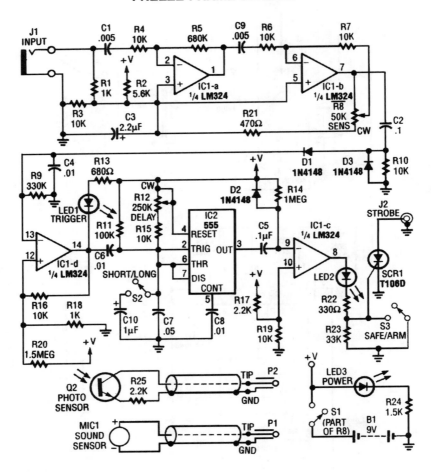

1994 ELECTRONIC EXPERIMENTERS HANDBOOK

Fig. 103-4

A sensor input from a mike or photodetector is amplified and used to trigger an SCR, which, in turn, triggers a camera strobe. This circuit is useful for freeze-frame photos of phenomena, such as drops of water, explosions, shots, etc.

SOLID-STATE STROBOSCOPE

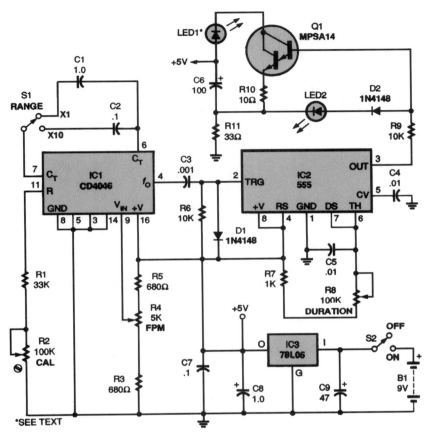

POPULAR ELECTRONICS

Fig. 103-5

The schematic diagram for the stroboscope is shown. When switch S2 is closed, a 9-V battery B1 is connected to IC3, a 78L05 voltage regulator, which provides a 5-V source to the circuit. Capacitors C8 and C9 filter the supply. A CD4046 phase-locked loop, IC1, provides a low-frequency square wave to the circuit. Potentiometer R4—a 5000-Ω, 10-turn, precision potentiometer with a built-in turns counter—provides a way to adjust and get a direct readout of the FPM rate from 0 to 1000. Switch S1 lets you select the multiplier affecting that rate. Setting S1 to ×1, therefore, gives you a 0- to 1000-FPM range, whereas setting S1 to ×10 results in a 0- to 10,000-FPM range. The values of the multiplier settings are determined by capacitors C1 and C2. A 555 timer, IC2, configured in a monostable mode, provides a pulse with a width that is adjustable via R8. That audio-taper potentiometer allows the duration or pulse width to be varied from 10 μs at the minimum setting to 1000 μs at the maximum setting. An MPSA14 Darlington transistor, Q1, is configured as a capacitive-discharge-type current sink, which provides an approximately 90-mA pulse (for the pulse duration set by R8) through LED1, the flash output. The super-bright LED specified for LED1 has a maximum rating of 100 mA.

931

104

Switching Circuits

The sources of the following circuits are contained in the Sources section, which begins on page 1043. The figure number in the box of each circuit correlates to the entry in the Sources section.

SHUNT PIN-DIODE RF SWITCH

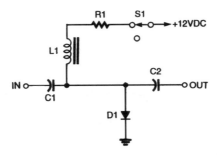

POPULAR ELECTRONICS

Fig. 104-1

D1 is a PIN diode (MV3404) used as a switch, with dc bias supplied via R1 and L1. This shunt PIN-diode switching circuit directs signals to ground when D1 is forward-biased.

BASIC PIN-DIODE RF SWITCH

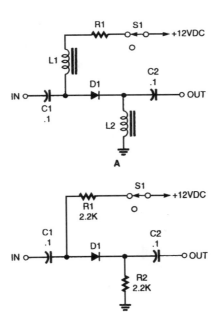

POPULAR ELECTRONICS

Fig. 104-2

D1 is a pin diode (MV3404) used as a switch, with dc bias supplied via R1. L1 and L2 are RF chokes. An alternative approach eliminating the RF chokes is also shown.

VHF TRANSCEIVER T/R SWITCH

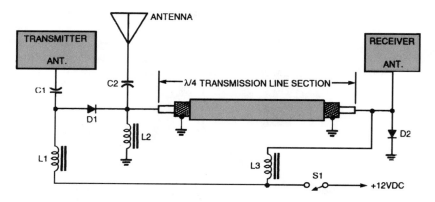

POPULAR ELECTRONICS

Fig. 104-3

This is a transceiver transmit/receive switch that uses a PIN diode instead of a relay. Diodes can be MV3404 or similar types.

SERIES-SHUNT PIN-DIODE RF SWITCH

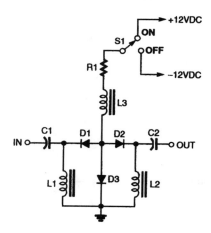

POPULAR ELECTRONICS

Fig. 104-4

A combination of series and shunt switching results in superior isolation between the input and output when in the OFF condition. Diodes can be MV3404 or similar types.

PIN-DIODE HIGH-ISOLATION SWITCH CIRCUIT

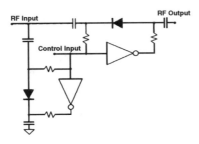

Fig. 104-5

This circuit is a basic PIN-diode switch using series- and shunt-diode building blocks. R is chosen from $I_{diode} = (V_{cc} - V_{diode}) / R$, where I_{diode} is the specified diode current and R is total dc circuit resistance.

SPDT PIN-DIODE HIGH-ISOLATION SWITCH CIRCUIT

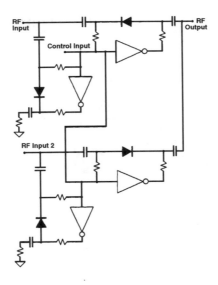

Fig. 104-6

This circuit is a basic SPDT PIN-diode switch using a series- and shunt-diode building block. Resistance R is chosen from $I_{diode} = (V_{cc} - V_{diode})/R$, where I_{diode} is the specified diode current and R is total dc circuit resistance.

PNP SWITCHES

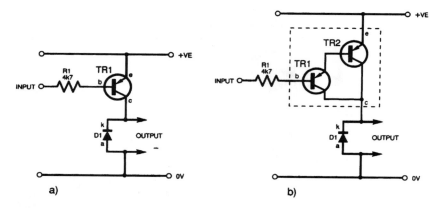

EVERYDAY PRACTICAL ELECTRONICS

Fig. 104-7

This figure shows two examples of a PNP switch: (*a*) a single transistor and (*b*) a Darlington transistor.

AC POWER CONTROLLER FOR VCR

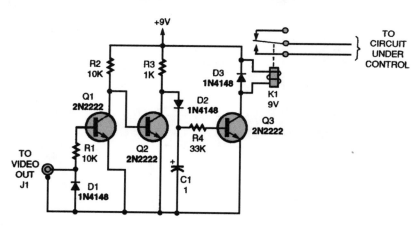

POPULAR ELECTRONICS

Fig. 104-8

This circuit can be used to switch on a manual-control TV when a VCR signal is present. When the tape starts, video is present at J1 from the VCR video output jack. This is detected, and the dc produced in the detector (D1) circuit turns on Q1, cutting off Q2, and allowing relay driver Q3 to actuate the relay K1.

3.3- OR 5-V ANALOG SWITCHES

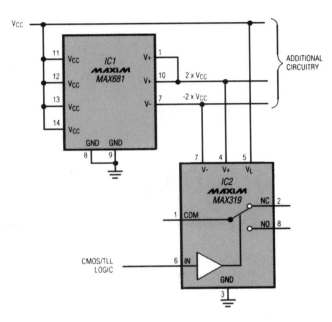

MAXIM

Fig. 104-9

By adding a single component to a 3-V-only or 5-V-only board, you can operate conventional CMOS analog switches with performance approaching that specified with ±15-V supplies. This means fast switching, low ON resistance, CMOS/TTL compatibility, low power consumption, and a signal range ($\pm V_{cc}$) that exceeds the input supply range (V_{cc} to ground). Simply add a charge-pump voltage converter (IC1), which produces ± 2-V_{cc} outputs from a V_{cc} input. These unregulated voltages ensure reliable switch operation for V_{cc} levels as low as 3 V. Logic thresholds for the switch remain unaffected. A V_{cc} of 3 V (for instance) produces ±6 V rails for the switch (IC2), resulting in ON resistance <30 Ω, switching times <200 ns, leakage <0.1 nA, and V_{cc} current <0.5 mA. Raising V_{cc} to 5 V produces ±10-V rails, resulting in ON resistance <20 Ω, switching times <150 ns, leakage <0.4 nA, and V_{cc} current <1.3 mA. IC1 can easily power additional switches and/or low-power op amps, but more than a few milliamperes of load current degrades performance by lowering the unregulated supply rails.

105

Telephone-Related Circuits

The sources of the following circuits are contained in the Sources section, which begins on page 1043. The figure number in the box of each circuit correlates to the entry in the Sources section.

PHONE-IN-USE INDICATOR

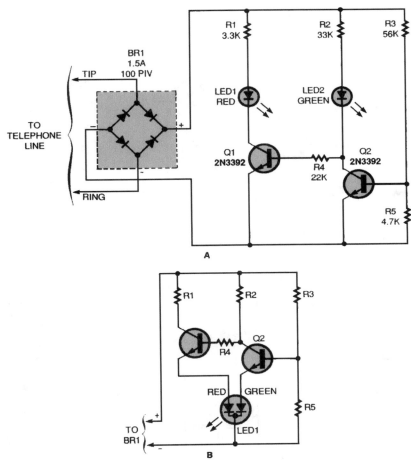

POPULAR ELECTRONICS

Fig. 105-1

Tired of having your phone calls interrupted by others picking up the receiver? This circuit (A) will show others when the phone is in use. If you would like to experiment with a tricolor LED, check out the optional schematic (B).

When all telephones are on-hook, Q2's base is turned on by a voltage-divider circuit, consisting of R3 and R5. (The value shown for R5 causes the device to switch over at about 9 V; it can be changed to facilitate other voltage levels.) Transistor Q2 allows current to flow through R2 and LED2, indicating that the phone line is not in use. It also effectively grounds the base of Q1 and forces LED1 to remain off. When the voltage drops because a telephone goes off-hook, Q2 stops conducting, which allows a little current to flow from R2, LED2, and R4 to Q1's base. When that occurs, Q1 conducts, energizing LED1 and LED2 is deprived of sufficient current to glow. The bridge rectifier compensates for a possible reversal between the tip and ring wires, and rectifies the ring signal.

REMOTE TELEPHONE BELL RINGER

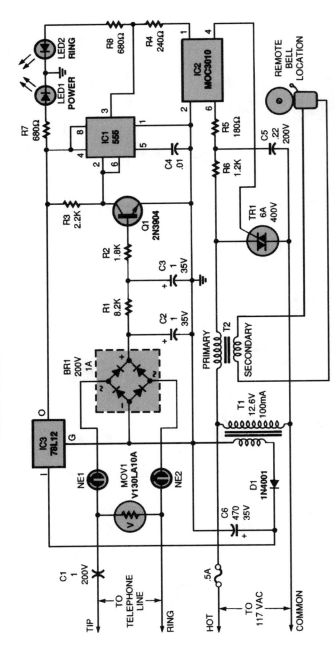

Fig. 105-2

The telephone-line extension bell ringer shown will enable you to add a remote ringer in your garage or some other area where a ringing telephone cannot be heard. Up to four ringers can be used on a single telephone line, and a remote bell can be used 100 feet or more away from the unit. By substituting a light bulb for T2 and dispensing with the bell, the circuit can be made useful for the hearing-impaired. About 50 to 60 V dc is present between the tip and ring (red and green) wires of an unoccupied telephone line. Capacitor C1 blocks that dc voltage. The MOV just shunts any dialing pulses generated by a rotary phone that might be on the same line. To make the phone ring, the tip and ring wires deliver an ac signal of between 90 and 130 V to the phone. That ac signal is coupled through C1 to the two neon lamps, NE1 and NE2. Those neon bulbs provide line isolation be-

REMOTE TELEPHONE BELL RINGER *(Cont.)*

tween the unit and the telephone line. They also neon fire (ionize) when more than 100 V is present on the phone line (in other words, during the ring signal). When they fire, they form a three-step voltage divider with the bridge rectifier. The voltage across the bridge is rectified, then filtered by R1, R2, C2, and C3, and causes Q1 to conduct. Then pins 2 and 6 of U1 go low, causing pin 3 of U1 to go high. The optoisolator and triac then turn on, applying power to the remote bell through a doorbell transformer (T2).

REMOTE TELEPHONE RINGER

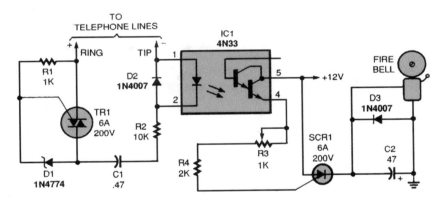

POPULAR ELECTRONICS

Fig. 105-3

When there is no ring signal, the phone line's on-hook voltage (about 50 V) keeps the triac (Radio Shack 276-1001 or equivalent) from switching on, so the optocoupler doesn't conduct. When the phone rings, terminal 4 of the optocoupler feeds pulses through R3 and R4 to the gate of SCR1. The activated SCR connects the bell to the 12-V supply. Pulses are about 400 Hz, so the bell might sound a bit rough. The opening of the bell's breaker points cuts the thyristor's holding current to stop it from conducting when the gate signal stops. A 12-V, 2-A power supply operates the unit. Adjust R3 high enough so that the SCR is not triggered between rings, but low enough to trigger it when the phone is ringing.

SUBSCRIBER-LINE INTERFACE-CIRCUIT POWER SUPPLY

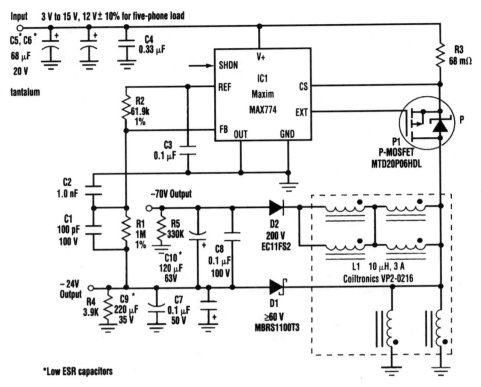

ELECTRONIC DESIGN

Fig. 105-4

A new multiwinding transformer (configurable by the user for a variety of applications) enables an inverting controller to produce the high negative voltages required by an ISDN board or other telephone line card. Such line cards use a subscriber-line interface circuit (SLIC), such as the 79R79 ringing SLIC from AMD. This IC generates the off-hook and on-hook signal transmission, ring-tone generation, and ring-tip detection that constitute an analog telephone interface. For off-hook signal transmission, it requires a tightly regulated −24 or −48 V; to generate ring tones, it requires a loosely regulated −70 V. The "five-ringer-equivalent" requirement demands 9 to 10 W from the −70-V output, which translates to a full-load I_{out} of about 150 mA. IC1 is an inverting switching regulator that usually converts a 3- to 16-V input to a fixed output of −5 V or an adjustable output. In the circuit shown, three pairs of windings in series (provided by a single off-the-shelf multiwinding transformer) enable IC1 to generate the high voltages needed by a SLIC IC1 (D1). Connecting a diode and output capacitors (C7 and C9) at the first or second pair of windings produces −24 V (as shown) or −48 V, respectively. Feedback to the IC via R1 and R2 achieves tight regulation at this output. The transformer turns ratios establish a loose regulation at the −70-V output. The circuit can service a five-

942

telephone load (10 W) from an input of 12 V, ±10 percent. It operates down to 3 V and produces about 2.4 W at 3.3 V and 3.9 W at 5 V. The −70-V output depends on cross-regulation with respect to the −24-V output. It is, therefore, affected by relative loading on the two outputs (i.e., whether one is heavily loaded and the other lightly loaded, or vice versa).

FM TELEPHONE TRANSMITTER

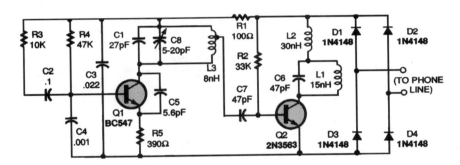

POPULAR ELECTRONICS

Fig. 105-5

The circuit connects in series with either the tip or the ring (green or red) telephone line. Power for the circuit is full-wave bridge-rectified from the phone line by diodes D1 through D4. Transistor Q1, capacitors C1 and C8, and inductor L3 form an FM oscillator that operates at a frequency of around 93 MHz. Variable capacitor C8 allows the oscillator frequency to be adjusted between 90 and 95 MHz. To move the tuning area up to the 98- to 105-MHz range, C1 must be replaced with a 10-pF capacitor. Audio from the phone line is coupled through R3 and C2 to the base of Q1, where it frequency-modulates the oscillator. Transistor Q3, inductor L1, and capacitor C6 form a power amplifier. The signal tapped off L3 in the oscillator circuit is fed to the base of transistor Q2, and the FM signal is transmitted from Q2's collector. Inductor L2 is a radio-frequency shunt that decouples power and audio from the amplifier circuit.

TELEPHONE-LINE SIMULATOR

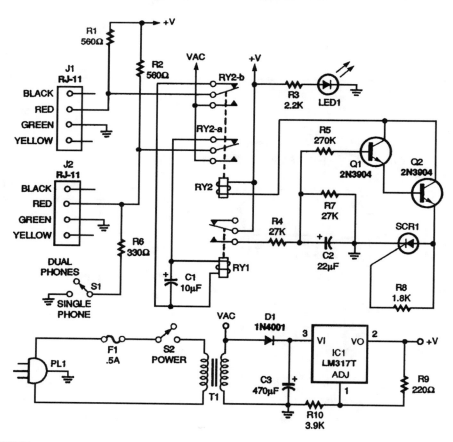

Fig. 105-6

When both handsets are on-hook, resistors R1 and R2 supply power to them. When both handsets are off-hook, they transmit and receive their own audio. All the TCB does is supply power to them through R1 and R2. If switch S1 is closed, resistor R6 simulates a telephone being plugged into

TELEPHONE-LINE SIMULATOR *(Cont.)*

J2, permitting the testing of only one telephone. Now consider the situation where one handset is on-hook and one is off-hook. That causes the on-hook line to go to 24 Vdc and the off-hook line to go to 7 Vdc. The coil of relay RY1 is connected across the two telephone lines, and the voltage difference between the two lines energizes it. When the contacts of RY1 are closed, C2 charges through R4; it takes about 1 second for C2 to charge to 12 Vdc. The 12 Vdc across C2 causes a voltage-controlled switch consisting of R5, Q1, Q2, SCR1, and R8 to close, thus energizing RY2. When RY2 is energized, RY1 is removed from the circuit and a 60-Hz, 37-V p-p sine wave is placed on the telephone lines, causing the telephones to ring. Because RY1 is removed from the circuit, capacitor C2 starts discharging through R7. It takes about 1 second for the capacitor to discharge to about 2.4 Vdc. That lower voltage level causes the voltage-controlled switch to disable RY2, removing the ring voltage from the telephone lines and putting relay RY1 back in the circuit. If one telephone is still off-hook and one is on-hook, the cycle is repeated.

TELEPHONE BELL INDICATOR CIRCUIT

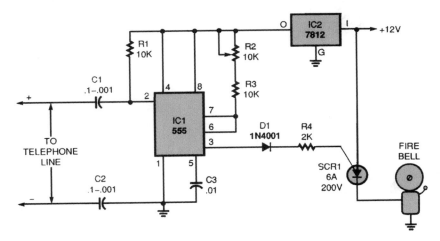

POPULAR ELECTRONICS

Fig. 105-7

This circuit uses a 555 timer to gate an SCR. The SCR passes 12 V to an alarm bell. The bell should be of the mechanical interrupter type so that it will reset and ring only when ring signals are on the telephone line.

PHONE CALL COUNTER

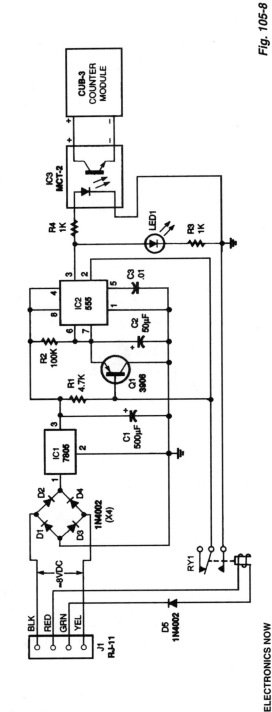

Fig. 105-8

This circuit counts any string of rings as one call, regardless of duration. If someone hangs up after 1 ring or 10, the counter advances by 1. Similarly, if someone reaches an answering machine and hangs up, or leaves a message, the counter still advances by one call. The counter circuit must ignore signals and react only to the 90-Vac that comes in when the phone is ringing. Because the counter circuit is isolated from the phone line, it needs its own power supply. With some phone systems, power can be provided by the yellow and black wires of the phone line, which typically are connected to a small transformer that provides about 8 Vac. The bridge rectifier D1 to D4, voltage regulator IC1, and filter capacitor C1 convert this to 5 Vdc to power the circuit. If your phone line does not have active yellow and black wires, 6.3 Vac must be connected to the input of the rectifier. Timer IC2 is triggered by relay RY1, which has a coil voltage of about 48 V. Because diode D5 is reverse-biased relative to the normal dc across the phone line, it will pass only partially rectified ac. When the phone rings, D5 passes a partially rectified 90 V, which is enough to energize the relay. Actually, the relay will chatter at about 20 Hz during a ring signal because the rectified voltage is not pure dc. This generates a rapid string of pulses to trigger the timer. But because the timer operates in a retriggerable mode, its output remains high for the duration of the rings. The NE555 timer (IC2) is wired in a retriggerable, monostable configuration. The output stays high for a length of time determined by the time constant, $1.1 \times R_2 \times C_2$. Because the timer is retriggered with each ring

PHONE CALL COUNTER *(Cont.)*

signal, its output will remain high for about 5.5 seconds after the phone stops ringing. The counter will advance by one count whenever the output goes high. Indicator LED1 shows when the phone is ringing and when the output of the timer is high. The high output activates optoisolator IC3 and advances the counter by 1. The counter can be any digital or electromechanical counter whose operation is not affected by the duration of the trigger signal.

SIMPLE PHONE-IN-USE INDICATOR

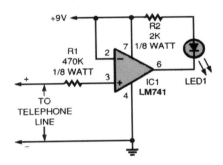

POPULAR ELECTRONICS

Fig. 105-9

The LM741 op amp is used as a voltage comparator, comparing the telephone-line voltage to the battery voltage. The telephone line drops below 9 V when it is in use. That drop turns the op amp on, which lights the LED. Resistor R1 prevents the circuit from loading the telephone line excessively. In my location, the in-use circuit drops the line voltage approximately 5 V. That leaves quite a bit of reserve because a telephone line operates at approximately 40 to 48 V. Resistor R2 limits the current to the LED. A 1000-Ω resistor will brighten the LED; however, the total circuit current draw will increase. The circuit (as shown) consumes 1.15 mA in standby and 3.80 mA when indicating that the line is in use.

ANSWERING MACHINE MESSAGE STOPPER

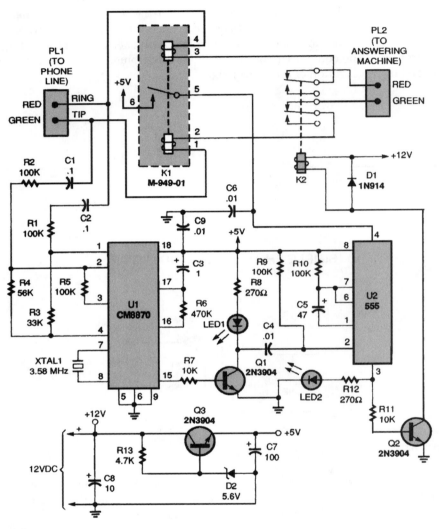

Fig. 105-10

The message stopper connects between your present answering machine and any convenient modular phone jack. When any extension telephone on the same line rings and the answering machine answers the phone first, you can easily stop the outgoing message and reset the machine for the next call simply by pressing a key on the tone-dialed telephone's keypad. The schematic diagram for the message stopper is shown. Power for the circuit is supplied by a 12-V source. It is fed to the junction of K2 and D1, also dropped to 5 V through a voltage-regulator stage composed of Q3, resistor R13, and zener diode D2, and fed to the balance of the circuit. Plug PL1 connects to the telephone line. The tip and ring conductors of that plug are connected in series to line-sense relay K1, line-disconnect relay K2, and finally plug PL2, which connects to the answering machine. Integrated circuit U1, a CM8870 DTMF receiver, monitors the phone line for the presence of DTMF signals. That chip contains an internal op-amp stage that allows it to be interfaced to the phone line using only a pair of capacitors (C1 and C2) and a few resistors. The voltage gain of the internal amplifier is unity in this circuit. When a connected answering machine answers the line and a DTMF signal is detected by U1, the output at pin 16 of U1 goes high. That causes capacitor C3 to begin discharging through resistor R6. The output at pin 15 goes high. A DTMF tone pair must be present before pin 15 goes high approximately $\frac{1}{3}$ second and is determined by the following formula $t = 0.67RC$, where t is the time in seconds, R is the value of R6 in ohms, and C is the value of C3 in farads. As the output of pin 15 goes high, Q1 is turned on, lighting LED1 and triggering the 555 timer U2, configured as a monostable multivibrator, into operation via C4. However, U2 can receive power only if the answering machine answers the line. That is accomplished by the relay contacts within line-sense relay K1 applying +5 V to pin 4 of U2. Once U2 is triggered, pin 3 goes high for approximately 5 seconds, LED2 illuminates, and Q2 switches on. Relay K2 switches on, disconnecting the answering machine from the phone line for approximately 5 seconds. That should give the typical answering machine time to detect the line disconnection and force it to reset for the next call. However, the length of the time period can be altered by changing the values of R10 or C5, as shown in the formula: $t = RC$, where t is the time in seconds, R is the value of R10 in ohms, and C is the value of C5 in farads.

TELEPHONE RING FLASHER

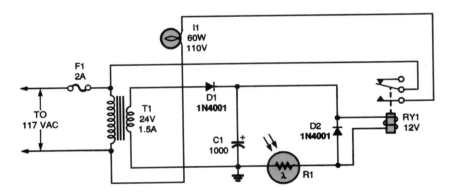

POPULAR ELECTRONICS

Fig. 105-11

When the phone rings, the flasher turns on and off, causing photocell R1 to conduct. Relay RY1 is then energized, which completes the circuit between the light and the ac line. Thus, the light flashes in step with the rings.

UNIVERSAL TELEPHONE HOLDING CIRCUIT

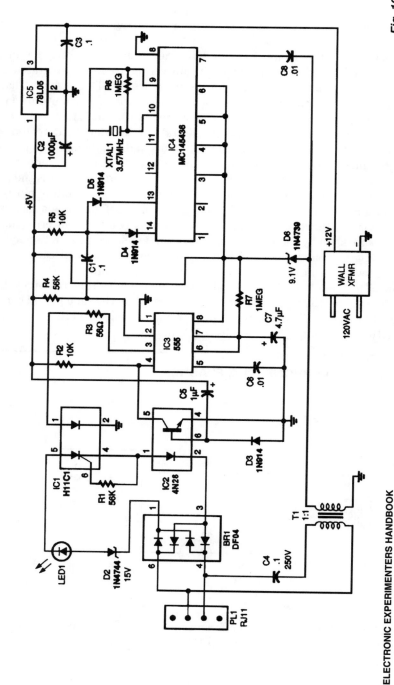

Fig. 105-12

ELECTRONIC EXPERIMENTERS HANDBOOK

If you have Touch-Tone telephone service, you can put a call on hold from any phone in your house by plugging this simple device into any telephone jack. The universal hold circuit works with any phone that has a keypad with a # key. To put a call on hold, press the # key and hang up the phone. A timer extends the #-key function while you hang up phones that have a keypad built into the handset. The universal hold circuit first detects the dual-tone, multifrequency (DTMF) signal that is generated when the # key is pressed. It then activates a circuit that partially loads the telephone line so that the central office "thinks" a phone is still off-hook—even after it is hung up. The hold circuit remains active for 5 seconds after the # key is released, so the key does not have to be held down while the phone is being hung up. When any phone is again picked up, the hold function is canceled.

106

Temperature-Related Circuits

The sources of the following circuits are contained in the Sources section, which begins on page 1043. The figure number in the box of each circuit correlates to the entry in the Sources section.

SIMPLE LED THERMOMETER

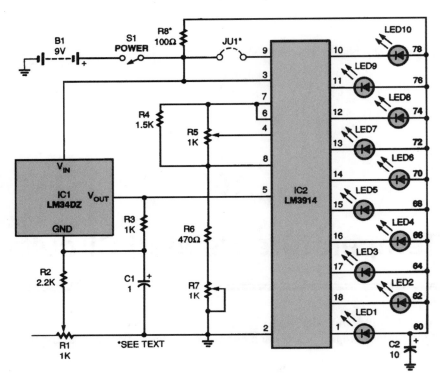

POPULAR ELECTRONICS

Fig. 106-1

As shown, the circuit is powered by a 9-V battery, B1, but it can easily run off any 7- to 10-Vdc power supply. At the heart of the thermometer is IC1, an LM34 temperature sensor. That device produces a voltage between the V_{out} and GND terminals that is linearly proportional to temperature. Although the output is usually 10 mV/°F, IC1 is connected in a resistor network (made up of R1 to R3) with a gain that provides an output of 40 mV/°F. Capacitor C1 is used as a noise bypass across R1 and R2. Because the voltage output by IC1 will be used by the rest of the circuit to "determine" what the temperature in the room is, potentiometer R1 will have to be calibrated exactly. The output of IC1 is fed to pin 5 of IC2, an LM3914 LED bar- or dot-graph driver, which is where the actual temperature-determining process occurs. IC2 has 10 internal comparators, the output pins of which are connected to LED1 to LED10. The voltage input to pin 5 is compared by IC2 to the voltages at pins 4 and 6; that process determines which LED or LEDs light. The LEDs can be set to light either one at a time (dot mode) or progressively (bar mode). When jumper JU1 is not installed, dot mode is enabled. When the jumper is installed, the chip is in bar mode. In dot mode, the LED that corresponds to the correct input voltage lights by itself. When the input voltage increases, an LED representing a higher

SIMPLE LED THERMOMETER *(Cont.)*

temperature will light and the LED previously lit will extinguish. In bar mode, the LED representing the temperature will light and all of the lower LEDs will stay lit. Each mode has its advantages—the dot mode uses less current because only one LED is lit at a time, but the bar mode is easier to read at a glance. Resistor R8 and capacitor C2 provide decoupling for the LED-supply circuit. If bar-mode operation is desired, it is recommended that you reduce the value of R8 to 15 Ω. For a range of 60 to 78°F to be displayed, pin 6 must have a reference of 3.345 V and pin 4 should have a reference of 2.545 V. Those values are obtained through adjustments of potentiometers R5 and R7.

TEMPERATURE-CONTROLLED FAN

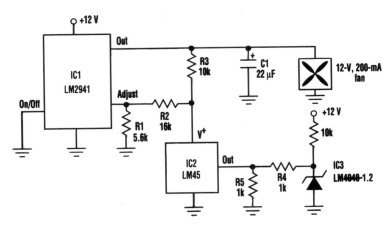

Fig. 106-2

The circuit described here needs only three ICs to smoothly increase fan speed as temperature rises above an easily set trip point. Low-dropout voltage regulator IC1 provides power to the fan and to temperature sensor IC2. IC2 drives the middle of the resistor divider across voltage reference IC3. Because the LM45's output stage can only source current, its output voltage will remain at about 610 mV until IC2's temperature rises above 61°C, IC2 will drive the 500-Ω Thevenin resistance of the divider, which will cause IC2's supply current to increase rapidly with rising temperature. As the temperature rises above the point where IC2's output voltage exceeds the Thevenin voltage set by R4 and R5, IC2's supply current increases. IC2's supply current flows through R1 and directly affects the regulator's output voltage. If the temperature rises 20°C above the nominal 61°C threshold, the regulator's output voltage will rise to 12 V and the fan will operate at full speed.

LINEAR TRUE-MEAN-SQUARE TEMPERATURE CONTROLLER

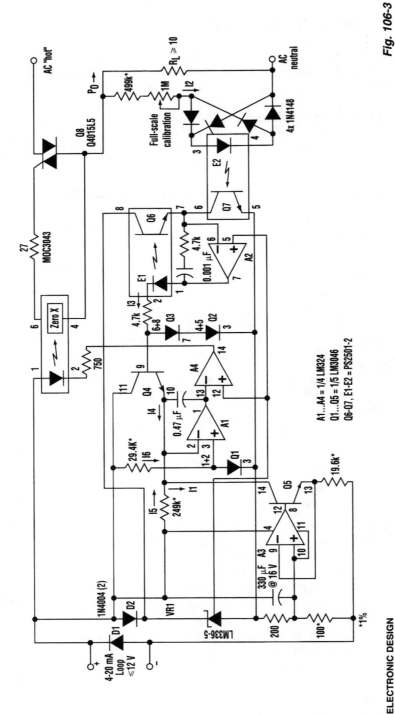

Fig. 106-3

ELECTRONIC DESIGN

The power-control circuit shown outputs true-mean-square power in direct proportion to a 4- to 20-mA current-loop control input. The delivered power is tightly regulated, thus canceling line-voltage variations. In addition, it is proportioned on a fast 8.33-ms timebase, which avoids thermal ripple. The circuit operates as follows: 4- to 20-mA control inputs are converted by A3 and Q5 to negative 20 to 100 μA (I_1). I_5 (fixed at ±20 μA) zero-corrects I_1, and the 0- to 80-μA difference is applied to the summing point of the A1 integrator. A4 compares the accumulated integral to a 2.5-V reference tapped from the ADJUST terminal of VR1 and, when the integral rises above that, it turns on the 3043 triac trigger optocoupler. Zero-cross switching of the Q8 triac

954

LINEAR TRUE-MEAN-SQUARE TEMPERATURE CONTROLLER *(Cont.)*

minimizes generated noise. Ac half-cycles through Q8 heat R1 and push load-monitor current I_2 through LED E2. To balance the resulting Q7 photocurrent, A2 produces I_3, which causes matching conduction in the E1/Q6 optocoupler. Close tracking between sections of the 2501-2 dual optocoupler assures good proportionality between I_2 and I_3. Because I_3 also biases series-connected Q2 and Q3, voltage applied to the base of Q4 will be $2[X \log(Y \times I_2) + Z]$, where X and Z are constants common to all five transistors in the 3046 monolithic array, and Y is set by the "Fullscale Cal" pot. As a result, antilog transistor Q4's emitter current is closely given by $(Y \times 12)^2/I_6$. When integrated by A1, it gives an accurate prediction of the true-mean-square power dissipated in R1. The resulting feedback loop acts to adjust Q8's duty cycle to regulate R1 power, allowing the temperature controller to accurately and linearly track the 4- to 20-mA control input signal. Operating power for the circuit is developed from the 4- to 20-mA loop current, eliminating any need for another power source.

TEMPERATURE ADAPTER FOR DVM

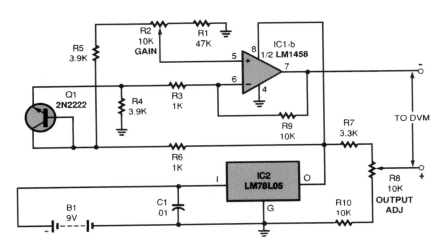

POPULAR ELECTRONICS

Fig. 106-4

A schematic for the temperature adapter is shown. The circuit outputs 0.01 V/°F. To determine temperature, you ignore the decimal point in the display, and the temperature readout is to the tenth of a degree. For example, if the temperature in a room is 75.5°, your DMM would read +0.755 V. The voltage input for op amp IC1 is obtained from a transistor Q1 used as a diode in this application. The linear voltage drop is measured and used to indicate temperature. The low-power regulator LM78L05 (IC2) provides a fixed 5-V source for the circuit. The voltage drop across Q1 is detected and amplified by IC1, an LM1458 op amp. The output of IC1 is then fed to the input terminals of a DMM. Put the positive lead of a voltmeter on pin 7 of IC1, and the negative lead on ground. Adjust R2 for a reading of 2.5 to 3 V. Connect the output of the adapter circuit to a voltmeter set to the 2-V dc scale. Adjust R8 so that the multimeter reads the same as a stabilized thermometer in the same vicinity as the temperature probe (ignore the decimal point).

TEMPERATURE-TO-FREQUENCY CONVERTER CIRCUIT

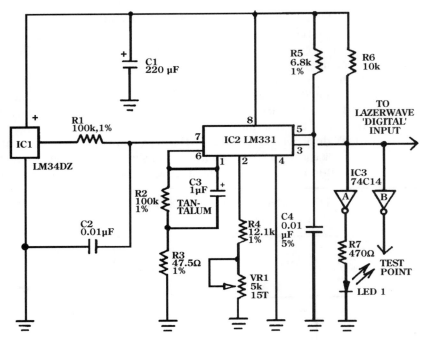

Fig. 106-5

IC1 produces the Fahrenheit-dependent voltage at its output. This is fed into the LM331 VFC through precision resistor R1. Inside IC2 are a switched current source, an input comparator, and a one-shot timer. The input comparator is designed to produce a trigger pulse if the voltage at pin 7 is greater than the voltage at pin 6. It does this by comparing the two voltages, one of which is generated internally by the switched current source, and the other of which is obviously the input. The switched current source charges capacitor C3. When the comparator trips and produces its trigger pulse, it activates the monostable multivibrator, which outputs a positive pulse that turns on the output transistor. The monostable multivibrator turns on the switched current source for a period of 1.1 times *RC,* where *RC* is the combination of resistor R5 and capacitor C4. C1 charges up to a point where it exceeds the input voltage, and the comparator resets itself; thus the frequency output transistor Q1 will be turned off. Resistor R2 discharges C3 in a time interval that depends on the values of both components. Then, the input comparator and monostable multivibrator are reset, and another cycle will be implemented as long as there is sufficient input voltage. Finally, resistors R4 and VR1 set the amount of current the constant-current source is able to inject into C3. Thus, VR1 can set the response "span" of the IC by adjusting it. IC3, two gates of a hex Schmitt trigger, are used to buffer IC2's output to light an LED and to make available a digital signal. Remember two things about IC3: Its logic 1 output will be a full 9 V, not the TTL level of 5 V, and all unused inputs must be tied to ground to prevent self-oscillation at RF frequencies and consequent malfunction of the circuit.

LINEAR READOUT CIRCUIT

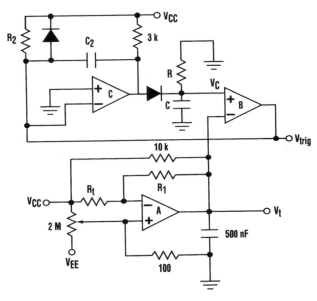

ELECTRONIC DESIGN

Fig. 106-6

Thermistors have found their way into a number of temperature-sensing applications. However, their resistance is a nonlinear function of temperature. In many applications, a digital readout of the thermistor also is desired. It is possible to produce a digital signal with a frequency that is linearly proportional to the temperature. The entire active portion of the circuitry can be implemented using $\frac{3}{4}$ of an LM339. Comparator A is configured to operate as an op amp that generates voltage (V_t) which is inversely proportional to the thermistor's resistance, R_t. The voltage V_t is used as the threshold for the relation oscillator formed by comparators B and C. Comparator B monitors the voltage on the integrating capacitor (V_c). When it crosses V_t, it triggers the one-shot formed by comparator C via the signal V_{trig}. The output of comparator C resets the integrating capacitor for a duration, t_o, where $t_o = (R_2 + 3 \, k\Omega)C_2$. Then the cycle begins again. The circuit achieves linear readout because the exponential nonlinearity of the thermistor is compensated by the exponential decay of the voltage in the relaxation oscillator.

PC TEMPERATURE INTERFACE

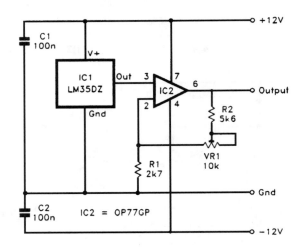

EVERYDAY PRACTICAL ELECTRONICS

Fig. 106-7

This simple temperature interface covers a range of 0 to 51°C. Dual balanced 12-V supplies were used to power the prototype, but dual 5-V supplies should just about suffice. IC1 is the temperature sensor. Over this range of temperatures, the LM35CZ should work well. The output from IC1 feeds into a noninverting-mode amplifier based on IC2. The closed-loop voltage gain of the amplifier is set by resistors R1 and R2 and preset VR1. The latter is adjusted to give a voltage gain of 5. VR1 is given the correct setting by first subjecting the sensor to a temperature, which is equal to about 50 or 100 percent of the full-scale value (i.e., about 20 to 51°C.). In most cases, the room temperature will be about 20 to 25°C, which will suffice. An accurate thermometer is used to measure the room temperature, and then VR1 is adjusted for the appropriate reading.

The following GW Basic or Q BASIC program reads the temperature sensor and prints the temperature on the screen:

```
10      REM BASIC TEMPERATURE PROGRAM
20      CLS
30      OUT &H37A,1
40      OUT &H37A,0
50      X=INP(&H379) AND 120
60      X=X/8
70      OUT &H37A,4
80      Y=INP(&H379) AND 120
90      Y=Y*2
100     Z=X+Y
110     Z=Z/5
120     Z=Z-11
```

958

PC TEMPERATURE INTERFACE (Cont.)

```
130    PRINT Z "Degrees C."
140    FOR DELAY=1 TO 20000
150    NEXT DELAY
160    A$=INKEY$
170    IF LEN(A$)=1 THEN END
180    CLS
190    GOTO 30
```

SOLID-STATE THERMOMETER

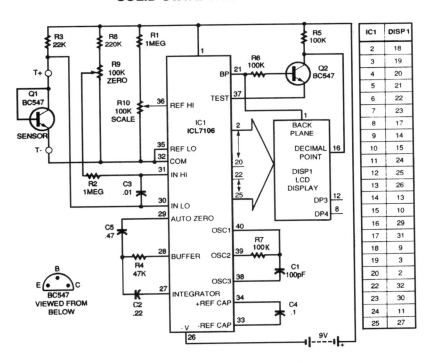

ELECTRONIC EXPERIMENTERS HANDBOOK

Fig. 106-8

The AM ICL7106 3½-digit analog-to-digital converter by Harris Semiconductor reads the −2.2-mV/°C change in the junction voltage drop of a BC547 transistor. R9 and R10 provide calibration. The basic chip has a 0- to 199.9-mV range. An LCD display made by Varitronix is used as a readout. The dc input voltage is correlated with the ambient temperature surrounding Q1.

TEMPERATURE MONITOR

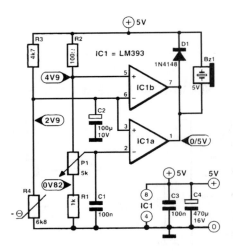

ELEKTOR ELECTRONICS

Fig. 106-9

The temperature sensor is a resistor, R4, with negative temperature coefficient (NTC); the resistance of such a resistor drops when its temperature rises. Resistors R4 and R3 are part of a resistance bridge whose variable branch consists of R1, R2, and P1. The metering diagonal is connected to the inputs of comparator IC1a. The voltage at the inverting input of IC1a is set with P1 to a level that with normal temperatures is a little lower than that at the noninverting input. When the temperature rises, the resistance of R4 falls. This results in the comparator's changing state (to low), which causes the piezo buzzer to sound. Care should be taken to ensure that the voltage across the metering diagonal does not drop below 3.5 V to prevent the common-mode dynamic range of the LM393 from being exceeded. The NTC can be a 5- or a 10-kΩ type. It must be in good thermal contact with the heat source. Preset P1 must be set to a position where, after the part or device being monitored has attained normal operating temperature, the buzzer just does not sound. The buzzer must be a 5-V type.

TEMPERATURE SHIFT CIRCUIT

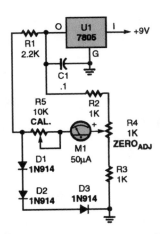

The temperature-shift meter is designed to detect small changes in temperature, rather than give exact temperature levels.

POPULAR ELECTRONICS

Fig. 106-10

TEMPERATURE-COMPENSATED MILLIVOLT REFERENCE

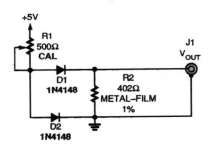

POPULAR ELECTRONICS

Fig. 106-11

Two identical diodes are used to derive a temperature-compensated voltage across R2. Select two diodes of the same manufacture with, for example, 10-mV difference in forward drop at the same current. Then the output will be compensated because both diodes will very likely have the same temperature coefficient (they are of identical manufacture).

107

Tesla Coil Circuits

The sources of the following circuits are contained in the Sources section, which begins on page 1043. The figure number in the box of each circuit correlates to the entry in the Sources section.

Tesla Coil
Solid-State Tesla Coil

TESLA COIL

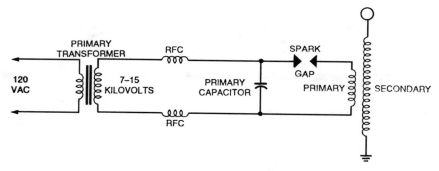

ELECTRONICS NOW

Fig. 107-1

The traditional Tesla coil is connected as shown. The primary capacitor, an HV ac type, typically 500 to 5000 pF, and the inductance of the primary winding should resonate around the resonant frequency of the secondary coil.

SOLID-STATE TESLA COIL

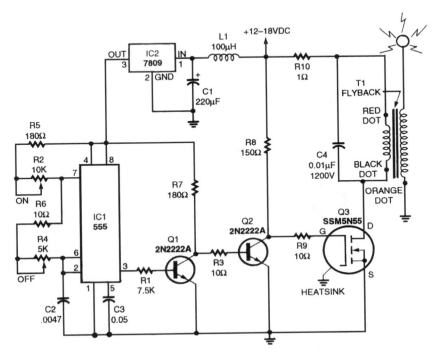

ELECTRONICS NOW

Fig. 107-2

A TV flyback transformer can double as a low-power Tesla coil. The Tesla circuit consists of a pulse generator, a driver circuit, and a high-voltage transformer. Resistors R1 and R2 determine the time duration that the output at pin 3 is off, while R3 and R4 along with R1 and R2 determine the ON time. Inductor L1 and regulator IC2 provide a clean, stable power source for the timer. Transistor Q1 acts as a buffer. Resistor R6 determines the rise time based on the time constant developed by R6 and the inherent gate capacitance of Q3. Resistor R8 limits current so that excessive current will not damage T1's primary winding. Capacitor C5 absorbs some of the back EMF generated in T1's primary. The pulse waveform from IC1 is applied to Q1, which provides the high current necessary to offset the high capacitance of Q3. Capacitor C5 partially absorbs the primary EMF, reducing the stress on Q3. The spike produced in the secondary creates a ringing oscillation. When this oscillation begins to decay, Q3 is once again switched into its ON state. This dumps the energy into C5 and builds the magnetic field in T1. If the timing of both the ON and OFF states of the pulse train is adjusted correctly, the secondary of T1 produces a nearly constant, high-frequency, high-voltage current.

108

Theremin Circuits

The sources of the following circuits are contained in the Sources section, which begins on page 1043. The figure number in the box of each circuit correlates to the entry in the Sources section.

VACUUM-TUBE THEREMIN VOLUME-CONTROL CIRCUIT

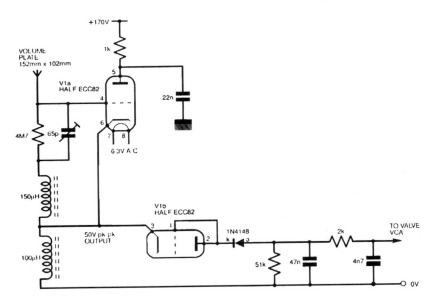

EVERYDAY PRACTICAL ELECTRONICS

Fig. 108-1

Gain is controlled by a variable negative grid bias in an audio amplifier stage. This oscillator produces a negative dc voltage by means of a rectifier. As a hand is brought up to the volume plate antenna, this loads the oscillator, reducing output and hence the negative dc output. This means increased gain for the controlled audio stage.

THEREMIN

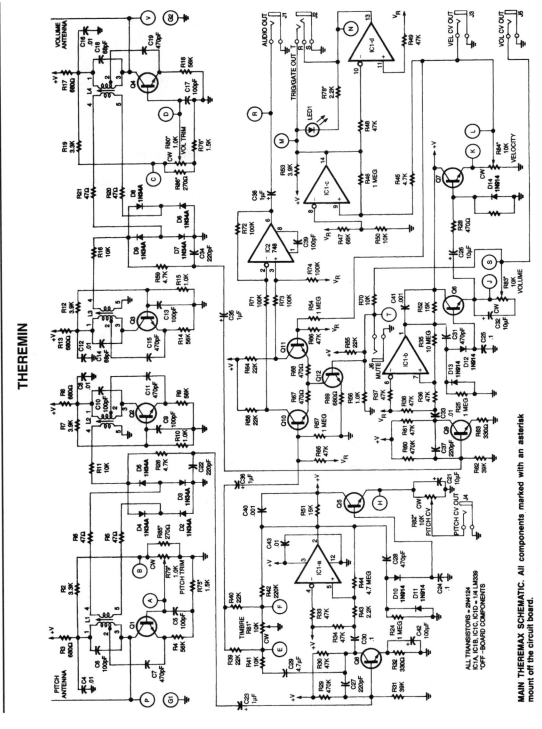

MAIN THEREMAX SCHEMATIC. All components marked with an asterisk mount off the circuit board.

Fig. 108-2

ELECTRONICS NOW

THEREMIN *(Cont.)*

The theremin uses several RF oscillators connected to antennas to both generate audio frequencies and vary the volume of the note produced. The pitch is generated by heterodyning two oscillators, one fixed and one variable, and the volume is controlled by another pair of oscillators in which the beat frequency that is generated is used to derive a control voltage for the volume-control circuit.

THEREMIN VOLUME-CONTROL CIRCUIT

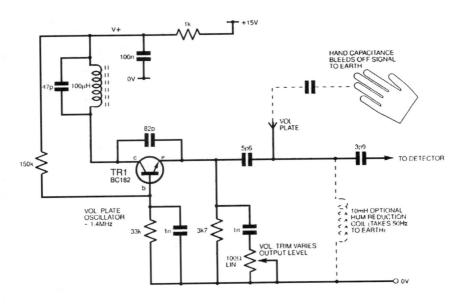

Fig. 108-3

The circuit shows a capacitive potential divider with volume trim on the fixed oscillator. The hand capacitance loads down the RF to a detector stage, reducing the detector output. This output is used to control the gain of an audio amplifier.

THEREMIN CIRCUIT

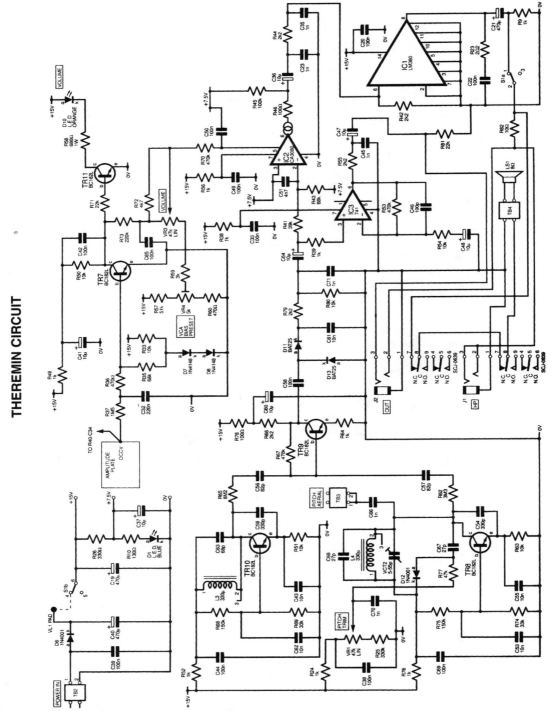

Fig. 108-4

EVERYDAY PRACTICAL ELECTRONICS

THEREMIN CIRCUIT *(Cont.)*

A theremin produces audio derived from mixing two high-frequency oscillators and using their audio-frequency difference as the note produced. Another oscillator is used to derive a volume-control signal, which varies the audio output level. Two or more oscillators are connected to short external rod or plate antenna electrodes; these are used as controls by bringing an object (one's hands) in proximity to them. By controlling hand movements, it is possible to create music or other sounds with this device.

ONE-CHIP THEREMIN CIRCUIT

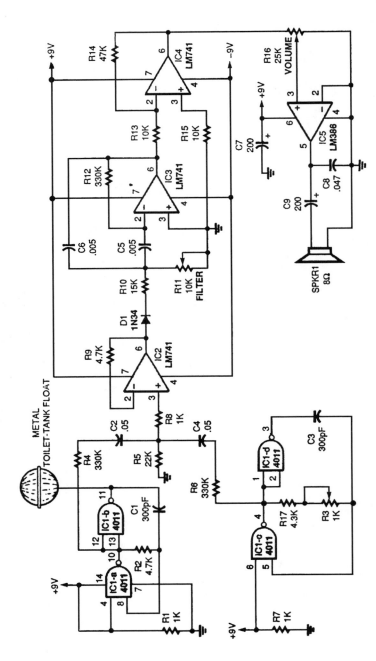

POPULAR ELECTRONICS

Fig. 108-5

Typically, a theremin uses two separate Colpitts LC oscillators, one of which can be slightly varied in frequency. The two frequencies are mixed together, and demodulated to reveal a beat frequency. If the two oscillators are at the same frequency, there is no beat or audio, but if they are off because of the proximity of your hand, a difference or beat frequency results, which is the audio output of the theremin. A 4011 quad gate, IC1, is the heart of this theremin. Two gates are used for each of the two required oscillators running at 250 kHz. For the aerial, use a metal toilet-tank float. It provides much better sensitivity than a length of wire. The two RF signals are mixed, then amplified by IC2, an LM741 op amp. Audio is detected by D1, a 1N34 diode. Another LM741, IC3, is set up as an adjustable bandpass filter; still another LM741, IC4, further amplifies the audio for IC5, an LM386 audio amplifier.

972

THEREMIN CIRCUIT VOLUME SECTION

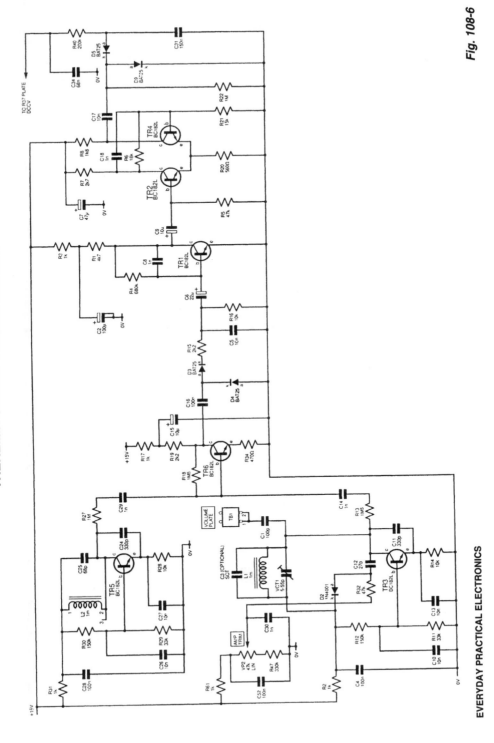

Fig. 108-6

The player's hand capacitance loads down the RF to a detector stage, reducing the detector output. This output is used to control the gain of an audio amplifier in the pitch section.

109

Timer Circuits

The sources of the following circuits are contained in the Sources section, which begins on page 1043. The figure number in the box of each circuit correlates to the entry in the Sources section.

CD4050 Timer
Low-Cost Timer
Long-Period Timer
CD4528 CMOS Timer
Audible Timer
Variable Duty-Cycle Timer
5-min Ac Timer
Dual Timer
Appliance Timer
555 Time Delay
Programmable Timer/Sequencer
Simple Very Long-Period Timer

CD4050 TIMER

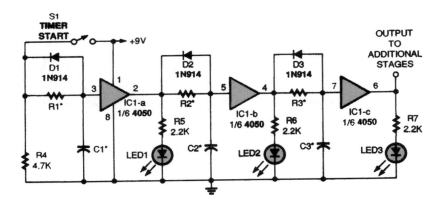

POPULAR ELECTRONICS

Fig. 109-1

Three stages are used to produce three consecutive time delays. The time delay for each stage is very close to $1.1 R_n C_n$. Diode D_n allows quicker reset of stage n.

LOW-COST TIMER

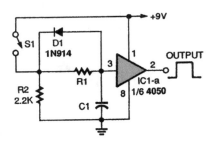

POPULAR ELECTRONICS

Fig. 109-2

Want a simple timer at practically no cost? Depending on what parts you have, you might be able to build this circuit for less than a dollar.

When S1 is closed, C1 begins to charge through R1 toward the 9-V supply. When the voltage at the input of the buffer, pin 3, reaches about 70 percent of the supply voltage, the output switches high for an ON-time delay function. As long as S1 remains closed, the circuit will maintain the high output at pin 2. Opening S1 resets the timer by allowing C1 to discharge through D1 and R2.

LONG-PERIOD TIMER

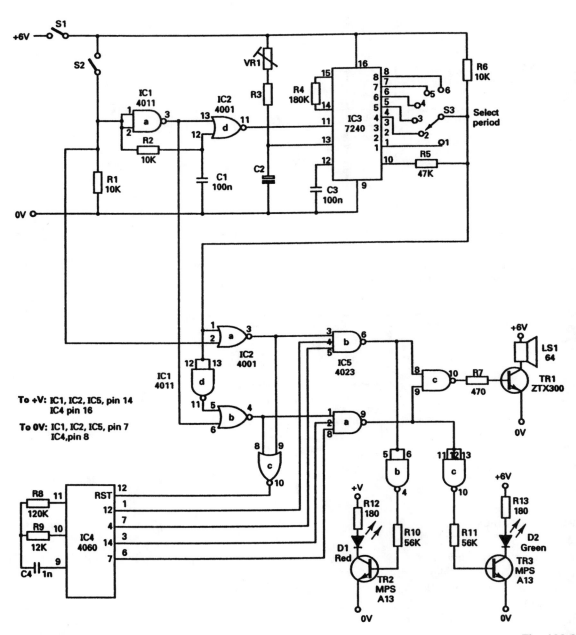

LONG-PERIOD TIMER *(Cont.)*

This circuit is based on the 7240 CMOS programmable timer chip (IC3), which has an accuracy of 0.5 percent. It contains a time-base generator, the frequency of which is decided by a resistor and a capacitor. The basic time period is RC seconds; so, given that the maximum value of R is 10 MΩ and the maximum value of C is 1000 μF, the maximum time period is 10,000 seconds, (2.8 hours) IC3 also has an eight-stage binary divider chain. The total period available is 27 times the above, which is just over 14 days. In this figure, S1 is an optional power switch. S2 is the switch which is closed to initiate timing. Closing S2 generates a brief low pulse that goes to pin 11 of the timer IC3 and starts the timing. The output of the timer is normally high, but it goes low for the whole of the timing interval. The length of the interval is selected by a rotary switch S3. Pins 1 to 8 of IC3 are the outputs from the eight-stage divider chain. When the counter is reset, they all go high; while timing, they go through an inverted binary sequence. R5 connects the output to the RESET terminal (pin 10), so the counter is reset at the end of the interval. The first stage of the logic consists of two NOR gates and a NAND gate wired as an inverter (part of IC1, 4011). These gates detect the two alarm states. A high output on either pin 8 or pin 9 causes a low output from gate IC2c (pin 10). This makes the RESET input of IC4 low. IC4 is a 14-stage counter with its own oscillator, which begins to oscillate when the RESET is made low. The oscillator has a period of about 25 kHz, which is divided down to produce 1.6 kHz at pin 7 (high-pitched note), 200 Hz at pin 6 (low-pitched note), 6 Hz at pin 1 (fast bleeping), and 1.5 Hz at pin 3 (slow beeping). The remainder of the logic consists of gates producing the fast high-pitched bleep signal, which goes to the red LED (D1) by way of transistor TR2, and the slower, low-pitched bleep, which goes to the green LED (D2) by way of TR3. Both signals go to the speaker LS1 by way of TR1.

CD4528 CMOS TIMER

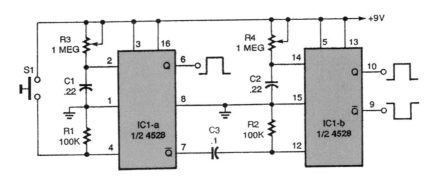

POPULAR ELECTRONICS

Fig. 109-4

The CMOS timer shown here can easily be cascaded with other similar 4528 circuits. R_1C_1 and R_2C_2 determine the timing.

AUDIBLE TIMER

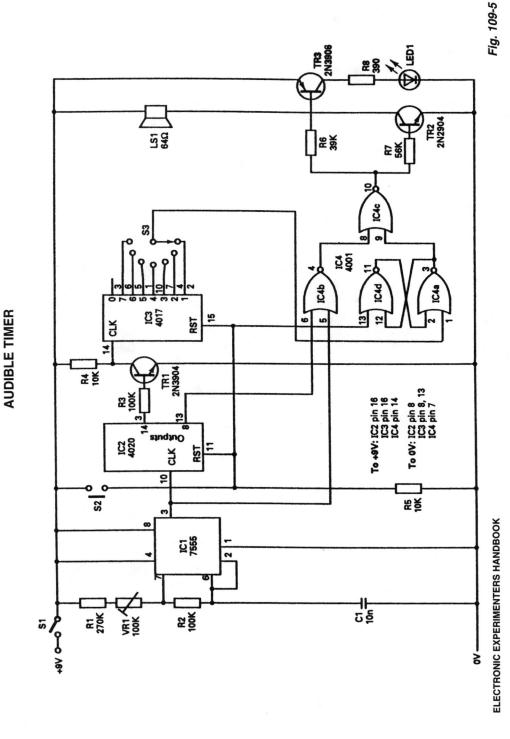

Fig. 109-5

The audible timer relies on the principle of obtaining a relatively long timing period by dividing down the frequency of a high-frequency astable multivibrator. The chip used for the multivibrator is the well-known 555 timer in its CMOS form. The preset

AUDIBLE TIMER *(Cont.)*

resistor VR1 allows the frequency to be set to 273 Hz. This provides the alarm tone. The 273-Hz signal is divided by the 14-stage counter IC2. At pin 8 of the counter, a signal at approximately 1 Hz is used to make the note intermittent. The timing frequency comes from pin 14 of the IC, at which the astable frequency is divided by 2^{14}, (16,384), giving a frequency of 1/60 Hz (1 count per minute). The output from pin 14 is inverted by transistor TR1 and fed to the clock input of a second counter (IC3). The outputs of this counter are normally at logical high, except that just one of the outputs is low at each stage of counting. When the counter is reset, output 0 goes high. On the next positive-going clock input, output 0 goes low and output 1 goes high. At each successive high-going clock input, the outputs from 0 to 7 go high, in turn, repeating. The rotary switch (S3) selects the output to be used to indicate the termination of the timing period. The circuit is reset by pressing button S2, which resets both counters and also the flip-flop formed by gates IC4a and IC4d. This turns off the NPN transistor TR2, which drives the loudspeaker, but turns on the PNP transistor (TR3), causing the LED to light. While the RESET button is held, the output from IC2 pin 3 is low, turning off TR1. The counter is incremented every minute until the output selected by S3 goes high. The high level from the selected output sets the flip-flop. The result is a note at 273 Hz, pulsing at the rate of 1 Hz. This is heard from the loudspeaker. The LED flashes on at the same time. Because the mark-space ratio of the astable output is high, the LED is turned on at almost full brightness during the ON periods.

VARIABLE DUTY-CYCLE TIMER

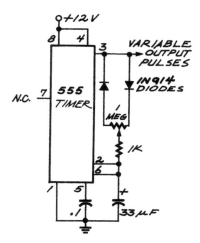

Fig. 109-6

This NE555 circuit produces a variable duty-cycle output.

5-min AC TIMER

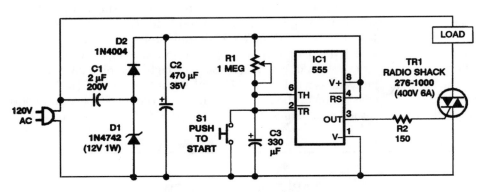

ELECTRONICS NOW

Fig. 109-7

This 555-based circuit will power a load for up to 5 minutes, depending on the setting of R1. Note the transformerless power supply made up of C1, C2, D1, and D2. The timer covers 0 to 5 minutes, or a little more, by measuring the time needed to charge C3 through R1, which is adjustable. To switch the ac power to the load, the timer uses a triac (a solid-state ac switch) instead of a relay. In the power supply, the capacitive reactance of C1 limits the current without generating heat. Use a metal-film capacitor for C1, not an electrolytic. Be sure to observe the voltage rating—use a capacitor rated for at least 200 V (preferably 600 V). Zener diode D1 limits the voltage, and D2 and C2 are the rectifier and filter.

DUAL TIMER

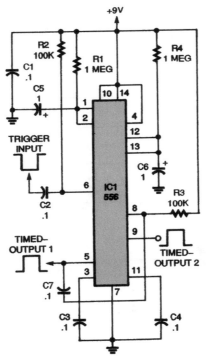

Fig. 109-8

Two are sometimes better than one. This dual timer contains a 556 IC, which contains two 555 timers in one package.

APPLIANCE TIMER

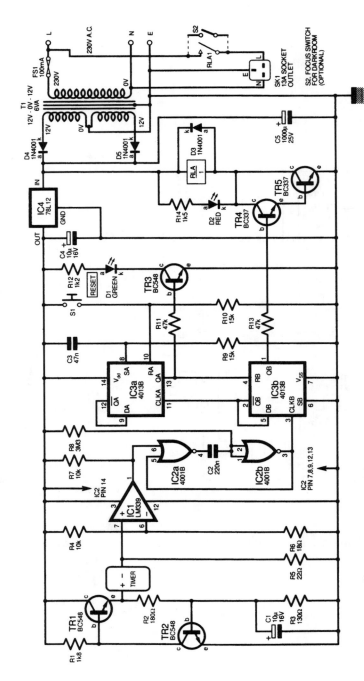

Fig. 109-9

EVERYDAY PRACTICAL ELECTRONICS

This design uses a low-cost digital kitchen timer to control appliances. No modification to the timer is needed; the only requirement is that it must beep both when the Start button is pressed *and* when the period has expired. This appliance timer detects the supply current flowing through the timer, and drives an external relay circuit, which operates loads for a period set by the kitchen timer. A low-voltage supply is formed by transistors TR1 and TR2, a ring-of-2 constant-current source adapted to provide about 1.5 V at the TR1 emitter (e). This powers the timer. At power on, capacitor C3 provides a positive pulse, which sets IC3a (pin 8), one-half of a dual D-type flip-flop. Output QA (pin 13) goes high, which resets IC3b, so that QB (pin1) is low. Tran-

982

APPLIANCE TIMER *(Cont.)*

sistors TR4 and TR5 are off, and so the relay RLA does not operate. However, transistor TR3 is driven on and illuminates the green LED D1 to indicate reset. The digital timer can now be set to the desired duration. Then switch S1 is pressed, which resets IC3a; QA goes low, which enables IC3b. When the timer's Start button is pressed, the increase in its supply current caused by the beep rises from a few microamperes to 5 mA or more. The voltage across resistor R5 rises, and comparator IC1, whose threshold is set by resistors R4 and R5, will output a brief positive pulse. This is cleaned up by the monostable formed from IC2a and IC2b, and clocks IC3b. Output QB (pin 1) now goes high and powers the relay via the Darlington transistor pair. At the end of the period, the first beep clocks IC3b again. Output QB (pin 2) goes high to clock IC3a (pin 11); QA goes high and resets IC3b so that the relay switches out.

555 TIME DELAY

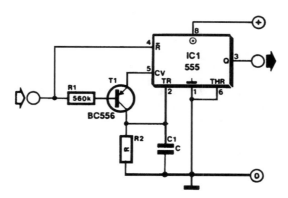

ELEKTOR ELECTRONICS

Fig. 109-10

Many electronic circuits frequently require the brief delay of a pulse. Such a delay, here between 100 μs and 100 seconds, is easily provided by a simple circuit based on the popular 555. That is more than adequate for most applications. The output of the 555 can go high only if the potential at pin 2 drops below a third of the level of the supply voltage, provided that the level at pin 4 is high. In quiescent operation, the level at pin 4 is low and C1 is charged via T1, so the output is low. When the input goes high, T1 is switched off and C1 is discharged via R1. In that condition, the RESET state is cancelled, and after a time delay that depends on the state of discharge of C1, the output of the 555 goes high. The time delay in seconds is calculated from $\tau = 0.69 R_1 C_1$, where R1 must be greater than or equal to 10 kΩ.

PROGRAMMABLE TIMER/SEQUENCER

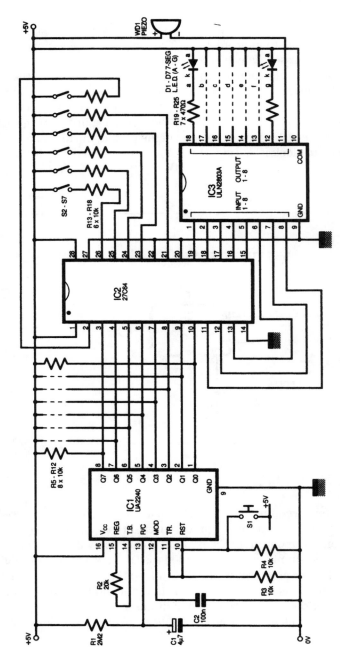

Fig. 109-11

EVERYDAY PRACTICAL ELECTRONICS

The circuit diagram of this programmable sequencer could be applied in a number of timing uses. Prior to the start of each interval, an audible tone is generated, and a seven-segment LED display shows the interval number. Prior to the end of that interval, the buzzer sounds again. IC1 is a 2240 timer/counter device that clocks up to 256 periods, the durations of which are determined by resistor R1 and capacitor C1. The 2240 has its RESET and TRIGGER pins (10 and 11) wired via a push switch S1. The timer is used to operate a memory chip, IC2. The first 8 addressed bits are used by the timer/counter to step through the 256 steps available for each program. The other 6 bits are selected by an external switch network S2 to S7 which permits up to 2^6 programs, each of 256 steps. The 27C64 has eight outputs, which was enough for a single-digit LED display plus a piezo buzzer.

SIMPLE VERY LONG-PERIOD TIMER

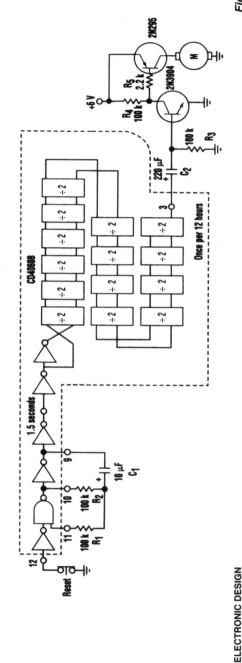

Fig. 109-12

ELECTRONIC DESIGN

This circuit is straightforward. An RC oscillator consisting of R1, R2, and C1, with the internal gates on the CD4060, generates a 1.5-second clock that is subsequently divided to a 12-h clock at the output. Other times are available for output or for ANDing with the longer-duration signals. The output is capacitively coupled to the two-transistor driver to provide a several-second pulse every 12 hours.

985

110

Tone-Control Circuits

The sources of the following circuits are contained in the Sources section, which begins on page 1043. The figure number in the box of each circuit correlates to the entry in the Sources section.

Guitar Treble Booster
Treble Booster Circuit

GUITAR TREBLE BOOSTER

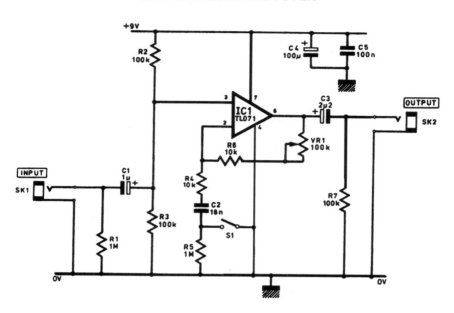

EVERYDAY PRACTICAL ELECTRONICS

Fig. 110-1

The circuit diagram shown was designed to increase the treble content of an electric guitar's output. It boosts frequencies above 1 kHz and helps to restore any higher-frequency losses. IC1 is a TL071 op amp connected as a noninverting amplifier, which is connected in the line between the guitar and the main amplifier. Potentiometer VR1 determines the level of treble boost applied; with the values of components shown, it will give between 3 and 20 dB. Resistor R6 and the potentiometer VR1 could be replaced by a fixed resistor, if required. The actual boost frequency is set by the values of resistor R4 and capacitor C2. With switch S1 open, the circuit acts as a voltage follower, with no effect on the guitar signal. Closing S1 grounds C2 to the 0-V line, which introduces the boost. Resistor R5 holds the "earthy" end of C2 at 0 V when switch S1 is open, and prevents switching noise appearing on the signal when S1 is operated. Resistors R2 and R3 set the input impedance of the circuit at 50 kΩ, which should suit most electric guitars. If the circuit is housed in a diecast case, S1 can be a foot-operated switch and SK1 could be a stereo jack socket with the sleeve and ring connections connected to the battery negative and 0-V rail, respectively. Thus, the unit will switch on automatically whenever the guitar is plugged in. Connect the case for the "earth" side of the circuit. Expect reasonable life from a PP3 battery.

TREBLE BOOSTER CIRCUIT

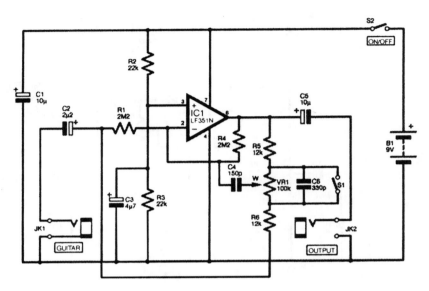

EVERYDAY PRACTICAL ELECTRONICS

Fig. 110-2

The circuit is basically a noninverting-mode amplifier. Resistors R2 and R3 bias IC1's noninverting input (pin 3), and R1 plus R4 act as a negative feedback network that sets the closed-loop voltage gain of IC1 at unity. However, R1 and R4 are used only to set the operating conditions of IC1 at dc and at low to middle audio frequencies. The high-frequency voltage gain is governed by a more-complex negative feedback network that is composed of resistors R5 and R6, potentiometer VR1, and capacitors C4 and C6. R5, VR1, and R6 form a conventional negative feedback network that enables the voltage gain of the circuit to be varied. With the wiper (w) of VR1 toward the top end of its track, the circuit provides losses of nearly 20 dB. Moving the wiper down to the bottom gives almost 20 dB of gain. It is coupled to the inverting input via capacitor C4, which has quite a low value. Consequently, C4 provides an efficient coupling only at high frequencies. At middle and bass frequencies, capacitor C4 has a very high impedance and provides only a loose coupling to IC1's inverting input. Therefore, control VR1 is ineffective at these lower frequencies, where the voltage gain of the circuit is largely governed by resistors R1 and R4. This permits the required boost and cut to be obtained at high frequencies, but keeps the gain at about unity at middle and low frequencies.

111

Touch-Control Circuits

The sources of the following circuits are contained in the Sources section, which begins on page 1043. The figure number in the box of each circuit correlates to the entry in the Sources section.

TOUCH SWITCH I

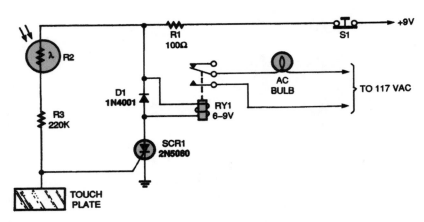

POPULAR ELECTRONICS

Fig. 111-1

This touch switch controls an ac-powered bulb, and requires only a few parts. Touching the touch plate triggers the SCR's gate, turning it on and allowing current to flow from the cathode to the anode of SCR1, thereby activating the relay.

TOUCH SENSOR

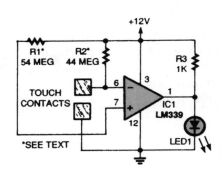

POPULAR ELECTRONICS

Fig. 111-2

This circuit is a basic two-contact touch-switch sensor circuit. The negative input (pin 6) of IC1 is tied to the positive supply through R2 (which is actually two 22-MΩ resistors connected in series), while the positive input (pin 7) is connected through R1 (made up of two 22-MΩ and one 10-MΩ resistor in series).

TOUCH SWITCH II

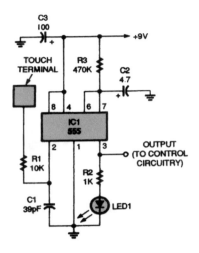

IC1, a 555 timer, is connected in a one-shot multivibrator circuit that is triggered by touching the touch terminal. The timed ON period is about 4 seconds with the component values given. To increase the ON time, increase the value of R3 or C2; to decrease the ON time, reduce the value of R3 or C2. The 9-V output at IC1 pin 3 can be used to drive an optocoupler, a power transistor, a hexFET transistor, CMOS circuitry, and more.

POPULAR ELECTRONICS

Fig. 111-3

LIGHT-OPERATED TURNOFF TOUCH SWITCH

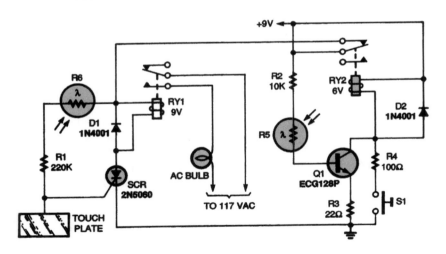

POPULAR ELECTRONICS

Fig. 111-4

When R5 is illuminated by a flashlight, its resistance lowers, leaving only 10,000 Ω as the minimum base resistance. That process gives a positive potential to the base of Q1, therefore turning on the transistor, activating RY2, and turning off the SCR. Or you can press S1 to turn the circuitry off.

SWITCH-OPERATED TURNOFF TOUCH SWITCH

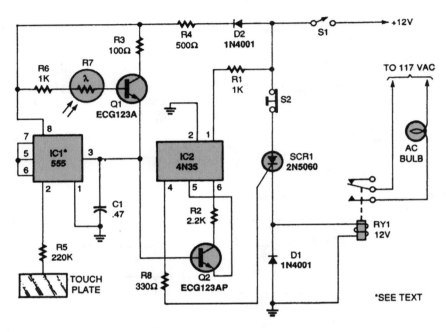

Fig. 111-5

When S1 is closed, the circuitry is on in standby mode. When the touch plate is touched, the output of the 555 timer IC1 goes high, supplying a positive potential to the base of transistor Q2, aiding its bias. The transistor is then on, allowing current to flow through pins 4, 5, and 6 of IC2. The gate of the SCR is then triggered, the relay is energized, and its contacts turn the ac bulb on. To turn the bulb off, just press S2 (RESET), which disconnects the anode of the SCR from the positive supply, turning it off and deenergizing the relay. The light-dependent resistor (R7) is used if you want to turn on the circuit remotely. Just point a flashlight at R7 to decrease its resistance, leaving only the 1000-Ω resistor as the base resistance. The resistance gives the base a positive potential, forward biasing the emitter-base junction of transistor Q1. Transistor Q2 is also turned on because its base is made positive by Q1, which triggers SCR1, energizing the relay.

ALTERNATIVE LIGHT-OPERATED TURNOFF TOUCH SWITCH

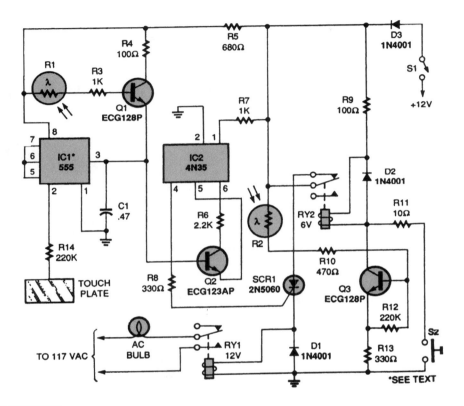

POPULAR ELECTRONICS

Fig. 111-6

When the touch plate is touched, it turns the output of IC1 on, giving a positive potential to the base of Q2. In turn, that turns on pins 4, 5, and 6 of IC2, triggering the SCR's gate and energizing RY1. When RY1 is energized, its contacts pull in, turning on the ac bulb. When R1 is hit by a strong light, it turns on the SCR and energizes RY1. To turn the ac bulb off, just point a strong flashlight at R2; this decreases the base resistance of Q3, making its base positive and, therefore, energizing RY2. When the contact of RY2 pulls in, it disconnects the anode of the SCR from the positive supply, turning it off. Or you can press S2 to energize RY2 and turn off the ac bulb.

SIMPLE TOUCH SWITCH

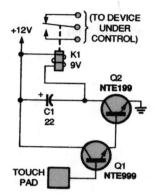

This circuit has two high-gain transistors. Operation occurs when the ambient 60-Hz ac field is impressed on the touch pad during the finger contact. The signal turns on Q1, causing Q2 to energize the relay. Capacitor C1 is used to prevent the relay from oscillating.

POPULAR ELECTRONICS

Fig. 111-7

TOUCH SWITCH III

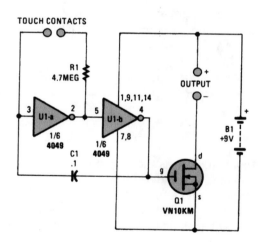

POPULAR ELECTRONICS

Fig. 111-8

The touch switch shown uses 9-Vdc operation, rather than the commonly used 120-Vac.

112

Transmitter Circuits

The sources of the following circuits are contained in the Sources section, which begins on page 1043. The figure number in the box of each circuit correlates to the entry in the Sources section.

LOW-POWER PLL-STABILIZED FM STEREO TRANSMITTER

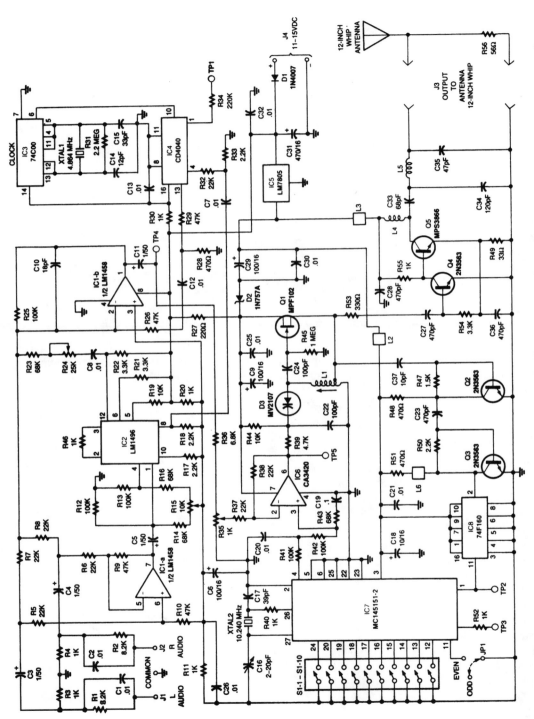

Fig. 112-1

LOW-POWER PLL-STABILIZED FM STEREO TRANSMITTER *(Cont.)*

The schematic shows a high-performance FM stereo transmitter, consisting of a stereo generator IC1 (audio matrixing), a balanced modulator IC2, a pilot and subcarrier generator IC3 and IC4, and a PLL synthesizer IC6, IC7, and IC8. Frequencies from 76 to 108 MHz are supported, with 100-kHz resolution. All oscillators are crystal-controlled for stability. The output is 0.5 to 0.7 V into 50 Ω. In the United States, in order to comply with FCC regulations, the transmitter is terminated in a 56-Ω resistor, which is nonradiating, and a small whip antenna (up to 12 inches long) is used as a radiating probe. The radiated field must be kept to 250 μV/m at 3 m from the antenna. In open locations, a 4-inch whip should give sufficient range (100 feet typical); for use inside buildings, where some losses are encountered, a 12-inch antenna might be needed. Do not use more than a 12-inch whip. In other nations, where regulations allow, a matched antenna can be used and the output amplifier can be run at 15 V or more, with over 200 mW output possible. Up to 24 V can be used if higher-voltage bypass capacitors are used on the supply line and R27 is changed to 680 Ω. IC5 might need heatsinking in this case.

A complete kit of parts, including PC board, is available from North Country Radio, P.O. Box 53, Wykagyl Station, New Rochelle, NY 10804-0053A.

INTERMEDIATE-FREQUENCY TRANSMITTER

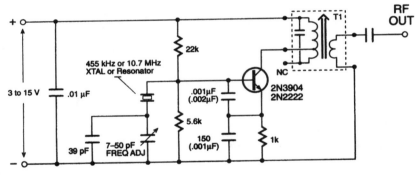

NUTS AND VOLTS

Fig. 112-2

The oscillator transmitter circuit is a variation of the Colpitts configuration. This particular circuit exhibits excellent frequency stability, good isolation between the frequency reference and the output, and isolated low output impedance. If you have the correct crystal on hand, that's fine. If not, a ceramic resonator will also work with the components shown. The tuned circuit T1 is an IF "can," salvaged from an old AM and FM receiver; be sure that it has a primary tap. The primary of the 455-KHz, IF can will measure 3 to 5 Ω, and the 10.7-MHz IF can will measure less than 1 Ω; these measurements will help you to identify the correct inductor. The components in parentheses should be used for the 455-kHz transmitter.

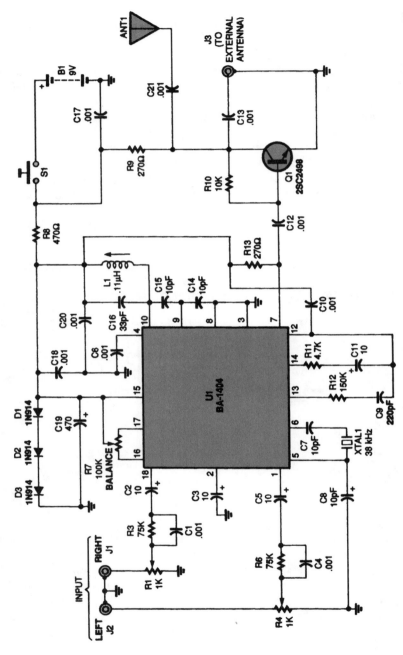

BA1404 STEREO TRANSMITTER

Fig. 112-3

ELECTRONIC EXPERIMENTERS HANDBOOK

At the heart of the schematic for the FM stereo transmitter is the BA1404 FM stereo transmitter IC. Note that there is no R2 or R5 circuit.

The circuit uses the BA1404 IC (Rohm semiconductor) to implement a complete low-power FM stereo transmitter in the 88- to 92-MHz band. Line-level audio is fed to U1, the BA1404. Xtal1 is a 38-kHz crystal, and the RF oscillator tank C16/L1 (and associated components) determine RF frequency. The circuit should be carefully constructed to minimize frequency drift and microphonics. Q1 is a buffer amplifier to reduce oscillator frequency pulling. A 9-V supply is used to power the circuit.

AM TONE TRANSMITTER

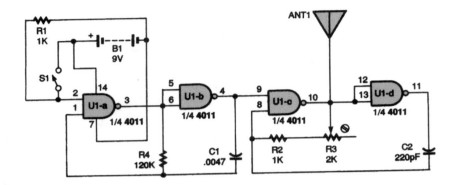

POPULAR ELECTRONICS

Fig. 112-4

Power for the circuit is supplied by B1, a 9-V battery. Two sections of a 4011 NAND gate, U1-c and U1-d, are used as part of a radio-frequency (RF) oscillator (carrier). The other two sections, U1-a and U1-b, are used in an audio-frequency (AF) oscillator (modulator). Switch S1 enables and disables the modulation to allow the transfer of an intelligent message with the transmitter. When you press S1, the AF oscillator composed of U1-a, U1-b, R4, and C1 starts generating an audio signal. That signal gates the RF oscillator composed of U1-c, U1-d, R2, R3, and C2 on and off. When on, the RF oscillator runs at 1 MHz. The resulting output is sent out ANT1 as an AM signal.

WIRELESS CAMCORDER MICROPHONE

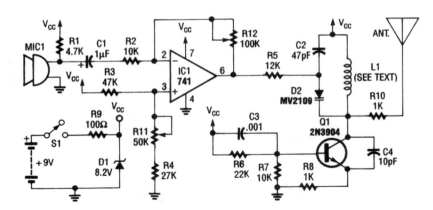

ELECTRONIC EXPERIMENTERS HANDBOOK

Fig. 112-5

A 741 op amp acts as an audio amplifier and frequency-modulates VHF oscillator Q1 via varactor D2. L1 is 2.5 turns of #18 wire on a $\frac{5}{16}$-inch-diameter form. The turn spacing can be varied to adjust the transmitter frequency.

VACUUM-TUBE TRANSMITTER

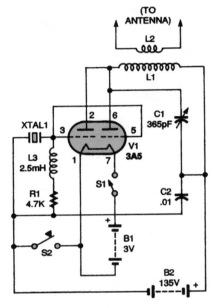

Fig. 112-6

The schematic for the transmitter is shown in the figure. In the circuit, power is supplied by a 3-V battery B1 and a 135-V battery B2. The latter battery, B2, can be replaced by a power-supply circuit. Switch S1 is used to turn the circuit on and off. The heart of the circuit is a 3A5 tube, V1. Both sections of V1 are connected in parallel in a conventional Miller oscillator. Resistor R1 sets the bias for the tube; the value of R1 might seem low to some because it is only 4700 Ω, but keep in mind that V1 is a transmitting triode operating at class C. A 2.5-mH RF choke coil (L3) is needed to keep the radio frequency where it belongs. Capacitor C1 and inductor L1 resonate at the frequency of a 40-m (7- to 7.3-MHz) amateur crystal, XTAL1. Inductor L2 couples RF energy to an attached antenna; C2 is a bypass capacitor. Switch S2 is a telegraph-type key switch. That is used to generate the content of any transmissions you make. Transformer T1 provides isolation from the power line. The isolated ac voltage is rectified by diode D1 and filtered by capacitor C1.

113

Ultrasonic Circuits

The sources of the following circuits are contained in the Sources section, which begins on page 1043. The figure number in the box of each circuit correlates to the entry in the Sources section.

ULTRASONIC MOTION DETECTOR

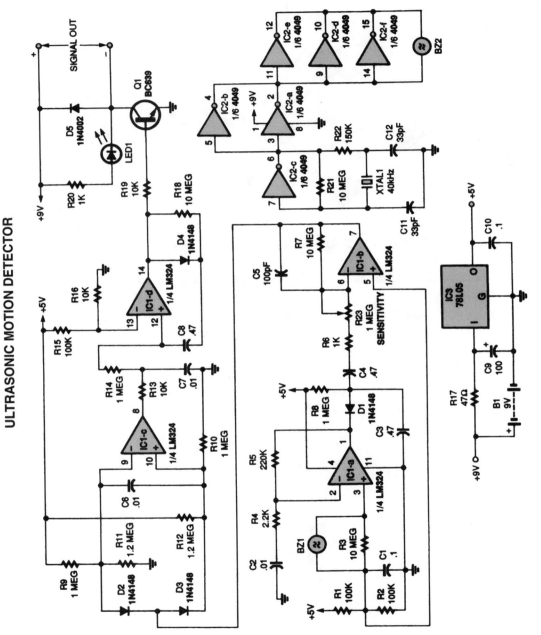

POPULAR ELECTRONICS

Fig. 113-1

ULTRASONIC MOTION DETECTOR *(Cont.)*

The receiver section of the circuit is made up of four ac-coupled stages, each built around one of four sections of an LM324 op amp, IC1. At the input to the third stage—a differential amplifier built around IC1-c—are two diodes, D2 and D3. They detect both positive and negative pulses. When there is no movement, the voltage at pin 7 of IC1-b is half the supply voltage and neither D2 or D3 can conduct. The voltage at pin 8 of IC1-c is then low. If the signal rises above +0.7 V, D3 conducts, causing the output on pin 8 to go high. If the signal falls below −0.7 V, D2 conducts, which also causes the output to go high. The fourth stage, built around IC1-d, is set up as a monostable flip-flop. That stage converts any signal that gets through the filter into a pulse substantial enough to turn on transistor Q1. When Q1 conducts, LED1 turns on and an output signal is provided to drive a separate relay or any other device connected to the circuit.

CMOS ULTRASONIC GENERATOR

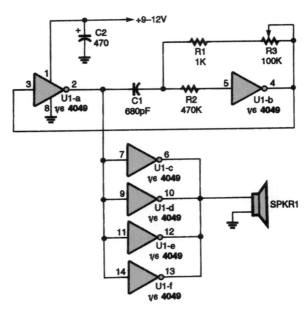

POPULAR ELECTRONICS

Fig. 113-2

This ultrasonic generator uses a single CD4049 IC. The frequency range is about 15 to 50 kHz.

TUNABLE ULTRASONIC AMPLIFIER

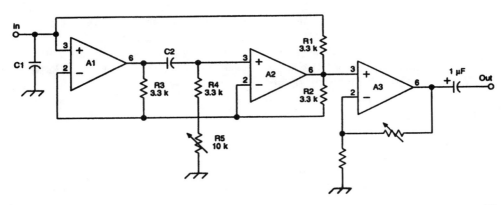

73 AMATEUR RADIO TODAY

Fig. 113-3

The RF portion of the VLF receiver is shown (a precise rectifier/filter circuit is needed following the output of the figure to generate the dc signal monitored by SID hunters). To tune the SID-hunting frequencies, C2 is 0.001 µF. The range of inductance simulated is 10.9 to 43.9 MHz. If C1 is a 0.002-µF unit, then the tuning range is 17 to 34 kHz.

PLL ULTRASONIC GENERATOR

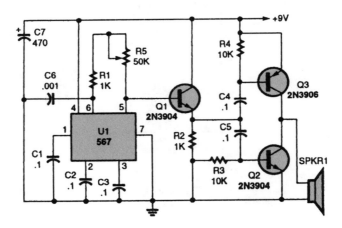

POPULAR ELECTRONICS

Fig. 113-4

This ultrasonic generator is built around a 567 PLL. By adding a telegraph key, as described in the text, it can be turned into an ultrasonic transmitter.

567 ULTRASONIC RECEIVER

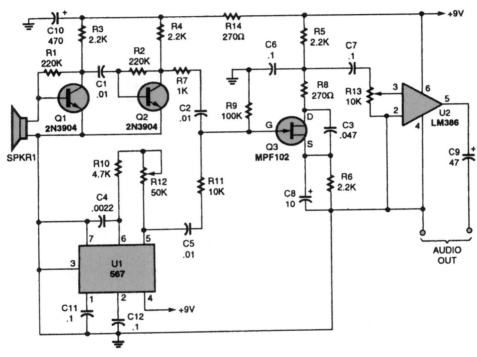

POPULAR ELECTRONICS

Fig. 113-5

This tunable ultrasonic receiver is ideal for use with the ultrasonic transmitter. It, too, is built around a 567 PLL.

114

Video Circuits

The sources of the following circuits are contained in the Sources section, which begins on page 1043. The figure number in the box of each circuit correlates to the entry in the Sources section.

Simple Video Receiver
Video Fader
Sync Stripper
Video Inverter
Video Monitor Terminator
Video Pattern Generator
RGB Sync Combiner Circuit
Monitor Video Amplifier
RGB-to-NTSC Converter
Video Negative Viewer
Multirate Video Sync Stripper
Video Amplifier with Sync Stripper and Dc Restorer
Video Distribution Amplifier
150-MHz Video Line Driver
Video-Operated Controller

SIMPLE VIDEO RECEIVER

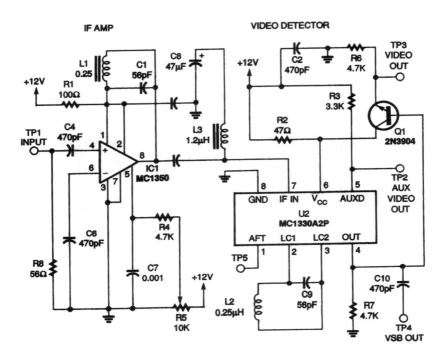

Fig. 114-1

The IF input from the tuner is fed into an MC1350 amplifier (IC1) through capacitor C4. The 56-Ω input resistor R8 approximates the impedance of the tuner's 50-Ω output. The IF amplifier is tuned by variable inductor L1 and capacitor C1, and its gain is controlled by potentiometer R5. The output of the IF amplifier is coupled into an MC1330 low-level video detector (IC2) through C5. The optional LC trap circuit, consisting of C8 and L3, can be adjusted to eliminate a specific signal, but the circuit will work well without the trap. The MC1330 is also tuned with a tank circuit, consisting of L2 and C9. The video outputs are biased with resistor networks. The noninverted video is fed into the base of a 2N3904 NPN transistor Q1 configured as a unity-gain, common-collector, buffer amplifier for impedance matching. The primary video output is taken from the top of Q1's emitter resistor R6, and the auxiliary video output is taken from pin 5 of IC2. The VSB output is an ac-coupled version of the primary output, and it can be used for any sound output you wish to use in the future. The switching carrier output can be used for automatic fine-tuning (AFT) circuits.

VIDEO FADER

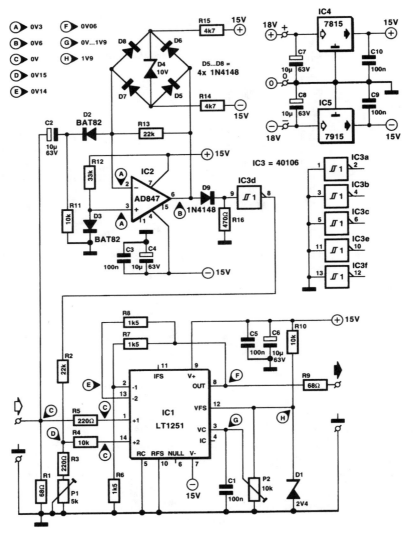

Fig. 114-2

The circuit shown here does an adequate job of allowing you to fade a video signal virtually down to the black level without losing sync on the monitor. The sync pulses are extracted from the composite video (CVBS signal) by op amp IC2 and its surrounding components. Components C2, R11, and D2 form a video-clamping circuit. Diode D3 provides a bias at the + input of IC2. Because of the rectifying action of D2, the op amp amplifies only the negative part of the CVBS signal. The clamping circuit in the feedback path of the AD847 (D5 to D8) prevents the op amp from going into satu-

VIDEO FADER *(Cont.)*

ration. The amplifier sync signal is digitized by a diode, D9, and a Schmitt-trigger gate, IC3d, before it is appied to the + input of one of the two fast op amps contained in the video fader IC, an LT1251. The sync level is set to the optimum level with the aid of preset P1. The LT1251 uses preset P2 to determine the level ratio between the sync channel and the video channel. The control voltage for the fader is derived from a reference voltage created by zener diode D1. The mixed output signal is available at pin 8 (dc-coupled), at an impedance of about 75 Ω. Current consumption of the circuit is less than 30 mA. The indicated test voltages are applicable with no input signal applied to the circuit. The fader also reduces the level of the color burst. Consequently, the picture can go black and white just before the black level is reached.

SYNC STRIPPER

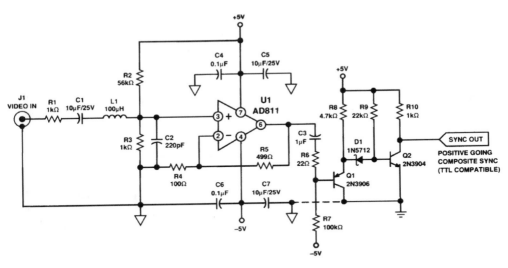

Fig. 114-3

This circuit provides a sync-stripping function. The output is positive-going TTL-compatible.

VIDEO INVERTER

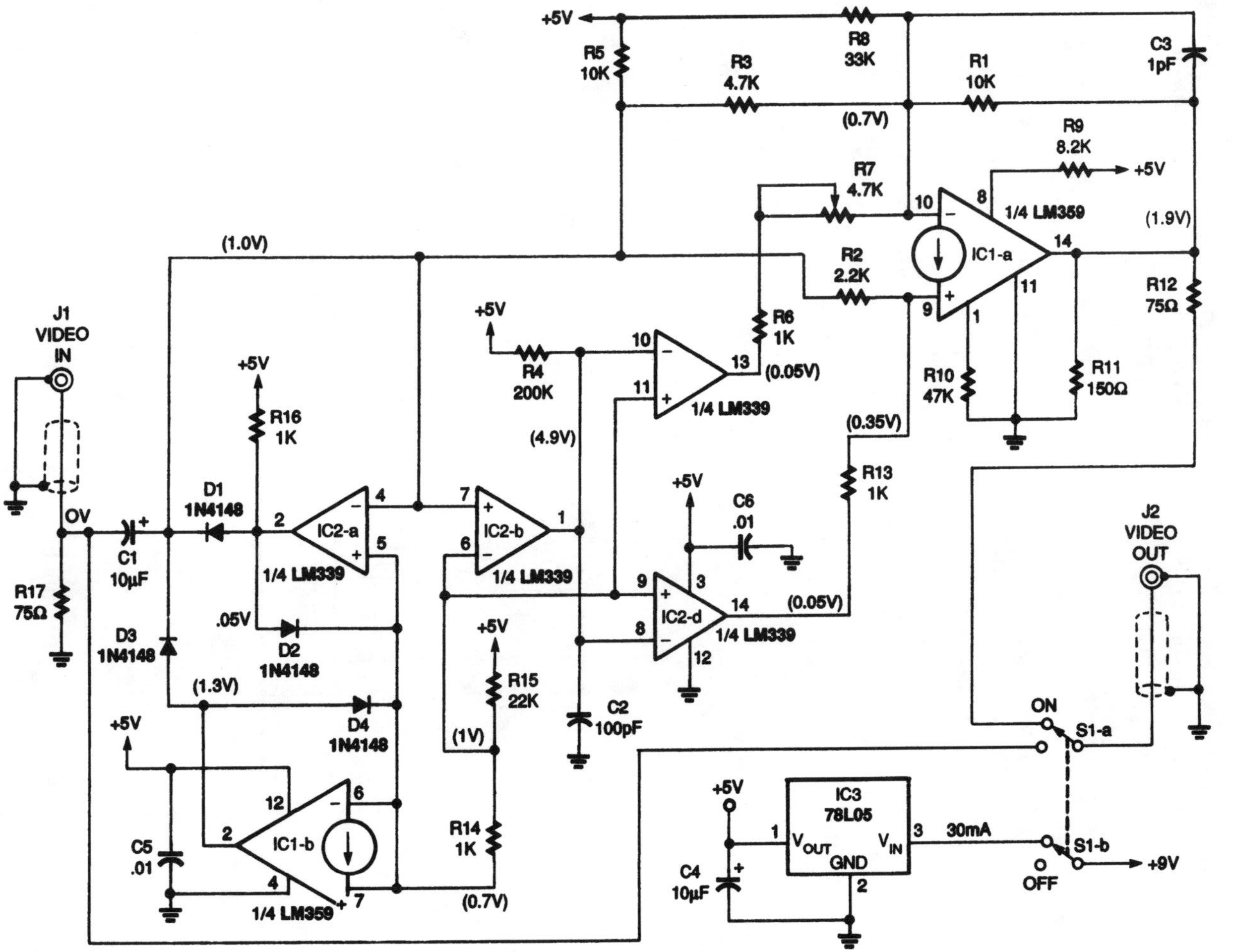

Fig. 114-4

ELECTRONICS NOW

VIDEO INVERTER *(Cont.)*

This video inverter circuit is based on an LM359 dual Norton op amp and an LM339 quad voltage comparator for dc restoration, sync detection and processing, and buffering and inversion of the video signal. The dc level of the video signal is restored to 0.7 V because the inputs of IC1-b operate at one diode voltage drop above ground. Comparator IC2-b acts as a threshold detector set to detect sync pulses 300 mV above the restored dc level of 0.7 V. Components C2 and R4 stretch the detected 4-μs sync pulses to 8 μs to include the duration of 8 cycles of 3.58-MHz color burst after the input sync pulse. Comparators IC2-c and IC2-d buffer the stretched sync pulse with open-collector outputs at pin 13 and 14 to gate the buffer/inverter action of the main amplifier IC1-a. Resistor R3 sets the gain of the inverting input of the main video amplifier IC1-a to 2 ($-R_1/R_3=-2$), and R4 sets the gain of the noninverting input of IC1-a to 4 ($R_1/R_4=+4$). These two inputs combine to achieve an overall gain of +2 as a buffer or −2 as an inverter. During the sync and color-burst portions of the input signal, the IC2-c and IC2-d outputs are open and the overall gain is +2. The rest of the time, a low at the output of IC2-d (pin 14) short-circuits the noninverting +4 signal path for an overall gain of −2, while a low at IC2-c pin 13 injects the necessary offset in the inverted video signal to keep it above the black level. Potentiometer R7 adjusts the black level at the output to correspond to the peak white level in the negative. Resistor R12 matches the output for 75 Ω. A 78L05 regulator (IC3) provides a stable 5-V supply from a 9-V battery or ac adapter.

VIDEO MONITOR TERMINATOR

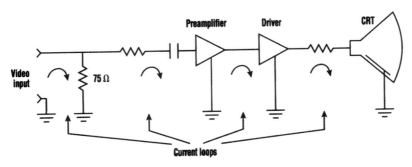

ELECTRONIC DESIGN

Fig. 114-5

The best method of terminating video output signals is to route the video cable directly to the video-amplifier input connector without connecting its signal or shield wires to other parts of the monitor.

VIDEO PATTERN GENERATOR

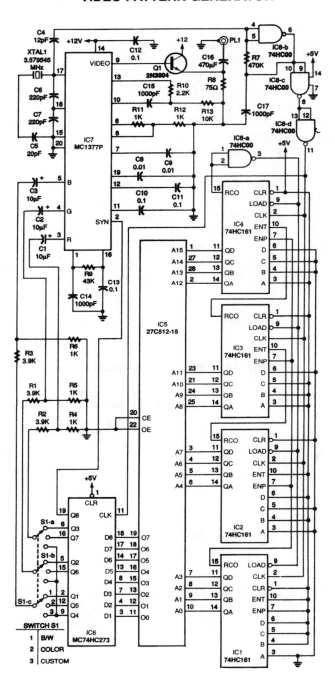

Fig. 114-6

VIDEO PATTERN GENERATOR *(Cont.)*

A 3.58-MHz clock drives a counter chain, which addresses a preprogrammed EPROM, to generate a complete NTSC sync and video waveform. The EPROM can be programmed for various test patterns as required. A switch provides a choice of a checkerboard pattern, a color bar pattern, or a custom pattern. The total number of pixels in the custom pattern is 43,605. Each pixel can be programmed to one of the eight colors in the color bar pattern. The color signal is formed in the MC1377 encoder chip from the RGB digital inputs. The composite video out at PL1 is 2.5 V p-p.

RGB SYNC COMBINER CIRCUIT

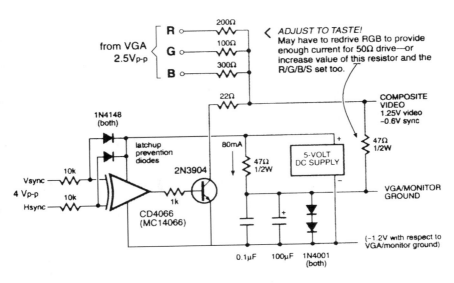

All resistors 1/4W unless otherwise specified.
This circuit is UNTESTED.

NUTS AND VOLTS

Fig. 114-7

This circuit shows one way of combining RGB and sync signals from a computer to form an NTSC B/W video signal that is viewable on an ordinary video monitor.

MONITOR VIDEO AMPLIFIER

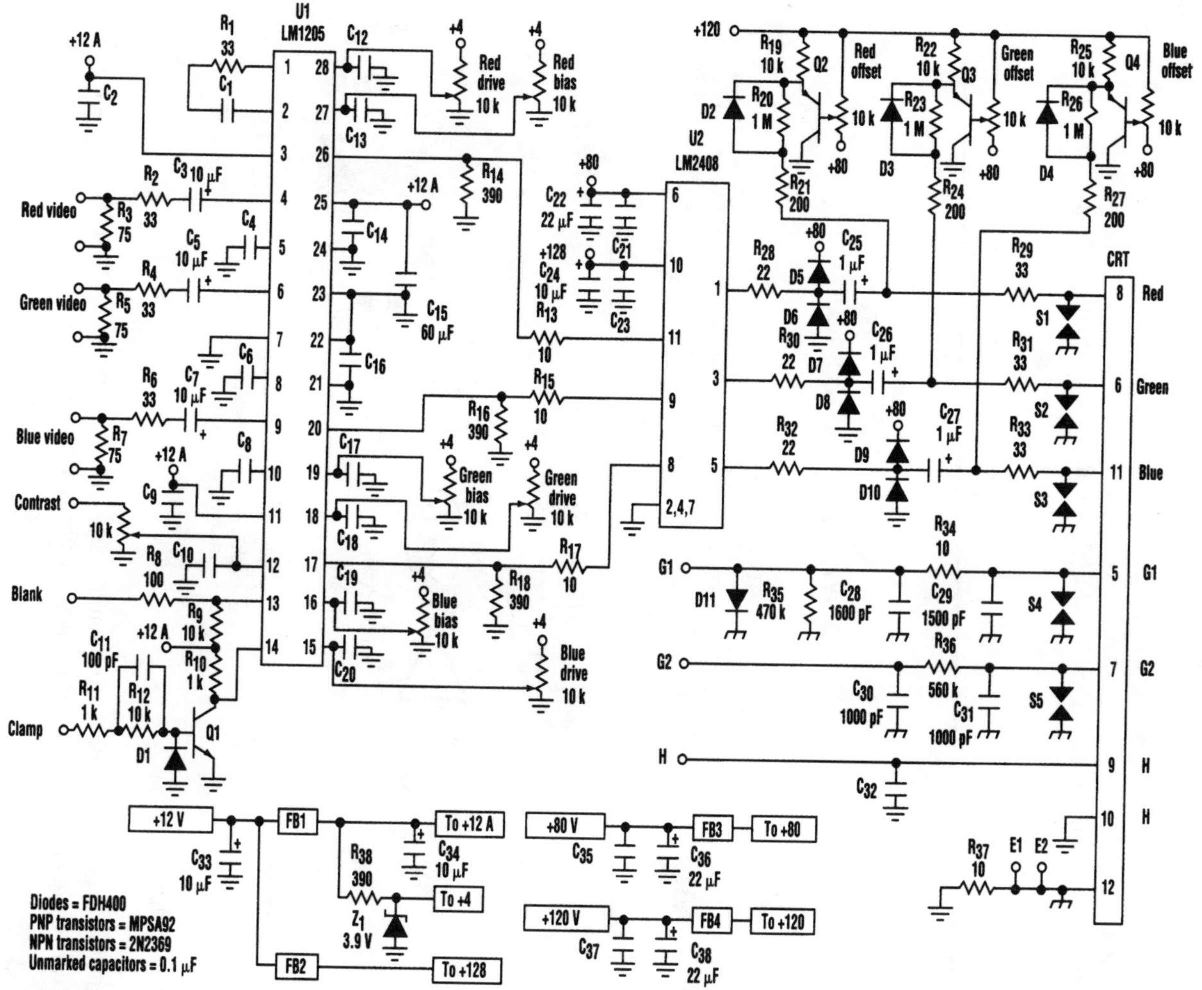

ELECTRONIC DESIGN

Fig. 114-8

Diodes = FDH400
PNP transistors = MPSA92
NPN transistors = 2N2369
Unmarked capacitors = 0.1 μF

1014

MONITOR VIDEO AMPLIFIER *(Cont.)*

This circuit uses a preamplifier and a video driver IC. The monitor video amplifier will affect all aspects of future computer monitors, including higher-resolution pictures and faster video speeds. Such an amplifier is usually split into a preamplifier section and a CRT driver section.

RGB-TO-NTSC CONVERTER

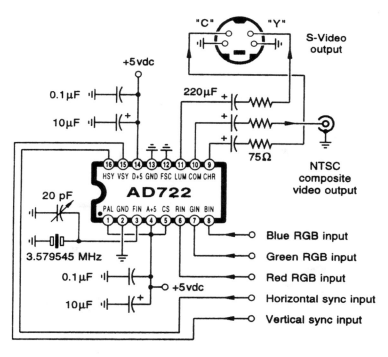

ELECTRONICS NOW

Fig. 114-9

This figure shows an RGB-to-NTSC converter circuit. Note that the input sweep rates, interlace, overscan, and program content must be TV-compatible if you want useful results. The AD722 is manufactured by Analog Devices.

VIDEO NEGATIVE VIEWER

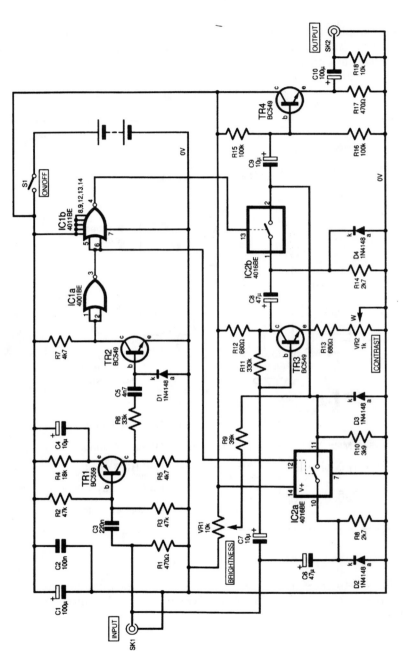

Fig. 114-10

EVERYDAY PRACTICAL ELECTRONICS

This circuit will invert the video while maintaining sync polarity. Therefore, it will produce a negative image of the video signal. The sync is stripped by TR1, TR2, and IC1a and IC1b. Then it controls analog switches IC2a and IC2b to maintain sync polarity while switching in the inverted video produced by inverter TR3. TR4 is an emitter-follower for low output impedance. The unit was originally intended for viewing photographic negatives on a video monitor.

MULTIRATE VIDEO SYNC STRIPPER

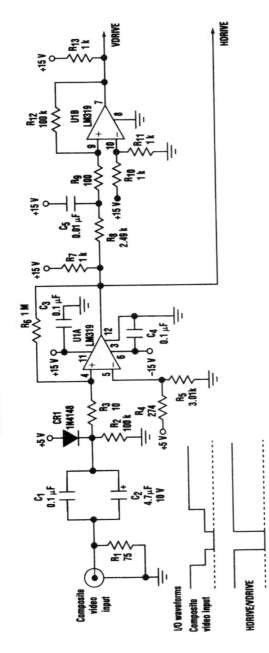

Fig. 114-11

ELECTRONIC DESIGN

This dual-comparator, multirate video sync stripper provides separate horizontal and vertical pulses that are consistent with the composite video-input-signal sync pulses. Using an LM319 dual-comparator integrated circuit and associated passive components, this circuit can strip horizontal (HSYNC) and vertical (VSYNC) sync pulses from standard RS170 video (525/2:1 interlaced) through an industry high-end video rate of 1280 × 1024/1:1 (noninterlaced). The composite video input signal is ac-coupled through capacitors C1 and C2 to the junction of resistors R2 and R3 and diode CR1. The video sync tips are clamped by CR1 at approximately 4.5 V and applied to noninverting input of comparator U1A (pin 4). Positive feedback from resistors R6 and R3 provides a hysteresis of about 150 µV to ensure a stable state change at the output of U1A (pin 12). R8 and C5 constitute a low-pass filter that prevents the filtered amplitude of the horizontal sync pulse from dropping below the threshold voltage of 7.5 V, established by R10 and R11 for the inverting input of U1B (pin 9). R9 and R12 supply a hysteresis of about 15 mV to the noninverting input of U1A (pin 10). The inherently longer vertical sync pulse width provides sufficient time for U1A (pin 12) to be near ground. As a result, C5 is adequately charged to exceed the threshold voltage and generate a vertical (VDRIVE) at U1B (pin 7).

1017

VIDEO AMPLIFIER WITH SYNC STRIPPER AND DC RESTORER

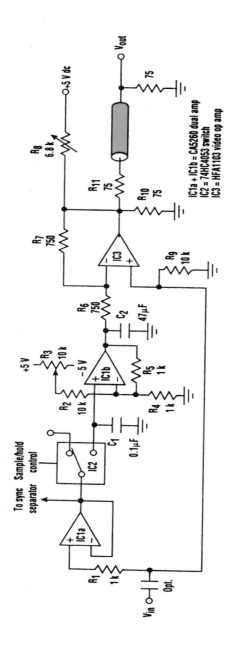

IC1a + IC1b = CA5260 dual amp
IC2 = 74HC4053 switch
IC3 = HFA1103 video op amp

Fig. 114-12

ELECTRONIC DESIGN

This circuit transmits 220-MHz (−3-dB bandwidth) video signals while stripping off the sync pulse and performing dc restoration. It is configured for a typical video cable-driver application driving a double-terminated 75-Ω load. In other words, the HFA1103 (IC3) is configured for a gain of +2 to ensure unity gain overall. The HFA1103 video op amp is specially designed to perform sync stripping. Its open-emitter NPN output forms an emitter-follower with the load resistor, and passes the active video signal while virtually eliminating the negative sync pulse. Residual sync, defined as the remainder of the original −300-mV sync pulse, referenced to ground, is only 8 mV at the cable output of the HFA1103. Because the HFA1103 contains no active pull-down, output linearity degrades as the signal approaches 0 V. To deal with this, a 6.8-kΩ pull-up resistor (R8) and a 75-Ω pull-down resistor (R10) on the output ensure a fixed positive offset voltage, in this case, +50 mV. This offset was arbitrarily chosen as a good compromise between linearity near the dc level and minimum residual sync. Increasing R_8 decreases residual sync at the expense of linearity. Conversely, lowering R_8 decreases linearity error, but increases residual sync. This circuit achieves dc restoration by using a CA5260 dual op amp (IC1a, IC1b) coupled with a sample-and-hold circuit, based on a 74HC4053 switch (IC2). V_{in}, consisting of the input video signal and a dc offset (V_{dc}), is connected to the noninverting input of

VIDEO DISTRIBUTION AMPLIFIER

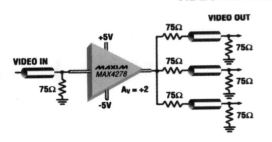

This video distribution amplifier uses a single MAX4278 with a gain of 2 (6 dB) so that three 75-Ω loads can be driven from one 75-Ω source. Each driven load has a source impedance of 75 Ω. Gain is flat to 0.1 dB up to 150 MHz.

MAXIM

Fig. 114-13

150-MHz VIDEO LINE DRIVER

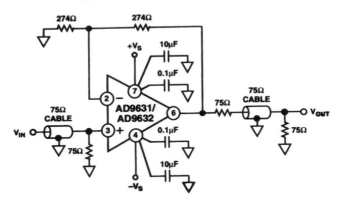

ANALOG DIALOGUE

Fig. 114-14

This line driver uses an Analog Devices P/N AD9631/9632 and has a bandwidth of better than 150 MHz.

VIDEO-OPERATED CONTROLLER

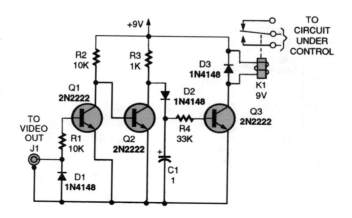

POPULAR ELECTRONICS

Fig. 114-15

If you need a timer that is programmable, precise, and will provide long delays, almost any VCR can be used without any internal modification. All that is required is a circuit that will detect the presence of the video output signal when the recorder turns on. The first diode (D1) clamps the negative video to ground. The rest of the circuit responds to the frame markers to charge up the capacitor and turn on the relay. A tape isn't even required (in most cases) if the recorder is tuned to an active channel because the video-out signal appears as soon as the recorder is turned on. Therefore, the circuit's ON time is not limited by tape length.

115

Voltage-to-Frequency Converter Circuits

The sources of the following circuits are contained in the Sources section, which begins on page 1043. The figure number in the box of each circuit correlates to the entry in the Sources section.

MICROPOWER VOLTAGE-TO-FREQUENCY CONVERTER

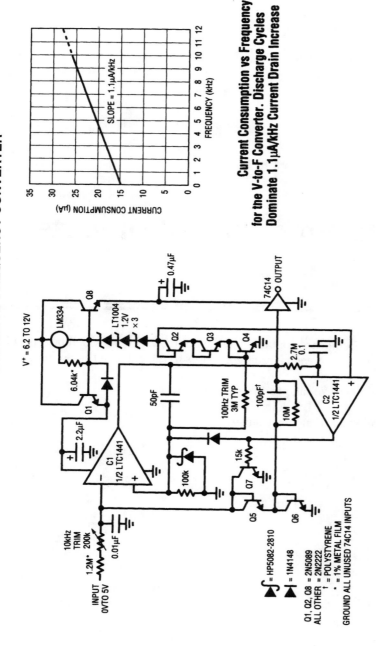

Current Consumption vs Frequency for the V-to-F Converter. Discharge Cycles Dominate 1.1μA/kHz Current Drain Increase

LINEAR TECHNOLOGY

Fig. 115-1

This voltage-to-frequency converter takes full advantage of the LTC1441's low power consumption under dynamic conditions. A 0- to 5-V input produces a 0- to 10-kHz output with 0.02 percent linearity, 60 ppm/°C drift, and 40 ppm/V supply rejection. Maximum current consumption is only 26 μA, 100 times lower than that of currently available circuits. C1 switches a charge pump, composed of Q5, Q6, and the 100-pF capacitor, to maintain its negative input at 0 V. The LT1004s and associated components form a temperature-compensated reference for the charge pump. The 100-pF capacitor charges to a fixed voltage; hence, the repetition rate is the circuit's only degree of freedom to maintain feedback. Comparator C1 pumps uniform packets of charge

MICROPOWER VOLTAGE-TO-FREQUENCY CONVERTER *(Cont.)*

to its negative input at a repetition rate precisely proportional to the input voltage-derived current. This action ensures that circuit output frequency is strictly and solely determined by the input voltage. Start-up or input overdrive can cause the circuit's ac-coupled feedback to latch. If this occurs, C1's output goes low; C2, detecting this via the 2.7-MΩ/0.1-μF lag, goes high. This lifts C1's positive input and grounds the negative input with Q7, initiating normal circuit action.

WIDE-BANDWIDTH VOLTAGE-TO-FREQUENCY CONVERTER

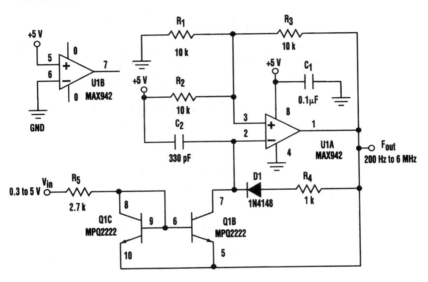

Fig. 115-2

Shown is the design for a low-cost, wide-bandwidth voltage-to-frequency converter (VFC). The core element of the prototype VFC is a Maxim MAX942 high-speed comparator. When operating, R1, R2, and R3 provide hysteresis with trip points from one-third to two-thirds of the supply voltage. This permits wide adjustability in the width of the output pulse. Q1C and Q1B form a current mirror that linearizes the charge potential across C2. As the voltage at pin 2 of U1 crosses the lower trip point, the comparator output turns on. The current mirror sink node is raised above the program voltage, which turns it off. D1 turns on, removing the charge on C2 through R4. When the voltage at U1 pin 2 crosses the upper trip point, the circuit resets. Upon observation, the prototype circuit linearity was approximately 2 percent over the tested range. Also, up to 6-MHz operation was possible using the MAX942 by carefully selecting C2 and the hysteresis resistors. Discrete transistors can be used instead of the Motorola MPQ2N2222 monolithic quad package. This can cause some instability at low frequencies. The transistors should be located in close thermal proximity. On the high side, a good, matched transistor pair can be used. Analog Devices' MAT-01 is a good choice.

VOLTAGE-TO-FREQUENCY CONVERTER

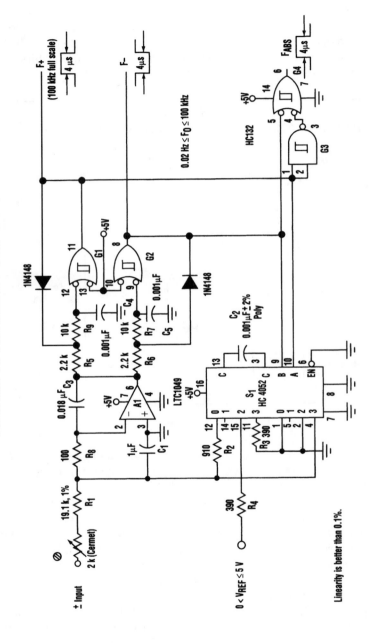

Linearity is better than 0.1%.

ELECTRONIC DESIGN

Fig. 115-3

The voltage-to-frequency converter draws <10 mA from one +5-V supply while producing three 0- to 100-kHz pulse outputs. One output is proportional to positive input voltages and inactive for negative inputs, another responds when the input is negative, and a third outputs a frequency that is proportional to the input's absolute value. The converter's unadjusted zero offset is less than 1 ppm of full scale and its linearity is better than 0.1 percent.

116

Waveform Generator Circuits

The sources of the following circuits are contained in the Sources section, which begins on page 1043. The figure number in the box of each circuit correlates to the entry in the Sources section.

555 RAMP GENERATOR

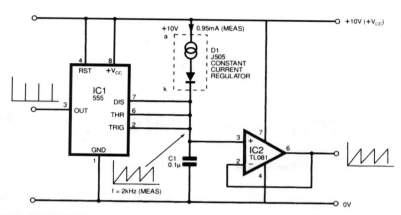

EVERYDAY PRACTICAL ELECTRONICS

Fig. 116-1

When the timing capacitor of an ordinary 555 circuit charges, an exponential curve across the capacitor results. Obviously, a fast-rising square wave is seen at the output. An accurate ramp voltage can be generated by the 555, though, by ignoring the square-wave output and sneakily using a constant-current source instead of the resistor charging network. The figure shows how a simple J505 constant-current diode can be used to create a reasonable ramp, which will be observed across the capacitor C1. D1 is specified at 1 mA nominal. Assuming that C1 is discharged initially, after power-up the 555 triggers ($V_{trig} < \frac{1}{3} V_{cc}$), and C1 starts to charge at a fixed rate through the constant-current source D1 until the threshold voltage of $\frac{2}{3} V_{cc}$ is reached. Then the 555 (IC1) will rapidly discharge the capacitor into pin 7 down to $\frac{1}{3} V_{cc}$, when the chip will trigger again. With a 12-V rail, the result is a sawtooth waveform of 4 V p-p. A fast series of spikes is seen at pin 3.

SIMPLE SIGNAL GENERATOR

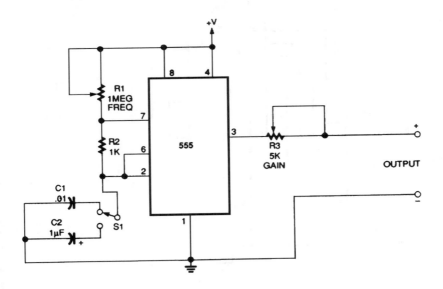

Fig. 116-2

The use of an NE555 allows generation of signals from 1 Hz to higher than 10 kHz. The output is a square wave.

ATTACK/DECAY RAMP GENERATOR

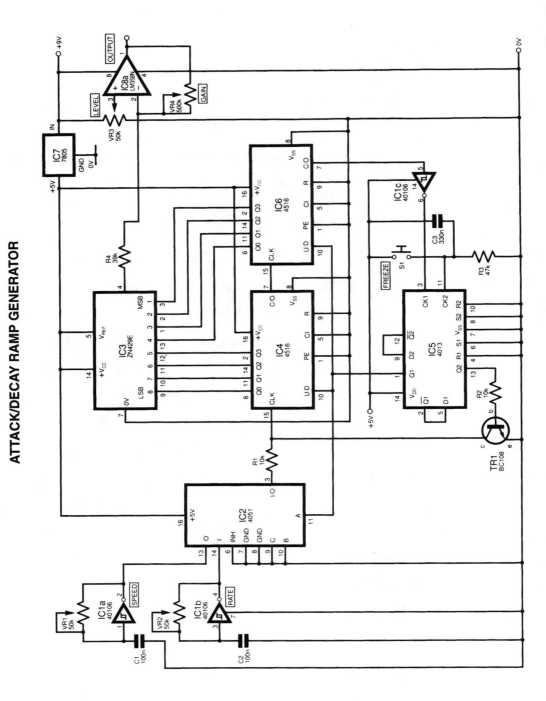

Fig. 116-3

EVERYDAY PRACTICAL ELECTRONICS

ATTACK/DECAY RAMP GENERATOR *(Cont.)*

In the circuit diagram shown, a digital-to-analog converter (DAC) IC is used to create an envelope waveform, the attack and decay of which are controllable. Typical applications can include audio and music synthesizer circuitry. Its operation is as follows. IC1a and IC1b form two variable-frequency oscillators, whose outputs are fed to IC2. This is configured as a demultiplexer, addressed by a signal appearing on pin 11. The signal from IC1a or IC1b appears at the output, pin 3. Initially pin 11 of IC2 will be high, so IC1a will be selected and fed through resistor R1 to IC4, a binary counter, which is cascaded to IC6. In total, this results in an 8-bit binary count (0 to 255) appearing at the counters' outputs. The speed of the count is adjustable with VR1. The counter output is fed directly to IC3, a ZN429E digital-to-analog converter. The ramp voltage appears at pin 4 of IC3, and will always be between 0 and 4 V. When the counter reaches its terminal count of 255, pin 7 of IC6 (CARRY OUT) will go low. This signal is inverted by IC1c and used as a clock pulse to IC5, a 4013 D-type flip-flop. The output at pin 1 of IC5 now goes low, so that a logic 0 is placed on pin 11 of IC2 and on pin 10 (UP/DOWN) of both counter ICs. Oscillator IC1b is now fed through the demultiplexer to the counters, which now count down from 255. The output voltage from IC3 will now fall at a rate determined by the setting of VR2. As the count passes zero and rolls back to 255, a pulse again appears at pin 7 of IC6, and the cycle continues. At any time, switch S1 (debounced by C3 and R3) can be momentarily activated. This causes a high output at pin 13 of IC5, which drives transistor TR1, shunting the clock signal to the counters. This will "freeze" the count at that point. A second operation of S1 restarts the count from where it left off. A buffer/amplifier formed by IC8a (one half of an LM358N) allows the output voltage of the DAC (IC3) to be increased or the dc level shifted.

SIMPLE TRIANGLE-WAVEFORM GENERATOR

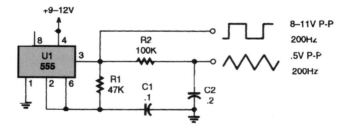

POPULAR ELECTRONICS

Fig. 116-4

This triangle-waveform generator was designed to give good results with as few parts as possible.

DESIRED WAVEFORM GENERATOR

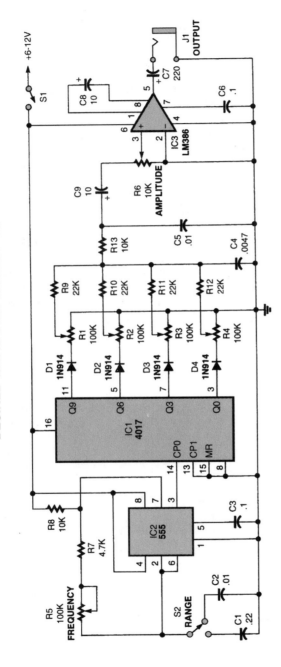

Fig. 116-5

POPULAR ELECTRONICS

This is a filtered step-function generator in which the wave amplitude at each step is set by a linear potentiometer (R1 to R4). With the potentiometers arranged in time order and side by side (the way they are in a graphic equalizer), the positions of the knobs describe the output waveform. The 555 oscillator, controlled by R5 and C1 or C2, sets the duration of each step, and (therefore) also the wave frequency. The 555 oscillator, controlled by R5 and C1 or C2, sets the duration of each step, and (therefore) also the wave frequency. Potentiometer R5 controls the step frequency and the capacitors define the frequency range. The output of the 555 is applied to the clock input for the 4017 decimal counter, which activates its outputs one at a time. The linear potentiometers divide the voltage of each output, and thus determine the output voltage at a given step point. The schematic shows potentiometers only in Q0, Q3, Q6, and Q9, but for smoother waveforms potentiometers should be at all of the outputs, except the CARRY-OUT pin. The composite waveform is now purified by the filter formed by R13, C4, and C5, so the output will be rounded. The amplitude is controlled by the LM386 audio amplifier's input. For high frequencies, replace the amplifier with a 741 or some other high-frequency op amp.

1030

SIMPLE PULSE GENERATOR

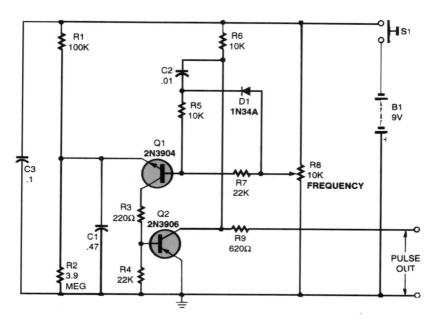

POPULAR ELECTRONICS

Fig. 116-6

Timing capacitor C1 is charged by battery B1 through R1 until Q1 is forward-biased and starts conducting. When this occurs, a forward base bias is applied to Q2 through Q1 and R3. As Q2 starts conducting, a pulse is made across collector load R6, which is coupled back to Q1's base through C2 and series resistor R5. That drives Q1 to rapidly discharge C1, and then the cycle starts over. In essence, Q1 and Q2 form a high-gain amplifiers, with C2 providing positive feedback. Diode D1 reduces C2's recovery time, while R8 establishes Q1's base bias level. Thus, it is the point at which timing capacitor C1 charges before feedback and capacitor discharge are initiated. The lower Q1's initial base bias, the higher the circuit's repetition rate. Resistor R8, therefore, serves as a frequency control, providing nine octaves of coverage. The circuit will work from 5 to 25 V, but a 9-V source is optimum. Output impedance is 600 Ω. A sawtooth signal suitable for use as a linear scope sweep is available across C1. For frequency-divider, sweep-generator, and time-marker uses, synchronization pulses can be applied to Q1's base through a small capacitor. The frequency range can be shifted by using other values for C1 and C2. With values of 100 μF for C1 and 3 μF for C2, the circuit will work down to about 1/20 Hz.

10-MHz FUNCTION GENERATOR

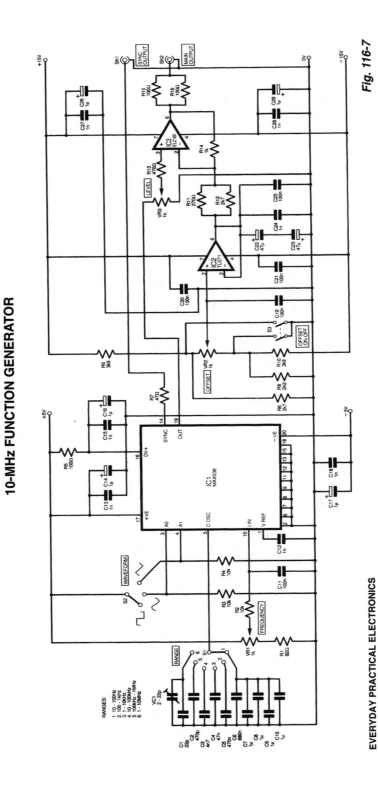

EVERYDAY PRACTICAL ELECTRONICS

Fig. 116-7

This circuit is built around the MAX038 IC and can provide sine, square, and triangle waveforms from 10 Hz to 10 MHz. PC-board layout techniques suitable for 50 MHz or higher should be used for best results with this circuit.

TRIANGLE-WAVEFORM GENERATOR

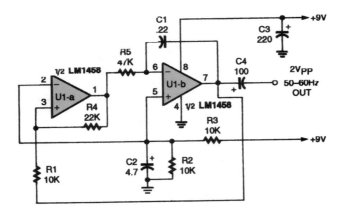

POPULAR ELECTRONICS

Fig. 116-8

A low-frequency triangle-waveform generator is made from a single LM1458 dual op amp and a few inexpensive components. The output frequency can be varied by changing the R_5. Changing the value of C_1 will shift the circuit's frequency range: Increasing the value will lower the oscillator's frequency and reducing the value will increase the frequency. Resistor R4 can be varied to change the output level. Powered from a 9-V supply, the circuit produces a 2-V p-p triangular output waveform.

PROGRAMMABLE PULSE GENERATOR

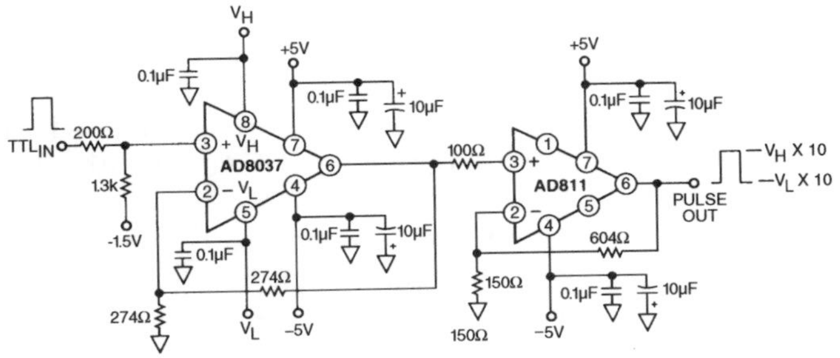

ANALOG DIALOGUE

Fig. 116-9

This circuit provides a programmable flat-pulse generator with TTL input and up to 24 V p-p output with a 2500-V-µs slew rate. Clamp amplifiers are useful as protective buffers for A/D inputs, as programmable flat-pulse-forming amplifiers, and as amplitude modulators. Because V_H and V_L clip in the input stage and have input bandwidths comparable to the signal inputs, they can also be used in simple circuitry to generate the positive or negative absolute value of input signals (i.e., full-wave rectify them).

HARMONIC GENERATOR

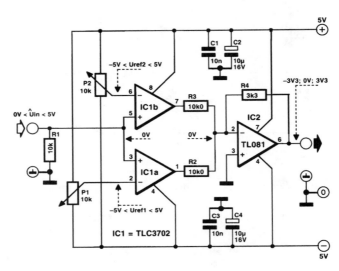

Fig. 116-10

This simple circuit acts as a frequency multiplier, which converts any sloped input waveform into a distorted output waveform, whose frequency spectrum is marked by a rich harmonics content. The input frequency does not disappear from the spectrum, however. When a pure sine-wave input signal is applied, the circuit generates odd harmonics only. The push-pull outputs of comparators IC1b and IC1a then supply differential signals to adder IC2, which cancels out even-numbered harmonics. For different mark/space ratios of the output signals, as set with P1 and P2, the harmonic spectrum changes. A duty factor of 0.25, for instance, will cause the circuit to supply the second, sixth, and tenth harmonics, but not the fourth one. The spectrum function is then written as sin x/x. The reference levels for the input comparators determine the output waveform. The desired degree of distortion of the input waveform is, therefore, adjusted with the two presets. R2 and R3 should be matched to within 1 percent.

STEPPED TRIANGLE-WAVEFORM GENERATOR

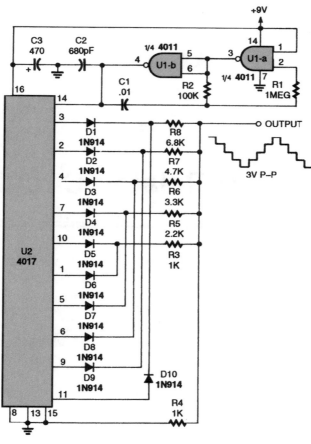

POPULAR ELECTRONICS

Fig. 116-11

An astable multivibrator (made up of U1-a and U1-b) drives a 1-of-10 counter. A resistor matrix acts as a rough D-A converter and produces the stepped triangle waveform shown. The waveform can be tailored to suit by adjusting R_3 through R_8.

117

Weather-Related Circuits

The sources of the following circuits are contained in the Sources section, which begins on page 1043. The figure number in the box of each circuit correlates to the entry in the Sources section.

Wind-Direction Sensor
Sferic Simulator Circuit
Anemometer
Weather Vane

WIND-DIRECTION SENSOR

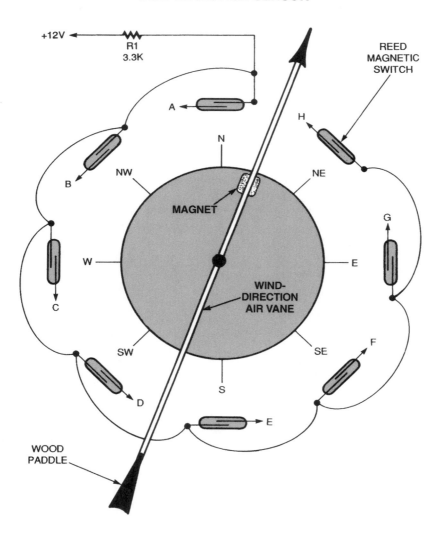

+12V

R1
3.3K

REED
MAGNETIC
SWITCH

A

B

NW

N

NE

H

MAGNET

W

E

G

WIND-
DIRECTION
AIR VANE

C

SW

SE

F

S

D

E

WOOD
PADDLE

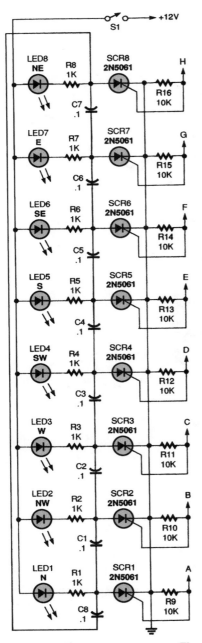

Fig. 117-1

To use the circuit you will need a wind vane like the one shown. It should have a weighted front end and an air paddle in the rear. Attach a small, strong magnet to the front part of the wind-vane arm. Then, position eight reed magnetic switches in a circle around a piece of plastic pipe and electrically connect them (as shown). Note that the reed switches are electrically connected to points labeled *A* through *H*. Those points correspond to points A through H in the wind-direction decoder circuit. The decoder uses eight low-current 2N5061 SCRs and eight LEDs to latch and display the wind vane's position. If the wind vane is pointing due north, reed switch A is closed, sending current into the gate of SCR1. That current turns the thyristor on, causing it to light LED1. If the wind shifts slightly to a position in between north and northwest or north and northeast, without activating any of the reed switches, LED1 will remain on, indicating that the last wind direction was north. With SCR1 turned on, one end of capacitors C8 and C1 is pulled to ground. The other end of both capacitors is tied to the +12-V bus through a resistor and LED; that means that both capacitors are charged to near 12 V. All other capacitors are not charged because both ends of each capacitor are returned to the +12-V bus through a resistor and LED. When the wind direction shifts to the northwest, reed switch B turns SCR2 on, thereby "taking" the positive end of C1, which is connected to its anode, to ground. This negative pulse turns SCR1 off as SCR2 turns on, lighting LED2 and turning off LED1.

SFERIC SIMULATOR CIRCUIT

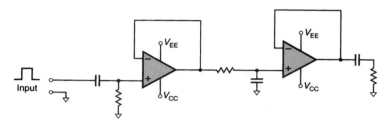

Notes: 1. Capacitors 20 nF   2. Resistors 1kΩ   3. Operational Amplifiers LF353

The **Sferic Simulator** generates a bipolar pulse signal like that radiated by lightning when triggered by a fast-rising square wave.

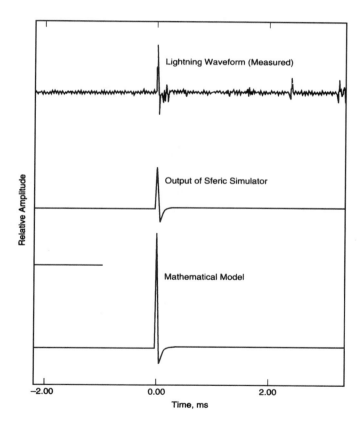

Lightning Waveform (Measured)

Output of Sferic Simulator

Mathematical Model

Relative Amplitude

−2.00   0.00   2.00

Time, ms

The **Output Waveform** generated by the sferic simulator resembles a real sferic and a sferic simulated by use of a mathematical model.

SFERIC SIMULATOR CIRCUIT *(Cont.)*

This circuit generates a signal that simulates the waveform of a sferic—the very-low-frequency electromagnetic signals generated by a lightning strike. This circuit is designed to test and calibrate lightning-detecting instruments. Typically, lightning-detecting instruments have been calibrated and tested using single-frequency signals or broadband noise sources, which do not have the spectral characteristics of sferics. This circuit generates a bipolar pulse waveform that closely approximates the main features of sferics. The circuit contains few components. This circuit assembly is light in weight and compact; even a "breadboard" version could fit in a 1.5-inch square (about 3.8 cm). To ensure that the output waveform has the desired shape, the trigger pulse fed to the input terminal of this circuit must be a square wave with rise and fall times less than 100 ns. To prevent coalescence of sequential output pulses, the square-wave pulse-repetition frequency should be kept at or below 3000 Hz.

ANEMOMETER

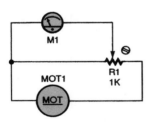

POPULAR ELECTRONICS   *Fig. 117-3*

Here is a really simple and inexpensive anemometer. You can use any small dc motor for MOT1. Meter M1 can be any milliampere or microampere meter with markings from 0 to 50 or 100. The parts, including the 1000-Ω, 15-turn potentiometer, are available at hobbyist sources. A three-paddle wheel for the motor can be made with golf-ball-sized hollow plastic balls. Cut two of them in half, and bolt three of the halves to a PVC hub, using 6-32 machine screws. Calibrate the circuit by holding it out of a car window while someone drives steadily between 20 and 30 miles per hour on a still day.

WEATHER VANE

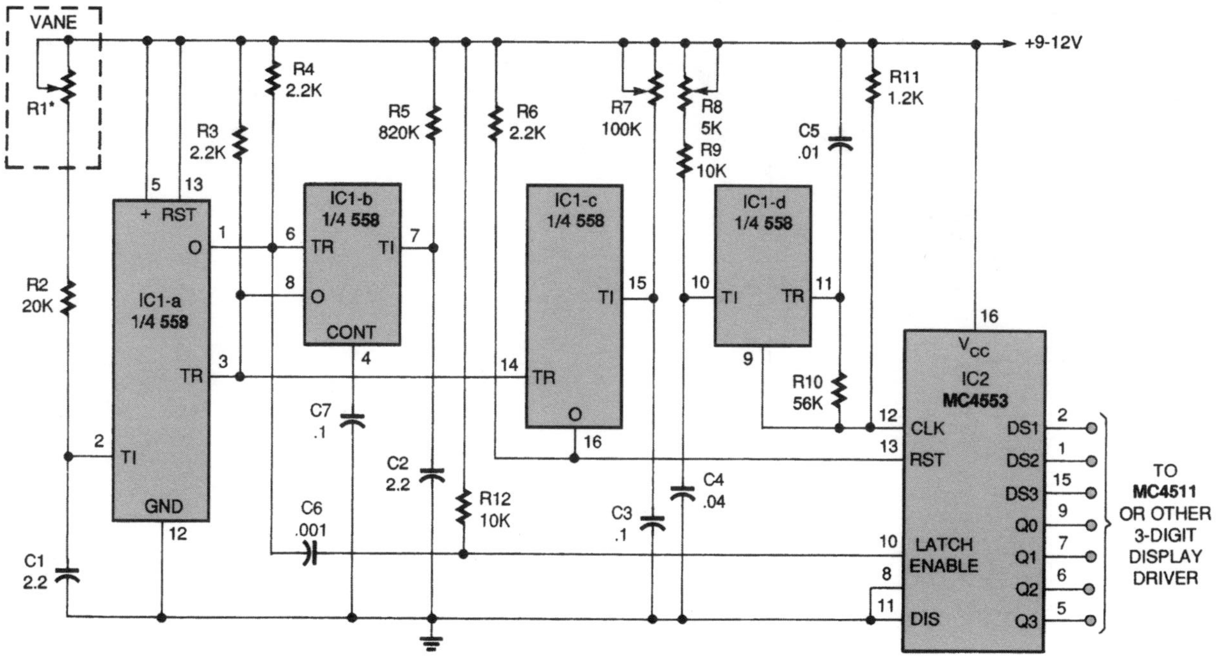

Fig. 117-4

POPULAR ELECTRONICS

1041

WEATHER VANE *(Cont.)*

IC1-a and IC1-b form an astable oscillator in which IC1-a produces a positive pulse width that is proportional to the resistance of the segmented potentiometer, R1. The negative pulse width, determined by IC1-b, is fixed. The third section, IC1-c, is triggered by the falling edge of IC1-b. Its pulse width is set by potentiometer R7 so that it is exactly equal to the width of the IC1-a's pulse when the vane is at zero resistance. Potentiometer R8 and IC1-d make up the frequency reference. An astable oscillator (formed out of IC1-d and its surrounding components) provides a clock signal to IC2, a 555 timer IC. The RESET pulse from IC1-c starts to count at zero, and the LATCH pulse from IC1-a freezes the instantaneous reading in the timer. So, U2 counts the time between the falling edge of IC1-a and the falling edge of IC1-c, which is determined by the position of the vane. The timer can feed its digital output to the three-digit display circuit of your choosing (such as an MC4511 display driver). To adjust the circuit, rotate R1 (the transducer) to its zero-resistance position (North) and adjust R7 until the readout is zero. Second, rotate R1 to its maximum-resistance position and adjust R8 until the readout is 350. That will give, theoretically, readings between zero and 350 in 10-degree increments as the vane is rotated.

Sources

Chapter 1

Fig. 1-1. Reprinted with permission from *Electronic Design*, January 6, 1997, pp. 164–166. Copyright © 1997, Penton Publishing Co.

Fig. 1-2. Reprinted with permission from *Electronic Design*, July 22, 1996, p. 112. Copyright © 1996, Penton Publishing Co.

Fig. 1-3. *Popular Electronics*, August 1997, p. 48.

Fig. 1-4. *Popular Electronics*, August 1997, p. 48.

Fig. 1-5. *Popular Electronics*, August 1997, p. 48.

Chapter 2

Fig. 2-1. *Electronics Now*, March 1997, p. 46.

Fig. 2-2. *Electronics Now*, February 1997, pp. 56–59.

Fig. 2-3. *Everyday Practical Electronics*, September 1996, p. 711.

Fig. 2-4. *Popular Electronics*, September 1996, p. 67.

Fig. 2-5. *Electronics Now*, October 1996, p. 15.

Fig. 2-6. *Popular Electronics*, October 1995, p. 74.

Fig. 2-7. *Popular Electronics*, April 1995, p. 31.

Fig. 2-8. *Electronics Now*, December 1996, p. 10.

Fig. 2-9. *Everyday Practical Electronics*, March 1996, p. 202.

Fig. 2-10. *Popular Electronics*, September 1994, p. 74.

Fig. 2-11. *Electronics Now*, May 1996, p. 10.

Chapter 3

Fig. 3-1. *73 Amateur Radio Today*, August 1996, pp. 46–47.

Fig. 3-2. *73 Amateur Radio Today*, February 1996, p. 32.

Fig. 3-3. *73 Amateur Radio Today*, January 1996, pp. 12–17.

Fig. 3-4. *QST*, December 1996, p. 68.

Fig. 3-5. *Popular Electronics*, July 1996, pp. 62–63.

Fig. 3-6. *Nuts and Volts Magazine*, January 1995, p. 61.

Fig. 3-7. *73 Amateur Radio Today*, February 1996, p. 32.

Fig. 3-8. *73 Amateur Radio Today*, August 1996, p. 54.

Chapter 4

Fig. 4-1. *QST*, December 1996, p. 41.

Fig. 4-2. *Electronics Now*, June 1996, pp. 38–75.

Fig. 4-3. William Sheets, 1995.

Chapter 5

Fig. 5-1. *Popular Electronics*, February 1997, p. 28.

Fig. 5-2. *Popular Electronics*, April 1996, p. 31.

Fig. 5-3. *Electronics Now*, August 1996, p. 10.

Fig. 5-4. *Popular Electronics*, July 1996, p. 66.

Fig. 5-5. Elektor Electronics, July–August 1996, p. 96.

Fig. 5-6. *Electronic Experimenters Handbook*, 1996, p. 25.

Fig. 5-7. *Popular Electronics*, October 1996, p. 6.

Fig. 5-8. *Popular Electronics*, December 1994, p. 31.

Fig. 5-9. *Electronic Hobbyists Handbook*, Fall 1995, pp. 40, 42–44.

Fig. 5-10. *Everyday Practical Electronics*, March 1996, p. 201.

Fig. 5-11. *Popular Electronics*, May 1994, p. 80.

Fig. 5-12. *Electronic Experimenters Handbook*, 1996, p. 63.

Fig. 5-13. *Everyday Practical Electronics*, March 1996, p. 194.

Fig. 5-14. *Analog Dialogue*, vol. 26, no. 2, 1992, p. 12.

Fig. 5-15. *QST*, December 1996, p. 69.

Fig. 5-16. *Popular Electronics*, December 1996, p. 62.

Chapter 6

Fig. 6-1. Reprinted with permission from *Electronic Design*, October 13, 1995, p. 101. Copyright © 1995, Penton Publishing Co.

Fig. 6-2. Reprinted with permission from *Electronic Design*, November 6, 1995, p. 118. Copyright © 1995, Penton Publishing Co.

Fig. 6-3. Reprinted with permission from *Electronic Design*, August 5, 1996, p. 99. Copyright © 1996, Penton Publishing Co.

Fig. 6-4. Reprinted with permission from *Electronic Design*, August 19, 1996, pp. 113–114. Copyright © 1996, Penton Publishing Co.

Fig. 6-5. *Popular Electronics*, April 1997, pp. 58–59.

Fig. 6-6. Reprinted with permission from *Electronic Design, Analog Applications*, November 20, 1995, p. 36. Copyright © 1995, Penton Publishing Co.

Fig. 6-7. Reprinted with permission from *Electronic Design, Analog Applications*, November 20, 1995, p. 25. Copyright © 1995, Penton Publishing Co.

Fig. 6-8. Reprinted with permission from *Electronic Design, Analog Applications*, November 20, 1995, p. 36. Copyright © 1995, Penton Publishing Co.

Fig. 6-9. *Elektor Electronics*, July–August 1996, p. 100.

Fig. 6-10. *Analog Dialogue*, vol. 28, no. 3, 1994, p. 7.

Fig. 6-11. *Popular Electronics*, October 1996, p. 68.

Fig. 6-12. *Popular Electronics*, July 1996, p. 51.

Fig. 6-13. *Popular Electronics*, July 1996, p. 50.

Fig. 6-14. *Electronics Now*, October 1995, p. 37.

Fig. 6-15. *Electronics Now*, May 1996, p. 56.

Fig. 6-16. *Electronics Now*, November 1995, p. 41.

Fig. 6-17. Reprinted with permission from *Electronic Design, Analog Applications*, November 20, 1995, p. 57. Copyright © 1995, Penton Publishing Co.

Fig. 6-18. Reprinted with permission from *Electronic Design, Analog Applications*, November 20, 1995, p. 53. Copyright © 1995, Penton Publishing Co.

Chapter 7

Fig. 7-1. *Communications Quarterly*, Winter 1996, pp. 57–63.

Fig. 7-2. *CQ*, January 1996, p. 60

Fig. 7-3. *RF Design*, June 1990, p. 71.

Fig. 7-4. *73 Amateur Radio Today*, November 1995, p. 34.

Fig. 7-5. *Communications Quarterly*, Winter 1996, pp. 93–94.

Fig. 7-6. *73 Amateur Radio Today*, January 1997, p. 26.

Fig. 7-7. *73 Amateur Radio Today*, August 1996, p. 48.

Fig. 7-8. *73 Amateur Radio Today*, August 1996, p. 48.

Fig. 7-9. *RF Design*, Intertec Publishing Corp., June 1994, p. 56.

Fig. 7-10. *RF Design*, Intertec Publishing Corp., June 1994, p. 54.

Fig. 7-11. *RF Design*, Intertec Publishing Corp., June 1994, p. 54.

Fig. 7-12. *RF Design*, Intertec Publishing Corp., June 1994, p. 54.

Chapter 8

Fig. 8-1. *Popular Electronics*, February 1996, p. 39.

Fig. 8-2. *Everyday Practical Electronics*, November 1996, p. 845.

Fig. 8-3. *Electronic Hobbyists Handbook*, Fall 1995, pp. 49–51.

Chapter 9

Fig. 9-1. Reprinted with permission from *Electronic Design*, October 13, 1995, p. 101. Copyright © 1995, Penton Publishing Co.

Fig. 9-2. *Electronics Now*, October 1996, p. 16.

Chapter 10

Fig. 10-1. *QST*, June 1995, p. 38.

Fig. 10-2. *73 Amateur Radio Today*, October 1995, p. 10.

Fig. 10-3. *Electronics Now*, September 1995, p. 7.

Fig. 10-4. *73 Amateur Radio Today*, August 1996, p. 53.

Fig. 10-5. *Popular Electronics*, November 1996, p. 75.

Fig. 10-6. *73 Amateur Radio Today*, August 1996, p. 55.

Fig. 10-7. *73 Amateur Radio Today*, May 1996, p. 78.

Fig. 10-8. *Electronic Experimenters Handbook*, 1996, p. 74.

Fig. 10-9. *73 Amateur Radio Today*, September 1995, p. 14.

Fig. 10-10. *Popular Electronics*, November 1995, p. 80.

Fig. 10-11. *Popular Electronics*, August 1996, p. 34.

Fig. 10-12. *Popular Electronics*, November 1995, p. 60.

Fig. 10-13. *73 Amateur Radio Today*, January 1996, p. 14.

Fig. 10-14. *Popular Electronics*, November 1996, p. 75.

Fig. 10-15. *Popular Electronics*, November 1996, p. 76.

Fig. 10-16. *Popular Electronics*, November 1995, p. 79.

Chapter 11

Fig. 11-1. *Popular Electronics*, August 1994, p. 44.

Fig. 11-2. Elektor Electronics, July–August 1996, p. 90.

Fig. 11-3. *Popular Electronics*, March 1994, p. 67.

Fig. 11-4. *Popular Electronics*, January 1996, p. 62.

Fig. 11-5. *Popular Electronics*, January 1996, pp. 63–64.

Fig. 11-6. *Popular Electronics*, January 1996, pp. 63–64.

Fig. 11-7. *Everyday Practical Electronics*, September 1996, p. 682.

Fig. 11-8. *Electronic Hobbyists Handbook*, Fall 1995, pp. 63–65.

Fig. 11-9. *Popular Electronics*, July 1996, pp. 42–44, 68.

Fig. 11-10. *Everyday Practical Electronics*, March 1995, p. 192.

Fig. 11-11. *Electronics Now*, July 1996, pp. 47–50.

Fig. 11-12. *Nuts and Volts*, August 1994, p. 98.

Fig. 11-13. *Nuts and Volts*, February 1995, p. 43.

Fig. 11-14. *Popular Electronics*, April 1995, p. 68.

Fig. 11-15. William Sheets.

Fig. 11-16. *Nuts and Volts*, August 1994, p. 98.

Fig. 11-17. *Popular Electronics*, April 1995, p. 68.

Fig. 11-18. *Electronics Now*, March 1997, p. 10.

Fig. 11-19. *Popular Electronics*, September 1996, pp. 39–40, 56.

Chapter 12

Fig. 12-1. *Popular Electronics*, June 1994, pp. 22–25.

Fig. 12-2. *Electronics Now*, March 1997, p. 31.

Fig. 12-3. *Popular Electronics*, June 1994, p. 26.

Fig. 12-4. *Electronics Now*, March 1997, p. 32.

Fig. 12-5. *Popular Electronics*, January 1996, p. 62.

Chapter 13

Fig. 13-1. *Maxim Engineering Journal*, vol. 24, p. 18.

Fig. 13-2. Reprinted with permission from *Electronic Design*, February 3, 1997, p. 153. Copyright © 1997, Penton Publishing Co.

Chapter 14

Fig. 14-1. Reprinted with permission from *Electronic Design*, September 3, 1996, p. 102. Copyright © 1996, Penton Publishing Co.

Fig. 14-2. Linear Technology, *Power Solutions*, 1996, p. 6.

Fig. 14-3. Linear Technology, *Power Solutions*, 1996, p. 7.

Fig. 14-4. Reprinted with permission from *Electronic Design*, June 10, 1996, p. 108. Copyright © 1996, Penton Publishing Co.

Fig. 14-5. Reprinted with permission from *Electronic Design*, May 15, 1995, p. 135. Copyright © 1995, Penton Publishing Co.

Fig. 14-6. *Popular Electronics*, June 1994, p. 25.

Fig. 14-7. *Maxim Engineering Journal*, vol. 23, pp. 14–15.

Fig. 14-8. Linear Technology, advertisement.

Fig. 14-9. *Everyday Practical Electronics*, July 1996, p. 504.

Fig. 14-10. *Maxim Engineering Journal*, vol. 23, p. 10.

Fig. 14-11. *Everyday Practical Electronics*, July 1996, p. 502.

Fig. 14-12. *Electronics Now*, September 1995, p. 8.

Fig. 14-13. Reprinted with permission from *Electronic Design*, December 4, 1995, p. 110. Copyright © 1995, Penton Publishing Co.

Fig. 14-14. Linear Technology, *Power Solutions*, 1996, p. 8.

Chapter 15

Fig. 15-1. *Everyday Practical Electronics*, March 1997, p. 180.

Fig. 15-2. Elektor Electronics, July–August 1996, p. 95.

Fig. 15-3. Reprinted with permission from *Electronic Design*, August 19, 1996, p. 118. Copyright © 1996, Penton Publishing Co.

Fig. 15-4. Reprinted with permission from *Electronic Design*, August 17, 1995, p. 124. Copyright © 1995, Penton Publishing Co.

Fig. 15-5. *Popular Electronics*, November 1996, p. 64.

Fig. 15-6. *Electronics Now*, April 1997, pp. 58–59, 74.

Fig. 15-7. Reprinted with permission from *Electronic Design, Analog Applications*, June 24, 1996, pp. 72–74. Copyright © 1996, Penton Publishing Co.

Fig. 15-8. *Nuts and Volts*, November 1994, p. 125.

Fig. 15-9. *Maxim Engineering Journal*, vol. 19, p. 15.

Fig. 15-10. Linear Technology, Design Note 117.

Fig. 15-11. *Electronics Now*, April 1997, pp. 58–59, 74.

Fig. 15-12. *Popular Electronics*, April 1996, p. 67.

Fig. 15-13. Elektor Electronics, July–August 1996, p. 88.

Chapter 16

Fig. 16-1. *Nuts and Volts*, Jan. 1995, p. 111.

Fig. 16-2. *Nuts and Volts*, Jan. 1995, p. 111.

Fig. 16-3. *Nuts and Volts*, Jan. 1995, p. 112.

Chapter 17

Fig. 17-1. Reprinted with permission from *Electronic Design*, September 3, 1996, p. 100. Copyright © 1996, Penton Publishing Co.

Fig. 17-2. *Electronics Now*, November 1995, p. 33.

Chapter 18

Fig. 18-1. *Popular Electronics*, July 1996, p. 65.

Fig. 18-2. *Popular Electronics*, April 1997, pp. 66–67.

Fig. 22-5. *Maxim Engineering Journal*, vol. 24, p. 4.

Fig. 22-6. Reprinted with permission from *Electronic Design*, November 20, 1995, p. 109. Copyright © 1995, Penton Publishing Co.

Fig. 22-7. Reprinted with permission from *Electronic Design*, July 22, 1996, p. 109. Copyright © 1996, Penton Publishing Co.

Chapter 23

Fig. 23-1. *Electronic Hobbyists Handbook*, Fall 1995, pp. 24–26.

Fig. 23-2. *Nuts and Volts*, September 1994, p. 97.

Fig. 23-3. *Popular Electronics*, March 1996, pp. 55–57, 88.

Fig. 23-4. Elektor Electronics, July–August 1996, p. 85.

Fig. 23-5. *Electronics Now*, October 1996, p. 72.

Chapter 24

Fig. 24-1. *RF Design*, Intertec Publishing Corp., December 1995, p. 50.

Fig. 24-2. *Popular Electronics*, June 1997, p. 64.

Fig. 24-3. *Popular Electronics*, June 1997, pp. 63–64.

Fig. 24-4. *Popular Electronics*, June 1995, p. 83.

Fig. 24-5. *Popular Electronics*, June 1995, p. 84.

Fig. 24-6. *Popular Electronics*, June 1995, p. 85.

Fig. 24-7. *Popular Electronics*, June 1995, p. 83.

Fig. 24-8. Linear Technology, Design Note 137.

Fig. 24-9. *Popular Electronics*, June 1997, p. 64.

Fig. 24-10. *Popular Electronics*, June 1997, p. 64.

Chapter 25

Fig. 25-1. *Popular Electronics*, November 1996, p. 77.

Fig. 25-2. *Popular Electronics*, November 1996, p. 74.

Fig. 25-3. *Popular Electronics*, November 1996, p. 77.

Fig. 25-4. *Popular Electronics*, November 1996, p. 74.

Fig. 25-5. *Popular Electronics*, November ~1996, p. 73.

Fig. 25-6. *Popular Electronics*, November 1996, p. 74.

Chapter 26

Fig. 26-1. *73 Amateur Radio Today*, August 1996, pp. 34–36.

Fig. 26-2. *RF Design*, Intertec Publishing Corp., December 1995, p. 58.

Fig. 26-3. *73 Amateur Radio Today*, August 1996, pp. 34–36.

Chapter 27

Fig. 27-1. Reprinted with permission from *Electronic Design*, May 15, 1995, pp. 138–140. Copyright © 1995, Penton Publishing Co.

Fig. 27-2. Reprinted with permission from *Electronic Design*, Feb. 3, 1997, p. 158. Copyright © 1997, Penton Publishing Co.

Chapter 28

Fig. 28-1. Reprinted with permission from *Electronic Design*, June 24, 1996, p. 104. Copyright © 1996, Penton Publishing Co.

Fig. 28-2. *Maxim Engineering Journal*, vol. 24, pp. 7–9.

Fig. 28-3. Linear Technology, *Power Solutions*, 1996, p. 29.

Fig. 28-4. *Electronics Now*, Dec. 1994, p. 71.

Fig. 28-5. Reprinted with permission from *Electronic Design, Analog Applications*, November 20, 1995, pp. 60–61. Copyright © 1995, Penton Publishing Co.

Fig. 28-6. Linear Technology, *Power Solutions*, 1996, p. 46.

Fig. 28-7. Linear Technology, *Power Solutions*, 1996, p. 46.

Fig. 28-8. Linear Technology, *Power Solutions*, 1996, p. 32.

Fig. 28-9. Linear Technology, *Power Solutions*, 1996, p. 33.

Fig. 28-10. Linear Technology, *Power Solutions*, 1996, p. 33.

Fig. 28-11. Linear Technology, *Power Solutions*, 1996, p. 38.

Fig. 28-12. Linear Technology, *Power Solutions*, 1996, p. 39.

Fig. 28-13. Linear Technology, *Power Solutions*, 1996, p. 39.

Fig. 28-14. Linear Technology, *Power Solutions*, 1996, p. 40.

Fig. 28-15. Linear Technology, *Power Solutions*, 1996, p. 40.

Fig. 28-16. Linear Technology, *Power Solutions*, 1996, p. 42.

Fig. 28-17. Linear Technology, *Power Solutions*, 1996, p. 43.

Fig. 28-18. Linear Technology, *Power Solutions*, 1996, p. 42.

Fig. 28-19. Linear Technology, *Power Solutions*, 1996, p. 25.

Fig. 28-20. Linear Technology, *Power Solutions*, 1996, p. 43.

Fig. 28-21. Linear Technology, *Power Solutions*, 1996, p. 22.

Fig. 28-22. Linear Technology, *Power Solutions*, 1996, p. 23.

Fig. 28-23. Linear Technology, *Power Solutions*, 1996, p. 21.

Fig. 28-24. Linear Technology, *Power Solutions*, 1996, p. 12.

Fig. 28-25. Reprinted with permission from *Electronic Design*, October 24, 1995, p. 102. Copyright © 1995, Penton Publishing Co.

Fig. 28-26. Linear Technology, *Power Solutions*, 1996, p. 59.

Fig. 28-27. *Electronics Now*, December 1994, p. 73.

Fig. 28-28. Linear Technology, *Power Solutions*, 1996, p. 59.

Fig. 28-29. Reprinted with permission from *Electronic Design*, October 13, 1995, p. 104. Copyright © 1995, Penton Publishing Co.

Chapter 29

Fig. 29-1. *Popular Electronics*, July 1994, p. 68.

Fig. 29-2. *Popular Electronics*, March 1997, pp. 40–45.

Fig. 29-3. *Electronics Now*, March 1995, p. 86

Fig. 29-4. *Electronics Now*, March 1995, p. 86.

Fig. 29-5. *Nuts and Volts*, June 1995, p. 99.

Fig. 29-6. *Nuts and Volts*, June 1995, p. 98.

Fig. 29-7. *Nuts and Volts*, June 1995, p. 99.

Fig. 29-8. *Everyday Practical Electronics*, October 1995, p. 804.

Fig. 29-9. *Electronic Experimenters Handbook*, 1996, p. 26.

Fig. 29-10. *Electronics Now*, January 1994, p. 83.

Chapter 30

Fig. 30-1. *Electronics Now*, February 1997, p. 53.

Fig. 30-2. *Electronics Now*, July 1995, p. 8.

Fig. 30-3. *Popular Electronics*, March 1994, p. 90.

Fig. 30-4. William Sheets.

Fig. 30-5. Reprinted with permission from *Electronic Design*, November 20, 1995, p. 110. Copyright © 1995, Penton Publishing Co.

Fig. 30-6. *Popular Electronics*, March 1994, p. 34.

Fig. 30-7. *Popular Electronics*, February 1997, p. 73.

Fig. 30-8. Reprinted with permission from *Electronic Design*, June 26, 1995, p. 103. Copyright © 1995, Penton Publishing Co.

Fig. 30-9. *Electronics Now*, April 1996, p. 34.

Fig. 30-10. *Everyday Practical Electronics*, September 1996, p. 706.

Fig. 30-11. *Electronics Now*, September 1995, p. 37.

Chapter 31

Fig. 31-1. *Electronics Now*, March 1995, p. 62.

Fig. 31-2. *Maxim Engineering Journal*, vol. 19, p. 13.

Fig. 31-3. *Everyday Practical Electronics*, July 1996, pp. 445.

Fig. 31-4. *Everyday Practical Electronics*, October 1996, p. 784.

Fig. 31-5. *Electronics Now*, November 1995, p. 154.

Fig. 31-6. *Popular Electronics*, January 1997, pp. 41–44.

Fig. 31-7. *Everyday Practical Electronics*, October 1996, p. 785.

Fig. 31-8. Elektor Electronics, July–August 1996, p. 75.

Fig. 31-9. Elektor Electronics, July–August 1996, p. 88.

Fig. 31-10. *Popular Electronics*, October 1995, p. 77.

Fig. 31-11. *Everyday Practical Electronics*, October 1995, p. 813.

Fig. 31-12. *Popular Electronics*, January 1996, p. 65.

Fig. 31-13. *Everyday Practical Electronics*, October 1995, p. 813.

Fig. 31-14. *Everyday Practical Electronics*, October 1995, p. 813.

Fig. 31-15. *Everyday Practical Electronics*, October 1995, p. 812.

Fig. 31-16. *Everyday Practical Electronics*, June 1996, p. 453.

Fig. 31-17. *Popular Electronics*, October 1995, p. 76.

Fig. 31-18. *Everyday Practical Electronics*, July 1996, pp. 444–447.

Fig. 31-19. *Electronics Now*, January 1996, pp. 29–31.

Fig. 31-20. *Electronics Now*, July 1996, p. 14.

Fig. 31-21. *Electronics Now*, January 1996, pp. 29–31.

Fig. 31-22. *Popular Electronics*, January 1996, p. 66.

Fig. 31-23. *Popular Electronics*, January 1996, p. 66.

Fig. 31-24. Reprinted with permission from *Electronic Design*, April 17, 1995, p. 95. Copyright © 1995, Penton Publishing Co.

Fig. 31-25. *Popular Electronics*, January 1996, p. 65.

Fig. 31-26. *Popular Electronics*, October 1995, p. 77.

Fig. 31-27. *Electronics Experimenters Handbook*, 1996, p. 23.

Fig. 31-28. *Popular Electronics*, January 1996, pp. 66–67.

Fig. 31-29. *Popular Electronics*, September 1995, p. 30.

Fig. 31-30. *73 Amateur Radio Today*, September 1995, p. 16.

Fig. 31-31. Linear Technology, *Power Solutions*, 1996, p. 22.

Fig. 31-32. Reprinted with permission from *Electronic Design*, December 4, 1995, p. 114. Copyright © 1995, Penton Publishing Co.

Fig. 31-33. *Everyday Practical Electronics*, October 1995, p. 811.

Chapter 32

Fig. 32-1. *Electronics Now*, April 1997, p. 14.

Fig. 32-2. *Electronics Now*, April 1997, p. 14.

Fig. 32-3. *Electronics Now*, September 1996, p. 14.

Chapter 33

Fig. 33-1. Reprinted with permission from *Electronic Design, Analog Applications*, June 24, 1996, p. 79. Copyright © 1996, Penton Publishing Co.

Fig. 33-2. *Popular Electronics*, March 1994, p. 70.

Fig. 33-3. *Electronics Now*, December 1995, p. 10.

Fig. 33-4. *Popular Electronics*, October 1995, p. 73.

Fig. 33-5. *Popular Electronics*, October 1995, p. 73.

Fig. 33-6. *Electronics Now*, Nov. 1995, p. 8.

Fig. 33-7. *Electronics Now*, Nov. 1995, p. 8.

Fig. 33-8. Reprinted with permission from *Electronic Design*, December 4, 1995, pp. 109–110. Copyright © 1995, Penton Publishing Co.

Fig. 33-9. *Popular Electronics*, March 1994, p. 70.

Chapter 34

Fig. 34-1. *73 Amateur Radio Today*, February 1997, p. 26.

Fig. 34-2. Linear Technology, vol. 7, no. 1, February 1997, p. 21.

Fig. 34-3. *Popular Electronics*, November 1995, p. 80.

Fig. 34-4. *Electronics Now*, October 1996, p. 63.

Fig. 34-5. *Popular Electronics*, February 1997, p. 73.

Fig. 34-6. *Popular Electronics*, November 1995, p. 80.

Fig. 34-7. *Electronics Now*, October 1996, p. 61.

Fig. 34-8. *Electronic Experimenters Handbook*, 1996, pp. 95–108.

Fig. 34-9. *Electronics Now*, October 1996, p. 62.

Chapter 35

Fig. 35-1. *73 Amateur Radio Today*, May 1996, p. 67.

Fig. 35-2. *Popular Electronics*, February 1997, p. 63.

Fig. 35-3. Linear Technology, vol. 7, no. 1, February 1997, p. 12.

Fig. 35-4. *Electronic Experimenters Handbook*, 1996, p. 57.

Fig. 35-5. Linear Technology, vol. 7, no. 1, February 1997, p. 11.

Fig. 35-6. Reprinted with permission from *Electronic Design*, August 7, 1995, p. 88. Copyright © 1995, Penton Publishing Co.

Fig. 35-7. *73 Amateur Radio Today*, November 1995, p. 67.

Fig. 35-8. Reprinted with permission from *Electronic Design*, December 16, 1996, pp. 116–118. Copyright © 1996, Penton Publishing Co.

Fig. 35-9. *Popular Electronics*, December 1994, p. 42.

Fig. 35-10. Reprinted with permission from *Electronic Design*, June 26, 1995, pp. 106–108. Copyright © 1995, Penton Publishing Co.

Fig. 35-11. Elektor Electronics, July–August 1996, p. 87.

Fig. 35-12. *73 Amateur Radio Today*, November 1995, pp. 67–68.

Fig. 35-13. Elektor Electronics, July–August 1996, p. 86.

Fig. 35-14. *Popular Electronics*, May 1995, p. 78.

Fig. 35-15. Reprinted with permission from *Electronic Design*, August 7, 1995, p. 90. Copyright © 1995, Penton Publishing Co.

Fig. 35-16. *73 Amateur Radio Today*, November 1995, p. 67.

Chapter 36

Fig. 36-1. *Popular Electronics*, October 1996, p. 67.

Fig. 36-2. *Electronics Now*, May 1996, p. 8.

Fig. 36-3. *Popular Electronics*, July 1996, p. 64.

Fig. 36-4. *Everyday Practical Electronics*, September 1996, p. 725.

Fig. 36-5. *Nuts and Volts*, April 1996, p. 52.

Fig. 36-6. *Everyday Practical Electronics*, October 1995, p. 793.

Fig. 36-7. *Everyday Practical Electronics*, October 1995, p. 794.

Fig. 36-8. *Everyday Practical Electronics*, October 1995, p. 794.

Fig. 36-9. *Nuts and Volts*, June 1995, p. 77.

Fig. 36-10. *Electronics Now*, January 1996, p. 9.

Fig. 36-11. *Everyday Practical Electronics*, October 1995, p. 793.

Fig. 36-12. *Popular Electronics*, January 1996, p. 67.

Fig. 36-13. *Popular Electronics*, April 1997, pp. 58–59.

Fig. 36-14. *Popular Electronics*, October 1995, p. 77.

Fig. 36-15. Elektor Electronics, July–August 1996, p. 69.

Fig. 36-16. *Electronics Now*, December 1995, p. 40.

Fig. 36-17. *Electronic Experimenters Handbook*, 1996, pp. 51–52.

Fig. 36-18. Delton T. Horn, *Sound, Light, and Music*, TAB Books, pp. 25–26.

Chapter 37

Fig. 37-1. Linear Technology, *Power Solutions*, 1996, p. 24.

Fig. 37-2. *Electronics Now*, December 1996, p. 11.

Chapter 38

Fig. 38-1. *Everyday Practical Electronics*, September 1996, p. 694.

Fig. 38-2. *Maxim Engineering Journal*, vol. 19, p. 8.

Fig. 38-3. *Maxim Engineering Journal*, vol. 19, pp. 3–4.

Fig. 38-4. *Electronics Now*, November 1995, p. 33.

Fig. 38-5. *Maxim Engineering Journal*, vol. 19, pp. 8–9.

Fig. 38-6. *Maxim Engineering Journal*, vol. 19, pp. 3–4.

Fig. 38-7. *Maxim Engineering Journal*, vol. 19, pp. 9–11.

Fig. 38-8. *Maxim Engineering Journal*, vol. 19, pp. 3–4.

Fig. 38-9. Elektor Electronics, July–August 1996, p. 85.

Chapter 39

Fig. 39-1. *Popular Electronics*, September 1996, p. 68.

Fig. 39-2. *Everyday Practical Electronics*, October 1996, p. 802.

Fig. 39-3. *Everyday Practical Electronics*, August 1996, p. 643.

Fig. 39-4. *Popular Electronics*, April 1997, pp. 59–61.

Chapter 40

Fig. 40-1. *Nuts and Volts*, January 1995, p. 86.

Fig. 40-2. *Nuts and Volts*, January 1995, p. 8.

Fig. 40-3. *Nuts and Volts*, January 1995, p. 85.

Chapter 41

Fig. 41-1. Reprinted with permission from *Electronic Design*, September 16, 1996, p. 100. Copyright © 1996, Penton Publishing Co.

Fig. 41-2. Reprinted with permission from *Electronic Design*, October 1, 1996, p. 102. Copyright © 1996, Penton Publishing Co.

Fig. 41-3. *Popular Electronics*, July 1996, p. 65.

Fig. 41-4. *Nuts and Volts*, March 1993, p. 31.

Fig. 41-5. Reprinted with permission from *Electronic Design*, May 1, 1997, p. 169. Copyright © 1997, Penton Publishing Co.

Fig. 41-6. *Popular Electronics*, September 1995, p. 32.

Fig. 41-7. *Popular Electronics*, March 1995, p. 40.

Fig. 41-8. *Everyday Practical Electronics*, March 1997, p. 204.

Fig. 41-9. Reprinted with permission from *Electronic Design*, August 5, 1996, p. 102. Copyright © 1996, Penton Publishing Co.

Fig. 41-10. *Electronics Now*, December 1994, pp. 53–61.

Fig. 41-11. Reprinted with permission from *Electronic Design*, July 24, 1995, p. 115. Copyright © 1995, Penton Publishing Co.

Fig. 41-12. *Nuts and Volts*, October 1995, pp. 43–44.

Chapter 42

Fig. 42-1. *Electronics Now*, July 1996, p. 12.

Fig. 42-2. *Popular Electronics*, September 1995, p. 75.

Chapter 43

Fig. 43-1. *Nuts and Volts*, June 1995, p. 98.

Fig. 43-2. Delton T. Horn, *Sound, Light, and Music*, TAB Books, 1993, pp. 142–143.

Fig. 43-3. *Everyday Practical Electronics*, May 1996, p. 390.

Fig. 43-4. Reprinted with permission from *Electronic Design*, December 2, 1996, p. 120. Copyright © 1996, Penton Publishing Co.

Fig. 43-5. *Everyday Practical Electronics*, October 1995, p. 787.

Fig. 43-6. *Everyday Practical Electronics*, August 1996, pp. 612–617.

Fig. 43-7. *Everyday Practical Electronics*, October 1995, p. 787.

Fig. 43-8. *Everyday Practical Electronics*, August 1996, pp. 612–617.

Fig. 43-9. *Electronics Now*, December 1995, p. 8.

Fig. 43-10. Linear Technology, vol. 7, no. 1, February 1997, p. 6.

Fig. 43-11. Linear Technology, Design Note 118.

Fig. 43-12. *Electronics Now*, August 1995, pp. 42–43.

Fig. 43-13. *Electronics Now*, August 1995, pp. 48.

Fig. 43-14. *Electronics Now*, August 1995, pp. 48.

Fig. 43-15. *Electronics Now*, August 1995, pp. 49.

Fig. 43-16. *Nuts and Volts*, June 1995, p. 99.

Fig. 43-17. Linear Technology, vol. 7, no. 1, February 1997, p. 6.

Fig. 43-18. *Electronics Now*, August 1995, p. 49.

Fig. 43-19. *Popular Electronics*, January 1997, p. 68.

Fig. 43-20. *Electronics Now*, August 1995, p. 49.

Fig. 43-21. *Electronics Now*, August 1995, p. 50.

Fig. 43-22. *Popular Electronics*, May 1994, pp. 69–70, 96.

Fig. 43-23. *Everyday Practical Electronics*, May 1996, p. 389.

Chapter 44

Fig. 44-1. William Sheets.

Fig. 44-2. Linear Technology, *Power Solutions*, 1996, p. 46.

Fig. 44-3. Elektor Electronics, July–August 1996, p. 86.

Chapter 45

Fig. 45-1. *Electronics Now*, September 1996, p. 77.

Fig. 45-2 *Electronics Now*, September 1996, p. 79.

Fig. 45-3. Gordon McComb, *Laser Cookbook*, TAB Books, 1988, p. 146.

Fig. 45-4. *Popular Electronics*, November 1995, p. 78.

Fig. 45-5. *Nuts and Volts*, June 1995, pp. 108–110.

Fig. 45-6. Reprinted with permission from *Electronic Design*, July 22, 1996, p. 114. Copyright © 1996, Penton Publishing Co.

Fig. 45-7. *Nuts and Volts*, June 1995, pp. 108–110.

Fig. 45-8. *Communications Quarterly*, Winter 1996, pp. 18–19.

Fig. 45-9. *Communications Quarterly*, Winter 1996, p. 13.

Fig. 45-10. *Electronics Now*, May 1996, p. 52.

Fig. 45-11. *Communications Quarterly*, Winter 1996, p. 14.

Fig. 45-12. *Communications Quarterly*, Winter 1996, pp. 13–14.

Fig. 45-13. *Communications Quarterly*, Winter 1996, pp. 17–18.

Chapter 46

Fig. 46-1. Reprinted with permission from *Electronic Design*, December 16, 1995, pp. 86–88. Copyright © 1995, Penton Publishing Co.

Fig. 46-2. *Nuts and Volts*, November 1993, p. 123.

Fig. 46-3. *Electronics Now*, March 1996, p. 8.

Fig. 46-4. *NASA Tech Briefs*, May 1995, p. 40.

Fig. 46-5. Reprinted with permission from *Electronic Design*, November 6, 1995, p. 120. Copyright © 1995, Penton Publishing Co.

Fig. 46-6. *Electronics Now*, March 1996, p. 8.

Chapter 47

Fig. 47-1. *Everyday Practical Electronics*, May 1996, p. 389.

Fig. 47-2. *Popular Electronics*, September 1995, p. 74.

Fig. 47-3. *Everyday Practical Electronics*, December 1996, p. 931.

Fig. 47-4. *Popular Electronics*, September 1995, p. 73.

Fig. 47-5. *Popular Electronics*, October 1995, p. 74.

Fig. 47-6. *Electronic Experimenters Handbook*, 1996, pp. 50–51.

Chapter 48

Fig. 48-1. *Electronics Now*, July 1995, p. 11.

Fig. 48-2. *Popular Electronics*, October 1996, pp. 31–33.

Fig. 48-3. *Electronic Experimenters Handbook*, 1996, pp. 55–56.

Fig. 48-4. *Popular Electronics*, February 1996, p. 77.

Fig. 48-5. *Popular Electronics*, October 1994, p. 83.

Fig. 48-6. *Popular Electronics*, October 1996, p. 64.

Fig. 48-7. *Popular Electronics*, January 1997, p. 68.

Fig. 48-8. *Popular Electronics*, January 1997, p. 71.

Fig. 48-9. *Popular Electronics*, April 1996, pp. 64, 77.

Fig. 48-10. *Electronic Experimenters Handbook*, 1996, p. 55.

Fig. 48-11. *Popular Electronics*, January 1997, pp. 68–71.

Fig. 48-12. *Popular Electronics*, January 1997, pp. 71–72.

Chapter 49

Fig. 49-1. Reprinted with permission from *Electronic Design*, April 17, 1995, pp. 98–100. Copyright © 1995, Penton Publishing Co.

Fig. 49-2. Reprinted with permission from *Electronic Design*, October 1, 1996, pp. 98, 100. Copyright © 1996, Penton Publishing Co.

Fig. 49-3. Reprinted with permission from *Electronic Design*, January 6, 1997, pp. 166–168. Copyright © 1997, Penton Publishing Co.

Chapter 50

Fig. 50-1. *Popular Electronics*, April 1996, p. 65.

Fig. 50-2. *Popular Electronics*, April 1996, p. 66.

Fig. 50-3. *Popular Electronics*, September 1995, pp. 74–75.

Fig. 50-4. *NASA Tech Briefs*, August 1994, p. 33.

Fig. 50-5. *Electronics Now*, March 1997, p. 60.

Chapter 51

Fig. 51-1. Reprinted with permission from *Electronic Design*, November 4, 1996, pp. 110–112. Copyright © 1996, Penton Publishing Co.

Fig. 51-2. *73 Amateur Radio Today*, March 1996, p. 71.

Fig. 51-3. *Popular Electronics*, January 1996, pp. 54–58.

Fig. 51-4. *Popular Electronics*, November 1996, pp. 47–49.

Fig. 51-5. *Everyday Practical Electronics*, October 1995, p. 797.

Fig. 51-6. *NASA Tech Briefs*, August 1994, pp. 34–35.

Fig. 51-7. *73 Amateur Radio Today*, November 1995, pp. 22–25.

Chapter 52

Fig. 52-1. *Popular Electronics*, March 1996, p. 67.

Fig. 52-2. *Electronics Now*, May 1995, p. 10.

Fig. 52-3. *Popular Electronics*, March 1994, p. 83.

Fig. 52-4. *Popular Electronics*, October 1996, p. 67.

Fig. 52-5. *Everyday Practical Electronics*, September 1996, p. 717.

Fig. 52-6. *Electronic Experimenters Handbook*, 1996, p. 82.

Fig. 52-7. *Popular Electronics*, October 1996, pp. 39–45.

Chapter 53

Fig. 53-1. Reprinted with permission from *Electronic Design*, December 16, 1995, p. 88. Copyright © 1995, Penton Publishing Co.

Fig. 53-2. *Nuts and Volts*, June 1996, p. 84.

Fig. 53-3. *Nuts and Volts*, October 1995, pp. 117–118.

Fig. 53-4. *Popular Electronics*, May 1997, p. 67.

Fig. 53-5. National Semiconductor, *Linear Applications Handbook*, 1991, p. 207.

Fig. 53-6. *Popular Electronics*, October 1996, pp. 39–45.

Chapter 54

Fig. 54-1. *Everyday Practical Electronics*, September 1996, p. 712.

Fig. 54-2. Elektor Electronics, July–August 1996, p. 68.

Chapter 55

Fig. 55-1. *Popular Electronics*, May 1997, p. 68.

Fig. 55-2. Elektor Electronics, July–August 1996, p. 93.

Fig. 55-3. *73 Amateur Radio Today*, February 1997, p. 22.

Fig. 55-4. *73 Amateur Radio Today*, June 1996, p. 56.

Fig. 55-5. *Electronics Now*, February 1997, p. 61.

Fig. 55-6. *QST*, June 1995, pp. 40–42.

Fig. 55-7. William Sheets.

Fig. 55-8. *Electronics Now*, December 1995, p. 74.

Fig. 55-9. *Popular Electronics*, February 1997, pp. 76–77.

Fig. 55-10. *Popular Electronics*, February 1997, pp. 73–74.

Fig. 55-11. *Popular Electronics*, June 1996, pp. 31–32.

Fig. 55-12. NASA Tech Briefs, May 1995, p. 34.

Fig. 55-13. *Popular Electronics*, March 1996, pp. 41–45.

Fig. 55-14. *Popular Electronics*, November 1995, p. 57.

Fig. 55-15. *Popular Electronics*, November 1994, p. 91.

Fig. 55-16. *Electronics Now*, November 1995, p. 112.

Fig. 55-17. *Electronics Now*, January 1996, pp. 29–31.

Fig. 55-18. *Popular Electronics*, April 1995, pp. 53–55.

Fig. 55-19. *Electronics Hobbyists Handbook*, Fall 1995, pp. 78–80.

Fig. 55-20. *Electronics Now*, July 1996, pp. 43–45.

Fig. 55-21. Reprinted with permission from *Electronic Design*, April 17, 1995, p. 98–100. Copyright © 1995, Penton Publishing Co.

Fig. 55-22. *Everyday Practical Electronics*, August 1996, pp. 628–629.

Fig. 55-23. *Everyday Practical Electronics*, October 1996, p. 801.

Fig. 55-24. *Electronics Now*, April 6, 1996, pp. 39–40, 75–78.

Fig. 55-25. *Popular Electronics*, March 1996, pp. 62, 73.

Fig. 55-26. *Popular Electronics*, April 1996, p. 63.

Fig. 55-27. *Electronic Experimenters Handbook*, 1996, p. 56.

Fig. 55-28. *Popular Electronics*, November 1996, p. 73.

Fig. 55-29. *Electronics Now*, April 1996, p. 70.

Chapter 56

Fig. 56-1. *Everyday Practical Electronics*, September 1996, p. 718.

Fig. 56-2. *Electronics Now*, November 1995, p. 9.

Fig. 56-3. *73 Amateur Radio Today*, November 1995, pp. 20–21.

Chapter 57

Fig. 57-1. *Popular Electronics*, May 1996, p. 65.

Fig. 57-2. *Popular Electronics*, October 1994, p. 84.

Fig. 57-3. *Everyday Practical Electronics*, November 1996, p. 841.

Fig. 57-4. *Popular Electronics*, July 1996, pp. 65–66.

Fig. 57-5. *Popular Electronics*, January 1997, p. 60.

Chapter 58

Fig. 58-1. *Electronics Now*, September 1996, pp. 61, 79.

Fig. 58-2. *Popular Electronics*, May 1994, p. 80.

Fig. 58-3. *Popular Electronics*, May 1994, p. 79.

Fig. 58-4. *Electronics Now*, March 1996, pp. 35–36.

Fig. 58-5. *Everyday Practical Electronics*, September 1996, p. 711.

Fig. 58-6. *Popular Electronics*, March 1997, pp. 50–57.

Fig. 58-7. Elektor Electronics, December 1996, p. 111.

Fig. 58-8. *Electronic Experimenters Handbook*, 1996, p. 24.

Chapter 59

Fig. 59-1. *NASA Tech Briefs*, May 1995, p. 36.

Fig. 59-2. *Everyday Practical Electronics*, October 1996, p. 801.

Fig. 59-3. Reprinted with permission from *Electronic Design*, May 1, 1997, p. 172. Copyright © 1997, Penton Publishing Co.

Fig. 59-4. *Popular Electronics*, October 1995, p. 50.

Fig. 59-5. *Electronics Now*, January 1997, p. 13.

Fig. 59-6. *Electronics Now*, November 1995, p. 9.

Chapter 60

Fig. 60-1. *73 Amateur Radio Today*, February 1997, p. 33.

Fig. 60-2. *73 Amateur Radio Today*, February 1997, p. 32.

Chapter 61

Fig. 61-1. *Maxim Engineering Journal*, vol. 24, p. 3.

Fig. 61-2. *Nuts and Volts*, November 1993, p. 122.

Fig. 61-3. Reprinted with permission from *Electronic Design*, May 1, 1995, p. 116. Copyright © 1995, Penton Publishing Co.

Fig. 61-4. Linear Technology, *Power Solutions*, 1996, p. 28.

Fig. 61-5. *73 Amateur Radio Today*, January 1997, p. 22.

Fig. 61-6. *Electronics Now*, Feb. 1996, p. 33.

Fig. 61-7. Reprinted with permission from *Electronic Design*, January 20, 1997, p. 153. Copyright © 1997, Penton Publishing Co.

Fig. 61-8. *RF Design*, Intertec Publishing Corp., November 1995, p. 56.

Fig. 61-9. *73 Amateur Radio Today*, November 1995, p. 68.

Fig. 61-10. Reprinted with permission from *Electronic Design*, April 17, 1995, p. 98. Copyright © 1995, Penton Publishing Co.

Fig. 61-11. Reprinted with permission from *Electronic Design*, January 20, 1997, pp. 153–154. Copyright © 1997, Penton Publishing Co.

Fig. 61-12. Reprinted with permission from *Electronic Design, Analog Applications*, June 24, 1996, p. 85. Copyright © 1996, Penton Publishing Co.

Fig. 61-13. *Electronics Now*, Jan. 1994, p. 83.

Fig. 61-14. *Maxim Engineering Journal*, vol. 23, p. 8.

Fig. 61-15. Reprinted with permission from *Electronic Design*, July 10, 1995, pp. 101–102. Copyright © 1995, Penton Publishing Co.

Fig. 61-16. *Electronics Now*, February 1997, p. 14.

Fig. 61-17. *Nuts and Volts*, November 1993, p. 122.

Fig. 61-18. *Popular Electronics*, June 1996, pp. 39–40, 75.

Fig. 61-19. *Popular Electronics*, June 1993, p. 70.

Fig. 61-20. Reprinted with permission from *Electronic Design*, November 4, 1996, p. 109. Copyright © 1996, Penton Publishing Co.

Fig. 61-21. Linear Technology, vol. 7, no. 1, February 1997, p. 22.

Fig. 61-22. Reprinted with permission from *Electronic Design*, November 4, 1996, p. 109. Copyright © 1996, Penton Publishing Co.

Fig. 61-23. *Popular Electronics*, March 1996, pp. 68–75.

Fig. 61-24. Reprinted with permission from *Electronic Design, Analog Applications*, November 20, 1995, p. 22. Copyright © 1995, Penton Publishing Co.

Fig. 61-25. *73 Amateur Radio Today*, November 1995, p. 68.

Fig. 61-26. *Electronics Now*, April 1996, p. 78.

Fig. 61-27. Reprinted with permission from *Electronic Design*, Nov. 4, 1996, p. 109. Copyright © 1996, Penton Publishing Co.

Fig. 61-28. *Nuts and Volts*, June 1995, p. 100.

Fig. 61-29. Reprinted with permission from *Electronic Design, Analog Applications*, June 24, 1996, p. 77. Copyright © 1996, Penton Publishing Co.

Fig. 61-30. Reprinted with permission from *Electronic Design, Analog Applications*, November 20, 1995, p. 32. Copyright © 1995, Penton Publishing Co.

Fig. 61-31. Reprinted with permission from *Electronic Design*, Oct. 16, 1996, p. 104. Copyright © 1996, Penton Publishing Co.

Fig. 61-32. *Everyday Practical Electronics*, June 1996, p. 752.

Fig. 61-33. *Everyday Practical Electronics*, June 1996, p. 452.

Fig. 61-34. *Popular Electronics*, October 1996, p. 68.

Fig. 61-35. *Nuts and Volts*, June 1995, p. 100.

Fig. 61-36. Reprinted with permission from *Electronic Design*, August 19, 1996, pp. 116–118. Copyright © 1996, Penton Publishing Co.

Fig. 61-37. Elektor Electronics, July–August 1996, p. 101.

Fig. 61-38. *Popular Electronics*, February 1995, p. 30.

Fig. 61-39. *Nuts and Volts*, Nov. 1993, p. 123.

Fig. 61-40. Reprinted with permission from *Electronic Design*, Aug. 7, 1995, p. 87. Copyright © 1995, Penton Publishing Co.

Fig. 61-41. Reprinted with permission from *Electronic Design*, Dec. 4, 1995, p. 110. Copyright © 1995, Penton Publishing Co.

Fig. 61-42. Reprinted with permission from *Electronic Design*, July 24, 1995, p. 118. Copyright © 1995, Penton Publishing Co.

Fig. 61-43 *RF Design*, Intertec Publishing Corp., June 1994, p. 56.

Fig. 61-44. Reprinted with permission from *Electronic Design, Analog Applications*, November 20, 1995, p. 34. Copyright © 1995, Penton Publishing Co.

Fig. 61-45. Reprinted with permission from *Electronic Design*, April 17, 1995, pp. 126–128. Copyright © 1995, Penton Publishing Co.

Fig. 61-46. National Semiconductor, *Linear Application Specific IC's Databook*, 1993, pp. 2–52.

Fig. 61-47. Reprinted with permission from *Electronic Design*, September 18, 1995, p. 97. Copyright © 1995, Penton Publishing Co.

Fig. 61-48. Reprinted with permission from *Electronic Design*, November 18, 1996, p. 120. Copyright © 1996, Penton Publishing Co.

Fig. 61-49. *Everyday Practical Electronics*, July 1996, pp. 520–524.

Fig. 61-50. *NASA Tech Briefs*, May 1995, p. 38.

Fig. 61-51. Reprinted with permission from *Electronic Design*, March 6, 1995, p. 91. Copyright © 1995, Penton Publishing Co.

Fig. 61-52. *Electronics Now*, Feb. 1997, p. 48.

Fig. 61-53. *Electronics Now*, February 1997, p. 49.

Fig. 61-54. *Popular Electronics*, May 1995, p. 78.

Fig. 61-55. *Popular Electronics*, May 1995, p. 76.

Fig. 61-56. *Popular Electronics*, May 1995, p. 77.

Fig. 61-57. *Radio-Electronics Experimenters Handbook*, 1992, p. 75.

Fig. 61-58. *Electronics Now*, February 1997, p. 49.

Fig. 61-59. *Popular Electronics*, May 1995, p. 77.

Fig. 61-60. *Radio-Electronics Experimenters Handbook*, 1992, p. 74.

Fig. 61-61. *Popular Electronics*, May 1995, p. 77.

Fig. 61-62. *73 Amateur Radio Today*, July 1996, p. 71.

Fig. 61-63. Reprinted with permission from *Electronic Design*, December 2, 1996, pp. 115–116. Copyright © 1996, Penton Publishing Co.

Fig. 61-64. Reprinted with permission from *Electronic Design*, Nov. 6, 1995, p. 113. Copyright © 1995, Penton Publishing Co.

Fig. 61-65. *Nuts and Volts*, June 1995, p. 99.

Fig. 61-66. *Electronics Now*, January 1996, pp. 29–31.

Chapter 62

Fig. 62-1. *Electronics Now*, Feb. 1997, p. 52.

Fig. 62-2. *RF Design*, Intertec Publishing Corp., June 1994, p. 56

Fig. 62-3. *Popular Electronics*, July 1996, p. 52.

Fig. 62-4. *Everyday Practical Electronics*, June 1996, p. 452.

Fig. 62-5. *Electronics Now*, Feb. 1997, p. 53.

Chapter 63

Fig. 63-1. *Popular Electronics*, March 1995, p. 29.

Fig. 63-2. *Everyday Practical Electronics*, August 1996, p. 642.

Fig. 63-3. *Everyday Practical Electronics*, September 1996, p. 712.

Fig. 63-4. *Everyday Practical Electronics*, November 1996, p. 841.

Fig. 63-5. Elektor Electronics, July–August 1996, p. 69.

Fig. 63-6. *Popular Electronics*, March 1995, p. 29.

Fig. 63-7. *Electronics Now*, Feb. 1996, p. 9.

Fig. 63-8. *Popular Electronics*, March 1995, p. 30.

Fig. 63-9. *Everyday Practical Electronics*, August 1996, p. 643.

Fig. 63-10. *Popular Electronics*, March 1995, P. 30.

Chapter 64

Fig. 64-1. Reprinted with permission from *Electronic Design, Analog Applications*, June 24, 1996, p. 72. Copyright © 1996, Penton Publishing Co.

Fig. 64-2. Reprinted with permission from *Electronic Design, Analog Applications*, June 24, 1996, p. 82. Copyright © 1996, Penton Publishing Co.

Chapter 65

Fig. 65-1. *Popular Electronics*, February 1996, p. 68.

Fig. 65-2. *Popular Electronics*, February 1996, p. 67.

Fig. 65-3. *Popular Electronics*, July 1996, p. 69.

Fig. 65-4. *Everyday Practical Electronics*, March 1996, p. 201.

Fig. 65-5. *Popular Electronics*, July 1996, p. 70.

Fig. 65-6. *Popular Electronics*, July 1996, pp. 70–71.

Fig. 65-7. Elektor Electronics, December 1996, p. 110.

Fig. 65-8. *Electronics Hobbyists Handbook*, Fall 1995, pp. 55–57.

Fig. 65-9. *Popular Electronics*, July 1996, p. 70.

Chapter 66

Fig. 66-1. *Electronics Now*, October 1996, p. 71.

Fig. 66-2. *Nuts and Volts*, November 1993, p. 123.

Fig. 66-3. *Electronics Now*, March 1996, p. 12.

Fig. 66-4. *Nuts and Volts*, August 1994, p. 83.

Fig. 66-5. *Electronics Now*, March 1996, pp. 8–12.

Fig. 66-6. *Electronics Now*, May 1995, p. 8.

Fig. 66-7. *Electronics Now*, February 1997, p. 14.

Fig. 66-8. *Electronics Now*, August 1996, p. 8.

Fig. 66-9. *Popular Electronics*, April 1997, pp. 58–59.

Fig. 66-10. *Popular Electronics*, September 1995, pp. 32, 88.

Fig. 66-11. Reprinted with permission from *Electronic Design*, November 18, 1996, pp. 115–116. Copyright © 1996, Penton Publishing Co.

Fig. 66-12. *Electronics Now*, January 1997, p. 30.

Chapter 67

Fig. 67-1. *Popular Electronics*, April 1995, p. 70.

Fig. 67-2. *Popular Electronics*, April 1995, p. 70.

Fig. 67-3. *Popular Electronics*, April 1995, p. 70.

Chapter 68

Fig. 68-1. *Everyday Practical Electronics*, September 1996, p. 670.

Fig. 68-2. *Electronics Now*, November 1995, p. 23.

Chapter 69

Fig. 69-1. Reprinted with permission from *Electronic Design*, July 10, 1995, pp. 104–106. Copyright © 1995, Penton Publishing Co.

Fig. 69-2. *Maxim Engineering Journal*, vol. 24, p. 17.

Chapter 70

Fig. 70-1. *Everyday Practical Electronics*, September 1996, p. 724.

Fig. 70-2. *Everyday Practical Electronics*, September 1996, p. 724.

Fig. 70-3. *Everyday Practical Electronics*, September 1996, p. 725.

Fig. 70-4. *Electronics Now*, February 1997, p. 13.

Chapter 71

Fig. 71-1. Reprinted with permission from *Electronic Design*, September 3, 1996, p. 98. Copyright © 1996, Penton Publishing Co.

Fig. 71-2. Reprinted with permission from *Electronic Design*, April 17, 1995, p. 128. Copyright © 1995, Penton Publishing Co.

Chapter 72

Fig. 72-1. *Popular Electronics*, October 1996, pp. 56–58.

Fig. 72-2. *Popular Electronics*, April 1997, p. 59.

Fig. 72-3. *Electronics Now*, October 1996, pp. 41–45.

Fig. 72-4. *Electronic Experimenters Handbook*, 1996, pp. 84–86.

Fig. 72-5. *Electronics Now*, June 1996, pp. 43–44, 85–88.

Fig. 72-6. *73 Amateur Radio Today*, January 1997, pp. 10–16.

Fig. 72-7. *Electronics Now*, June 1995, p. 46.

Chapter 85

Fig. 85-1. *Electronics Now*, December 1994, p. 73.

Fig. 85-2. Reprinted with permission from *Electronic Design*, June 24, 1996, p. 106. Copyright © 1996, Penton Publishing Co.

Fig. 85-3. *Electronics Now*, December 1994, p. 74.

Fig. 85-4. Reprinted with permission from *Electronic Design*, May 1, 1995, p. 113. Copyright © 1995, Penton Publishing Co.

Fig. 85-5. *Electronics Now*, December 1994, p. 74.

Fig. 85-6. Reprinted with permission from *Electronic Design*, April 17, 1995, p. 130. Copyright © 1995, Penton Publishing Co.

Fig. 85-7. *Electronics Now*, December 1994, p. 74.

Fig. 85-8. Reprinted with permission from *Electronic Design*, July 10, 1995, p. 106. Copyright © 1995, Penton Publishing Co.

Fig. 85-9. Reprinted with permission from *Electronic Design*, October 14, 1994, p. 102. Copyright © 1994, Penton Publishing Co.

Fig. 85-10. Linear Technology, *Power Solutions*, 1996, p. 23.

Fig. 85-11. Reprinted with permission from *Electronic Design*, November 4, 1996, p. 112. Copyright © 1996, Penton Publishing Co.

Fig. 85-12. Reprinted with permission from *Electronic Design*, January 20, 1997, pp. 154–155. Copyright © 1997, Penton Publishing Co.

Fig. 85-13. Reprinted with permission from *Electronic Design*, September 3, 1996, p. 102. Copyright © 1996, Penton Publishing Co.

Fig. 85-14. *Electronics Now*, December 1994, p. 73.

Fig. 85-15. Reprinted with permission from *Electronic Design*, May 1, 1995, p. 118. Copyright © 1995, Penton Publishing Co.

Chapter 86

Fig. 86-1. *Popular Electronics*, July 1996, p. 52.

Fig. 86-2. *Popular Electronics*, May 1996, p. 57.

Fig. 86-3. *Popular Electronics*, November 1995, pp. 84–85.

Fig. 86-4. *Nuts and Volts*, November 1995, p. 123.

Fig. 86-5. *Everyday Practical Electronics*, March 1996, p. 220.

Fig. 86-6. *Popular Electronics*, July 1996, p. 52.

Chapter 87

Fig. 87-1. *Popular Electronics*, November 1996, pp. 41–43.

Fig. 87-2. *Electronics Now*, December 1995, p. 69.

Fig. 87-3. *Nuts and Volts*, March 1993, p. 31.

Fig. 87-4. *Popular Electronics*, May 1997, pp. 69–70.

Fig. 87-5. *Electronic Experimenters Handbook*, 1996, p. 26.

Fig. 87-6. *Electronics Now*, August 1996, p. 51.

Fig. 87-7. *Electronics Now*, May 1996, p. 8.

Fig. 87-8. *Popular Electronics*, March 1996, p. 68.

Fig. 87-9 *Popular Electronics*, February 1997, p. 77

Fig. 87-10. *Popular Electronics*, February 1996, pp. 33–34, 78.

Fig. 87-11. *Popular Electronics*, March 1994, p. 89.

Chapter 88

Fig. 88-1. *Electronics Now*, December 1995, p. 10.

Fig. 88-2. *Nuts and Volts*, June 1996, p. 83.

Fig. 88-3. *Electronics Now*, Sept. 1995, p. 39.

Fig. 88-4. Reprinted with permission from *Electronic Design, Analog Applications*, June 24, 1996, p. 77. Copyright © 1996, Penton Publishing Co.

Fig. 88-5. *Nuts and Volts*, Dec. 1994, p. 137.

Fig. 88-6. *Popular Electronics*, November 1994, p. 91.

Fig. 88-7. *Popular Electronics*, May 1996, p. 73.

Fig. 88-8. *Everyday Practical Electronics*, September 1996, p. 681.

Fig. 88-9. *NASA Tech Briefs*, July 1996.

Fig. 88-10. *Maxim Engineering Journal*, vol. 24, p. 19.

Fig. 102-6. *Popular Electronics*, January 1996, pp. 35–36.

Fig. 102-7. *Everyday Practical Electronics*, November 1995, p. 879.

Chapter 103

Fig. 103-1. Elektor Electronics, July–August 1996, p. 90.

Fig. 103-2. *Electronics Now*, May 1996, p. 8.

Fig. 103-3. *Popular Electronics*, November 1995, p. 77.

Fig. 103-4. *Electronic Experimenters Handbook*, 1994, p. 68.

Fig. 103-5. *Popular Electronics*, September 1996, pp. 51–52, 56.

Chapter 104

Fig. 104-1. *Popular Electronics*, December 1994, p. 41.

Fig. 104-2. *Popular Electronics*, December 1994, p. 41.

Fig. 104-3. *Popular Electronics*, December 1994, p. 41.

Fig. 104-4. *Popular Electronics*, December 1994, p. 41.

Fig. 104-5. *RF Design*, Intertec Publishing Corp., November 1995, p. 56.

Fig. 104-6. *RF Design*, Intertec Publishing Corp., November 1995, p. 56.

Fig. 104-7. *Everyday Practical Electronics*, June 1996, p. 453.

Fig. 104-8. *Popular Electronics*, May 1996, p. 74.

Fig. 104-9. *Maxim Engineering Journal*, vol. 19, p. 14.

Chapter 105

Fig. 105-1. *Popular Electronics*, November 1996, p. 65.

Fig. 105-2. *Popular Electronics*, November 1996, pp. 66–67.

Fig. 105-3. *Popular Electronics*, November 1996, p. 66.

Fig. 105-4. Reprinted with permission from *Electronic Design*, November 4, 1996, p. 114. Copyright © 1996, Penton Publishing Co.

Fig. 105-5. *Popular Electronics*, January 1996, p. 37.

Fig. 105-6. *Electronics Now*, September 1995, p. 43.

Fig. 105-7. *Popular Electronics*, November 1996, p. 66.

Fig. 105-8. *Electronics Now*, June 1995, p. 48.

Fig. 105-9. *Popular Electronics*, November 1996, p. 65.

Fig. 105-10. *Popular Electronics*, September 1995, pp. 57–58.

Fig. 105-11. *Popular Electronics*, November 1996, p. 64.

Fig. 105-12. *Electronic Experimenters Handbook*, 1996, pp. 36–38.

Chapter 106

Fig. 106-1. *Popular Electronics*, July 1996, pp. 45–46, 74.

Fig. 106-2. Reprinted with permission from *Electronic Design*, August 19, 1996, pp. 114–116. Copyright © 1996, Penton Publishing Co.

Fig. 106-3. Reprinted with permission from *Electronic Design*, January 20, 1997, pp. 156–158. Copyright © 1997, Penton Publishing Co.

Fig. 106-4. *Popular Electronics*, January 1997, pp. 48–50.

Fig. 106-5. *Nuts and Volts*, September 1994, p. 97.

Fig. 106-6. Reprinted with permission from *Electronic Design*, October 24, 1995, p. 104. Copyright © 1995, Penton Publishing Co.

Fig. 106-7. *Everyday Practical Electronics*, July 1995, p. 551.

Fig. 106-8. *Electronic Experimenters Handbook*, 1996, pp. 39–42.

Fig. 106-9. Elektor Electronics, July–August 1996, p. 84.

Fig. 106-10. *Popular Electronics*, March 1994, p. 90.

Fig. 106-11. *Popular Electronics*, September 1996, p. 69.

Chapter 107

Fig. 107-1. *Electronics* Now, November 1994, p. 61.

Fig. 107-2. *Electronics Now*, November 1994, p. 61.

Chapter 108

Fig. 108-1. *Everyday Practical Electronics*, November 1996, p. 845.

Fig. 108-2. *Electronics Now*, February 1996, pp. 31–73.

Fig. 108-3. *Everyday Practical Electronics*, November 1996, p. 845.

Fig. 108-4. *Everyday Practical Electronics*, November 1996, p. 849.

Fig. 108-5. *Popular Electronics*, August 1996, pp. 66–67.

Fig. 108-6. *Everyday Practical Electronics*, November 1996, p. 845.

Chapter 109

Fig. 109-1. *Popular Electronics*, September 1996, p. 58.

Fig. 109-2. *Popular Electronics*, September 1996, p. 58.

Fig. 109-3. *Electronic Experimenters Handbook*, 1996, pp. 77–78.

Fig. 109-4. *Popular Electronics*, September 1996, p. 58.

Fig. 109-5. *Electronic Experimenters Handbook*, 1996, pp. 75–76.

Fig. 109-6. *Electronics Now*, September 1995, p. 14.

Fig. 109-7. *Electronics Now*, January 1997, p. 12.

Fig. 109-8. *Popular Electronics*, September 1996, pp. 57–58.

Fig. 109-9. *Everyday Practical Electronics*, December 1996, p. 930.

Fig. 109-10. Elektor Electronics, July–August 1991.

Fig. 109-11. *Everyday Practical Electronics*, May 1996, p. 390.

Fig. 109-12. Reprinted with permission from *Electronic Design*, June 26, 1995, p. 108. Copyright © 1995, Penton Publishing Co.

Chapter 110

Fig. 110-1. *Everyday Practical Electronics*, March 1995, p. 193.

Fig. 110-2. *Everyday Practical Electronics*, October 1995, p. 782.

Chapter 111

Fig. 111-1. *Popular Electronics*, October 1996, p. 65.

Fig. 111-2. *Popular Electronics*, January 1997, p. 72.

Fig. 111-3. *Popular Electronics*, April 1996, p. 66.

Fig. 111-4. *Popular Electronics*, October 1996, p. 65.

Fig. 111-5. *Popular Electronics*, October 1996, p. 65.

Fig. 111-6. *Popular Electronics*, October 1996, p. 66.

Fig. 111-7. *Popular Electronics*, September 1995, p. 30.

Fig. 111-8. *Popular Electronics*, Fact Card 206.

Chapter 112

Fig. 112-1. *Electronics Now*, June 1997, p. 31.

Fig. 112-2. *Nuts and Volts*, Jan. 1995, p. 61.

Fig. 112-3. *Electronic Experimenters Handbook*, Fall 1995, pp. 30–32.

Fig. 112-4. *Popular Electronics*, November 1995, p. 43.

Fig. 112-5. *1994 Electronic Experimenters Handbook*, p. 104.

Fig. 112-6. *Popular Electronics*, February 1996, p. 31.

Chapter 113

Fig. 113-1. *Popular Electronics*, March 1996, pp. 52–55.

Fig. 113-2. *Popular Electronics*, December 1994, p. 73.

Fig. 113-3. *73 Amateur Radio Today*, November 1995, p. 68.

Fig. 113-4. *Popular Electronics*, December 1994, p. 74.

Fig. 113-5. *Popular Electronics*, December 1994, p. 74.

Chapter 114

Fig. 114-1. *Electronics Now*, December 1995, pp. 77–78.

Fig. 114-2. Elektor Electronics, July–August 1996, p. 74.

Fig. 114-3. *Analog Dialogue*, vol. 26, no. 1, 1992, p. 13.

Fig. 114-4. *Electronics Now*, July 1995, p. 43.

Fig. 114-5. Reprinted with permission from *Electronic Design, Analog Applications*, November 20, 1995, p. 56. Copyright © 1995, Penton Publishing Co.

Fig. 114-6. *Electronics Now*, September 1996, pp. 51–73.

Fig. 114-7. *Nuts and Volts*, January 1995, p. 61.

Fig. 114-8. Reprinted with permission from *Electronic Design, Analog Applications*, November 20, 1995, p. 55. Copyright © 1995, Penton Publishing Co.

Fig. 114-9. *Electronics Now*, January 1996, p. 39.

Fig. 114-10. *Everyday Practical Electronics*, March 1997, pp. 167–172.

Fig. 114-11. Reprinted with permission from *Electronic Design*, July 10, 1995, pp. 102–104. Copyright © 1995, Penton Publishing Co.

Fig. 114-12. Reprinted with permission from *Electronic Design, Analog Applications*, November 1995, p. 58. Copyright © 1995, Penton Publishing Co.

Fig. 114-13. *Maxim Engineering Journal*, vol. 23, p. 16.

Fig. 114-14. *Analog Dialogue*, vol. 28, no. 3, 1994, p. 7.

Fig. 114-15. *Popular Electronics*, December 1994, pp. 30–31.

Chapter 115

Fig. 115-1. Linear Technology, Design Note 137.

Fig. 115-2. Reprinted with permission from *Electronic Design*, November 20, 1995, p. 112. Copyright © 1995, Penton Publishing Co.

Fig. 115-3. Reprinted with permission from *Electronic Design*, September 18, 1995, pp. 100–102. Copyright © 1995, Penton Publishing Co.

Chapter 116

Fig. 116-1. Linear Technology, Design Note 137.

Fig. 116-2. Reprinted with permission from *Electronic Design*, November 20, 1995, p. 112. Copyright © 1995, Penton Publishing Co.

Fig. 116-3. Reprinted with permission from *Electronic Design*, September 18, 1995, pp. 100–102. Copyright © 1995, Penton Publishing Co.

Chapter 117

Fig. 117-1. *Popular Electronics*, February 1997, pp. 74–75.

Fig. 117-2. *Nasa Tech Briefs*, January 1997, pp. 41–43.

Fig. 117-3. *Popular Electronics*, February 1996, pp. 67–68.

Fig. 117-4. *Popular Electronics*, February 1996, p. 69.

Index

1119

ABOUT THE AUTHORS

Rudolf F. Graf has 45 years of engineering, sales, and marketing experience in the electronics field. He has written more than 30 books, with a total of around 3 million copies in print, and well over 100 articles. A senior member of the IEEE and a licensed amateur radio operator (KA2CWL), he has a BSEE degree from Polytechnic Institute of Brooklyn and an MBA from New York University. He is self-employed.

William Sheets is a self-employed circuit design engineer with more than 25 years of experience in RF, analog, and digital electronics. He has written numerous articles for electronics publications and coauthored five books with Graf. The designer and builder of a satellite TV system, many transmitters and receivers, and a computer, he has an MEE degree from New York University. His interests include amateur radio (K2MQJ).